Smart Grid

Smart Grid

Integrating Renewable, Distributed & Efficient Energy

Edited by
Fereidoon P. Sioshansi
Menlo Energy Economics

AMSTERDAM • BOSTON • HEIDELBERG • LONDON
NEW YORK • OXFORD • PARIS • SAN DIEGO
SAN FRANCISCO • SINGAPORE • SYDNEY • TOKYO
Academic Press is an imprint of Elsevier

Academic Press is an imprint of Elsevier
225 Wyman Street, Waltham, MA 02451, USA
The Boulevard, Langford Lane, Kidlington, Oxford, OX5 1GB, UK

Notices
Knowledge and best practice in this field are constantly changing. As new research and experience broaden our
understanding, changes in research methods, professional practices, or medical treatment may become necessary.

Practitioners and researchers must always rely on their own experience and knowledge in evaluating and using
any information, methods, compounds, or experiments described herein. In using such information or methods
they should be mindful of their own safety and the safety of others, including parties for whom they have a
professional responsibility.

To the fullest extent of the law, neither the Publisher nor the authors, contributors, or editors, assume any liability
for any injury and/or damage to persons or property as a matter of products liability, negligence or otherwise, or
from any use or operation of any methods, products, instructions, or ideas contained in the material herein.

Library of Congress Cataloging-in-Publication Data
Smart grid : integrating renewable, distributed & efficient energy / [edited by] Fereidoon P. Sioshansi.
    p. cm.
  Includes bibliographical references and index.
  ISBN 978-0-12-386452-9 (hardback)
    1. Smart power grids. 2. Electric power distribution–Energy conservation. 3. Electric utilities–Cost
effectiveness. 4. Renewable energy sources. I. Sioshansi, Fereidoon P. (Fereidoon Perry)
  TK3105.S53 2011
  621.319–dc23   2011027601

British Library Cataloguing-in-Publication Data
A catalogue record for this book is available from the British Library.

For information on all Academic Press publications
visit our Web site at *www.elsevierdirect.com*

Typeset by: diacriTech, Chennai, India

Contents

Part I
Setting the Context: The What, Why, How, If, and When of Smart Grid

Part II
Smart Supply: Integrating Renewable & Distributed Generation

The present electric power delivery infrastructure was not designed to meet the needs of a restructured electricity marketplace, the increasing demands of a digital society, or the increased use of renewable power production. In addition, investments in expansion and maintenance are constantly being challenged, and the existing infrastructure has become vulnerable to various security threats. Today's power system is largely comprised of large central-station generation connected by a high voltage network of grid to local electric distribution systems, which, in turn, serve homes, business, and industry. In today's power system, electricity flows predominantly in one direction using mechanical controls.

The smart grid still depends on the support of large central-station generation, but it includes a substantial number of installations of electric energy storage and of renewable energy generation facilities, both at the bulk power system level and distributed throughout. In addition, the smart grid has greatly enhanced sensory and control capability configured to accommodate these distributed resources as well as electric vehicles, direct consumer participation in energy management, and efficient communicating appliances. This smart grid is hardened against cyber security while assuring long-term operations of an extremely complex system of millions of nodes.

As a result, there is now a national and international imperative to modernize and enhance the power delivery system. The smart grid is envisioned to provide the enhancements to ensure high levels of security, quality, reliability, and availability (SQRA) of electric power; to improve economic productivity and quality of life; and to minimize environmental impact while maximizing safety and sustainability. The smart grid will be characterized by pervasively collaborative distributed intelligence, including flexible wide band gap communication, dynamic sharing of all intelligent electronic devices, and distributed command and control.

As further described in this volume, achieving this vision will require careful policy formulation, accelerated infrastructure investment, and greater commitment to public/private research, development, and demonstration (RD&D) to overcome barriers and vulnerabilities.

Clark W. Gellings
Fellow
Electric Power Research Institute

In his seminal book *The Third Wave* published in 1980, **Alvin Toffler** talked about the coming *information revolution*. What is remarkable about his book is that he predicted many of the things that have happened since, even before there was the Internet as we know it.

The information age that he predicted is, of course, upon us. It has transformed many industries in ways that few could have imagined in 1980. Incidentally, if you searched for the term *smart grid*—there was no Google back in 1980—you would probably not have come up with very many hits. Smart grid was not a household word yet.

Consider a few examples of the pervasive impact of the information age, which we take for granted today.

- The plastic credit card in your wallet—With few exceptions, you can use it virtually anywhere around the globe—and at the end of the month it will tell you what you bought, when, where, for how much, in what currency—which it will convert to your local currency for a small fee.
- Your mobile phone—Based on rich levels of data on individual consumers and an analysis of their usage patterns, your local service provider can now deliver customized programs and offerings to customer segments down to an individual level. In turn, you as a consumer now have tremendous flexibility in reviewing your usage, adjusting plans, and adding services almost on demand.
- The express package you ship or the item you order online—You can track the package from the beginning to the end of its journey.
- Purchasing your airline tickets—airlines today are able to adjust pricing structures, almost in real time, in response to supply and demand and a number of other variables too numerous to mention. As travelers we are able to check in online or get an alert on our mobile phone if a flight is delayed, even before the gate agent knows.

From finance to delivery services to a vast range of products, nearly every industry around the globe has been transformed and revolutionized by information technology. The electric power sector—not to offend anyone—is a laggard in fully adopting information technology. For the most part, it still bills customers—or ratepayers in many places—based on monthly or quarterly volumetric consumption times a flat rate, with a few exceptions, of course. Most consumers are not aware of how much they are consuming or how much it is really costing them—or the service provider.

A typical electric utility's knowledge of the state of its distribution network is minimal. In many places, customers call to report that the lights are out, and then someone has to be sent to the field to figure out what happened, when, and why. In short, the industry has yet to embrace the information age.

The *smart grid*, a catch-all term, promises to change this. Taken as a whole, it has the potential to usher the electricity sector into the information age. As the chapters of this book explain, the smart grid, along with smart meters, smart devices, and smart prices, can catapult the electric power industry into a new era, allowing the service providers to deliver new and improved services at lower costs and with higher levels of reliability. These benefits can be had while reducing the sector's considerable emissions of greenhouse gases by increasing the penetration of renewable energy resources and distributed and on-site generation. The list of what can be done is endless, limited only by our imagination.

While the smart grid's great potentials are exciting, its challenges are equally daunting. Capturing its many purported benefits will not be easy or necessarily cheap. But as other industries have made the transition to the information age, the electric power sector too must embrace the concept and bring its full potential benefits to customers and society at large.

This book, with contributions from a number of experts, practitioners, and scholars, describes the advantages of the smart grid while addressing the obstacles to its achievement.

Guido Bartels
General Manager, Energy & Utilities, IBM
Chairman, Global Smart Grid Federation

Jérôme Adnot is a Full Professor in charge of the Demand Side Management research team of the Center of Energy and Processes of Mines-ParisTech. He also leads several master's degree programs at the school including the program on Mobility and Electric Vehicles.

Since 1991 he has focused on Demand Side Management working on the evaluation of new energy technologies. He is the co-author of numerous reports on the subject, for example for the European Commission and for the French environmental agency, ADEME. He has authored more than 100 scientific papers on these topics and is a member of the editorial board of the *Energy Efficiency Journal*.

Professor Adnot graduated in mechanical engineering from the Ecole des Mines de Nancy and holds PhDs from Mines-ParisTech and Université Paris VI.

Graeme Ancell is Planning and Development Manager at Transpower New Zealand Limited. He is involved in the long-term development of the transmission grid in New Zealand. He has worked in both grid development and system operations divisions in Transpower.

Dr. Ancell has interests in the integration of large-scale wind generation in island power systems, power system economics, and smart grids.

Dr. Ancell received his BE (Honors) and PhD degrees from the University of Auckland, New Zealand. He received a master's degree in Information System Management from Victoria University, New Zealand.

Margaret Armstrong is a professor at the Cerna, Mines-ParisTech, where she is the co-leader of the quantitative finance group. Her research focuses on modeling electricity markets, especially the impact of derivatives such EDF's virtual power plants on day-ahead prices, and on the evaluation of projects such as power plants, mines, and oil fields that are subject to technical and financial uncertainty. The quantitative finance group and the CEP have a joint project to study the impact of electric vehicles on the French market.

Professor Armstrong has a master's degree in mathematical statistics from the University of Queensland, Australia, and a PhD in geostatistics from Mines-ParisTech.

Guido Bartels leads IBM's energy and utilities business globally, which includes the Intelligent Utility Network initiative, the company's portfolio of solutions for the smart grid. He is also a member of IBM's senior leadership team. From 2007–2010 he was the Chairman of the GridWise™ Alliance during which the membership of the smart grid advocacy organization grew eightfold. He has been instrumental in several other countries in setting up their national smart grid organizations, leading to the Global Smart Grid Federation in 2010, where he is the current Chairman.

He also serves as a member of the U.S. Department of Energy's Electricity Advisory Committee and as Vice Chairman of the New York State Smart Grid Consortium. As an advocate of the smart grid, Mr. Bartels believes it to be the ultimate enabler for continued economic growth, national security, and a sustainable energy future.

Mr. Bartels earned an MBA degree in business economics from the University of Amsterdam in the Netherlands.

Adam Berry is a Research Scientist with the Energy Transformed Flagship, Commonwealth Scientific and Industrial Research Organization (CSIRO). He has previously been a post-graduate researcher at the University of Tasmania, Australia.

Berry draws on a background in artificial intelligence, evolutionary computation, and machine-learning research to drive his current work on the optimal planning and operation of electrical networks in an era of increasingly distributed generation. He is particularly focused on microgrids and their role in future electricity networks, publishing numerous peer-reviewed articles and reporting to the Australian government on this area.

Berry graduated from the University of Tasmania, Australia, with first-class science honors in 2003 and received his PhD from the same institution in 2008.

John Boys is Professor of Electronics at the University of Auckland where he was the Head of Department and is now a Research Professor. Previously, he was a Lecturer/Senior Lecturer at The University of Canterbury. Prior to that, he worked in Europe developing new business opportunities for international companies.

Professor Boys' main interests include power electronics and AC motor control, leading to dynamic demand control of micro hydro systems, and more recently to inductive power transfer (IPT), which entails transfer of electrical power without wires. This technology is now licensed worldwide in factories and particularly in clean manufacturing for computer chips and flat-panel displays. Research with inductive systems is ongoing for inductively powered EVs charged dynamically through the road surface. He is the holder of more than 25 U.S. patents, principally in the IPT area. He is a Distinguished Fellow of IPENZ and a Fellow of the Royal Society of New Zealand.

He received an ME (Dist) and a PhD from the University of Auckland.

Steven Braithwait is a Vice President at Christensen Associates Energy Consulting, in Madison, Wisconsin, where he specializes in market-based retail electricity pricing and demand response (DR), with an emphasis on measuring customer price response, estimating DR load impacts, and evaluating demand response program benefits and costs. Most recently, he managed statewide ex post impact evaluations of the major California utilities' critical-peak pricing, demand bidding, and aggregator DR programs for large business customers.

Dr. Braithwait has also assisted clients in designing market-based service offerings including real-time pricing, critical-peak pricing, and demand response programs. He has delivered papers at numerous industry conferences on the topics of market-based pricing, customer price response, demand response programs, and load forecasting.

Dr. Braithwait earned an undergraduate degree in mathematics at Occidental College in Los Angeles, and a doctorate in economics at UC Santa Barbara.

Christine Brandstätt is a researcher at the Bremer Energie Institut, a scientific institute investigating and advising on issues concerning energy economics and energy politics. She contributes to the institute's research focus on energy economics and market design and is currently engaged in integration of renewable energy and smart grids.

As a collaborator in the research project Innovative Regulation for Intelligent Networks, which receives funding from the German Federal Ministry of Economics and Technology, she pursues research on incentive structures to efficiently support the development of smart grids in energy networks. Her interests include sustainable energy supply, energy policy, market design, and regulation. She previously collaborated in the EU project, Sustainable Energy Management Systems, investigating the requirements for sustainable energy supply at the community level.

Mrs. Brandstätt received her undergraduate degree in industrial engineering and environmental planning from the University of Applied Sciences in Trier, Germany, and holds masters in management and engineering of the environment and energy from the Universidad Politécnica de Madrid, Spain; Ecole des Mines de Nantes, France; and Kungliga Tekniska Högskolan in Stockholm, Sweden.

Gert Brunekreeft is Professor of Energy Economics at Jacobs University Bremen in Germany and Director of the Bremer Energie Institut. Before joining Jacobs University, he was a Senior Economist at EnBW AG and held research positions in applied economics at Tilburg University, the University of Cambridge, and Freiburg University. He is associate researcher to a number of research centers and is associate editor for the journal *Competition and Regulation in Network Industries*.

Professor Brunekreeft's main research interests are in industrial economics, regulation theory, and competition policy of network industries, especially electricity and gas markets. Current research includes the economics of vertical unbundling and the relation between regulation and investment, and he is the project leader for IRIN—Innovative Regulation for Intelligent Networks. He has authored several books and numerous publications including in *Journal of Regulatory Economics*, *Utilities Policy*, *Oxford Review of Economic Policy*, and *Energy Journal*.

He holds a degree in economics from the University of Groningen and a PhD from Freiburg University, both in economics.

Judy W. Chang is Principal at The Brattle Group engaged in evaluating the potential impact of integrating renewable energy resources in power systems. She recently completed several resource planning projects that involve the development of scenarios and strategies for electric systems in the United States to meet long-range electric demand while considering the growth of renewable energy, energy efficiency, other demand-side resources, and the potential retirement of existing generation facilities due to various environmental regulations. She is the founding executive director of a non-profit organization, New England Women in Energy and the Environment.

She holds a master's in Public Policy from Harvard University's Kennedy School of Government.

Jeong-Gon Choi is a Manager of the Main Office Design Team in the General Management Department of the Korea Power Exchange, in Seoul, Korea. Over the past 10 years, Choi has played an instrumental role in fostering the inclusion of demand-side resources in the Korean energy market, including involvement in projects to develop the first Korea Power Exchange (KPX) demand response program and to establish required IT systems for Korea power system operation. He also provides advice to public officials on demand-response-enabling policies.

He currently manages the design of KPX's new main office complex, which is expected to be the first smart building in the country with information-based technologies and operational strategies that will enable the KPX office to mitigate energy costs through participation in demand response programs.

Mr. Choi received a degree in Electrical Engineering at Hanyang University in Seoul and is a certified Project Management Professional.

David J. Cornforth is a Research Scientist with the Energy Transformed Flagship, Commonwealth Scientific and Industrial Research Organization (CSIRO). He has been an educator and researcher at the University of Newcastle, Charles Sturt University, and the University of New South Wales in Australia.

Dr. Cornforth leads research in local power systems, microgrids, and the integration of renewable energy into the grid. He has a background and interests in artificial intelligence, multiagent simulation, and optimization. He has published widely in academic conferences and journals on microgrids, the application of artificial intelligence to build smarter power systems, optimization techniques, and computer security.

Dr. Cornforth has a BSc in Electrical and Electronic Engineering from Nottingham Trent University, UK, and a Ph in Computer Science from the University of Nottingham, UK.

Susan Covino is the Senior Consultant, Market Strategy, at PJM. She previously served as Manager for Demand Side Response. Prior to joining New

Power she was responsible for gas and electric matters for Enron in the mid-Atlantic and northeast regions of the United States. She also previously served KCS Energy Marketing as Vice President and General Counsel.

Ms. Covino has been actively involved in the development of demand response in PJM and the mid-Atlantic region. She has served as PJM's liaison to the Mid-Atlantic Distributed Resources Initiative (MADRI) and has led the development of multiple iterations of MADRI's Demand Response Road Map developed in cooperation with PJM. Ms. Covino has also led PJM stakeholder outreach in demand response, organizing two demand response symposia over the last three years.

Ms. Covino earned a Bachelor of Arts degree with a double major in economics and history from the University of Connecticut and a Juris Doctor degree from Dickinson School of Law.

Kelly B. Crandall is the Special Projects Coordinator in the Local Environmental Action Division (LEAD) of the City of Boulder, CO. She assists the city with regulatory and programmatic issues related to smart grid, demand-side management, and renewable energy. Boulder is the site of a major smart grid project sponsored by Xcel Energy.

Prior to joining the City of Boulder staff, Ms. Crandall was a Smart Grid Analyst with the National Renewable Energy Laboratory, where she consulted with the National Science & Technology Council's Subcommittee on Smart Grid. Her focus was on consumer issues, including data privacy, demand response, and education. She previously presented before the Colorado Public Utilities Commission on feed-in tariffs, and has written and presented on renewable energy credit market reform and green building legal challenges.

She received a BA with highest honors in History and Political Science from the University of Florida and a JD from the University of Colorado Law School. She is a licensed attorney in Colorado.

Ahmad Faruqui is a Principal at The Brattle Group where he is focused on assessing the economics of dynamic pricing, demand response, advanced metering infrastructure, and energy efficiency in the context of the smart grid. He pioneered the use of experimentation in understanding customer behavior, and his early work on time-of-use pricing experiments is frequently cited in the literature.

He has assisted FERC in the development of the National Action Plan on Demand Response and led the effort that resulted in A National Assessment of Demand Response Potential. He co-authored EPRI's national assessment of the potential for energy efficiency and EEI's report on quantifying the benefits of dynamic pricing. He has assessed the benefits of dynamic pricing for the New York Independent System Operator, worked on fostering economic demand response for the Midwest ISO and ISO New England, reviewed demand forecasts for the PJM Interconnection, and assisted the California Energy Commission in developing load management standards.

Dr. Faruqui holds a PhD in economics from the University of California, Davis.

Frank A. Felder is Director of the Center for Energy, Economic, and Environmental Policy and Associate Research Professor at the Bloustein School of Planning and Public Policy at Rutgers University.

Dr. Felder directs applied energy and environmental research. Ongoing and recent projects include energy efficiency evaluation studies, economic impact of renewable portfolio standards, and power system and economic modeling of state energy plans. He is also an expert on restructured electricity markets. He has published widely in professional and academic journals on market power and mitigation, wholesale market design, reliability, transmission planning, market power, and rate design issues. He was a nuclear engineer and submarine officer in the U.S. Navy.

Dr. Felder holds undergraduate degrees from Columbia College and the School of Engineering and Applied Sciences and a master's and doctorate from M.I.T in Technology, Management and Policy.

Nele Friedrichsen is a researcher at the Bremer Energie Institut, a scientific institute investigating and advising on issues concerning energy economics and energy politics. She contributes to the institute's research on energy economics and market design, including smart grids, regulation, and network pricing.

Her research interests include sustainability, energy policy, regulation, and vertical governance in the electricity system and market design. She studies arrangements that safeguard efficient coordination in liberalized markets. Contributing to the project IRIN—Innovative Regulation for Intelligent Networks—she analyzed incentive structures that efficiently support the evolution of smart grids, including intelligent network pricing and its potential for enhancing efficiency.

Mrs. Friedrichsen received a Diploma from the University of Flensburg, Germany, in Energy and Environmental Management and is currently pursuing a PhD in Economics at Jacobs University in Bremen, Germany.

Alain Galli is a Professor at the Cerna, Mines-ParisTech, where he is the co-leader of the quantitative finance group. His research focuses on modeling commodities especially oil and CO_2, and on the evaluation of projects such as power plants, oil fields, and mines that are subject to technical and financial uncertainty. The quantitative finance group and the CEP have a joint project to study the impact of electric vehicles on the French market.

Professor Galli has a PhD in mathematics from the University of Grenoble, France.

Clark W. Gellings is a Fellow at the Electric Power Research Institute, responsible for technology strategy in areas concerning energy efficiency, demand response, renewable energy resources, and other clean technologies.

Mr. Gellings joined EPRI in 1982, progressing through a series of technical management and executive positions, including seven vice president positions.

Prior to joining EPRI, he spent 14 years with Public Service Electric & Gas Company in New Jersey. He is the recipient of numerous awards, has served on numerous boards and advisory committees both in the United States and internationally, and is a frequent speaker at industry conferences.

Gellings has a Bachelor of Science in electrical engineering from Newark College of Engineering in New Jersey, a Master of Science degree in mechanical engineering from New Jersey Institute of Technology, and a Master of Management Science from the Wesley J. Howe School of Technology Management at Stevens Institute of Technology.

Bruce Hamilton is CEO of Adica LLC a Chicago-based global provider of smart grid software, consulting services, and workforce development training. He is a founding Director of the Institute for Sustainable Energy Development (ISED) and previously served as Head of Energy Modeling, Databanks, and Capacity Building at the International Atomic Energy Agency.

His current responsibilities include development and application of advanced analytics for evaluating the impact of smart grid technology and providing consulting services for smart grid roadmap development and business case evaluation. He is lead author of a *Feasibility Report on a Smart Grid and Green Technology Development Initiative Between the State of Illinois and Korea* and provides program management support for the Illinois-Korea Smart Grid Partnership. He served as guest editor of a recent edition of the Institute for Electrical and Electronics Engineers (IEEE) *Power & Energy Magazine* and is the IEEE Power & Energy Society's designated contributor to the IEEE *Smart Grid Newsletter*.

Mr. Hamilton earned an undergraduate degree at Benedictine University in Illinois and an MS in Computer Science at Illinois Institute of Technology.

Daniel Hansen is a Vice President at Christensen Associates Energy Consulting, in Madison, Wisconsin, working in a variety of areas related to retail and wholesale pricing in electricity and natural gas markets. He has used statistical models to forecast customer usage, estimate customer load response to changing prices, and estimate customer preferences for product attributes.

He has developed mathematical models that analyze risks associated with weather and price uncertainty, forecast market shares for products and providers, and simulate customer demand response to alternative rates. These capabilities have been used to develop and price new product options, evaluate existing pricing programs, evaluate the risks associated with individual products and product portfolios, and develop cost-of-service studies.

Dr. Hansen earned an undergraduate degree in economics and history at Trinity University in San Antonio and a doctorate in economics at Michigan State University.

Philip Q. Hanser is a Principal at The Brattle Group where he is engaged in projects ranging from renewables integration to retail and wholesale rate design to transmission system issues. Prior to joining his present job, he was with the Electric Power Research Institute (EPRI) in charge of the Demand-Side Management Program.

He has served as an expert witness before the Federal Energy Regulatory Commission, various state commissions, and federal and state courts on topics including market design, transmission tariffs, generation contracts, demand-side management, energy efficiency, electricity demand forecasting, and rate design. He served as one of the state of California's witnesses in the FERC's hearings on Enron's market manipulations during the California electricity crisis.

He has advanced degrees from Columbia University in economics and mathematical statistics.

Patti Harper-Slaboszewicz is a Business Strategy Senior Specialist Consultant with CSC/Utility Industry Division, a global leader of technology-enabled business solutions and services. She actively participates in the firm's smart grid consulting work for utility business clients as they plan and implement smart metering, dynamic pricing, and energy efficiency programs. Her current focus is on planning products and services for utility retail customers.

She previously led the AMI MDM Working Group to assist utilities in understanding the need for meter data management (MDM) and to successfully implement MDM. She was project manager for the award-winning Power-CentsDC dynamic pricing pilot program in the District of Columbia and presented the results of the program to the Executive Office of the President, the Department of Energy, and the National Science and Technology Council along with the rest of the program team. She started her utility career at Pacific Gas and Electric Company in the IT and Rate Departments, where she fielded the first time-of-use study for small commercial customers.

Ms. Harper-Slaboszewicz received her BA in Mathematics and MA in Economics from UC Berkeley.

Steven G. Hauser is Vice President, Grid Integration, at the National Renewable Energy Laboratory in Golden, CO, and serves as President Emeritus of The GridWise® Alliance, which he was instrumental in establishing. He is responsible for new strategies to lead national efforts to create a smarter grid and is a recognized expert on transforming the power sector to meet future economic, environmental, and energy security mandates. He has been active in clean and renewable energy technology developments and is a frequent speaker on the smart grid.

Mr. Hauser was the driving force behind the creation of the GridWise® Alliance and has been instrumental in bringing a broad coalition of companies to collaborate to implement its vision and in creating the new Global Smart Grid Federation. Previously, he held senior management positions at GridPoint, Battelle, and SAIC. During his career he has served as an advisor to numerous clean energy organizations.

He received a BS in engineering physics from Oregon State University and an MS in chemical engineering from the University of Washington.

Jennifer Hayward is a Research Scientist in the Energy Transformed Flagship, Commonwealth Scientific and Industrial Research Organization (CSIRO). She leads research projects on technology cost projections and has made economic modeling contributions to renewable energy projects.

Dr. Hayward's focus is on developing new methods and modeling approaches to provide robust projections of the capital costs of existing and emerging electricity generation technologies. She is a computational chemist who has worked in interdisciplinary environments to provide analysis of issues of future strategic significance.

She has a BSc (Honors) from the University of Newcastle, a PhD in Chemistry from the University of Sydney, and a Graduate Diploma in Biotechnology from the Australian National University.

Stephen Healy is a Senior Lecturer in the School of History and Philosophy and Co-Coordinator of the Environmental Studies Program of UNSW's Faculty of Arts and Social Science. He is also Research Coordinator of the Centre for Energy and Environmental Markets.

Dr. Healy has worked for Greenpeace International, the NSW EPA, and Middlesex University, UK, where he led the Science, Technology, and Society Program. His research interests include climate change, energy, risk and uncertainty, and public participation. He is interested in the historical development of contemporary systems of energy provision, energy institutions, governance, and politics and in re-envisioning energy consumption. He has published across a number of fields and co-authored *Guide to Environmental Risk Management*, published in 2006.

He has a BSc (Honors) in Physics and an Electrical Engineering PhD, in Photovoltaics, from UNSW.

Theodore Hesser is the **Energy Smart Technologies** analyst for Bloomberg New Energy Finance's research team based in New York. His work focuses on market trends and strategy within the emerging smart grid, energy efficiency, battery storage, and advanced transportation sectors.

Prior to joining Bloomberg he worked as a MAP fellow at NRDC where he was engaged in research focusing on the role of direct load control in integrating renewable resources onto the grid.

Ted received a BA in physics from The Colorado College and a MS in Civil and Environmental Engineering from Stanford University.

Magnus Hindsberger is Senior Manager Market Modeling at the Australian Energy Market Operator (AEMO). He is responsible for market simulations of future electricity and gas market outcomes supporting the long-term electricity planning process and cost-benefit assessments of new investments. Previously, he worked at Transpower New Zealand Limited, the Nordic consulting company

ECON, as well as Elkraft System now merged into Energinet.dk, which is the Danish transmission system operator.

He has a strong background in electricity market modeling and its application for long-term scenario planning and cost-benefit assessments. He is interested in new technology and has participated in international working groups including the IEA, CIGRE, and EU Framework Programme on distributed generation, demand-side participation, and electric vehicles.

Dr. Hindsberger holds MS and PhD degrees in Engineering (Operations Research) from the Technical University of Denmark.

Valeriia Iezhova holds an engineering degree from Kiev Polytechnic Institute and a Masters in Energy Engineering and Policies from Mines-ParisTech.

Her research at the Cerna, Mines-ParisTech concerned the evolution of power production in France and the consequent changes in prices in the liberalized power market. She also had an extended internship at Finance & Investments Department of EDF Groupe, one of the world's biggest electricity producers, where she participated in the development of an investment project in electricity generation and transmission.

Amira Iguer is a Research Engineer at the Center of Energy and Processes of Mines ParisTech. She has participated in energy and environmental research projects, including energy and price modeling, environmental impact assessment and management plans for power plants, and business models for load control. The objective of her most recent project was to assess the energy potential offered by electric vehicles with a V2G configuration to the grid.

Ms. Iguer holds a degree in Process and Environmental Engineering from the Ecole Polytechnique in Algiers and a Masters in Energy Engineering and Policies from the Mines-ParisTech.

Warren Katzenstein is a Principle Consultant with KEMA, formerly an Associate with The Brattle Group and whose expertise includes renewable energy. He recently developed new power spectrum analysis techniques to characterize intermittent renewable variability and quantify reductions in variability due to geographic diversification or storage.

He received a PhD in Engineering and Public Policy from Carnegie Mellon University and a BS in Engineering from Harvey Mudd College.

Chris King is President, Strategic Consulting at eMeter, a leading provider of smart grid software. He is responsible for strategic activities, including market analysis, product strategy, and policy, working with regulators and legislators worldwide.

Mr. King has 30 years of related experience as rate design director at Pacific Gas & Electric, VP – Sales & Marketing at CellNet Data Systems, CEO at electricity retailer Utility.com, and at eMeter. His research interests include smart grid economics, regulatory policy, technology, and consumer benefits. He chairs the European Smart Energy Demand Coalition and is a Director of the Demand

Response and Smart Grid Alliance and the Association of Demand Response and Smart Grid Professionals. He has testified before the U.S. Congress, several U.S. state bodies, and international regulatory bodies.

Mr. King holds bachelors and masters degrees in Biological Sciences from Stanford University, a masters in Management Science from Stanford's Graduate School of Business, and a doctorate in Law from Concord University.

Michael Koszalka is Vice President at ICF Inc., responsible for the firm's energy efficiency (EE) practice in the Western Region of the United States. He has more than 30 years of experience in the utility industry, both within utilities and in a consulting role. He has led all aspects of demand-side management (DSM) programs from business case development and integrated resource planning to developing and evaluating request for proposals for implementation services; managed EE and demand response (DR) programs; conducted program evaluations, measurement, and evaluation; and developed regulatory reports and filings. He has also developed and implemented strategic marketing programs and has experience with integrated resource planning (IRP).

Mr. Koszalka's main interests include using smart grid technologies to improve the effectiveness of demand response and energy efficiency efforts. He is a certified Measurement and Verification Professional (CMVP) and an accredited Lead GHG Verifier. He has taught both graduate and undergraduate marketing and management courses.

He has a BS in electrical engineering from the University of California at Berkeley and an MBA in Marketing Management from Golden Gate University in San Francisco.

Lorenzo Kristov is a Principal in the Market and Infrastructure Development Division of the California ISO, where he is engaged in both market design and infrastructure policy initiatives to enhance the efficiency and competitiveness of the markets and to facilitate state and federal environmental policy goals. He was a principal designer of the ISO's locational marginal pricing (LMP) market structure launched in 2009 and led the redesign of the ISO's transmission planning process, approved by FERC in 2010, to better support the integration of renewable energy resources. Prior to joining CAISO he worked at the California Energy Commission representing the Commission in CPUC proceedings to develop the rules for retail direct access. Before that, he was a Fulbright Scholar in Indonesia working on creating a commercial and regulatory framework to advance private power development.

Dr. Kristov is currently working to better integrate grid planning, interconnection policy, and market design from a whole-system perspective to enable the ISO to facilitate and adapt to the momentous technology- and environmental policy-driven changes occurring in California's power sector.

He received a bachelor's in mathematics from Manhattan College, a master's in statistics from North Carolina State University, and a PhD in economics from the University of California at Davis.

Peter Langbein is the Manager of Demand Response Operation for the PJM Interconnection, with responsibility for demand response participation that represents over 1 million customers and $500 million in revenue in 2010 from PJM's Energy, Capacity, and Ancillary Service Markets. He is responsible for the various demand response business processes and associated systems from entry into the market to the coordination of the final settlements for demand response activity.

Prior to joining PJM, he was engaged in competitive retail electricity market across North America in various capacities. He has over 17 years experience in the wholesale and retail power markets from retail sales and marketing, commodity risk management and trading, to customer care and settlements.

Mr. Langbein holds a BA in economics with honors from Montclair State College and an MA in Applied Microeconomics and Economics from New York University.

William Lilley is a Principal Research Scientist at Energy Transformed Flagship, Commonwealth Scientific and Industrial Research Organization (CSIRO), Australia's national science agency. Dr. Lilley designs and leads large integrated modeling programs investigating possible energy futures. In 2009 Dr. Lilley was seconded as a principal advisor to the Australian federal government's Smart Grid Smart City initiative. He is a scientist with expertise in distributed energy, smart grids, and transport and environmental analysis. He is currently designing an analytical framework to optimize the development of generation assets and associated transmission and distribution networks to meet Australia's greenhouse gas reduction targets.

Dr. Lilley recently released a comprehensive report detailing the value of distributed energy, broadly covering distributed generation, demand management, and energy efficiency for Australia. He has studied the impact of motor vehicle and industrial emissions on human exposure, emission measurements from motor vehicles and freight locomotives, inverse modeling techniques for determining fluxes from complex area sources, impacts of intelligent transport systems, and the development of a near field chemically reactive dispersion model.

Dr. Lilley holds undergraduate degrees in Science and a PhD in Environmental Science from the University of Newcastle, Australia.

Iain MacGill is an Associate Professor in the School of Electrical Engineering and Telecommunications at the University of NSW, and Joint Director (Engineering) for the University's Centre for Energy and Environmental Markets (CEEM). Iain's teaching and research interests include electricity industry restructuring, sustainable energy technologies, and energy and climate policy.

CEEM undertakes interdisciplinary research in the analysis and design of energy and environmental markets and their associated policy frameworks. It brings together UNSW researchers from the Faculties of Engineering, Business, Science, Law and Arts and Social Sciences. Dr. MacGill leads two of CEEM's three major research programs. The Sustainable Energy Transformation program is researching sustainable energy technology assessment, renewable energy integration and frameworks for sustainable energy transitions. The Distributed Energy Systems program is undertaking research on 'smart grids' and 'smart' homes, and frame-works for facilitating distributed generation and demand-side participation in electricity markets. Iain has published widely in these and related areas.

Dr. MacGill has a Bachelor of Engineering (Electrical) and a Master's of Engineering Science (Biomedical) from the University of Melbourne, and a PhD on electricity market modeling from the University of NSW.

Kamen Madjarov is an Associate with The Brattle Group. His work focuses on the analysis of integrating renewable generation onto the power grid. He was recently part of a team that designed and developed a computational tool that quantifies the operational impact of variable resources on the grid.

He is currently completing his PhD in Economics at Northeastern University.

Todd McGregor is the AMI Program Manager at Pepco Holdings, Inc (PHI), one of the largest energy delivery companies in the mid-Atlantic region, serving 1.9 million customers in Delaware, the District of Columbia, Maryland, and New Jersey. Mr. McGregor's responsibilities include procuring and deploying the Advanced Metering Infrastructure (AMI) and Meter Data Management System, redesigning the affected business processes, managing the change to the new business processes, realizing the operational and customer benefits associated with AMI, and supporting regulatory filings in the context of PHI's smart grid initiative. He was also recently elected as Co-Chair of the Implementation Working Group of the Gridwise Alliance.

He previously served at PHI as the process manager for meter-to-cash, where he was responsible for the design, execution, and improvement of PHI's "order to cash" process.

Mr. McGregor received his BS in Accounting and Business Administration from Illinois State University and an MBA in Finance and Accounting from the University of Illinois at Urbana-Champaign.

Se Jin Park is a Manager of the Demand Market Team in the Electricity Market Department of the Korea Power Exchange in Seoul, Korea. His long career as an established Electricity Engineer in Korea includes years of industry experience related to demand response, market operation, and power system planning.

Mr. Park provides leadership and support for an International Collaboration Program on Distributed Energy Resources. In doing so, he collaborates with U.S. universities, research centers, and businesses on research aimed to strengthen the planning, analysis, and implementation of distributed energy resources.

Mr. Park received a degree in Electrical Engineering at Yeungnam University.

Glenn Platt leads the Local Energy Systems Theme within the Energy Transformed Flagship, Commonwealth Scientific and Industrial Research

Organization (CSIRO). Prior to CSIRO, Platt worked in research and engineering roles with a variety of organizations, from Nokia Mobile Phones to small engineering consulting firms, covering the telecommunications, engineering, and industrial automation fields.

The Local Energy Systems Theme within CSIRO aims to apply the latest state-of-the-art technologies to reducing energy consumption through energy management systems, improved building design and modeling, new devices, and smart grid technologies.

Dr. Platt holds a PhD, MBA, and an electrical engineering degree, all from the University of Newcastle, Australia.

William Prindle is a Vice President with ICF International, a global energy-environment consultancy. He helps lead the firm's energy efficiency work for government and business clients, including major support work for U.S. EPA's ENERGY STAR® and related efficiency programs, as well as utility efficiency program development. He also supports the firm's corporate energy management, carbon management, and sustainability advisory services.

He was previously Policy Director at the American Council for an Energy-Efficient Economy (ACEEE), where he led research and advocacy work on energy and climate policy for federal and state governments. Prior to that, he directed buildings and utilities programs for the Alliance to Save Energy, and was a management consultant before that.

Mr. Prindle received his B.A. in Psychology from Swarthmore College and M.S. in Energy Management and Policy from the University of Pennsylvania.

Luke Reedman is Energy Modeling Stream Leader in the Energy Transformed Flagship, Commonwealth Scientific and Industrial Research Organization (CSIRO). Dr. Reedman is an economist with expertise in energy market analysis and economic modeling of the transport and stationary energy sectors. He leads energy modeling research with ongoing research in multiscale energy modeling, integrated assessment of the development of energy markets and networks, sustainable aviation fuels, alternative road transport analysis, and investment under uncertainty.

Dr. Reedman has published on energy futures, examining the effects of carbon constraints on the deployment of distributed generation, the impact of policy uncertainty on the deployment of low-emission electricity generation technologies, and the possible implications of convergence between the electricity and transport sectors.

Dr. Reedman holds undergraduate degrees in Economics and Applied Finance and a Ph in Economics from the University of Newcastle, Australia.

Philippe Rivière is an Associate Professor at Mines ParisTech and the head of the MSc Energy Strategies.

He has been working on the Demand Side Management research team since 2004, focusing mainly on HVAC of buildings. He led several studies on this topic for the French Energy Agency ADEME and the European Commission. His expertise extends to modeling energy systems in buildings, power generation, and the life-cycle analysis of these systems.

Professor Rivière graduated in mechanical engineering from the Ecole Nationale Supérieure des Techniques Avancées and holds a PhD in Energy from Mines-ParisTech.

Mark A. Rothleder is Director of Market Analysis and Development at the California Independent System Operator Corporation (CAISO) and leads the renewable integration initiative. He has previously held the positions of Principal Market Developer and Director of Market Operations and was part of the CAISO start-up team. Prior to joining CAISO, he worked in the Electric Transmission Department of Pacific Gas & Electric Company, where his responsibilities included Operations Engineering, Transmission Planning, and Substation Design.

Since joining the CAISO he has worked extensively on implementing and integrating the approved market rules for California's competitive Energy and Ancillary Services markets and the rules for Congestion Management, Real-Time Economic Dispatch, and Real-Time Market Mitigation into the operations of the CAISO Balancing Authority Area. He was instrumental in the implementation of market rules and software modifications to CAISO's Market Redesign and Technology Upgrade project (MRTU) in 2009.

Rothleder is a registered Professional Electrical Engineer in the state of California and earned a BS in Electrical Engineering from the California State University, Sacramento, and an MS in Information Systems from the University of Phoenix.

Heather Sanders is the Director of Smart Grid Technologies and Strategy at the California Independent System Operator (CAISO), where she is responsible for research and promotion of smart grid and other technologies supporting ISO reliability, market efficiency, and transmission utilization objectives.

Ms. Sanders recently led an effort to develop a Smart Grid Roadmap for the ISO that resulted in a high-level approach to integrating advanced and smart grid technologies to support the California state and ISO objectives.

She graduated from South Dakota School of Mines and Technology with a BS in Electrical Engineering and has an MBA from the University of Utah.

Fereidoon P. Sioshansi is President of Menlo Energy Economics, a consulting firm based in San Francisco, California, serving the energy sector. Dr. Sioshansi's professional experience includes working at Southern California Edison Company (SCE), the Electric Power Research Institute (EPRI), National Economic Research Associates (NERA), and most recently, Global Energy Decisions (GED), now called Ventyx. He is the editor and publisher of *EEnergy Informer*, is on the Editorial Advisory Board of *The Electricity Journal*, and serves on the editorial board of *Utilities Policy*.

Dr. Sioshansi's interests include climate change and sustainability, energy efficiency, renewable energy technologies, regulatory policy, corporate strategy, and integrated resource planning. His four recent edited books, *Electricity Market Reform: An International Perspective*, with W. Pfaffenberger; *Competitive Electricity Markets: Design, Implementation, Performance; Generating Electricity in a Carbon Constrained World;* and *Energy, Sustainability, and the Environment: Technology, Incentives, Behavior*, were published in 2006, 2008, 2009, and 2011, respectively.

He has degrees in Engineering and Economics, including an MS and PhD in Economics from Purdue University.

Paul M. Sotkiewicz is the Chief Economist in the Market Services Division at the PJM Interconnection where he provides analysis and advice on PJM's design and performance, including the implications of federal and state policies on various PJM markets. Prior to his present post he served as the Director of Energy Studies at the Public Utility Research Center, University of Florida, and prior to that he was an economist in the Office of Economic Policy and later served on the Chief Economic Advisor's staff at the Federal Energy Regulatory Commission.

At PJM, Dr. Sotkiewicz has led initiatives to reform scarcity pricing as mandated by FERC Order 719 and has also led efforts examining transmission cost allocation and the potential effects of climate change policy on PJM's energy market. He has also been involved in developing proposals for compensating demand resources in PJM's energy market prior to the recent FERC NOPR on demand response compensation and is involved in the development of price-responsive demand.

Dr. Sotkiewicz received a BA in History and Economics from the University of Florida and an MA and PhD in Economics from the University of Minnesota.

James Strapp is a Partner in IBM's Energy and Utilities Strategy and Change Consulting Practice, based in Toronto, Canada, where he leads a global team of smart grid specialists providing expertise on IBM's largest and most strategic engagements around the world. In this capacity, he is actively engaged in a variety of smart grid projects in North America, Australia, and Europe. He has been involved in Ontario's electricity market for nearly 15 years, including engagements with the Independent Electricity System Operator, the Ontario Energy Board, and local distribution companies.

He started his consulting career 20 years ago focusing on greenhouse gas abatement strategies and the evaluation of demand-side management programs, including engagements in Ontario, British Columbia, the United Kingdom, and Australia.

Mr. Strapp has a BA from the University of Western Ontario and an MSc in Geography from the University of Victoria.

Samir Succar is a Staff Scientist at the Natural Resources Defense Council's (NRDC) Center for Market Innovation based in New York. His work focuses on the integration of renewable energy and the role of T&D infrastructure upgrades, demand resources, energy storage, and other enabling technologies.

Prior to joining NRDC, he was a member of the research staff of the Energy Systems Analysis group at the Princeton Environmental Institute at Princeton University, where his research focused on integration issues associated with utility-scale renewable energy and on enabling technologies for intermittent generation. His interests include reliance on energy storage to enhance transmission infrastructure utilization and mitigating the intermittency of renewable energy, specifically compressed air energy storage, and other bulk storage technologies.

Dr. Succar received a BA from Oberlin College and a PhD in Electrical Engineering from Princeton University.

Steve Sunderhauf is Manager of Customer Programs at Pepco Holdings, Inc. (PHI), the holding company for Atlantic City Electric Company, Delmarva Power & Light Company, and Potomac Electric Power Company.

Mr. Sunderhauf is responsible for the design and evaluation of customer demand-side management programs for PHI's three regulated utilities. His responsibilities include the design, development, and evaluation of new smart-grid-enabled customer product offerings, which include advanced metering infrastructure (AMI) enabled dynamic pricing and AMI-enabled demand response equipment.

Mr. Sunderhauf earned an economics degree from Bucknell University, an M.S. degree in public policy and management from Carnegie-Mellon University, and a J.D. from the George Washington University. He is a member of the Maryland Bar.

Chris Thomas is Policy Director at the Citizens Utility Board (CUB) in Chicago, where he is engaged as a utility economist and energy policy consultant providing technical support for the development of comprehensive energy market action plans and encouraging innovation in energy markets to benefit customers. He is responsible for developing and overseeing implementation of CUB's policy agenda, which advocates for inclusion of demand-side resources into the energy supply portfolio.

His international experience includes leading research and training on energy markets and smart grid development for government agencies and businesses in Asia, Europe, and the Middle East. He currently provides leadership for the Illinois Smart Building Initiative and is a member of the program management office for the Illinois-Korea Smart Grid and Green Economy Committee. He is a frequent speaker at regional and national conferences.

Mr. Thomas received an undergraduate degree in Business Administration at Truman State University in Missouri and an MS in Economics and Finance at Southern Illinois University, Edwardsville.

Dan Ton is Program Manager of Smart Grid R&D within the U.S. Department of Energy (DOE) Office of Electricity Delivery and Energy Reliability. He is responsible for developing and implementing a multi-year R&D program for next-generation smart grid technologies to transform the electric grid in the United States through public/private partnerships.

Previously, he managed the Renewable Systems Integration program within the DOE Solar Energy Technologies Program.

Mr. Ton received a BS in Electrical Engineering and an MS in Business Management from the University of Maryland.

W. Maria Wang is a Technology Manager for Energy & Environmental Resources Group, LLC (E2RG), where she provides technical consulting services to government agency programs related to the smart grid, renewable energy, and energy storage. Dr. Wang supports the U.S. Department of Energy (DOE) Smart Grid R&D Program, GridApp utility consortium, and the Batteries for Advanced Transportation Technologies (BATT) Program.

She has also supported the DOE Solar Energy Technologies Program culminating in two reports, "High Penetration of Photovoltaic Systems into the Distribution Grid" and "Solar Energy Grid Integration Systems – Energy Storage." In addition, Dr. Wang has completed a comprehensive summary of the Renewable Systems Interconnection Study for Sandia National Laboratories.

Dr. Wang has a BS from the Massachusetts Institute of Technology and MS and PhD degrees in Chemical Engineering from Stanford University.

Jianhui Wang is a Computational Engineer at the Center for Energy, Environmental, and Economic Systems Analysis (CEEESA) at Argonne National Laboratory.

Dr. Wang is the chair of the IEEE Power & Energy Society power system operation methods subcommittee and co-chair of an IEEE task force on integration of wind and solar power into power system operations. He has authored/co-authored more than 60 journal and conference publications. He is an editor of the *IEEE Transactions on Smart Grid* and an associate editor of the *International Journal of Renewable Energy Technology*. His research interests include large-scale modeling and simulation, electric power systems optimization, electric vehicles, electricity restructuring, computational intelligence, climate change, and the smart grid.

He received a BS and MS from the School of Business Administration in North China Electric Power University and a PhD in Electrical Engineering from the Illinois Institute of Technology.

THE INCEPTION OF THE SMART GRID

With no warning, over 50 million people in the Northeast United States and Ontario, Canada, lost power in August 2003. Everyone's first reaction was, how could this happen and why was it so widespread? But once the initial investigations of the massive failure of the grid were conducted and appropriate blame was placed on the guilty parties, the focus of attention shifted to how can we avoid a similar accident in the future.

The first obvious answer was to find ways to make the grid more intelligent, more reliable, and more secure. Among the suggestions made were to make the grid self-detecting and self-healing, so that future accidents could be detected at an early stage, resolved, or isolated to avoid another catastrophic cascading failure of the scale of the August 2003 blackout. This line of thinking naturally led to the term *smart grid*, which has become a household word.

Since 2003, a flurry of research, vast effort, and large investments of capital and resources have gone into the *smart grid*. The investments to date, however, are likely to be dwarfed by the additional investments expected in the coming years. There is near unanimous support for the smart grid concept, even during the current economic downturn. A significant amount of money in the US stimulus funding, for example, was specifically targeted towards smart grid investments. Internationally, the smart grid is equally popular as indicated in a recent study attempting to describe its universal appeal and growing footprint (Figure 1).

Yet, despite the widespread embrace of the smart grid concept, there is no universal definition of what it encompasses. It is probably safe to say that there are as many definitions of "smart grid" as there are smart grid projects, experts, or practitioners.[1] This poses a challenge, especially for an edited book such as this, which consists of contributions from a diverse number of experts, scholars, academics, and practitioners from different walks of life, representing different disciplines and perspectives. For these reasons, it is helpful to provide a definition and scope for smart grid at the outset.

WHAT IS THE SMART GRID?

For the purpose of the present volume, smart grid is defined as any combination of enabling technologies, hardware, software, or practices that collectively make the delivery infrastructure or the grid more reliable, more versatile,

---

[1]Hauser et al. provide further perspective on the inception and evolution of smart grid.

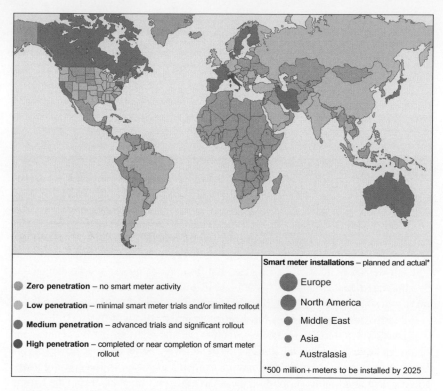

FIGURE 1 The growing global footprint of smart grid. *(Source: Smart from the Start, PricewaterhouseCooper, 2010)*

more secure, more accommodating, more resilient, and ultimately more useful to consumers.

Moreover, a useful definition of the smart grid must encompass its ultimate applications, uses, and benefits to society at large. In this sense, the smart grid must include a number of key features and characteristics including, but not limited, to the following:

- It must facilitate the integration of diverse supply-side resources including increasing levels of intermittent and non-dispatchable renewable resources;
- It must facilitate and support the integration of distributed and on-site generation on the customer side of the meter;
- It must allow and promote more active engagement of demand-side resources and participation of customer load in the operations of the grid and electricity market operators;
- It must allow and facilitate the prices-to-devices[2] revolution that consists of allowing widespread permeation of dynamic pricing to beyond-the-meter

---

[2]This word, to the editor's knowledge, was first coined by EPRI.

applications, enabling intelligent devices to adjust usage based on variable prices and other signals and/or incentives;

- It must ultimately turn the grid from a historically one-way conduit that delivers electrons from large central stations to load centers, to a two-way, intelligent conduit, allowing power flows in different directions, at different times, from different sources to different sinks;
- It must allow for broader participation of energy storage devices on customers' premises or centralized devices to store increasing levels of energy when it is plentiful and inexpensive, to be utilized during times when the reverse is true, and accomplish this intelligently and efficiently;
- It must allow distributed generation as well as distributed storage to actively participate in balancing generation and load;
- It must encourage more efficient utilization of the supply-side and delivery "network" through efficient and cost-effective implementation of dynamic pricing and similar concepts;
- It must allow storage devices on customer side of the meter, including electric batteries and similar devices, to feed the stored energy back into the grid when it makes economic sense to do so;
- It must facilitate any and all concepts and theories that encourage greater participation by customers and loads in balancing supply and demand in real time through concepts broadly defined as demand response;
- It must make the "grid"—the vast generation, transmission, and distribution network—more robust, more reliable, and more secure to interruptions, and less prone to accidents or attacks of any kind; and
- Of course, it must accomplish all these while reducing the costs of the operation and maintenance of the "network," with commensurate savings to ultimate consumers.

If this sounds like an all-encompassing definition, it is intentional. If it sounds like a tall order, that is also intentional as a glance at the titles of the chapters of this book suggests. Before leaving this subject, I must also point out what my definition of the smart grid and this book do *not* cover, and that is technical engineering detail. If, for example, you are looking for a discussion of dynamic loading of transmission lines made possible through the application of synchronous phasers[3] and capacitors, you will be disappointed. This book is not technical and its intended audience is not engineers.

WHAT'S WRONG WITH THE EXISTING GRID?

The short answer, as pointed out by Gellings in the book's Foreword, is that the current grid is a relic of the past, designed to meet the needs of a different industry in a bygone era with outdated technologies that are incapable of meeting today's requirements, let alone those of the future.

---

[3]Chapter 6 actually does mention synchronous phasers, but only in passing.

The existing grid was designed essentially as a one-way conduit to transmit vast amounts of power generated primarily at a limited number of large central power stations to major load centers.[4] The original thinking behind the entire network, from generation to transmission and distribution, was based on the outdated premise that customer load is a given, which requires generation to be adjusted to meet. The balancing of supply and demand in real time was routinely accomplished by adjustments on the supply side. Until recently, customer demand was not subject to control or manipulation, with virtually no means or incentives for load to play an active role.

In a *dumb grid*—in contrast with the emerging smart grid—environment managing, influencing or controlling customer demand, was problematic or simply impossible mostly due to the limitations of the technology, primarily in two key areas:

- First, until recently, primitive electromechanical meters in use to measure consumption for all but the largest customers were only capable of registering volumetric usage not differentiated by time of use, voltage, wattage, or anything else. More sophisticated meters are needed to be able to influence the pattern and amount of customer consumption.
- Second, severe limitations in communications between the suppliers and end-users have been another major impediment to managing consumption until recent times. As further described in numerous chapters of this book, suppliers as well as the grid operator need much more robust means of communication to send price signals and receive feedback from consumers in real time if there is any hope of influencing demand.

As further described below, the metering limitations placed severe restrictions on how electricity could be priced—resulting in flat rates, undifferentiated by time or location of use, application, or anything else. Moreover, limitations on communications meant that revenue collection could be nothing but primitive. Until recently, most consumer bills were calculated based on only one piece of data—a monthly or quarterly kWh consumption figure—and in many cases even that could not be achieved, resulting in an estimate.[5] Even today, the vast majority of consumers, including many in developed countries, are billed on a flat multiplier—the cents/kWh—multiplied by the volumetric consumption for the period.

---

[4]The existing grid was designed to meet the needs of large and vertically integrated utilities operating their networks largely in isolation as independent entities. Organized wholesale markets, third-party generators, and increased penetration of renewable energy generation are relatively recent developments that were not part of the grid's original design.

[5]Customers in many parts of the world are billed quarterly, in some cases based on a single annual reading that attempts to reconcile the estimated bills with the actual reading.

These technological limitations plus the low and falling per-unit costs of generation meant that consumers would be billed on simple tariffs that remained flat for all hours of the day, week, month, and year and only infrequently adjusted. Moreover, customers were charged postage stamp rates that did not vary across wide geographic areas.

Viewed in this context, it is probably a small exaggeration to say that for much of its history, the electric power industry did its best to literally *disengage* customers from the upstream side of the meter simply because the technology did not allow anything else (Box 1).

Box 1: How Would the Smart Grid Reengage the Disengaged Customers?[6]

It was probably not deliberate but for much of its history, the electric power sector paid scant attention to what consumers did to electricity once it reached their meter. Beyond the meter was not only beyond the industry's technical reach, but figuratively and literally beyond its control, influence, or interest. In the process, customers were virtually disengaged from the upstream side of the business.

This was accomplished through massive investment in an extensive and ubiquitous distribution network, the so-called poles and wires, which connected virtually every consumer in developed parts of the world to the grid. Within their premises, customers are never more than a few feet away from a switch or a plug—and for all practical matters, that is all they need.

Consumers can turn on the switch or plug in any device at anytime, anywhere, and draw as much as they want, for as long as they want it. The industry has made it easy and simple for them.

The switches and the plugs, of course, are connected to a transformer, which is connected to a distribution network, which is connected to high-voltage transmission lines, which are connected to a network of central power plants *somewhere*. And power plants, in turn, are fed by primary energy sources that go even further upstream of the plug or the switch. But the average consumer need not know anything about any of this. For all practical purposes, any plug or any switch *anywhere* is connected to an *infinite* supply of energy.

Making matters even more convenient for consumers, over the years, utilities invested massive amounts in ever larger and more efficient power plants. For much of the 20th century, the unit cost of electricity *declined* in real terms. For most consumers, electricity bills were a negligible and diminishing percentage of their disposable income. Some utilities even encouraged their consumers to use *more* by offering *declining block tariffs*, where unit costs fall with increased consumption. Even today, electricity prices in many parts of the world are a relative bargain as exemplified in the following map of current average retail electricity rates in the United States.

---

[6]Excerpted from *EEnergy Informer*, Feb. 2011.

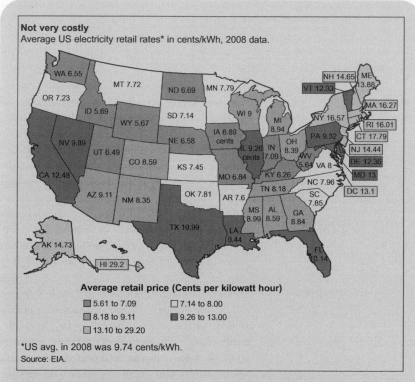

Not very costly
Average US electricity retail rates* in cents/kWh, 2008 data.

Average retail price (Cents per kilowatt hour)

- 5.61 to 7.09
- 7.14 to 8.00
- 8.18 to 9.11
- 9.26 to 13.00
- 13.10 to 29.20

*US avg. in 2008 was 9.74 cents/kWh.
Source: EIA.

Starting in the mid-1970s—the date varies in different parts of the world—the industry's miracle of *economies of scale* and declining per-unit costs came to an end. Building new capacity gradually became *more* expensive relative to the average cost of the existing capital stock. Adding new capacity began to *increase*, rather than *decrease*, the average generation costs. For example, nuclear power, once promised to be too cheap to meter, turned out to be a rather expensive proposition as a generation of new units entered service in the 1970s and 1980s.

Starting in the 1970s, fuel prices also began to rise, to varying degrees in different parts of the world. Oil, the universal benchmark for energy costs, jumped to $25 a barrel during the first oil crisis in 1973, is now hovering around $100, and is projected to become more expensive over time.

In the 1980s, new concerns about security of supplies, fuel diversity, and price volatility were added to the list of woes. In the 1990s and 2000s, additional issues including fears about the scarcity of finite supplies of fossil fuels,[7] effect of anthropogenic greenhouse gas emissions on climate, and sustainability became pronounced. Gradually but surely, the industry's single-minded focus on the supply side has given

---

[7]The subject of scarcity and price of fossil fuels, of course, is controversial. As prices rise, more reserves become economically feasible or "recoverable," in industry parlance. Moreover, improved technology, such as horizontal drilling for non-conventional shale gas, has the potential to vastly increase what can be considered recoverable.

way to a new appreciation of customer demand and increased efficiency of energy utilization, including new ways of balancing supply and demand in real time.

Many within and outside the industry are now convinced that the cheapest kWh is the one we do not consume. Energy efficiency—once given mostly lip service—is now considered a major energy resource, and a cost-effective option.

These same visionaries are now in favor of turning things around by *reengaging* the disengaged consumers. Even though these initiatives are in early stages of their evolution, they are nevertheless moving in the direction of making consumers more active in the market of supply and demand.

Among the most stunning examples of this *paradigm change* is the evolution of thinking at the Federal Energy Regulatory Commission (FERC), the closest thing in the US to a federal energy regulator. In a recent interview with the *New York Times*,[8] current FERC chairman Mr. Jon Wellinghoff said, "The energy future of the US looks radically different from its past," partly because consumers will become, "… active parts of the grid, providing energy via their own solar panels or wind turbines, a system called distributed generation; stabilizing the grid by adjusting demand through intelligent appliances or behavior modification, known as demand response; and storing energy for various grid tasks." Mr. Wellinghoff is not only supportive of such schemes but believes that "consumers should get paid to provide these services."

Wellinghoff's Quotable Quotes—Excerpted from *New York Times* Article November 29, 2010

- "I believe that for **markets to be competitive**, we need to have as many different types of resources in those markets as possible."
- "We're doing what we can to the extent that we have jurisdiction to ensure that there are no barriers to **distributed generation** becoming part of whole-sale markets."
- "To the extent that you can put **demand response** in the system—that is, have consumers control their loads at times when the system is stressed—you can reduce substantially the amount of fossil fuel generators that are needed to relieve that stress."
- "If a **battery** or a dishwasher or a water heater or an aluminum pot or a compressor in a Wal-Mart can respond on a microsecond basis, and it takes the generator a minute to respond, that faster response should be rewarded a higher payment because, in fact, it's providing a better service."
- "We're reviewing the **economic benefits of storage** and how storage should be compensated for the various services it can provide to the grid."

While such ideas are not necessarily new or novel, coming from FERC's chairman, they get noticed (see quotes from the NYT article above). Moreover, FERC has taken the unusual step of actively promoting these ideas, not just through interviews and public pronouncements, but through published reports, studies, and surveys as well as a number of *orders*, which in no uncertain terms oblige the Regional

---

[8]*Making the consumer an active participant in the grid*, New York Times, 29 Nov 2010.

Transmission Organizations (RTOs) and Independent System Operators (ISOs) to implement the concepts.[9]

In the past couple of years alone, FERC has published several seminal studies documenting the substantial potential for demand response (DR) in the United States (see the accompanying graph) and has issued a number of orders that are prompting fundamental changes in organized U.S. electricity markets.

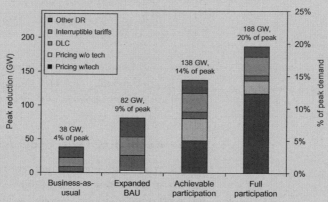

What's the scope for DR?
US DR potential by 2019 from different types of programs and under alternative scenarios.

Source: National assessment of DR potential, FERC, June 2009.

As a result of FERC's incessant prodding in the past several years, most organized wholesale market operators in the US have now developed various forms of DR programs, have incorporated demand-side bidding into their predominantly supply-focused auctions, and are broadly—although not always effectively or successfully—supportive of the growing significance of the demand side of the market.

It sounds simple and trivial today—but the idea of getting consumers to become active participants in the market is still novel to many in the industry and even more so to the average consumer who has been successfully trained to be a passive user.

One of the two most obvious examples of this paradigm shift is the effort to make price signals more transparent and visible to consumers, including the rising interest in dynamic pricing.[10]

The second is the growing interest in demand response (DR) programs, broadly defined as anything that influences consumers to reduce load during peak demand periods and/or shift load to off-peak periods, usually in response to incentives or price signals.[11]

---

[9]Chapter 17 provides further elaboration of a number of pertinent FERC orders.

[10]Chapter 3 expands on the concept of dynamic pricing. Chapter 12 provides evidence on how large commercial and industrial customers have responded to dynamic pricing in California.

[11]Several chapters in this volume cover demand response in more detail.

Dynamic pricing, at its core, is nothing more than alerting customers that a kWh consumed at 2 PM on a hot summer afternoon is *not* the same as a kWh consumed at 2 AM. The former costs a lot more to generate and deliver. Under a flat tariff regime, a kWh is a kWh, no matter when and where it is consumed.

Getting consumers to accept and digest this simple and rather obvious message is going to take a lot of effort. The utility industry only has itself to blame for not making this obvious point sooner, say, starting 30 years ago when the industry's peak demand period became pronounced with the increased penetration of air conditioning load in most summer-peaking utilities.

Flat rates are *flat wrong* because they give the *wrong* signal to consumers that it is okay to use as much as they want, any time they want it. The result is needle-sharp peak loads experienced by summer-peaking utilities, such as those in California (see graph below).

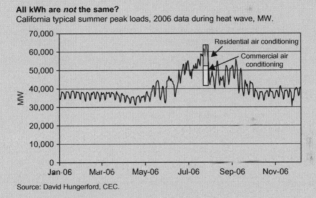

All kWh are *not* the same?
California typical summer peak loads, 2006 data during heat wave, MW.

Source: David Hungerford, CEC.

The growing interest in rising block rates is another manifestation of this concern. Under a rising block rate tariff, such as those in effect in California (see the table below), consumers pay progressively higher per-unit prices as they consume larger volumes of electricity. The subtle—or perhaps not so subtle—message conveyed by rising block tariffs is to punish those consumers with prodigious energy consumption while rewarding the frugal ones. Most consumers are barely aware of their total bills—most pay little attention to the details.

Anecdotal observations and discussions with friends and neighbors suggest that most residential consumers are totally unaware of the existence of the rising block tariffs even when they vehemently complain about their high utility bills. The simple message that the more you consume, the higher the prices, does not appear to have been communicated to the average consumer in California despite the fact that these tariffs have been in place for some time. This, by the way, appears to be a universal problem. Most consumers in liberalized European markets do not appear to understand their electric bills either.

The More You Use, the Higher the Price

Example of the 5-tier price scheme for residential customers in SCE service area,* in effect Oct 09

Tier	Price Cents/kWh[1]	Baseline Allowance[2]
Tier 1	11.808	0–100%
Tier 2	13.741	101–130%
Tier 3	23.334	131–200%
Tier 4	26.833	201–300%
Tier 5	30.334	>300%

*Baseline allowance is determined by applicable climate zone; higher allowances apply to high temperature zones, lower for mild coastal zones.
[1]For low-income customers, applicable prices for the first three tiers are 8.533, 10.668, and 18.051 cent/kWh, respectively, with the tier 3 rate applied to all usage above 130% of baseline allowance.
[2]Link to SCE's Baseline Allocation table: http://www.sce.com/CustomerService/billing/tiered-rates/baseline-chart-map.htm
Source: Southern California Edison Company.

Which brings us to the growing emphasis being placed on demand response (DR). Huge amounts of resources and effort are going into conveying another simple message that any 3rd grader intuitively understands. Cut back all but your most essential usage during periods when electricity is scarce and expensive to generate and deliver—namely during peak demand periods—or else be prepared to pay a premium.

Reducing peak demand through incentives? What a novel idea
Estimates of the potential scale of demand response under alternative scenarios over a decade.

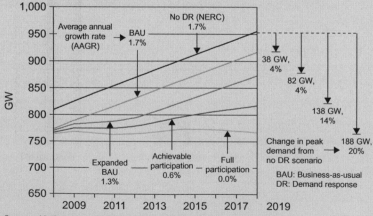

Source: National assessment of DR potential, FERC, June 2009.

> What makes this simple message complicated is also historical. For over a century, consumers were encouraged to use as much as they wanted, whenever they wanted it, and with flat tariffs that told them all kWh are the same. Now, we are trying to tell them, "Forget about what we told you before and why we did not charge you differentiated rates. Starting tomorrow, we are going to charge you more when it costs us more to generate and deliver the juice." It may be a simple message, but try explaining it to the average consumer and see how far it goes.
>
> Yet, the time has finally arrived to deliver the message—the bitter medicine—that all kWh are not the same. As the earlier graph from a recent FERC survey suggests, the industry's peak demand is not a given. It can be managed, modified, and reduced by aggressive implementation of dynamic pricing with a host of enabling technologies that are now being tried in a variety of smart grid implementation projects around the world.[12] It may sound fantastic and futuristic, but it is no longer a theory or a pipe dream.

Aside from the limitations in metering and communications, the systematic disengagement of customers described in Box 1 can be traced to another important characteristic of the industry, which no longer applies.

For much of its history, the electric power industry was blessed with enormous economies of scale. As the industry grew and expanded, electricity could be generated and supplied to consumers at falling real costs. The industry, dominated by engineers, was preoccupied with rapid expansion necessitated by a rapid rise in demand, which the falling prices further encouraged. The industry, it could be said, was supply-focused. Under prevailing rate of return regulations, still in effect in many parts of the world, suppliers benefited from more investments on which they could earn a regulated rate of return. And the same regulation gave them every financial incentive to encourage more consumption—not less.

Under such circumstances, why would a supplier bother with promoting energy efficiency or fuss with managing demand? In fact, consumers were encouraged to use more, in some cases with falling block tariffs, which meant the more they used the lower the per-unit cost. With the introduction of commercial nuclear power there was even talk that electricity would simply be "too cheap to meter,"[13] with the implication that consumers may simply be billed a flat and low monthly fee—forget about metering altogether.

In the beginning, none of this much mattered since consumers used little and meter reading, billing, and revenue collection were considered a necessary nuisance, with the best and brightest industry people concentrating on building bigger power plants and more transmission and distribution lines. But in recent

---

[12]Part IV of the book provides a number of examples of efforts to better manage peak demand using the smart grid, smart meters, smart prices, and smart devices.

[13]The now famous quote is attributed to the chairman of the Atomic Energy Commission during President Eisenhower's administration.

years, as consumer loads, especially in the commercial and residential sectors, have grown, flat undifferentiated tariffs have turned into something of a curse.

With the phenomenal growth of air conditioning globally over the past 3 decades, flat undifferentiated tariffs have resulted in highly pronounced spiky peaks such as those exhibited in California (see Box 1). The air conditioning load, driven by hot summer temperatures, has become a noticeable nuisance, sometimes referred to as "the load from hell," due to its spiky and highly unpredictable nature. Serving this load requires significant investment in peaking generation, results in price spikes in the market, and has generated interest in ways of managing it.

A number of chapters of this book are focused on solutions to address the peak load problem and the inherent inefficiencies and inequities associated with flat rates in networks where costs of service are highly dependent on time of use and, to a lesser extent, its location.[14]

There are, of course, a number of other reasons suggesting that our existing grid is out of line with our current needs and, more important, future requirements. Among the most compelling reasons is the increasing significance of renewable energy resources. While some of these, such as geothermal, may operate as base-load or can be controlled, such as hydro, others, particularly wind and solar energy, are intermittent by nature and largely non-dispatchable.

Our existing grid is ill equipped to deal with these types of renewable resources.[15] First, system operators who manage generation and transmission networks dislike non-dispatchable resources that are not under their direct command and control. Second, the intermittency and variability play havoc with their traditional means of balancing supply and demand, which is based on scheduling generation to meet variable but highly predictable load. This philosophy does not bode well going forward in many parts of the globe where intermittent renewables are projected to make up a third or more of the resource base within a decade, if not earlier. As shown in Figure 2, more than 30 states in the US now have mandatory targets for renewable energy resources going forward.

In the European Union, there is a target to meet 20% of the energy requirements by 2020 from renewable resources. Germany, following the Fukushima accident in Japan in March 2011, is considering a total phase-out of its existing nuclear fleet, mostly to be replaced by renewable energy—as much as 80% of the generation needs by 2050.

Another major phenomenon affecting the electric power sector is the emergence of distributed generation (DG).[16] With rapid technological development and an equally rapid drop in prices, many DG technologies already are or will

---

[14]Brunekreeft et al. discuss the implications of the prevailing tariffs that do not vary by location, giving inefficient price signals to both load and generation.

[15]Chapter 6 describes the challenges faced by California's grid operator as the state is marching toward a 33% renewable target by 2020.

[16]See Chapter 7.

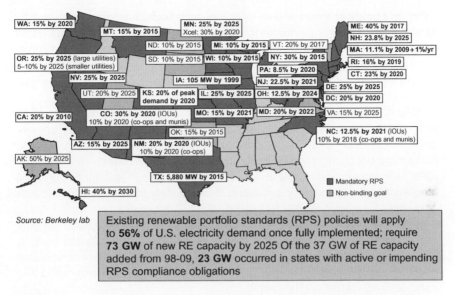

FIGURE 2 US RPS mandates. *(Source: Supporting Solar Power in RPS Standards: Experience from the US, Lawrence Berkeley National Laboratory (LBL), Oct 2010, available at http://eetd.lbl .gov/ea/ems/reports/lbnl-3984e-ppt.pdf)*

soon become commercially viable. Many industrial, commercial, and even residential customers can, or will soon be able to, generate a growing percentage of their internal needs from localized or on-site DG resources. Rooftop-mounted solar photovoltaics (PVs), solar hot-water collectors, ground-source geothermal heat pumps, fuel cells, and a number of other emerging technologies mean that consumers can become self-sufficient or, in some cases, be in a position to sell their excess generation back into the grid.[17]

This has created not only technical challenges for utilities, long used to selling to consumers, but has also raised a number of thorny issues such as how much they should pay for the juice when it is fed into the grid.[18] Moreover, the fact that an increasing number of customers may conceivably become *net producers* over time poses challenges to utilities and network operators who, after all, must maintain the network even in the face of flat or falling net volumetric consumption.

Making matters more complicated is the growing interest in energy efficiency, promoted in a number of jurisdictions. Since in many cases the cheapest

---

[17]California Governor Jerry Brown has set a target of 12 GW of DG, although it is not clear how and when this goal may be reached. If achieved, that would be roughly equivalent to building ten nuclear power plants, giving new meaning to DG.

[18]In many jurisdictions, utilities and suppliers are allowed to charge a higher tariff when feeding customers and a lower one when excess electricity is fed into the grid. There are also restrictions on how much can be fed into the grid depending on the local retail regulations.

kWh may be those not consumed, utilities with high and rising rates may be facing something totally unprecedented: *demand elasticity*. As they raise retail rates, consumers may find it attractive to cut back consumption, either through investing in energy efficiency and/or self-generation.[19] Utilities can no longer assume that consumers will consume what they always have. This is as much a surprise to suppliers as it is to newly empowered consumers.

The skeptics should take a look at regulatory and policy developments in a number of bellwether jurisdictions including California and within the European Union where the concept of *zero net energy* (ZNE) or passive homes has been proposed or is under serious consideration.[20] At its core, ZNE requirements will result in dramatic declines in per-customer volumetric consumption as an increasing number of households meet the new standards over time.[21]

Since an average home or office cannot generate a lot of power, the only conceivable way to meet the ZNE requirement is to *dramatically* cut back on consumption through increasing the efficiency of energy use. This means higher efficiency appliances, improved lighting, better shell insulation, and more judicious and sparing use of energy.

Moreover, because on-site power generation is unlikely to be correlated with on-site consumption, the average consumer will be feeding its excess generation into the grid part of the time, while taking power off the grid at other times. The grid of the future will therefore be required to work much harder than today, while net volumetric consumption per customer declines over time due to ZNE-type requirements. This has important implications on tariffs, which must not only be dynamic, but reflect the two-way nature of transactions.

One interesting result of ZNE-type mandates is that in a not too distant future, a portion of customers will approach little or zero net volumetric consumption levels. Traditional tariffs that are based on a multiplier times volumetric consumption will no longer apply in this environment.

A similar phenomenon has already occurred in the mobile telephone industry. As more consumers use mobile devices to send and receive text messages, share files and photos, watch videos, listen to music, and access the Internet—as opposed to talking—mobile phone companies have gradually migrated to a new pricing paradigm. Most mobile consumers are now charged based on bandwidth and data transfer speeds, not minutes talking on the phone. The definition

---

[19]In areas with existing high retail rates such as in Hawaii, many distributed generation options are already cost effective. In states with rates in multiple tiers such as California, consumers in top tiers are facing marginal rates in the 40–50 cents/kWh, which makes PVs at 16 a relative bargain.

[20]In California, all residential homes built after 2020 are to meet the ZNE standard, which will apply to new commercial establishments starting in 2030. The definition of ZNE is that it must generate as much energy as it consumes.

[21]ZNE requirements are typically imposed on new construction, sparing existing premises. For this reason, their impact will take time to become noticeable.

of *service* has changed from what it used to be. The time has arrived for utilities to consider alternatives to billing customers based on flat rates and volumetric consumption.

Still, other technical developments are likely to further complicate matters for the traditional grid and traditional utility service paradigm. A new generation of hybrid and full electric vehicles (EVs), for example, not only promises to serve as a potentially huge storage medium for grid operators to "dump" vast amounts of unneeded intermittent energy when network demand is low, but also offers the possibility that some of the stored energy may be fed back into the grid during high demand periods, the so-called vehicle-to-grid (V2G) technology.

Several chapters in the book cover a number of interesting features of EVs and the critical role that the smart grid must play in turning these fictional scenarios into reality.[22] Energy storage, whether in EVs or in centralized plants or in the form of energy stored in other media, water, ice, compressed air, etc., is recognized as a central requirement of any smart grid of the future.[23]

WILL CUSTOMERS BUY IT?

While discussing such exciting developments, it is not unusual to lose sight of the central role of customers in the rapidly evolving smart grid paradigm. Engineers, technicians, economists, and everyone else, it seems, all have grand designs for the smart grid, smart meters, smart prices, and smart devices. Others envision grand schemes where all these components will be coordinated and integrated so that the inherent synergies of the smart grid can be captured and the vision realized. But will consumers buy any of this? Will they do as the experts hope? Will they behave as the models predict? Will they be engaged as many expect them?

As a number of chapters in Part III of this book point out, customer acceptance and participation, by all accounts, are among the biggest and most challenging pieces of the smart grid puzzle. Consumers, most social scientists and anthropologists insist, have their own needs and priorities. With increasing demands on their limited disposable time and even more limited attention span, they *may* or *may not* behave as the experts and engineers want them to.

Moreover, consumers are wary of the data privacy issues and increasingly concerned about "big brother" looking over their shoulders, remotely monitoring and controlling their devices, and managing their electricity consumption. A number of smart metering implementation projects have already encountered significant consumer backlash, and this may be the tip of the iceberg.

---

[22]In particular see Chapters 18 and 19.
[23]Chapter 5 provides further details on storage.

In California, for example, where utilities are engaged in an ambitious rollout of smart meters, a significant number of consumers have expressed concerns about the smart meters, for a number of reasons. During a recent trip to Europe, I was equally surprised at the level of distrust and skepticism I heard from many I spoke with.

Data access and privacy issues aside, all indicators are that a majority of consumers may take limited interest in many of the features of the smart meters, smart devices, and dynamic pricing espoused by the experts in this book and elsewhere. In the end, many of the purported benefits of the smart grid may not come to fruition if consumers do not perceive them as benefits in ways that are meaningful and practical to them. And that would be a great tragedy.

But as further described in several chapters in this volume, there may be ways for consumers to participate in the smart grid environment without taxing them too much.[24] Clearly, much more research is needed in this critical area.

ARE THE BENEFITS WORTH THE COSTS AND THE RISKS?

I hope that the preceding discussion has highlighted some of the key applications of the smart grid, which explain the high level of interest and investment going into it. However, implementing these schemes and achieving their purported benefits will not be easy nor can be taken for granted. For every great opportunity, there are multiple challenges, obstacles, and pitfalls. Part IV of the book, which presents a number of case studies, offers anecdotal examples of some of these difficulties that lie ahead and must be confronted and successfully addressed.

But the benefits are certainly real and potentially significant, supporting the business case for pursuing the investments (Box 2). Among the benefits is that the smart grid offers great potential to address the challenges posed by climate change, something that may not be immediately obvious or generally acknowledged. As already mentioned, the smart grid promises to integrate a growing percentage of clean renewable energy resources as well as a better balancing of generation and load, both of which will go a long way to address concerns about climate change.

> **Box 2: What Are the Costs and Benefits of the Smart Grid[25]?**
>
> In 2004, the Electric Power Research Institute (EPRI), the collaborative research arm of the electric power industry, estimated that to implement the smart grid would cost around $165 billion for the US. That, many said, is a lot of money, and questioned if the benefits would outweigh the costs.

---

[24]Chapters 14–16 provide further insights.
[25]This box is excerpted from *EEnergy Informer*, May 2011.

How smart is smart enough?
Tomorrow's power system: A smart grid.

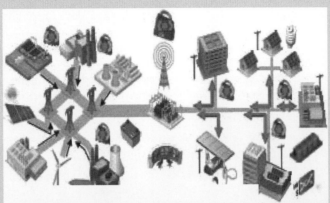

Source: Estimating the costs and benefits of the smart grid – A preliminary estimate of the investment requirements for a fully functioning smart grid, EPRI report # 1022519, April 2011.

Fast forward to 2011 and an updated EPRI assessment, which puts the cost figure for a fully functioning smart grid for the US electric power sector somewhere in the range of $338–476 billion.

If you think that is a lot, the US Interstate Highway System that began under President Eisenhower, cost roughly $500 billion if measured in 2011 dollars. It was a lot of money, but by most measures, it was a real bargain. EPRI claims that the same will be true for the smart grid, but only better. It estimates the potential benefits of the smart grid somewhere around $1.3–2 trillion.

Can we trust the numbers?
Summary of estimated cost and benefits of the smart grid.

	20-year total ($billion)
Net investment required	338–476
Net benefit	1,294–2,028
Benefit-to-cost ratio	2.8–6.0

Source: Estimating the costs and benefits of the smart grid, EPRI, April 2011.

The purported benefits of smart grid would come from multiple sources, some more difficult to assess or put a value on, including the following:

- More reliable power delivery and quality, with fewer and shorter outages;
- Enhanced cyber security;
- A more efficient grid with reduced energy losses and a greater capacity to manage peak demand, lessening the need for new generation;
- Environmental and conservation benefits including enhanced capabilities to integrate renewable energy and electric vehicles (EVs); and
- *Potentially* lower costs for customers through greater pricing choices and improved access to energy information.

As useful and timely as it is, the EPRI report is a "high level" exercise with ball-park guesstimates of what the costs and benefits may be. While this is helpful, for many technical people working in the trenches, the numbers are fluffy and the estimated cost/benefits suspect. Moreover, EPRI takes a holistic perspective of smart grid from a national point of view. This may be helpful for a policy maker but not necessarily for someone trying to justify the cost effectiveness of a particular smart grid, smart metering, or dynamic pricing scheme.

But as is often the case, one must start somewhere, and this is as a good place to start as any. A study by the International Energy Agency (IEA) provides another assessment of the benefits of the smart grid at an even higher level of abstraction.[26]

Explaining the increased price tag of the updated study, Mark McGranaghan, EPRI's VP of Power Delivery and Utilization, pointed out, "This cost assessment factors in new technologies and customer benefits that create a more resilient, self-healing and interactive grid that were not available when the 2004 analysis was completed."

Costs of the smart grid
Total smart grid costs.

Costs to enable a fully functioning smart grid ($M)		
	Low	High
Transmission and substations	82,046	90,413
Distribution	231,960	339,409
Consumer	23,672	46,368
Total	**337,678**	**476,190**

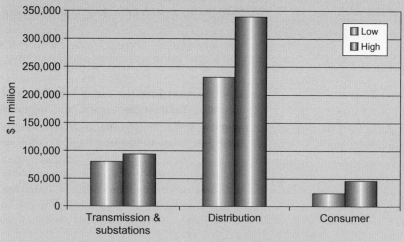

Source: Estimating the costs and benefits of the smart grid, EPRI, April 2011.

---

[26]The Technology Roadmap: Smart Grids, IEA, 2011.

EPRI considered projected costs over the next 20 years in four key areas, transmission, substation, distribution, and customer interface. Broadly, the costs fall into two categories:

- Investment required to meet load growth and to address existing deficiencies—such as power flow bottlenecks and high-fault currents that damage critical equipment—through equipment installation, upgrades, and replacement; and
- Investment needed to develop and deploy advanced technologies to achieve "smart" functionality of power delivery systems.

As is usually the case, it is harder to put a value on the benefits than the costs, but that does not prevent people from trying.

The reality on the ground, however, is rather different, at least as reported in a recent survey conducted by Microsoft Corp. and presented at the CERA Week Conference in Houston in mid-March 2011. According to a poll of 210 utility executives and smart grid experts from around the world, only 8% of the companies reported that they have passed from the planning into implementation phase—which means that 92% are still thinking about it, if that.

What about the benefits?
Estimated benefits of the smart grid.

Attribute	Net present worth (2010) $B	
	Low	High
Productivity	1	1
Safety	13	13
Environment	102	390
Capacity	299	393
Cost	330	475
Quality	42	86
Quality of life	74	74
Security	152	152
Reliability	281	444
Total	**1294**	**2028**

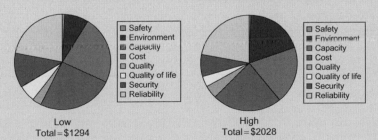

Low
Total=$1294

High
Total=$2028

Source: Estimating the costs and benefits of the smart grid, EPRI, April 2011.

Why hesitate when the benefits of the smart grid are so overwhelming, according to EPRI's assessment and others like it? The main barriers often mentioned include uncertainties on the financial, regulatory, technology, and return on investment aspects of smart grid initiatives – that covers the entire utility business spectrum.

Microsoft's managing director for Worldwide Power & Utilities Industry, Jon Arnold, tried his best to put a positive spin on all this when he said, "Our study clearly indicates the hype cycle is over, and more utilities today are planning smart grid implementations." But 64% of Microsoft's survey respondents said they did *not* have a clear view of the enterprise-wide information and technology infrastructure they will use to structure current and future smart grid deployments.

Mr. Arnold explained, "We're seeing a normal phenomenon occur in terms of the evolution of thinking about these projects. Utilities are finding out what they don't know, and they are, naturally, exerting some caution before making big investments, even though the willingness to spend is there."

Microsoft's own survey results as well as other studies can lead us to different conclusions. There is a lot *more* that the industry does *not* know about the how, what, when, where, and how much of the smart grid puzzle. And companies that have embarked on large-scale implementation projects are, by and large, finding it tough going.

Business school gospel says there are clear advantages to be the *first mover* in any business, but many in the industry are waiting to see what happens to those that jump in the fire first. This explains why so many utilities are engaged in smart grid pilot projects. It is a much safer bet. One cannot be accused of inaction or indecisiveness, while keeping the financial and technical risk exposure to a tolerable level.

The implementation of dynamic pricing in conjunction with the introduction of a host of enabling technologies is expected to lead to the "prices-to-devices" revolution, which in turn, promises to unleash the full potential of smart prices delivered to smart devices through smart meters resulting in more efficient utilization of electricity.[27]

There are already clear indications that electricity markets and grid operators are beginning to take appropriate steps in response to concerns about climate change. The rules of conduct in the electricity market in the UK, for example, are being modified in response to government's desire to meet future low carbon emission targets. In the case of California, the Independent System Operator (CAISO) has developed a future smart grid vision in response to the state's ambitious climate law as well as the 33% renewable target mandated by 2020.[28] Viewed in this context, investments in the smart grid will be doubly rewarding—as in killing two birds with one stone.

---

[27]Chapter 16 provides further detail.
[28]Chapter 6 covers this topic in more detail.

OUTLINE OF THE BOOK

This book, being an edited collection of chapters, covers a broad range of topics in the growing smart grid space. Moreover, it benefits from a diversity of perspectives inherent in the diversity of its contributors. Not only are they from different geographical parts of the world, they represent a wide spectrum of views, disciplines, and backgrounds.

As the editor, I have allowed maximum flexibility to the contributing authors to develop and present their perspectives as they see fit. This means that—despite my best attempts to present a closely intertwined and integrated message—the book by its collective nature offers a chorus of views and perspectives. On the flip side—and I must say that in my view this more than compensates for any shortcomings of an edited collection—this volume offers the reader a range of views that is often missing when a single author or a small group of authors write a book of this nature and scope.

The book is organized into 4 interrelated parts:

- I: Setting the context: The what, why, how, if, and when of the smart grid;
- II: Growing role of renewable and distributed generation;
- III: Smart infrastructure, smart prices, smart devices; and
- IV: Case studies, applications, and pilot projects

Part I provides the definition, sets the context, and describes the scope and the evolution of the smart grid concept, including discussion of its many potential benefits, challenges, and obstacles as well as policy and regulatory issues and alternative visions of what the smart grid may ultimately deliver.

In **Chapter 1, Smart grid is a lot more than just "technology," Steve Hauser** and **Kelly Crandall** present a vision of the "smart grid," an assessment of its potential impacts including a discussion of the challenges and obstacles to the achievement of this vision.

The authors describe the evolution of the smart grid concept from its inception to the present and present views of how it may transform the power sector over the next decade. As the title of the chapter suggests, this book is not a technical engineering book, but rather is focused on policy, regulatory, institutional, economic, and strategic aspects of the smart grid. This chapter provides an overview of the topics covered in the following chapters.

The chapter's main insight is to set the context of the changes the smart grid represents and provide brief answers to the what, why, how, if, and when of the smart grid. As with transformations in other industries, the changes are more organic and inevitable than legislated and engineered. The chapter cites specific projects that support the claims and refers to the evidence provided by the scholars and experts who have contributed to this effort.

In **Chapter 2, From smart grid to smart energy use, Stephen Healy** and **McGill** claim that smart energy use involves moving beyond monitoring, communications, smart meters, and dynamic pricing to more fundamentally empowering energy end-user decision-making.

They point out that although large-scale integrated grids and centralized generation have effectively delivered affordable energy services, they have also resulted in supply-side imperatives shaping the character of customer expectations, behaviors, and practices while, at the same time, removing customers from involvement in decisions about generation and supply.

The authors argue that growing demand, energy security, and climate change underscore a requirement to revisit "smart energy use." They show that one of the great contributions of emergent smart-grid and distributed energy options is to put end-users back at the heart of energy decision-making. The chapter concludes with a discussion of what technologies, changes in market structures, and governance processes might facilitate such a transition.

In **Chapter 3, The ethics of dynamic pricing, Ahmad Faruqui** examines arguments for and against the implementation of dynamic pricing based on its economic efficiency and ethical implications.

The author points out that fixed, flat tariffs and simple bills using dumb meters made sense and could be justified in the context of a declining cost industry reaping the benefits of massive economies of scale in generation and declining fuel costs. Those days, however, are over as the industry faces rising costs and pressures to invest in energy efficiency, fuel conservation, curbing greenhouse gas emissions, and so on.

The chapter's main conclusion is that dynamic pricing should be viewed in the context of today's realities and those that are likely to emerge in the future. Equity requires that prices reflect costs, and thus customers who cost more to serve—such as those who have peakier-than-average load shapes—should pay more and those who cost less to serve—such as those with flatter-than-average load shapes—should pay less. Efficiency requires that price signals convey the scarcity value of electricity to all customers and encourage curtailment of peak loads and their possible shifting to off-peak hours.

In **Chapter 4, The equity implications of smart grid: Questioning the size and distribution of smart grid costs and benefits, Frank Felder** points out that if smart grid technologies and policies are to achieve their potential to revolutionize the use of the electric power grid, then legitimate issues of equity need to be addressed to avoid delaying if not halting smart grid implementation.

Compounding the difficulty in analyzing the equity implications of smart grid is defining equity in the first place. The chapter uses multiple definitions of equity to analyze the potential implications of the smart grid, efficiently implemented, to determine whether there is an equity-efficiency conflict or synergy.

The chapter's main insight regarding the implications of equity for policy makers and regulators is that legitimate equity concerns exist and even with multiple and somewhat conflicting definitions of equity the smart grid can enhance equity if coupled with the right policies.

Part II is primarily focused on the upstream or supply-side of the smart grid, including the expected evolution of renewable energy resources, distributed generation and microgrids, and various options to match intermittent resources on the supply side with storage and flexible loads on the demand side.

In **Chapter 5, Prospects for renewable energy: meeting the challenges of integration with storage, W. Maria Wang, Jianhui Wang,** and **Dan Ton** discuss the issues and potential solutions in integration of large amounts of renewable energy into the power grid—a key requirement for the smart grid.

The chapter provides an overview of relevant technological developments focusing on solar power since wind integration is covered in other chapters. Since these resources are inherently intermittent and variable in their energy output, current industry practices need to be altered to accommodate high penetration of renewable generation. The authors discuss the role of various energy storage technologies, such as batteries and mechanical systems, in addressing this challenge and highlight projects supported by the US Department of Energy to demonstrate grid-scale storage technologies.

The authors conclude that the optimal technologies for addressing renewable energy integration will be application-specific.

In **Chapter 6, The smart grid vision and roadmap for California, Heather Sanders, Lorenzo Kristov,** and **Mark Rothleder** point out that the "smart grid" is a broad umbrella that encompasses emerging technologies covering the entire value chain, from electricity production and transportation to ultimate consumption. In this context, the California Independent System Operator, CAISO, envisions a future that captures the full potential benefits of smart grid technologies while meeting California's ambitious environmental, renewable, and energy efficiency policy goals.

The authors describe CAISO's roadmap for promoting the evolution of the smart grid in all its dimensions by focusing on ISO's core functions, namely reliable grid operation, efficient spot markets, open-access transmission service, grid planning, new generation interconnection, and integration of renewable resources to achieve California's ambitious environmental goals.

The successful evolution of the smart grid requires a consistent vision of the future capabilities of these technologies in supporting the ISO's core responsibilities, which must be coordinated with the requirements of the key stakeholders.

In **Chapter 7, Realizing the potential of renewable and distributed generation, William Lilley, Jennifer Hayward,** and **Luke Reedman** describe the role of distributed and renewable generation in the context of the evolving smart grid.

The authors believe that distributed and renewable technologies will evolve in response to external factors including changing fuel prices and the imposition of carbon pricing. Using an international economic analysis, the authors show that smart grid technologies and techniques can play an important role in integrating and managing the power produced by these options to ensure a safe, reliable, and secure energy supply.

The chapter's main contribution is to show that it is cost effective for smart grid technologies to be developed to enhance the integration and management of two-way flows that will result from the large-scale uptake of distributed and large proportions of intermittent renewable generation in both transmission and distribution systems as plant is deployed in response to climate change. The modeling demonstrates that the potential savings that can be gained globally by the year 2050 by the use of intermittent generation and smart grid technologies are of the order of $14 trillion dollars.

In **Chapter 8, What role for microgrids? Glenn Platt, Adam Berry,** and **David Cornforth** claim that microgrids hold great potential for developed electricity systems as a transitionary path between current methods of system operation and the truly dynamic smart grids often envisioned.

While microgrids are often seen as a technology for remote power provision, the authors point out that they are equally suitable to developed electricity systems, providing a hierarchy of operation that isolates the wider network control system from the low-level issues often associated with large numbers of renewable generation and storage devices. The authors detail these benefits through examples of deployed microgrids around the world, while also discussing the challenges associated with their operation and deployment.

The chapter's main conclusion is that microgrids, by *internally* addressing issues such as generation intermittency, storage management, reactive power flows, and load control, offer a fantastic route to a smart grid with a high penetration of renewable generation and dynamic load management, without requiring dramatic changes to the wider distribution system or its operation.

In **Chapter 9, Renewables integration through direct load control and demand response, Theodore Hesser** and **Samir Succar** point out that demand response (DR) has the potential to become the glue of a decarbonized grid. Specifically, direct load control (DLC) can be utilized to bolster the grid's reliability and flexibility to accommodate variable energy resources (VER).

The authors describe how regionally clustered wind development is expected to result from high-penetration scenarios. Heterogeneous distributions of wind resources will impose significant reliability concerns on balancing authorities in certain parts of the country. These expected reliability constraints can be ameliorated with the additional firming capacity of novel flexibility resources such as DLC.

The chapter's main contribution is a topology of DLC resources that includes wind integration costs on the regulation, load following, and unit commitment timescales. An upper and lower bound for DLC's contribution to growing ancillary service requirements are also presented.

In **Chapter 10, Riding the wave: using demand response for integrating intermittent resources, Philip Hanser, Kamen Madjarov, Warren Katzenstein,** and **Judy Chang** point out that the impact of wind's variability and unpredictability on system operations is becoming a growing concern with the continued penetration of wind resources. Traditional supply-side options will no doubt play a key role in integrating intermittent and non-dispatchable resources, but the smart grid can also play a role.

The authors review recent wind integration studies and examine tools to assess the smart grid's potential, particularly power spectrum analysis of the output of wind resources in combination with techniques such as optimal inventory methods including deploying demand-side resources for ancillary services and the benefits of substituting load for generation.

The chapter concludes that the potential for the smart grid as an integrating resource is large, but its success will critically depend on the ability to recruit and maintain customers' participation.

Part III is primarily focused on the growing role of smart infrastructure, smart meters, smart prices, and smart devices that, taken together, can result in smart and efficient energy utilization, broadly defined. It includes theoretical and applied topics.

In **Chapter 11, Software infrastructure and the smart grid, Chris King** and **James Strapp** address a number of major information technology challenges facing utilities, energy retailers, and other market participants in implementing smart grid initiatives.

The authors focus on two universal challenges: managing large and imperfect data streams and integrating a plethora of communicating smart grid devices with multiple legacy and new software applications. Resolving these difficulties requires overcoming IT silos built around individual business applications and supporting diverse devices from smart meters to smart appliances. The authors describe success strategies, including case studies, to address these problems. The strategies include flexibility to handle new versions of technology, new technologies, new applications, and new standards.

The chapter's main contribution is in providing strategies and specific best practice examples of how different utilities are addressing these problems in different parts of the world including California, Texas, Canada, Scandinavia, and the UK.

In **Chapter 12, How large commercial and industrial customers respond to dynamic prices—the California experience, Steven Braithwait** and **Daniel Hansen** point out that while a number of residential smart pricing pilot studies have been conducted in various regions of the US in support of smart grid business plans, few have focused on large commercial and industrial (C&I) customers. An important exception has been in California, where utilities have installed hourly interval meters on C&I customers and have experimented with a variety of dynamic pricing and demand response (DR) programs.

This chapter describes the results of load-impact evaluations of critical peak pricing (CPP) tariffs for large C&I customers. The CPP tariffs, which began in 2004 as voluntary programs, are transitioning to default tariffs.

The authors describe the potential role that C&I customers can play in linking wholesale and retail electricity markets by responding to dynamic pricing and DR program payments. The chapter concludes that CPP customers respond in significant ways, but load reductions tend to be concentrated in a relatively small fraction of the customers, and many price-responsive C&I customers appear to participate in DR programs rather than CPP.

In **Chapter 13, Smart pricing to reduce network investment in smart distribution grids—experience in Germany, Christine Brandstätt, Gert Brunekreeft,** and **Nele Friedrichsen** point out that integrating large amounts of decentralized resources into future smart grids requires significant distribution network investment. Smart pricing could substantially reduce this investment need through coordination of network, generation, and load.

Locational signals can be implemented in general tariff plans for network or energy, but also in individual smart contracts. The chapter gives a brief overview of theoretical concepts and international experience with locational pricing with a focus on distribution networks. The authors then present an in-depth analysis of the German case and the details of current developments towards smart pricing.

The authors provide insights on how small steps towards tariff flexibility and locational signals can reduce distribution network investment while encouraging efficient network development and utilization.

In **Chapter 14, Succeeding in the smart grid space by listening to stakeholders and customers, William Prindle** and **Michael Koszalka** examine a number of critical program design questions that smart grid technology poses, focusing on the customers, their needs, and their priorities.

The authors examine how technology, information, and behavior interact in the smart grid space; what level of energy savings can smart-grid-based programs produce; what kinds of product/service/program bundles can utilities and their partners offer that would meet both DR and energy efficiency goals; and what kinds of program models need to be developed/tested/evaluated to satisfy regulatory requirements.

The chapter's main contribution is to ask what might the future look like when viewed from the perspective of customers at the receiving end of the smart grid.

In **Chapter 15, Customer view of smart grid—set and forget? Patti Harper-Slaboszewicz, Todd McGregor,** and **Steve Sunderhauf** point out that while some see "smart homes" as the inevitable application of new technology to energy, there is a wide gap between what consumers *want* and what industry experts and vendors are *expecting* customers to do.

The authors suggest that convenience, ease of use, multifunctionality, and privacy will strongly influence how and when the smart grid is invited into our living rooms and becomes part of our daily lives. This transition requires an understanding of how to turn the industry's *need* for consumer engagement into reality on the ground.

The chapter's main conclusions are to determine practical means of engaging customers to interact with the smart home of the future.

In **Chapter 16, The customer side of the meter, Bruce Hamilton, Chris Thomas, Se Jin Park,** and **Jeong-Gon Choi** point out that utilities are in the process of installing smarter, and more expensive, meters in homes around the globe, offering the promise of improving the reliability of the grid while delivering cost savings to consumers. However, this promise can really only be achieved by going beyond the meter and affecting the way energy is consumed within customers' premises.

The authors examine how advanced technology, including programs that send prices to devices, can enhance customer participation and stimulate significant behavior change including the potential from developments still on the horizon.

Using concrete examples from around the globe, the chapter concludes that there are many sustainable business models that can enable consumers to bring home the anticipated benefits of the smart grid.

Part VI includes a number of experiments, case studies, or pilot projects that cover one or more interesting aspects or applications of the smart grid concept described in earlier sections of the book and presents the results that have been achieved or are expected from such applications.

In **Chapter 17, Demand response participation in organized electricity markets: a PJM case study, Susan Covino, Peter Langbein,** and **Paul M. Sotkiewicz** describe how PJM has succeeded in developing a significant role for demand participation in the largest organized US electricity market.

The authors examine the evolution of DR participation in PJM's markets achieved by implementation of early vintage smart grid technologies such as interval metering, which facilitates the necessary measurement and verification of demand reductions in conjunction with market design enhancements that monetize the value of DR. Customers can participate in PJM's energy markets, capacity market, synchronized reserve market, or the regulation market – with room for considerable growth with continued enhancements in the interface between customer load and PJM's demand-focused markets.

The chapter's main contribution is to present empirical evidence that the theory works in practice and results in win-win outcomes. Since 2002 DR participation in PJM has grown significantly to the point where the RPM capacity market now accounts for roughly 7% of PJM's all-time system peak, providing a benchmark for others.

In **Chapter 18, Perfect partners: Wind power and electric vehicles— a New Zealand case study, Magnus Hindsberger, John Boys,** and **Graeme Ancell** examine the integration of large-scale wind generation and large numbers of EVs into the New Zealand power system, a rather smallish and isolated network posing unique challenges.

The authors point out that the NZ system exhibits behavior not seen in larger, interconnected networks and examine how wind generation and EVs can complement each other, enabled by a smart grid, to support rather than challenge the operation of the power system.

The chapter's main contribution is to show how market integration costs can be lowered and investment in generation, transmission, and distribution deferred through linking the charging of EVs to system frequency as well as price signals.

In **Chapter 19, Impact of smart EVs on day-ahead prices in the French market, Margaret Armstrong, Amira Iguer, Valeriia Iezhova, Jerome Adnot, Philippe Rivière,** and **Alain Galli** examine the impact that the introduction of electric cars would have on day-ahead electricity prices over the next 15 years, using the French grid as a case study.

The authors are particularly interested in statistically predicting the impact of EVs on the network by simulating the offers to buy and sell electricity on the day-ahead market. Different scenarios for recharging the EVs are considered including a case where the grid is drawing power from the batteries when prices are high—the vehicle-to-grid (V2G) concept. Most studies on the impact of wind power on day-ahead prices make two strong assumptions, namely that the demand is inelastic and/or supply is the same as the merit order.

The major contribution of the chapter is to propose a method for modeling the offers to buy and sell power on the wholesale market in the absence of such restrictive assumptions.

The book's **Epilogue** offers the editor's concluding thoughts on how the smart grid vision espoused by the contributors to the book may ultimately be achieved.

Fereidoon P. Sioshansi
Menlo Energy Economics

Part I

Setting the Context: The What, Why, How, If, and When of Smart Grid

Chapter 1

Smart Grid Is a Lot More than Just "Technology"

Steve G. Hauser and Kelly Crandall

INTRODUCTION

The electricity industry has arguably been the primary driver for the world's growing economy from the early twentieth century to the present. The correlation between electricity use and availability with standard of living is strong, as shown in Figure 1.1, and it continues to be a critical element in developing economies around the world. Nevertheless, critical barriers are emerging that may prevent or challenge the continuation of the business-as-usual trends in the future. While the mandate of the power industry in the twentieth century was to provide any amount of power consumers required, anywhere, anytime, and at an affordable price—as eloquently described in Chapter 2 by Healy and MacGill—the mandate as we enter the second decade of the twenty-first century is much more complex, demanding not only reliable power, but cleaner and more efficient power with increasingly diverse consumer requirements.

As highlighted by Bartels in the book's Preface, one clear and common theme animates the vision of the twenty-first-century grid: Information. Hence, the smart grid is often used as the term for the twenty-first-century

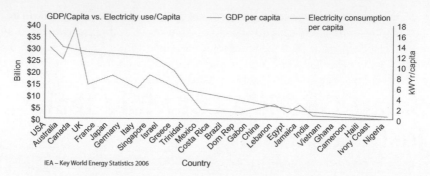

FIGURE 1.1 Global electricity consumption vs. gross domestic product.

grid. Over the past two decades, we've seen almost every industry change radically. From package delivery, to consumer banking, to retail sales, to airline travel, all have seen substantial changes in their infrastructure—the way they deliver services, the way they interact and engage consumers, and the way they conduct business. The changes in these industries rely not just on more information, but on better information—more accurate, more timely, and two-way communication that allows the "system operator" at the enterprise level to plan, design, and operate faster, smarter, and more efficiently. The end result is improved services, more convenience, lower costs, and happier customers. Customers' engagement with the service provider—whether it is the bank, the airline, the shipping company, the online retailer—frequently results in improved relationships that are win-win. The result is a stronger, more productive, and more robust economy. Despite the occasional misuse of information, whether intentional or inadvertent, the basic underlying changes have been a driving force fueling economic growth and prosperity.

The same principles apply to the electric power sector. With better information comes the opportunity for rapid innovation in utility operations. As described in other chapters of this volume, utilities can plan future requirements much more accurately, can design the system more efficiently and with closer tolerances, and can operate it in entirely new and more efficient ways. It also enables the option of empowering consumers by providing information on when and how they use energy, and gives them the ability to align their use with optimal system performance and diverse personal preferences.

Of course with greater information comes the need for a massive new information infrastructure to collect, transport, manage, store, analyze, and display this information. Such a nontrivial undertaking and investment require new thinking and new processes, raising issues that the power industry has only begun to understand and appreciate. Fortunately, other industries have already addressed and solved much of these issues for their own purposes, and the power industry can build on this experience.

This chapter offers a brief overview of the changes we can expect to see in the power industry over the next decade or two. The section "An Inevitable Evolution" provides a glimpse at the current state of the grid followed by the section "Envisioning the Smart Grid: What Do We Want From It?" laying out a vision of what a smart grid should include. The section "Path to the Smart Grid" describes some of the key efforts that are enabling the change and discusses efforts going forward. The "Conclusion" provides a final view of the expected changes.

AN INEVITABLE EVOLUTION

The electricity industry has done a superb job of building, operating, and maintaining a complex infrastructure that transports trillions of kWh of electricity across thousands of miles at the speed of light. Adequate and reliable electricity has been essential to creating economic prosperity. Worldwide, it has made us healthier, safer, and more comfortable by supporting innovations like indoor lighting and refrigeration [1]. The National Academy of Engineering named electrification of the United States the greatest twentieth-century engineering achievement—above the airplane, the internet, the computer, and television [2]. The same is true across the globe where economies and societies have prospered in large part due to the availability of electricity.

It's difficult to argue for a fundamental change to a system that has served so many so well for so long. Yet for decades, grid operators have run the system based on good estimates from years of experience, rather than on real-time information. This is not a criticism—it is the practical and economic reality of operating such a complex yet delicate system. Estimates of demand are often based on surrogate measurements of production, accurate in the aggregate but with little detail. At best consumer-level information was, and in many places still is, measured monthly by manual readings on the dials of analog meters.

Many studies over the last few years have described the need to transform the electricity industry. These studies often talk about the lack of investment in basic infrastructure, including the transmission and distribution network, the need for innovation, the aging equipment and workforce, the cost of power interruptions, the need for emission reductions, and more. All paint a dire picture in the hope that a "burning platform" argument will create a mandate for action among politicians and industry leaders. Undoubtedly, our needs for electricity are shifting in certain ways. We rely on electricity to fuel our increasingly digital economy and society—our massive data centers, our industrial processes, our gadgets, and the way we communicate with each other. Our demand for energy is increasing worldwide [3] as shown in Figure 1.2, and that energy must be reliable to a higher degree than ever before to meet our manufacturing, information and communication needs [4]. However, when the lights, heat, air conditioning, and television all come on nearly 100% of the time, it's hard to feel a sense of urgency for change.

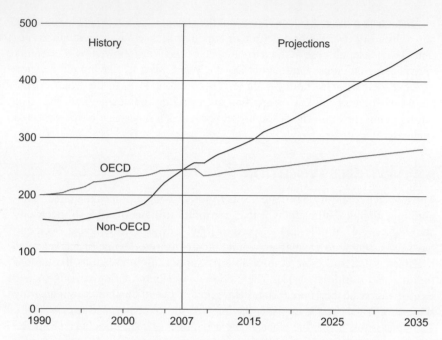

FIGURE 1.2   World energy consumption 1990–2035 (Quads). *(Source: International Energy Outlook, 2010)*

In fact, we are at a stage where the global movement toward a smarter grid has already begun. Change is coming, perhaps not so much by national mandates and policy changes as by focused technology evolution and market forces. While various forms of government mandates have forced technology changes, they are often short-lived and can have unintended negative effects. Technological progress continues apace—entrepreneurs and researchers are eager to transform the power sector with innovation [5]. Technology and manufacturing advances are now making it cheaper and easier to install sensors throughout the electricity infrastructure with inherent communication capabilities—providing more, better, and closer to real-time information. The energy efficiency of household appliances like refrigerators and air conditioners has increased considerably in the last few decades [6]. The cost per watt of installed photovoltaics (PV) continues to fall—as shown in Figure 1.3—in many locations approaching grid parity with traditional power sources [7].

These incremental improvements in the efficiency and cost effectiveness of using electricity are bolstered by rare but influential disruptive technologies—and occasionally new business models—that change markets and expectations, sometimes meeting unknown needs and creating entirely new markets [8]. Electric vehicles might be considered a disruptive innovation, and cost-effective, large-scale energy storage—when commercially viable—most certainly

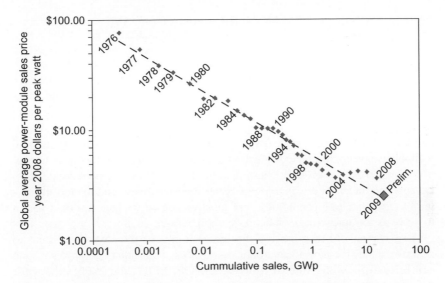

FIGURE 1.3   Decrease in PV module price and cumulative manufactured capacity, 1976–2009. *(Source: Mints 2009 and Mints 2010)*

will be.[1] The combination of incremental and radical improvements will lead to sweeping changes across the power sector. Moore's Law [9] and the Law of Accelerating Returns [10] both predict exponential growth in evolutionary processes, as the products of each stage contribute to the next. In other words, innovations like microgrids, covered in Chapter 8, by Platt et al., and electric vehicles, may be only the beginning. Our power ecosystem is evolving, inevitably, to offer us more information, more options, and more choice. We can only dimly see the future state from where we are now.

In the coming years, aging grid components will need to be replaced, and new generation and transmission infrastructure will be built to meet increasing demand and clean energy goals [11]. Unconfirmed stories tell of utilities special-ordering discontinued parts at high cost in lieu of investing in newer technologies. Yet shifting priorities, evolving technologies, and the experience of other industries suggest that we won't replace old systems with old systems. Additionally, new innovative companies are finding increasing value in meeting consumer diversity with customization [12]. Industries from package delivery to airline travel are being infused with more and better information. And pilot projects worldwide are beginning to demonstrate both a vision for, and the potential of, the twenty-first-century grid [13]—exemplified by one such pilot project for the Sydney metropolitan area in NSW, Australia.

---

[1]Chapter 5 by Wang and Wang covers storage in more detail.

ENVISIONING THE SMART GRID: WHAT DO WE WANT FROM IT?

As described in the following chapters of this book, the twenty-first-century grid—hereafter referred to as the smart grid—will include a complex network of technologies and systems, hardware and software, communications, and controls that taken together will provide both producers and consumers a high level of visibility and control. Being inherently flexible and adaptable to unanticipated future changes, it will include technologies that we already know and technologies yet to be created. Much of our efforts to date have focused on technologies and not on broader questions about energy management solutions—because we understand technologies better, because both vendors and utilities tend to gravitate to technologies, and mostly because we lack a robust vision to serve as a framework for meeting comprehensive goals.

It is less useful to prescribe the technologies of the smart grid than to describe what this new infrastructure and ecosystem must accomplish. One can refer to previous studies by NETL, DOE, EPRI, and many others for detailed descriptions of technology solutions. It's difficult to predict details of what changes will take place, in what time frame, and with what results, but it's extremely important for us to have a vision of where we want to go. By focusing on these priorities, we can create concrete goals and metrics while providing vital directional and correctional guidance to early deployments.

As consumers we all have a rich set of expectations for our electric power service. We want it to be always available and inexpensive. We want it not to pollute our cities or neighborhoods and would prefer not to see power lines. If we see ourselves as "green" consumers, we also want our electricity use to have limited impact on the global ecosystem. If we are on fixed incomes, we want costs to stay the same or go down. And if we are operating a business, we want 100 percent reliability without exceptions.

In other words, if we envision a grid that will meet our needs in the twenty-first century, we would want it to be:

- Affordable
- Clean
- Reliable
- Capable of supporting our evolving economy and society

Exactly how these priorities will shape the smart grid will vary based on factors like geography, generation profiles, building and user characteristics, economics, innovation, and policies.

We Would Want the Smart Grid to Be Affordable

Across decades and continents, the electric industry has effectively created an implied contract that obligates utilities to provide as much electricity as consumers desire—at a reasonable cost. Consumers and their advocates worry about who will bear the risk of smart grid and clean energy investments: utilities, governments (taxpayers), or ratepayers [14]. The chapter by Felder (Chapter 4) covers these issues at length. The question is, how do we add information and communications technologies, new energy technologies, and better service to the grid, but still make power less expensive? The short answer is by using those new technologies to provide more capabilities optimized for energy efficiency, operational efficiency, and demand management.

Energy Efficiency

From sweaters in the 1970s, to double-pane windows and heat pumps in the 1980s, to LEED standards and Energy Star appliances in the 1990s, conservation and energy efficiency have been with us since President Nixon declared his goal for "Energy Independence" in his 1974 State of the Union address [15]. And yet the typical U.S. consumer is still very wasteful in the use of electricity. Many of us now have more efficient appliances and relatively well-insulated homes, but most still use incandescent lights and have added a variety of power-hungry digital appliances that are often left on even when not being used. A recent study by the American Council for an Energy Efficient Economy (ACEEE) showed that average or better economic growth could be supported for the next 20 years and beyond with little or no increase in energy consumption [16].

While the concept of energy efficiency is understood by many consumers, their use of energy tends to be driven more by their desire for comfort and convenience than by concern for its effect on system costs or emissions reductions. Various campaigns for higher efficiency in homes, businesses, and transportation, like Energy Star [17], have had noticeable impact, but there are still millions of homes, businesses, and public buildings that use significantly more energy than is needed to maintain their comfort, health, security, and productivity [18]. Most consumers are still unlikely to pay $5 for a light bulb when they can get one for $1, let alone pay thousands of dollars for a heat pump when they can get a good furnace plus air conditioning for much less. And they are certainly more likely to pay to re-carpet their house or add a new deck than to add insulation to their walls and ceilings.

McKinsey & Company [19] found that nontransportation energy use could be reduced by more than 20% over the projected demand by 2020 (Figure 1.4), and that energy management systems are low-cost, high-impact players in the efficiency supply curve. Smarter technologies and a smarter grid are critical to unlocking this great potential—the smart grid should actively identify and mine energy efficiency opportunities throughout the entire system, using better information for better decisions. The ACEEE estimates that smart technologies saved 775 billion kWh in 2006 alone [20]. The same study shows that existing and near-commercial semiconductor-based appliances would support an economy in 2020 that is 35% larger than it is today, while using 7% less energy.

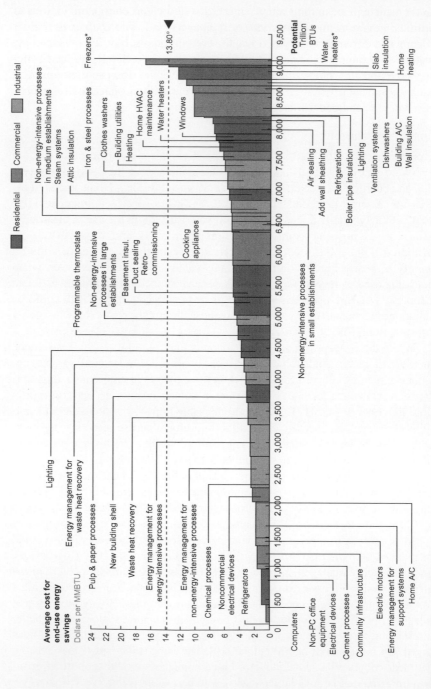

FIGURE 1.4 Energy efficiency supply curve. *(McKinsey & Company 2009)*

Moreover, the increasing capabilities of building energy systems will make continuous commissioning more feasible for residential consumers as well as large commercial and industrial ones. Smart homes and buildings will be able to monitor equipment performance, optimize comfort and efficiency, and convey environmental performance [21]. They will be able to target opportunities for efficiency with stunning accuracy [22]. In the smart grid, energy efficiency will become intrinsic to how we operate our buildings and live our lives.

Operational Efficiency

Modernizing equipment and automating operations can significantly improve performance, reduce losses, and reduce the cost of operating the grid. A recent EPRI report [23] indicates that most of the cost and benefits of the smart grid will occur in the distribution network (Figure 1.5). While most utilities have automated their business processes, many have not automated their field operations. The oft-repeated anecdote is that utilities know when the power is out at some locations—especially at the end of radial lines—when consumers call and tell them. They then often diagnose the problem by dispatching crews to drive the distribution lines with paper maps in hand until they spot the problem. At the transmission level, of course, operators may have sophisticated sensors, software, and controls at their disposal. Nevertheless, there are significant opportunities for improvement throughout the grid.

New technologies and systems are needed to implement proactive maintenance and repairs, reducing failures and increasing the life of aging equipment. Significant energy savings are possible through techniques such as voltage

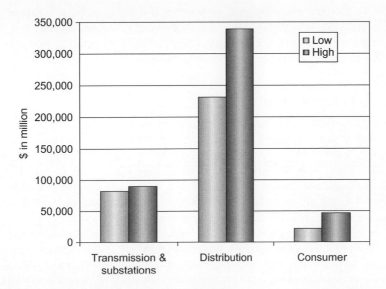

FIGURE 1.5   Total smart grid costs. *(Source: EPRI, 2011)*

optimization [24] and dynamic line rating. Many utilities have begun to address this issue by installing sensors, communications, and controls at critical points in their system. Simply knowing the status of various devices in the field in near-real time can enhance the ability of operators to optimize power flow and quality. Digital meters equipped with the ability to communicate at each account location can reduce the cost of reading the meters, automatically report local outages, allow operators to remotely connect and disconnect accounts, reduce theft, monitor voltage levels, and provide many other benefits. More progressive utilities have installed automated systems with dynamic digital maps and sophisticated management tools. These features not only improve the efficiency of utility operations, but also open up new levels of service and can make working conditions safer for field workers.

Given their different starting points and driving motivations, the full benefit of automating operations will vary substantially from utility to utility, with highly customized applications for each. Some, for example, may emphasize automatic fault detection because of the high cost of repairing complex underground systems. Others may emphasize automatic switching to reduce the duration of an outage to remotely located consumers. Eventually, the smart grid may provide ancillary services like frequency regulation by automatically adjusting demand with storage devices or by varying certain loads. In this vein, the Pacific Northwest National Laboratory, through its parent laboratory manager, recently licensed a frequency sensor/controller chip that can be installed inexpensively in various appliances and used to adjust loads in real time, thereby moderating frequency exertions on the grid automatically with little or no central control [25].

While full and complete grid automation is a noble goal, in isolation it is neither mission critical nor will it likely be cost effective. In designing and building the smart grid, it will be important not only to optimize performance but also to optimize the cost of operating the grid.

Demand Management

When consumers decide to buy more computers, heaters, air conditioners, or more and larger televisions, they expect their electric utility to make sure that electricity is available to power the devices. This is true whether it's the hottest day of summer or in the middle of a winter storm. Consumers do not perceive the grid-related consequences or costs of their decisions. As further described in Chapter 2 by Healy and MacGill, the industry essentially disengaged the consumers from the activities up-stream of the meter.

The easiest way to please the most consumers in the past has been to build the system big enough and robust enough to meet all of their needs, especially at times of peak consumption. According to the Edison Electric Institute, in 2008 U.S. consumers used 4,110,259,000,000 kWh [26], just over 4 trillion kWh, of electricity [27] (Figure 1.6). That's nearly 14 MWh per person

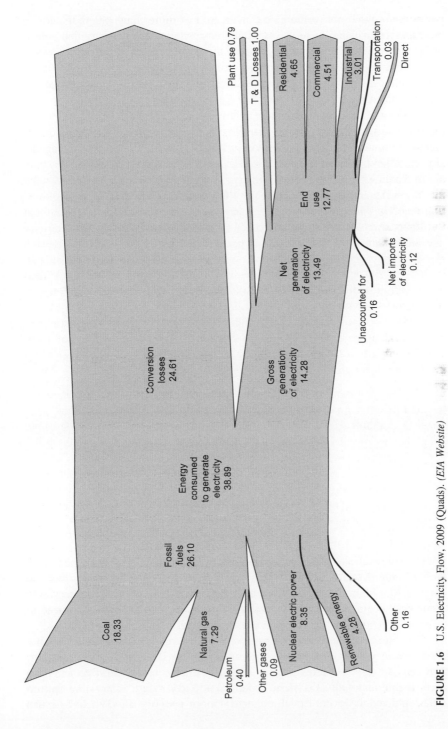

FIGURE 1.6 U.S. Electricity Flow, 2009 (Quads). (*EIA Website*)

per annum, much above the global average. For more than 100 years, we've built an infrastructure to serve essentially any amount of electricity that consumers request. We ask utilities to estimate the future consumer demand for electricity and build the production and delivery infrastructure—power plants, regional transmission systems, and local distribution systems—to accommodate this growth. And since we cannot store electricity—at least not much—production has to match consumer demand at any instant. Moreover, we ask utilities to build the system 10–20% larger than the anticipated demand to make sure we can accommodate unexpected growth or loss of critical generation or transmission lines [28].

In North America, the Federal Energy Regulatory Commission (FERC) and the North American Electric Reliability Corporation (NERC) and its affiliated regional reliability groups oversee system plans and operations to ensure that the lights stay on. Among many other things, they track trends in electricity use and ensure that the industry is properly planning for and building the power production and transmission capacity that are needed to meet demand. In 2009, NERC predicted that production capacity for electricity in the United States will grow from just over 1,000,000 MW, or 1 Billion kW, in 2009 to just over 1,400,000 MW in 2018, a 40% increase in nine years [29].

NERC also projects the corresponding consumed energy to grow from just over 4 TWh at present to just over 4.5 TWh in 2018—a total increase of about 15%. There are at least two reasons why production capacity is expected to grow more than twice as fast as electricity demand.

- The first is that more than 50% of the expected capacity increase is attributed to wind power, the availability of which is rarely coincident with peak demand (typically less than 25% of the time), whereas baseload plants—generally coal-fired or nuclear—are typically available during peak demand.
- The second reason is that, since the 1960s, our electricity use has been increasingly weather-dependent and peaky. Businesses tend to use more electricity in the mid-to-late afternoon, while residences use more electricity in mornings and evenings [30]. More than 10% of our total national production capacity is required less than 100 hours each year (Figure 1.7).

The net result is that in many regions, peak demand is growing much faster than average demand, a trend that is unhelpful if costs are to be kept low and plant factors high. Similar trends are experienced globally. This explains why reducing consumer electricity use during peak times—commonly referred to as demand response—has become a high priority.

There are varied technologies and programs that can make demand response relatively easy and cost effective for consumers. In 2009, FERC assessed current demand response programs state-by-state to project the improvements that are possible nationally [31]. The report identified a range of possible options that could reduce overall peak demand 4%, or roughly 38 GW, to as much

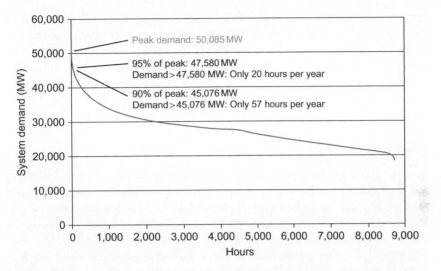

FIGURE 1.7   Typical demand response potential in CA. *(Source: Data derived from California ISO (CAISO))*

as 20%, or roughly 188 GW, over time, as further described in the book's Introduction. The latter figure would require that utilities be much more aggressive in seeking improvement in load factor and that consumers be induced to participate more widely in various demand response programs. In 2010, FERC issued a companion report that lays out a national plan for making demand response a priority [32]. Most recently, in March 2011 FERC released Order 745, which establishes rules by which demand response resources should be paid for participating in organized wholesale markets. While industry reaction to FERC's initiative is mixed at best, the ensuing debate will no doubt clarify many remaining questions about the true benefits of demand response [33].

In the smart grid, the exact amount of demand response is less important than the concept of creating a grid that allows the electricity system to reduce loads readily, when necessary, diverting electricity from noncritical to more critical and essential loads. This must be done both automatically—such as lights dimming in warehouses during peaks or refrigerator temperatures or thermostats shifting a few degrees—and with substantially enhanced consumer participation. Such elegant energy management systems might involve consumers setting their preferences and varying them when prices or conditions change.[2] New technologies and systems might provide ways to induce demand to follow changes in production—and production costs—instead of production

---

[2]Harper-Slaboszewica et al. in Chapter 15 titled "Set and Forget" describe relatively convenient ways for consumers to become active participants in balancing supply and demand.

responding to changes in load.[3] This approach should logically incorporate dynamic rates so that consumer behavior is linked to the cost of delivering energy at any particular time.[4] One of the important benefits of this change to the grid would be the potential to sharply reduce the reserve margin requirements for utilities and grid operators. This requirement, absolutely critical in our current grid system, could eventually become obsolete, with huge savings in capital investments that are reduced or rendered unnecessary.

We Would Want the Smart Grid to Be Clean

In spite of the growing political and societal concern about climate change, renewable energy in the United States until now has been a small contributor to power production, approaching just 4% of the total generation in 2008, excluding hydropower. There are many reasons for this: historically high costs for renewable energy, inconsistent federal and state incentives, and a lack of consumer acceptance. For various reasons, this is now changing. Even without clear federal mandates, dozens of states and cities have made carbon-reduction commitments that continue to drive markets and more will likely follow suit. Renewable energy and clean tech companies are at last able to forecast both stability and growth in long-term markets for their products and services. New technologies, new companies, and new business models are emerging.

While the expectations for national or global commitments to carbon reduction targets are not currently encouraging, other factors, such as the prospect for more stringent regulations, such as those proposed by the U.S. Environmental Protection Agency (EPA), will continue to put pressure on the industry to switch to a cleaner generation mix. Figure 1.8 shows the forecasted growth in renewable energy from the EIA (*Energy Outlook* 2011). On January 29, 2010, President Obama issued an executive order mandating certain reductions in carbon emissions by federal facilities—a 28% reduction over 2005 levels by 2020 [34]. Even without clear federal mandates, dozens of jurisdictions have made commitments that continue to drive markets [35]. Almost no matter what the final specific targets are and what form national and international policy takes, it is clear that the smart grid must accommodate the integration of massive amounts of carbon-free production technologies.[5]

Renewable Resources

While the amount of solar generation in the United States more than tripled between 2000 and 2008 [36], it still represents only a fraction of 1% of the

---

[3]Chapter 10 by Hanser et al. describes the role of load as a means of balancing supply and demand as opposed to the traditional method of relying entirely on adjustments on the supply side.
[4]Chapter 3 by Faruqui describes the role of dynamic pricing in the future smart grid.
[5]The chapter by Lilley et al. (Chapter 7), for example, examines the implications of a global push towards a cleaner energy mix in the electric power sector and its projected benefits.

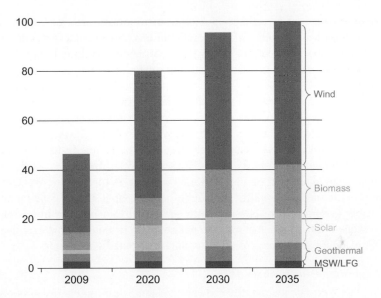

FIGURE 1.8   Projections of renewable energy generation capacity in the United States, 2009–2035 (GW). *(Source: Annual Energy Outlook 2011, EIA, Reference Case)*

total U.S. generation. Specific utilities have recently launched projects to explore the impact of high-penetration renewable scenarios by installing 30% or more solar PV in a particular neighborhood and assessing the grid impacts. These projects are an attempt to anticipate the market impact of very inexpensive solar equipment: while predictions on the future cost of PV systems are difficult, there is general agreement that costs will continue to come down— possibly to less than half the current cost by 2020.

A recent NERC reliability assessment [28] forecasts that nearly 230 GW of new wind generation may be added across the United States in the next decade—almost ten times the total current installed capacity. The same report estimates that over 30,000 miles of high voltage transmission will be needed to support all new power plant additions, largely driven by wind. NERC identifies the need for this transmission as one of the biggest challenges facing the industry, along with operational challenges such as the need for additional ancillary services to manage the variability of wind resources. New technologies will help to optimize the design, siting, and operation of transmission lines to maximize utilization. Studies are underway to explore methods for using both local load control and storage to reduce the problems of wind power's variability.

The power industry is beginning to carefully analyze the barriers to rapid scale-up of wind, solar, plug-in electric hybrid vehicles, and other clean technologies and the barriers to managing distributed (e.g., rooftop solar) power sources. Parallel impacts on transmission and distribution operations must be evaluated to determine their true value in reducing carbon emissions. Most studies indicate that when the percentage of renewable sources on the grid reaches 15% or more, substantial operational changes become necessary [37]. Reaching national goals of more than 30% still seems daunting, but grid operators in other countries are already gaining experience operating under these conditions.

The National Renewable Energy Laboratory (NREL) is working with the U.S. power industry to explore these issues through formal stakeholder engagement with groups such as the Western Governors Association, the U.S. Offshore Wind Collaborative, the Global Wind Energy Council, the Great Lakes Wind Collaborative, and many others. Recent reports summarized comprehensive and detailed analysis of wind and solar integration for both the Western and Eastern interconnections [38]. The studies indicate that higher penetrations of renewable generation are possible with additional transmission infrastructure and various operational changes. Several chapters in this volume examine the impact of renewable energy generation and distributed generation.

Distributed Energy Resources

During this next decade the drive toward consumer-centric energy systems, as well as expected cost reductions in PV, will accelerate installation of rooftop solar systems. This may also drive the move toward direct current "personal grids" that, combined with local network batteries, will give consumers the opportunity for clean critical power systems. A recent study analyzes the full benefits of locating PV systems at the edge of the grid, integrating them with smart grid systems and capturing these additional benefits by increasing predictability, reducing capacity requirements, and improving distribution level operations [39].

For the smart grid, the ability to store electricity may be as significant as the availability of information. For decades the holy grail of engineers—the practical and cost-effective storage of electricity—has been as elusive as any aspect of the industry. However, recent breakthroughs in materials and materials processing have shown promise in reaching this challenging goal. With the investments currently being made globally, storage, both mechanical and thermal, will likely be a foundational component of the smart grid.

Although a few years old now, the Pearl Street report on energy storage [40] is one of the most comprehensive descriptions of both the technologies and markets for electricity storage. The Pearl Street report looks at each of these applications in detail and estimates a value for each. It shows clearly how important it is to capture multiple value streams. Storage devices that are networked and flexibly dispatched for a variety of applications have the potential both to drive down costs and increase value. This is one of the main reasons why plug-in vehicles have received so much attention: they eventually should be able to provide electricity storage for their own use, as well as the grid, providing opportunities to create other markets and applications. PJM, the largest grid operator in the United States, has recently proposed a demonstration project that will "grid-connect" up to 1,000 electric-hybrid school buses, allowing them to be charged off-peak and potentially to discharge to the grid while sitting idle during the middle part of the day [41].

Historically, storage has been focused on individual applications such as vehicles or bulk storage to ease the variability of renewable resources. The economic value has always been difficult to demonstrate, even with costs of the technology projected to go down. The real value of storage is much broader to grid operations: it may lead to improved asset utilization for generation, improved provision of ancillary services, better integration of renewable resources, and congestion relief for transmission and distribution. Many studies are exploring the value of storage [42].

Electric Vehicles[6]

Depending on the design of the charging stations, a typical Plug-in Hybrid Electric Vehicle (PHEV) or Electric Vehicle (EV) could require almost as much power as a typical home, meaning that neighborhoods that have several electric vehicles may require upgraded infrastructure, especially in densely populated urban and suburban areas with older infrastructure. New, smarter infrastructure will give utility operators more freedom to manage these new loads optimally and minimize negative impacts. In late 2009, a coalition of companies that support an aggressive move toward electrified transportation published a report [43] outlining the benefits of and barriers to rapid electrification of transportation in the United States. While their position may not be entirely neutral, the report gives a fairly complete and balanced analysis of the benefits and impacts. They propose a realistic scenario that shows 25% of all new light-duty vehicle sales comprised of either EV or PHEV by 2020, with sales ramping up quickly by that time (Figure 1.9). Their plan targets approximately 90% of new vehicle light-duty sales being EV or PHEV by 2030, which would require consideration of serious policy changes to incentivize rapid switching by consumers. This would also have a huge impact on utilities, increasing the amount of energy required in 2030 by 10–15% for transportation alone. In this scenario, the smart grid will allow utilities to use primarily off-peak power for battery charging, increasing asset utilization, and absorbing excess renewable energy.

We Would Want the Smart Grid to Be Reliable

Despite its shortcomings, the existing grid—at least in developed countries— provides nearly ubiquitous power with reasonable quality and reliability to every consumer and every load. This universal standard of service has been essential over the last several decades, as consumers increasingly added devices

---

[6]Chapters by Hindsberger et al. (Chapter 18) and Armstrong et al. (Chapter 19) cover the issues involving penetration of large number of EVs and intermittent renewable generation.

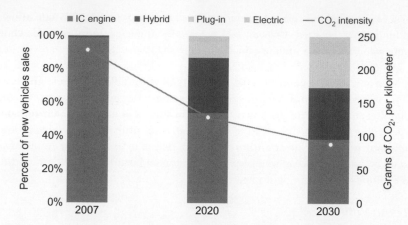

FIGURE 1.9 U.S. passenger vehicle sales by technology. *(Source: Electrification Coalition, 2009)*

and equipment that are designed and built to this standard. Our increasingly digital economy already requires higher levels of reliability and power quality [44].

Yet the cost of providing this service will be prohibitive if applied universally, as is common. To continue this universal service paradigm, and to provide the level of power quality and reliability needed to meet future loads, will necessitate the construction of expensive "super grids" [45]. These "super grids" will substantially increase the sophistication of transmission and distribution equipment to maximize power quality and minimize disruptions. A few countries, such as Singapore and Japan, have moved in this direction more rapidly than the United States and as a result have much more expensive electricity. China now touts its "strong grid" national strategy, which is similar in its objective to provide universal quality and reliability to its rapidly growing economy [46].

The alternative is a more market-driven paradigm in which reliability is modulated to match load: high levels of power quality and reliability are not universal, but are focused on a very small percentage of consumer loads that specifically require this level of service. In the smart grid, reliability represents the initial foray into highly differentiated services to consumers, which no longer represent broad ratepayer classes, but rather are based on sophisticated classification using diverse factors. This heterogeneous grid can provide highly differentiated service to consumers, with value and costs better aligned. Power quality and reliability could even vary by location,[7] with each utility finding the appropriate level for its varied groups of customers. Of course, electricity prices would need to be structured to correspond to the various levels of power quality and reliability that consumers desire. One would also expect a

---

[7]Brandstatt et al. (Chapter 13) discuss varying network access prices by location and its implications for the future smart grid.

basic level of power quality to be provided. In this future world, consumers would pay for these services as we now do in the highly differentiated telecommunications industry that we've become accustomed to, where everyone's "plan" is different.

Microgrids are increasingly being discussed as one innovative technology that could support such a shift.[8] The Departments of Defense and Homeland Security are also beginning to drive more serious efforts to increase the reliability of loads that are critical to national security. Certain communications systems, for example, require essentially 100% reliability—which is impossible to guarantee with the current grid infrastructure. A grid that operates with heterogeneous reliability might ensure that traffic lights and emergency facilities would still function during a typical power outage or a natural disaster.

We Would Want the Smart Grid to Continue to Support Our Economy and Society

The smart grid must enable every consumer—large and small—to participate in optimizing the cost and performance of their own use and of the entire grid. As with other industry transformations, a fundamental challenge is engaging the consumer in the transformation. The typical residential consumer today adds new loads to the grid with little understanding or regard for the production and delivery required to supply the electricity for that new load. Whether to power a new TV, computer, or electric car, consumers view their electricity demand to be incidental (at best) to their local utility's business or plans for expansion. Therein lays one of the basic paradoxes that plague the power industry: no single consumer, except for a few large industrial consumers, has much impact on how the system operates, but consumers collectively dominate the system design and operation. In other words, a utility cannot afford to pay much to affect one consumer's behavior, yet it is only by affecting each consumer that the system can be changed.

As the power requirements evolve with the economy of the twenty-first century, the simple distinction between residential, commercial, and industrial consumers becomes both less accurate and less convenient. Businesses are run from homes with increasingly sophisticated electronics and associated power needs. Health care occurs in homes demanding higher reliability. Commercial buildings become more sophisticated in their design and operation, demanding more customized power supply and often including their own local power supplies. The result is the emergence of new markets for electricity technologies, solutions, and services, and economic opportunities for innovative companies ready to meet these diverse requirements.

Fortunately, technologies that already engage consumers in other aspects of their lives can easily be adapted to manage their electricity use. This is

---

[8]Platt et al. (Chapter 8) discuss microgrids.

evidenced by companies like Google, IBM, Cisco, and others finding new markets in the power sector. Gadgets that are commonplace for communication and entertainment can also inform us about options for reducing our electricity use. Availability of information on consumer use will also allow utilities to analyze and segment users to better frame service and price offerings.[9] Some predict that hundreds of market segments will emerge with unique and highly customized requirements—and hence highly customized products and services. No longer would residential consumers be viewed as one generic and homogeneous market. Rather, one can imagine consumers being differentiated, for example, by whether one has a preponderance of energy-efficient appliances; whether one conducts business at home; or whether one has critical medical devices in the home. The line between commercial consumers and residential consumers may become blurred as homes become more differentiable, like businesses.

In the transition to the smart grid, consumers will have the opportunity to move from being passive participants to active, interactive, and even transactive customers, as their level of sophistication increases. This does not mean that they will be forced to make numerous decisions about energy use, but that they will possess intelligent software that understands their preferences and acts on their behalf. With richer information that is easy to interpret, consumers will be able to tailor their energy use to their needs and save money.

THE PATH TO THE SMART GRID

Smart grid concepts and technologies are increasingly being considered as tools that can be used to achieve important national, state, and local goals for energy efficiency, renewable energy, and demand response while supporting economic and societal needs. Changes will be based more on experiment than on edict, more on emergent processes than on engineering practices. Change—particularly change as sweeping and promising as the smart grid offers—is neither easy nor free. While these forces for change are largely organic and uncontrolled, it does not excuse any of us from thoughtful planning, careful anticipation, and assertive leadership. Creating a clear vision for our electricity system for decades ahead will force strategic decisions—ones that are essential to our prosperity. Results will point to the best paths—ones that enable power system builders and operators to provide, and consumers to use, electricity more efficiently, with large savings for the economy and the environment.

A spark of innovation has become more and more obvious across the industry. New ideas, new technologies, and new projects have emerged to push this

---

[9]Hamilton et al. (Chapter 16) cover the consumer side of the meter, while Harper-Slaboszewicz et al. (Chapter 15) discuss behavioral aspects of getting consumers to engage with variable prices and react to information provided through the smart grid.

innovation toward real impacts. Traditional power sector vendors such as GE, Siemens, ABB, Alstom, and Schneider Electric all have announced smart grid initiatives as evidenced by a significant increase in marketing efforts in the popular media—highlighted by GE's scarecrow dancing along the wires during a 2010 Super Bowl advertisement. Similarly, IT companies such as IBM, Accenture, Cisco, Google, and others are expanding traditional back-office products and services into more comprehensive smart grid solutions, often acquiring new companies and launching new product offerings. And new startups such as Tendril, GridNet, Silver Spring Networks, Smart Synch, GridPoint, eMeter, and others are growing steadily. Of course many other companies—both old and new—are providing storage devices, roof-top solar panels, wind turbines, electric vehicles and charging infrastructure, building automation, and on and on. All have thoughtful business plans, good technologies, and staff with a wealth of experience. They also have strong marketing programs that reach utility executives and the average consumer alike.

IBM is an early pioneer, launching their smart grid efforts more than a decade ago. Their global "Intelligent Utility Network" of several large utilities around the world has set the pace in many ways for the industry on various key issues. Their leadership has been obvious in many industry organizations and events around the world.[10] IBM's Smarter Cities Challenge, launched in late 2010, is indicative of their leadership on smart grid and related issues globally.[11]

In the past year, Commonwealth Edison launched a group of five smart grid pilot projects that will incorporate 10 Chicago communities. The projects included in the ComEd Smart Grid Innovation Corridor are intended to evaluate the smart grid technologies and implementation approaches for residential solar power, ComEd's first intelligent substation, distribution automation, EV charging stations, and dynamic voltage regulation. The project is designed to build on the information gathered from smart meters installed in 130,000 homes in the area.

The GridWise Alliance (www.gridwise.org) has emerged as the industry's leader on advocating for a smart grid. A broad coalition of more than 150 companies, it represents the interests of utilities, vendors, and other key stakeholders. It has been a key driver for new legislation and new initiatives in the United States and has spawned sister organizations in several other countries that similarly drive change in the power industry in their own countries.

Since June 2010, the National Science and Technology Council under the Executive Office of the President of the United States has engaged various federal agencies in developing policies and guidelines for accelerating smart grid development in the United States. During this process they engaged more than 100 different companies and organizations to provide expert input and

---

[10]This book includes contributions from IBM.
[11]See www.smartcititeschallenge.org.

released a report on their findings in mid-2011. Other federal organizations such as the Federal Energy Regulatory Commission, the Department of Energy, the National Institute of Standards and Technology, have active programs addressing various aspects of smart grid development.

While some argue that the roots of the present smart grid efforts can be traced back to various pilot-scale projects in the late 1980s, there is no doubt that whether measured by the number of digital meters deployed, by government and industry expenditures, or simply by the number of conferences and papers on the topic, we are now witnessing an explosion of activity. These efforts and many more like them around the world build on one another through complex interactions and exchanges of information, forming a new global smart grid industry. This new industry is still maturing—learning from its successes and from its mistakes—and collectively building a path to a smarter grid.

In the United States alone several hundred projects have been started in recent years, due in large part to the American Recovery and Reinvestment Act of 2009, which authorized the expenditure of $4.5 billion matched by an additional $5.5 billion from the project participants. Most of the 50 states have active projects somewhere in the state, and certain states like Texas, California, New York, and Vermont have comprehensive state-wide efforts underway. The latest details of the Recovery Act funded projects can be found at www.smartgrid.gov.

Likewise, major smart grid projects are now underway in Australia, Ireland, India, China, Korea, Japan, Canada, Britain, France, Italy, Spain, and numerous other countries. In 2010 several of these countries formed a new organization called the Global Smart Grid Federation (www.globalsmartgridfederation.org) with the purpose of sharing best practices and advocating for policies that encourage smart grid development around the world. This group encourages public agencies and private companies to collaborate on issues that are critical to rapid smart grid development. A recent report by the World Economic Forum titled "Accelerating Successful Smart Grid Pilots" [47] summarizes the key drivers and issues from interviews with smart grid pilot projects underway around the world. This report discusses the commonalities and differences of the projects in various countries.

The Electric Power Research Institute released a report in early 2011 that provides a broad assessment of the costs and benefits of deploying a smart grid [23]. The report titled "Estimating the Costs and Benefits of the Smart Grid—A Preliminary Estimate of the Investment Requirements for a Fully Functioning Smart Grid" factors in a wide range of new technologies, applications, and consumer benefits. The analysis shows that the investment needed to implement a fully functional smart grid in the United States ranges from $338 billion to $476 billion and can result in benefits between $1.3 trillion and $2 trillion. A few excerpts from the EPRI study including the estimated costs and benefits are included in the Introduction. While many assumptions went into EPRI's assessment, it suggests that the projected benefits to consumers far outweigh the costs.

Formal international agreements have also been recently negotiated. Growing out of the Major Economies Forum and the UN Climate Change Conference, smart grid was adopted in 2009 as a major new thrust for technology partnership. As a result, at the first Clean Energy Ministerial held in Washington, DC, in July 2010, the United States announced the launch of the International Smart Grid Action Network (ISGAN). ISGAN is an international partnership that was created to be a mechanism for multilateral, government-to-government collaboration to advance the development and deployment of smarter electric grid technologies, practices, and systems. ISGAN activities are focused on those aspects of the smart grid where governments have regulatory authority, expertise, convening power, or other leverage, with a focus on six principal areas: policy; standards and regulation; finance and business models; technology and systems development; user and consumer engagement; and workforce skills and knowledge. In April 2011, ISGAN was formally established with 19 countries and the European Commission as participants, as an Implementing Agreement under the International Energy Agency's *Framework for International Technology Cooperation*.[12]

CONCLUSIONS

Plans, strategies, and policy frameworks are gradually emerging around the globe. Cities, states, utilities, countries, and regions are all developing a vision and a path toward a twenty-first-century grid—a smart grid. Thoughtful leadership by some organizations and companies is resulting in common themes and even an informal consensus beginning to emerge in obvious and not so obvious ways. Add to this the rapidly growing practical experience of designing and operating various smart grid systems, and the result is not just new ideas, concepts, and impacts, but also new challenges that fuel research, analysis, and even new paradigms.

While these myriad efforts can sometimes seem unfocused, unstructured, slow, and less productive than desired, changes are already occurring. Consumers are beginning to understand and make small but important decisions to manage their energy use differently, based on timely and accurate information. Eventually consumers will embrace these changes the way they have in other markets, and they will demand solutions that go far beyond today's. Utilities are sensing the paradigm shift, and many are responding with their own plans. Some are embracing the pressure for better information. Some are embracing the pressure to engage consumers. Some are even thinking ahead toward better systems, models, operations—changes that may radically change their business over time.

Change is hard for any organization and even more so for an industry. While innovative technologies will motivate change, rules, regulations, contracts,

---

[12]IEA recently published a smart grid roadmap that captures, at a high level, the purported benefits of the smart grid in meeting future climate change targets.

models, systems and standards are more difficult and more complicated, often involving stakeholders with opposing goals and intent. The changes we've seen to date have been largely motivated by engineers and technologists with others acting primarily as observers. True change will require full participation from economists, regulators, lawyers, policy experts, financial experts, and many others.

While predicting the landscape of the power industry in 2020 and beyond is a challenge, the changes will be more radical, intense, and impactful than most can imagine. Technologies, systems, and institutions alike will change. Those that lead these changes will be those that determine the future of the industry.

While this volume does not pretend to be a comprehensive treatment of all the global efforts in smart grid progress, the authors and editors believe it to be indicative of the both the breadth and depth of these efforts. The goal is to inform readers, catalyze new thinking, and challenge old paradigms. With the fast-moving pace of the changes taking place across the industry, better information and better concepts will quickly emerge to build on those we present here. However, the contributions contained herein guide the progress taking shape and will play a significant role in accelerating this progress.

REFERENCES

[1] P. Fox-Penner, *Smart Power: Climate Change, the Smart Grid, and the Future of Electric Utilities* 1-4 (2010); *Independent Evaluation Group for the World Bank, The Welfare Impact of Rural Electrification: A Reassessment of the Costs and Benefits* 39-46. http://lnweb90. worldbank.org/oed/oeddoclib.nsf/DocUNIDViewForJavaSearch/EDCCC33082FF8BEE8 52574EF006E5539/$file/rural_elec_full_eval.pdf, 2008.

[2] National Academy of Engineering, *Greatest Engineering Achievements of the 20th Century.* http://www.greatachievements.org/.

[3] NERC, *Long-Term Reliability Assessment.* http://www.nerc.com/files/2010_LTRA_v2-.pdf. 2010. EIA, *International Energy Outlook.* http://www.eia.doe.gov/oiaf/ieo/electricity.html, 2010.

[4] EPRI, *The Cost of Power Disturbances to Industrial and Digital Economy Companies.* http:// www.onpower.com/pdf/EPRICostOfPowerProblems.pdf, 2001.

[5] The Cleantech Group, *Smart Grid Vendor Ecosystem Report.* http://www.energy.gov/media/ Smart-Grid-Vendor.pdf, 2010. GTM Research, *The Smart Grid in 2010: Market Segments, Applications, and Industry Players.* Available from http://www.gtmresearch.com/report/ smart-grid-in-2010, 2009.

[6] http://apps1.eere.energy.gov/buildings/publications/pdfs/corporate/regulatory programs mypp. pdf (pp. 30–33).

[7] G. Barbose, N. Darghouth, R. Wiser, LBNL-4121e, *Tracking the Sun III: The Installed Cost of Photovoltaics in the U.S. from 1998–2009*, p. 10 http://eetd.lbl.gov/ea/ems/reports/lbnl-4121e .pdf, 2010.

[8] J.L. Bower, C.M. Christensen, "Disruptive Technologies: Catching the Wave," *Harvard Business Review*, p. 43 (Jan.–February 1995).

[9] G.E. Moore, "Cramming more components onto integrated circuits," *Electronics* 38 (8) (1965) 114–117.

[10] R. Kurzweil, *The Law of Accelerating Returns*. http://www.kurzweilai.net/the-law-of-accelerating-returns/, 2001 (accessed 07.03.01).

[11] Public Utility Commission of Texas, *Need for Generation and Transmission Capacity in Texas; Renewable Energy Implementation and Costs*. https://www.puc.state.tx.us/electric/reports/renew/2006_Renew_Energy_Imp_Costs.pdf, 2006.

[12] C. Anderson, *The Long Tail: Why the Future of Business is Selling Less of More* (2006).

[13] World Economic Forum, *Accelerating Successful Smart Grid Pilots*. http://www3.weforum.org/docs/WEF_EN_SmartGrids_Pilots_Report_2010.pdf, 2010.

[14] AARP, NASUCA, Consumers Union, NCLC, & Public Citizen, *The Need for Essential Consumer Protections: Smart Metering Proposals and the Move to Time-Based Pricing*. http://www.nasuca.org/archive/White%20Paper-Final.pdf, 2010.

[15] R. Nixon, *State of the Union Address*. http://www.cs.duke.edu/courses/cps100/fall03/code/trie/data/sotu/1974rn.html, 1974 (accessed 30.01.74).

[16] S. Laitner, *Assessing the Potential of Information Technology Applications to Enable Economy-Wide Energy-Efficiency Gains*, August 17, 2007.

[17] EPA, Energy Star® and Other Climate Protection Partnerships: *2009 Annual Report*. http://www.energystar.gov/ia/partners/publications/pubdocs/2009%20CPPD%20Annual%20Report.pdf, 2009.

[18] M.C. Fuller, et al., LBNL-3960E, *Driving Demand for Home Energy Improvements*. http://eetd.lbl.gov/EAP/EMP/reports/lbnl-3960e-web.pdf, 2010.

[19] McKinsey & Company, *Unlocking Energy Efficiency in the U.S. Economy*. http://www.mckinsey.com/clientservice/electricpowernaturalgas/us_energy_efficiency/, 2009.

[20] J.A. ("Skip") Laitner, et al., *Semiconductor Technologies: The Potential to Revolutionize U.S. Energy Productivity* (ACEEE, May 2009).

[21] E. Mills, P. Mathew, LBNL-1972E, *Monitoring-Based Commissioning: Benchmarking Analysis of 24 UC/CSU/IOU Projects*. http://eetd.lbl.gov/emills/pubs/pdf/MBCx-LBNL.pdf, 2009.

[22] J. Heimbuch, *EcoFactor Helps Customer Discover Botched Home Improvement Project*, TREE-HUGGER.COM. http://www.trehugger.com/files/2010/11/ecofactor-helps-customer-discover-botched-home-improvement-project.php (accessed 23.07.10). J. Steinberg, *Optimizing HVAC Programming Behavior Remotely to Enhance Energy Efficiency and Demand Response: A Residential Field Study*, 2008 Behavior, Energy and Climate Change Conference. http://www.stanford.edu/group/peec/cgi-bin/docs/events/2008/becc/presentations/19-6C-03%20Optimizing%20HVAC%20Programming%20Behavior%20Remotely%20to%20Enhance%20Energy%20Efficiency%20and%20Demand%20Response%20-%20A%20Residential%20Field%20Study.pdf, 2010 (accessed 12/19/08).

[23] Electric Power Research Institute, Estimating the Costs and Benefits of the Smart Grid, Technical Report, 2011.

[24] K. Forsten, et al., "Green Circuits Through Voltage Control," *T&D World*. http://tdworld.com/overhead_distribution/voltage-optimization-energy-reduction/index.html (accessed 01.05.10). K. P. Schneider, et al., PNNL-19596, *Evaluation of Conservation Voltage Reduction (CVR) on a National Level*. http://www.pnl.gov/main/publications/external/technical_reports/PNNL-19596.pdf, 2010. R.W. Beck, *Northwest Energy Efficiency Alliance Distribution Efficiency Initiative Project Final Report*. http://www.comedamifuture.com/Resources/DEI%20Final%20Report%201207.pdf, 2007.

[25] PNL, Press Release, *Battelle Licenses Grid Friendly™ Appliance Controller*. http://www.pnl.gov/news/release.aspx?id=856 (accessed 3/22/11).

[26] Edison Electric Institute. See http://www.eei.org/Pages/default.aspx.

[27] 1 kWh is the amount of energy a 100-watt light bulb burns in 10 hours. On a per-capita basis this amounts to about 14,000 kWh per person, about double the usage of most other developed countries and many times higher than developing countries.

[28] NERC, *2010 Long-Term Reliability Assessment*. http://www.nerc.com/files/2010_LTRA_v2-. pdf, 2010.

[29] North American Electric Reliability Corporation. *2009 Scenario Reliability Assessment.* October 2009.

[30] These peak hours, of course, are not just for air conditioning but are the sum of lights, computers, motors, and many other loads—i.e., they include both extremely critical loads (like traffic lights and life-support equipment) and non critical loads (such as air conditioning and lighting in unoccupied buildings).

[31] Federal Energy Regulatory Commission, A *National Assessment of Demand Response Potential* (Staff Report. June 2009).

[32] Federal Energy Regulatory Commission, *National Action Plan on Demand Response*, June 2010, Staff Report.

[33] *EEnergy Informer*, May 2011, p. 8.

[34] "Obama Makes Good on Executive Order to Reduce the Federal Government GHGs," *The Green Market*, February 3, 2010.

[35] One example is the action that New Mexico, California, and other western states are taking to create the Western Climate Initiative, a regional cap-and-trade organization. http://www. westernclimateinitiative.org/.

[36] DOE – The Department of Energy. *2009 Renewable Energy Book.*

[37] Germany, Spain, and Denmark are already well past 15%. In 2009, the Spanish national grid set a variety of records, eventually topping 50 percent of its instantaneous total power demand from wind power.

[38] D. Lew, et al., *How Do Wind and Solar Power Affect Grid Operations: The Western Wind and Solar Integration Study*. NREL, May 2010; D. Corbus et al., *Eastern Wind Integration and Transmission Study – Preliminary Findings* (NREL, September 2009).

[39] Navigant Consulting, *The Convergence of the Smart Grid with Photovoltaics: Identifying Value and Opportunities* (December 2008).

[40] R. Baxter, J. Makansi, *Energy Storage: The Sixth Dimension of the Electricity Value Chain* (Pearl Street Executive Briefing Report, January 22, 2002). http://www.energystoragecouncil.org/1%20-%20Jason%20Makansi-ESC.pdf.

[41] "How Electric Cars Could Become a Giant Battery for Renewable Energy," *Mother Jones*, October 15, 2010.

[42] R. Sioshansi, P. Denholm, T. Jenkins, "Comparative analysis of the value of pure and hybrid electricity storage," *Energy Economics* 33 (1) (2011) 56–66.

[43] Electrification Coalition, *Electrification Roadmap: Revolutionizing Transportation and Achieving Energy Security* (November 2009).

[44] As a result, niche markets are developing for devices that improve the power quality and reliability for home computers and TVs with the promise of extending their life.

[45] Chris Marney, *Microgrids and Heterogeneous Power Quality and Reliability* (Lawrence Berkeley National Laboratory [LBNL-777E], July 2008).

[46] *A Strong Smart Grid: The Engine of Energy Innovation and Revolution*, State Grid of China, 2009.

[47] World Economic Forum, *Accelerating Successful Smart Grid Pilots*, 2010.

Chapter 2

From Smart Grid to Smart Energy Use

Stephen Healy and Iain MacGill

INTRODUCTION

This chapter first outlines in the following section the evolution of the electricity industry to show how the current emphasis on some specific smart grid applications and technologies, such as smart meters and dynamic pricing, represents only a modest change in the underlying historical trajectory of the industry. In particular, we suggest that such an emphasis does not adequately address the way contemporary consumer choice has come to be predominantly conditioned by the low-cost and seemingly boundless electricity availability many take for granted today. We also argue that this situation is reflected in current supply-side-oriented industry structures and arrangements, quite unlike the early days of the industry when institutional structures and arrangements commonly emerged as shared enterprises between end-users and generating plant developers. While electric lighting and some industrial applications were enthusiastically embraced when distributed electricity first emerged in the late nineteenth and early twentieth centuries, economies of supply soon had system engineers looking to stimulate new loads. Many new "energy services" were devised as a result, centering not only on new appliances and "white goods" but also upon previously unheralded "energy services" such as the domestic provision of chilled, dehumidified air. While most of these energy services are now taken for granted and assumed routine and necessary, this section will show that their mass uptake was the not the result of straightforward consumer choice but rather involved intensive marketing and a number of further institutional and political initiatives targeting intensive, mass electrification.

Indeed, the evolution of the electricity industry over the last century has largely centered upon extending the grid and scaling up centralized generation. Reasons for this included remotely located yet highly concentrated energy sources (particularly hydro and coal), economies of scale in generating plant, and the aggregation of diverse and highly distributed forms of supply and demand. In many regards this has been highly successful. One of its greatest successes has been to allow end-users to disengage from the electricity industry other than through paying their bills. Almost the entire population of the developed world now has access to a relatively low-cost and highly reliable electricity grid that appears to effortlessly supply just about any amount of electricity end-users choose to consume. For the most part there are no rules controlling the way small end-users install new loads, such as air conditioners, with the industry typically focused upon meeting projected future demand through the augmentation of supply financed by charging end-users on a cost-recovery basis. As a result electricity has become, rather than the "scarce resource" of economics textbooks, a cheap source of just about any energy service that end-users conjure up or have suggested to them.

Currently, however, there are growing pressures for change. One source for these was the wave of electricity industry restructuring across many countries in the 1990s centering upon the introduction of competition into electricity markets. The explicit principles behind the introduction of competitive electricity markets, and the associated corporatization, or privatization, of previously government-owned or regulated private monopoly industry participants, included improved efficiency and innovation, as well as improved customer choice. Outcomes to date have been mixed, and meanwhile, other, arguably more pressing, drivers for change have emerged. These include seemingly endless demand growth, growing energy security concerns (particularly with respect to the ongoing supply of fossil-fuel primary energy resources), and climate change. These are discussed in the section "Pressures for Change: Rising Demand, Energy Security, and Climate Change" of this chapter.

There is a promising range of existing and emerging technologies capable of helping address these challenges, with some that fit well within current centralized supply style arrangements. Many of the most promising options, however, center upon distributed energy options, including energy efficiency and demand management arrangements. These latter can help reduce demand growth, thereby improving energy security and reducing emissions. However, utilizing their full potential will almost certainly require a very different relationship between end-users and the electricity industry from that established historically. Here lies the promise and challenge of smart grids that hold the potential to facilitate such engagement.

However, the predominant current emphasis on means such as smart meters and dynamic pricing tends to conflate consumer choice and engagement with the collective response of aggregate demand to real-time prices. This chapter argues that this reflects the historical supply-side emphasis of the industry and substantively underplays the potential for further constructive end-user engagement. The chapter labels such more substantively empowered users "smart users" and explores how we might move further in this direction. The emergence of more localized smart grids and distributed energy options, often owned and operated by consumers, effects a return to the situation operating before the consolidation of large-scale integrated grid supply. In these emergent circumstances users are able to not only provision the suite of energy services they have inherited as efficiently and cheaply as possible but also to reflect upon the sustainability of these services and whether or not they might better be either substituted or, more simply, ceased altogether. This can, and does, result in the prioritization of more sustainable alternatives such as, for example, passive cooling rather than energy-intensive air conditioning. These matters make up the content of the section "Putting End-Users Back at the Heart of Energy Decision-Making: Emergent Smart Grid and Distributed Energy Options" of this chapter.

While emergent smart energy use commonly involves smart energy users reframing energy services and using a different set of energy technologies, this can only be effective in a market and regulatory environment favorable to the effective exercise of consumer sovereignty. This is a significant challenge because the emergence of large-scale integrated grids and centralized generation witnessed the disempowerment of consumers as supply-side imperatives came to frame and shape demand-side choice in numerous ways. The chapter concludes, in the section "Market and Regulatory Innovations for Smart Energy Use," with a discussion of market and regulatory innovations favorable to the facilitation of smart energy use and develops a number of recommendations for their further development and institutionalization.

DISEMPOWERING USERS—THE EMERGENCE OF SUPPLIER SOVEREIGNTY

Many have long argued that energy policy should focus on demand-side services such as heat, entertainment, light, mobility, and so on, delivered by fuels and electricity, rather than the supply-side criteria, such as the electricity-generating capacity within a particular jurisdiction, which currently dominate much energy policy thinking (e.g., [1]). While such a change in focus has the potential to facilitate improved efficiencies in the delivery of energy services, and help ensure that these better reflect demand, this position assumes that—given a blank slate—consumers would, essentially, choose a set of energy services analogous to those that dominate today. This section demonstrates that the historical emergence of the suite of contemporary demand-side services predominant today was as much about the choices of system designers and engineers as it was about unconstrained consumer choice.

The 1888 historical extract in Figure 2.1, compares the use of conventional AC generation and distribution to meet electrical lighting loads against the potential application of DC generation and distribution with local battery (accumulator) energy storage to better utilize generation and network assets. The extraordinary peaky character of expected electrical loads at that time (almost exclusively lighting) is highlighted in the graph showing the estimated load profile of lighting for London. The author suggests that peak lighting loads might best be met through a mix of centralized DC generation and local battery-sourced supply.

The interchangeability of electrical and mechanical energy was first demonstrated in Michael Faraday's 1831 discovery of electrical induction [2, p. 169]. Many built upon this discovery over the course of the following century with the contributions of some, notably Thomas Edison, particularly prominent. While Edison was instrumental in the development of specific technologies such as light bulbs and large generators, it was his vision, organizational acumen, and broader business skills that made him stand out. Edison's company, for example, was responsible for the first electricity-generating plant at Holborn Viaduct in London in January 1882, which was followed some months later by the first American power plant in New York [2, p. 170]. The initial application of these was to lighting. Although inefficient by modern standards early electric light was "about ten times brighter than gas mantles and hundred times brighter than candles" [2, p. 170]. As a result, electric light dominated the street lighting of industrial cities by the first decade of the twentieth century, and these, and related, developments were instrumental in the emergence of the modern city and industrialized society more generally. Electricity, an energy vector or carrier rather than an energy source, is the ultimate pollution displacer because there are virtually no direct environmental impacts associated with its use. The flexibility and high efficiency with which electricity can be used also make it exceedingly attractive for many applications, something that was soon put to widespread effect.

However, the benefits granted by electrification also had a downside. Urban sprawl, for example, previously constrained by horse- and steam-powered transport from suburban rail stations, was significantly facilitated from 1890, but increasingly from 1900, onward by the advent of electrified

JOURNAL

OF THE

SOCIETY OF

Telegraph-Engineers and Electricians.

Founded 1871.   Incorporated 1883.

Vol. XVII.	1888.	No. 73.

CENTRAL STATION LIGHTING :

TRANSFORMERS *V*. ACCUMULATORS.

By R. E. Crompton, Member.

The present paper is the outcome of the discussion which took place on Messrs. Kapp's and Mackenzie's papers on transformers, recently read before this Society.   I was asked to give facts and figures in support of the statement I then made, that I believed the distribution of electricity by transformers offered no special advantages over other methods, particularly over distribution by means of accumulators used as transformers.

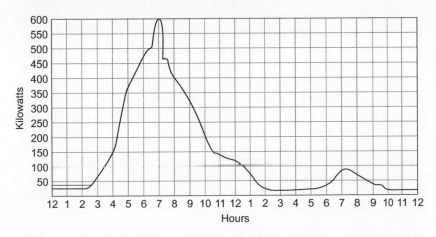

FIGURE 2.1   An early "smart grid" proposal published in 1888. Crompton, R. Central station lighting: Transformers v Accumulators, *Journal of the Society of Telegraph-Engineers and Electricians*, 1888, 17, 349–371.

tramways or trolleybuses [3, p. 148].[1] These were faster, cheaper, cleaner, and more convenient, and they significantly encouraged the expansion of city boundaries. This shaping of certain "forms of life" by electricity was not only structural, such as with the case of urban sprawl, but also intimate as in the case of emergent domestic energy services. In the domestic sphere where electric lighting was vigorously promoted and enthusiastically embraced from the beginning, this was not the case for other energy services. While there was some uptake, and promotion of these energy services in richer households it was only with economic collapse at the end of the 1920s that utility managers turned to the domestic sector as a serious source of potential new load [4]. Much of the groundwork for this had been done however. For example:

in 1920…General Electric's "advertising introduced a new objective: the creation and fostering throughout America of a new electrical consciousness which would normally express itself in a certain fundamental 'want' – the desire of individual families to make their homes into electrified dwelling places" [5, p. 265 (quoting the GE Publicity Department)].

This involved not only mass marketing but also "door-to-door selling, poster contests, parades…other locally focused advertising drives…[and involved] utilities often provid[ing] inexpensive wiring services, payable in monthly instalments," because "[u]nlike industrial sales, the domestic market had to be built literally unit by unit" [5, p. 265]. Many of the strategies used at the time underscore how these services were viewed very differently than today, something underlined by the then-common strategy of suggesting that electric appliances could replace domestic servants [5, p. 271]. That this went on for some time is illustrated by how electric irons, which replaced heavy cumbersome "flatirons" that required manual heating, were the second most successful energy service after lighting until the 1950s [4, p. 7]. Refrigerators, which displaced electric irons in terms of domestic "appliance saturation"[2] around 1950, provide an insightful case study of the character of many of these changes.

"[S]o rare in 1922 that it was not even covered in most standard home economics texts…there were only 20,000 refrigerators in the entire United States" in 1923 [5, p. 275]. Domestic refrigerators only became cost effective in the 1930s (276) as a result of design improvements from the late 1920s onward that eventually delivered self-contained machines substantially equivalent to those used today. An interesting twist to this story is that while this effort focused upon the optimization of electric compression machines, gas powered absorption refrigerators were more advanced from the beginning of this period and are still today the machine of choice where efficiency and reliability count

---

[1]The age of electrified tramways was short-lived, however, with the rise of automobile ownership from the following decade onward tending to thenceforth dominate the dynamic growth of urban sprawl.

[2]"[T]he number of appliances in stock per one hundred households" [4, p. 7].

most, such as in remote applications [6, pp. 128–143]. Needless to say electric utility companies, such as General Electric, were prime movers in the optimization of the electric compression machines that were to play such a significant role in domestic load building through the twentieth century and beyond.

Even the widespread assumption that the uptake of electrified energy services was, in some sense or other, "natural" because they were inherently superior in terms of the comfort, convenience, and time saving they delivered does not always stand up to scrutiny. As the title of Edith Cowan's classic *More Work for Mother* [6] declares sometimes the situation was the inverse. Vacuum cleaners, for example, whose uptake was only second to that of electric irons until around 1950 [4, p. 7], were expected typically to be used, by mothers, about once a week, whereas the far more infrequent rug beating they replaced was generally regarded as men's work up until this time. This was not the story told, however, by both appliance manufacturers, most of whom were also utilities, and various other societal interests with which they were closely aligned, such as the home economics movement. For them a very particular model of electrification took pride of place, focused by contemporary ideals such as that of progress, but the numbers tell a different story. Women spent as much time on housework late in the twentieth century as they did at the beginning [5,6], although what they did changed considerably.

Homes have, arguably, become cleaner and more comfortable (although see the discussion of the invention of the notion of thermal comfort below), but they have also tended to be more self-contained, self-sufficient, and isolated from neighbors and community. Doing the laundry, for example, once a communal activity, became a matter that a housewife shared solely with a machine, although automatic machines, which saved the chores associated with wringing and changing water, were only developed in the late 1930s and became common after World War II (WWII) [6, p. 94]. The emergence of air conditioning detailed below graphically illustrates both the artifice involved in these developments and how consumer choice took a back seat to the preferences of others as electrification grew apace.

A Case Study: The Emergence of "Thermal Monotony"

A debate over the character of healthy indoor environments, particularly that of schools, between fresh air and air conditioning (AC) advocates proved particularly pivotal to the development of domestic AC. This debate culminated in the establishment of an expert New York State Commission on Ventilation in 1913 whose final report came out in favor of window ventilation for classrooms "much to the indignation of the engineering community" [7, p. 68].[3] The report

---

[3]Cooper's *Air-Conditioning America* (1998) is the acknowledged authoritative history of the emergence of air conditioning, with Ackermann's *Cool Comfort* (2002) the only comparable, but narrower, study.

completed in 1917 wasn't officially released until 1923 giving engineers time to counter. The American Society of Heating and Ventilating Engineers (ASH&VE)[4] established the ASH&VE Research Laboratory in 1919, which "[a]s it took shape...attempted to address both technical issues and the larger problem of public confidence" [7, p. 69]. Specific attention was given to the establishment of numerical standards and to the development of improved, more accurate, measuring instruments. A contemporary journalist, quoted by the leading historian of AC, Gail Cooper [7, p. 70], testifies to the benefits so granted:

One of the engineers' most effective safeguards against loss of public confidence is the development of fixed, rational standards, and the determination of precise finite values, that he may thus express himself with an exactness which ensures understanding.

Just before the New York State Commission on Ventilation's final report was circulated, the engineers building upon earlier work published the "Comfort Chart" that "graph[ed]...the combinations of temperature and humidity at which most people felt comfortable" [7, p. 71]. The current ASHRAE Standard 55 is a straightforward elaboration of this work. The provision of these parameters, and instruments for their precise measurement, by ASH&VE's Research Laboratory was the first step in allowing the indoor conditions defined by the Comfort Chart to be generated and maintained, in different places, on demand.

While, today, many take it for granted that this particular definition of thermal comfort is somehow "natural," the marketing activities that its promotion involved suggest otherwise. The engineers voted "at the 1923 ASH&VE annual meeting...to delete the term [fresh air] from the society's proposed ventilation code" [7, p. 73], a term central to their opponent's rhetoric, and mounted a public relations campaign whose mouthpiece was the journal *Aerologist* founded in 1925. This rhetorical engineering of future debate regarding the relative merits of natural ventilation vis-à-vis AC, today commonly labeled "spin," was successful with use of the term 'fresh air' becoming indicative of "support for the open-air crusaders and an antitechnical spirit" (73). Ultimately in this "contest between competing professionals...the scientification of ventilation bolstered engineers' status and authority," turning what had been "the 'art of ventilation' into a science" and significantly enhancing the prospects of the emerging AC industry [7, p. 78]. These standards broke the, previously generally assumed, link between human comfort and natural climate and proved to be a key to the success of the Comfort Chart and its successors.

---

[4]In December 1958, members of ASH&AE and ASRE (the American Society of Refrigeration Engineers) voted to merge the two societies into the American Society of Heating, Refrigerating, and Air-Conditioning Engineers, Inc. (ASHRAE), which is the relevant current body (see: http://www.ashraeno.com/history.htm).

Cooper argues that "the ideals of the engineering profession were literally built into the systems that they sold" [7, p. 109]. While users may not always reflect these ideals their static character is reflected in a homogeneity common to air-conditioned environments and cultures, something reflected in the use of the term "thermal monotony" that commentators commonly use to describe air-conditioned environments and spaces. That AC does carry an homogenizing imperative is borne out by how, as it became an essential feature of the post-WWII U.S. manufacturing environment, it facilitated the decline of a distinctive southern U.S. culture. Before WWII "air-conditioning evolved from a bally-hooed novelty into a standardized expectation among middle-class users of public facilities" in the United States, with movie houses the most notable trailblazer [7; 8, p. 46]. Gail Cooper argues that the post-WWII success of AC was, however, less a matter of consumer approval and more a result of how "[a]rchitects and builders made the decision to air-condition American homes and offices, then lenders and regulators stamped the change with institutional approval" [7, p. 142; 8]. Although ideas, promotion, and marketing played key roles in the growth of AC,

the triumph of air conditioning in the 1950s was based upon its integration into building design, construction and financing. Architects, builders, and bankers accepted air conditioning first, and consumers were faced with a fait accompli that they had merely to ratify [7, p. 142].

Advertising specifically helped in "domesticating and normalizing the use of technologies that might at first seem frightening…[while] also ultimately marginalizing those who still refuse[d] to use them" [8, p. 125]. Domestically central AC made possible cheap, lightweight, thermally inferior construction techniques:

The peculiarities of tract housing and modern design made air-conditioning virtually essential. Tract houses were hot houses. Lightweight construction made them cheap and large picture windows gave them a modern look [7, p. 157].

New post-WWII office construction offered opportunities to air condition office space far cheaper than the costs of retrofitting older buildings [7, p. 159]. The escalating cost of urban land resulted in an increasing number of simple block shaped buildings, whose cheapest configuration left interior spaces that, without windows for ventilation or light, required AC and fluorescent lighting to render them habitable [7, p. 160]. Thus, as with domestic housing, "an air-conditioning system facilitated the construction of a less costly building style. The office block was a design dependent on air conditioning to make it habitable" [7, p. 161]. Government support was central and evidenced by, among other things, a 1955 change in federal government policy ensuring that "virtually all new government buildings would be air-conditioned" [7, p. 163]. The resultant proliferation of "[b]oth the small picture window house and the glass-fronted office block…heavily dependent on air conditioning to make their economical designs habitable" [7, p. 163] influenced not only the form and content of the built environment but also the "forms of life" it supported.

This is graphically illustrated by Arsenault's account of how "air conditioning has affected nearly every aspect of southern [US] life" [9, p. 628]. As in the United States more generally AC was first found in the movies, and by WWII most movie theatres in the South were air conditioned [9, pp. 603–4]. Further developments took some time to become established, and while by 1950 many federal buildings were air conditioned "most of the region's state buildings and county courthouses did not become air-conditioned until the 1960s" [9, p. 605]. The domestic uptake of AC followed, but more intensely than in the North, and the trajectory of cheap post-war tract home developments in the United States was more steep. "By 1955 one out of every twenty-two American homes had some form of air conditioning. In the South the figures were closer to one in ten" [9, p. 610]. This escalated such that "by the end of the [1960s] half of the homes and apartments in the South were air-conditioned" [9, p. 610].

By the 1970s,

the air conditioner invaded the home and the automobile, there was no turning back…
The South of the 1970s could claim air-conditioned shopping malls, domed stadiums,
dugouts, green-houses, grain elevators, chicken coops, aircraft hangers, crane cabs,
off-shore oil rigs, cattle barns, steel mills and drive-in movies and restaurants
[9, p. 613].

Arsenault underscores how industry, government, and finance institutions not only structurally facilitated these developments, but were also actively involved in the conditioning of consumer preferences and dispositions. Particularly intriguing examples of this include the issuing of a 'Discomfort Index' by the National Weather Bureau in 1959, calibrated to highlight how outdoor temperature and humidity diverged from indoor thermal comfort standards, and the meteorological invention of the term "heat storm" in the 1960s [9, p. 615].

While Arsenault notes that the mostly black poor are still often excluded from "thermal monotony" he argues that

[a]ir-conditioning has changed the southern way of life, influencing everything from
architecture to sleeping habits… contribut[ing] to the erosion of several regional
traditions: cultural isolation, agrarianism, poverty, romanticism, historical consciousness,
an orientation toward non-technological folk culture, a preoccupation with kinship,
neighborliness, a strong sense of place, and a relatively slow pace of life [9, p. 616].

While AC is evidently only one influence among many, and some of these changes are clearly positive, Arsenault convincingly argues that AC was a significant agent of such changes. For example, the demise of a vernacular architecture— open porches, high ceilings, breezeways, and so on—that AC helped bring about not only facilitated the replacement of front porch conversations by a retreat behind closed doors and windows, but "also influenced the character of southern family life" [9, p. 624] in more subtle and fundamental ways. In tandem with other, sometimes linked, developments, such as air-conditioned shopping malls and TV, Arsenault concludes that "even if, on balance, residential air conditioning strengthened the nuclear family, the impact on wider kinship networks probably went in the opposite direction" [9, p. 625]. That AC affected fundamental aspects of southern culture more generally is underlined by how it "modulated the daily and seasonal rhythms that were once an inescapable part of southern living…assault[ing]…the South's strong 'sense of place'" [9, pp. 627–8].

Figure 2.2 shows how recent statistics bear out this trajectory. From 1978 to 1997 centrally air-conditioned households increased from 23% to 47% across the United States as a whole while in the South 93% of households were air-conditioned by 1997 [10]. Figure 2.3 highlights how this broader influence of AC was not simply a function of its availability but also of the degree of use, with AC use in the South more than double that in the rest of the country.

Table 2.1 shows how, across the three main residential categories of U.S. electricity end use (AC, space heating, and water heating) AC demonstrated a small but notable increase in share from 1987 to 2009, up from 15.8% to 17.9%, while the share of both space heating and water heating declined from 10.3% to 9.1% and from 11.4% to 9.1%, respectively [11].

Similar patterns are now evident across the globe. In the authors' home state of New South Wales, for example, 50% of households were air conditioned by October 2006 [12], while the rapid growth of AC in Western Sydney, particularly the use of split mini-systems, is putting a considerable strain on local electricity substations [13, p. 5].

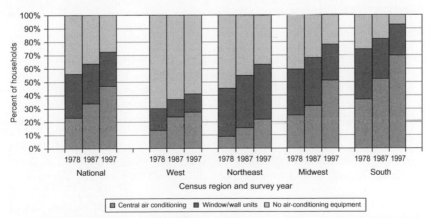

Sources: Energy Information Administration; 1978, 1987, and 1997 Residential Energy Consumption Surveys.

FIGURE 2.2   Type of air-conditioning equipment by census region, 1978, 1987, and 1997.

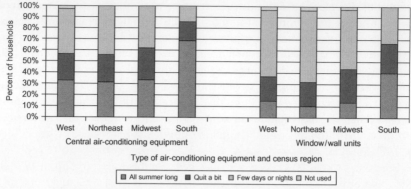

Sources: Energy Information Administration; 1997 Residential Energy Consumption Survey.

FIGURE 2.3 Frequency of air-conditioning usage by census region, 1997.

TABLE 2.1 Percent of U.S. Electricity Consumption by End Use

End Use	1987	1990	1993	1997	2001	2009
AC	15.8	15.9	13.9	11.8	16.0	17.9
Space Heating	10.3	10.0	12.4	11.4	10.1	9.1
Water Heating	11.4	11.2	10.3	11.0	9.1	9.3
Total Appliances	62.5	63.0	63.4	65.9	64.7	63.7

Source: EIA – http://www.eia.gov/emeu/recs/recs2001/enduse2001/enduse2001.html; http://www.eia.doe.gov/tools/faqs/faq.cfm?id=96&t=3.

In a highly stylized article Prins [14] indicted Americans as "condis" addicted to "coolth." "Condis"

abhor the heat and the wet. They inhabit civilization: Their natural habitat is the civis… They fear and detest the things that live in the Wild Woods. They insulate themselves from the Wild. In summer, they take Coolth with them to do this, to tame it. They seek predictability. They hate the spontaneous. Their lives celebrate the linear and overthrow of the cyclical. Day is as Night and Night is as Day to them. Time is Digital [14, p. 252].

While Prins's article has been subject to considerable criticism[5] it's notable that others reinforce the charge that AC-related practices and behaviors meet standard definitions of addiction [15, pp. 183–85]. Prins ultimately argues that

---

[5]Prins's [14] highly stylistic piece was published in a special volume of the journal *Energy and Buildings* on the social and cultural aspects of cooling and was contested by a number of invited commentaries (the commentaries and Prins's rejoinder can be found in volume 18, 1992, of *Energy and Buildings*: 259–268).

"fracturing the right to Coolth…[requires] the need to reconstruct America's contemporary cultural norms" [14, p. 258]. Interestingly many of Prins's critics charge that he overlooks key structural issues, such as how AC emerged in parallel with networked electricity supply. This underscores how "condis" addicted to "coolth" are not only the result of the psychological makeup of consumers but also the systemic imperatives embedded in networked electricity supply systems. Today, the emergent opportunities presented by DG and smart grids put such historical imperatives under a particularly interesting spotlight.

The Lessons of History

As electricity grids grew rapidly in the early decades of the twentieth century it became clear that optimal use of the thermal generating plant then dominating supply provision required significant additional loads, beyond those of industry. As a result a number of energy service options were developed reflecting, primarily, corporate, institutional, and other professional preferences rather than those of consumers. These were then aggressively marketed by those set to profit from them. This provision and structuring of both energy service options and of the means by which they might be obtained were not therefore the result of unconstrained consumer choice, as economic perspectives still tend to assume, but rather gave the consumers exercising that choice an ever-broadening range of services to choose from. The current dominant suite of energy services and the means by which they are delivered are thus as much a result of the decisions and actions of utility managers, engineers, and regulators as they are the result of the decisions and actions of consumers. To characterize contemporary decisions regarding these services as simply a straightforward matter of consumer choice, with "choice" narrowly conceived as little more than the responses of end-users to their own perceived needs, electricity prices, and information is, therefore, misleading and does not square with the empirical evidence. Some of the most telling relevant contemporary examples are some of the best known, such as the cheap off-peak, and highly inefficient heat and hot water storage systems embedded throughout centralized grid systems today. In other words the evidence suggests that our current energy-intensive lifestyles are as much about the systemic framing and conditioning of the choices available to consumers as they are about the exercise of individual consumer choice. Consequently, addressing this systemic framing and conditioning of the choices available to consumers requires that those currently responsible for overseeing and delivering current energy provision reinstate a substantive decision-making role for end-users.

PRESSURES FOR CHANGE: RISING DEMAND, ENERGY SECURITY, AND CLIMATE CHANGE

It is interesting to note that one of the stated objectives of the electricity industry restructuring undertaken in numerous countries over the last two decades was to facilitate customer choice [16]. However, restructuring centering upon the development of competitive electricity markets, and the associated corporatization, or privatization, of previously government-owned or regulated private monopoly industry participants, has had mixed results in this regard. While nominally focused upon improved efficiency, innovation, and consumer choice there have been some highly counterfactual outcomes such as the collapse of the Californian system, and the experience of the energy user has not yet been markedly changed in many cases (see, for example, [17]).

Now, however, current industry arrangements are under a growing range of pressures that hinge critically, on the nature and scale of desired energy services. They include growing energy demand and particularly peak demand, rising energy security concerns centering upon energy resource depletion and the increasingly concentrated distribution of what remains, and the threat of dangerous climate change.

Growing Demand

As documented in the previous section, growing demand has been the normal state for the electricity industry since its inception. While conventionally conceived as being driven by aggregate growth in population and wealth, it was rather the systemic intersection of this growth with the entrepreneurial exploitation of emerging technological possibilities that has been pivotal to growth in electricity demand. This remains the case today; consumer electronics, for example, is a sector particularly dynamic in proliferating new and/or augmented energy services (e.g., flat screen TVs; gaming consoles with enhanced functionality such as control via bodily movement; ever greater integration and augmentation of the features of hand-held digital devices, etc.), which is illustrated in Table 2.1 in terms of an ongoing growth in appliance penetration.

As a result end-users continue to find, and are indeed encouraged to adopt, new energy services. Some of these, such as air-conditioned homes and offices, require appliances and equipment with high energy and, particularly, peak power demands, and while there has been, and continues to be, energy efficiency improvements, peak demand has been less affected by these improvements. This steady growth in electricity demand, relative to other domestic energy sources, is shown in Figure 2.4.

The economics of end-user equipment has also changed with respect to supply. For example, electric kettles can cost as little as $10 while drawing 2 KW, and portable air conditioners can now be purchased at a cost of around $300 while drawing 1KW. In addition, the electricity industry has generally relied on diversity—that is on people not running toasters or kettles at same time—yet air conditioning is generally a highly correlated use. If your neighbor is running their air-conditioning unit then there's a reasonable chance you are running yours because use is driven by the weather. On the supply side

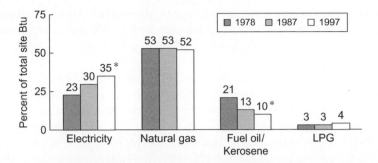

* The difference between the 1978 and 1997 estimates is statistically
significant at the 95-percent confidence level.
    Sources: Energy Information Administration; 1978, 1987, and 1997
Residential Energy Consumption Surveys.

FIGURE 2.4   Percent of total U.S. residential site energy consumption by energy source.

about the cheapest generation technology available is the Open Cycle Gas
Turbine at around $1000/KW [18]—1 to 2 orders of magnitude greater
expense per KW than these electrical loads. Furthermore existing electrical
networks don't have low-capital cost-peaking options as seen with generation
(plants that are cheap to build but expensive to run and, as a result, are better
for infrequent peaks than expensive to build but cheap to run baseload options).
Many industries around the world, including in Australia, spend more on networks than on electricity generation. One result of this is a potentially significant imbalance, with private investment by end-users in particular new
appliances potentially requiring the supply side of the electricity industry to
invest equivalent levels in generation and network infrastructure if reliability
is to be maintained. Air conditioning is a good case in point, having stressed
some local sub stations beyond capacity in recent years in the authors' home-
town of Sydney [13, p. 5].

Energy Security

This has both short and longer-term dimensions. While the longer term is
focused upon concerns with primary-fuel security, the short term is focused
upon peak demand, which tends to drive a longer term focus on investment
in generation and networks. The impact of the oil crisis in the early 1970s
provides an exemplary example of primary-fuel energy security concerns in
the electricity industry. Many countries at the time had significant oil-fired
generation, which is low cost, easy to transport, and easy to use, and the
1970s price shocks resulted in electricity industries globally turning to alternatives. Some countries, France and Japan for example, gave nuclear a significant
role from this time, while the R&D attention and funding this granted

renewables, most notably in the United States, significantly spurred the further development of some renewable technologies such as PV and wind. Furthermore, it had a very significant impact on improving the energy efficiency of many countries.

In some ways the electricity industry has lower energy security concerns than sectors like transport where alternatives to oil are more limited. By comparison a very wide range of potential primary energy sources can be turned into electricity although the industry worldwide remains very dependent on fossil fuels. Oil in particular and gas, to a lesser extent, have likely constraints on supply in the near to medium term, with likely upcoming price impacts. However, the geopolitics of significant fossil fuel reserves is a significant factor with more countries with gas than oil, for example. Coal is far more abundant and geographically dispersed but is also witnessing tightening demand and increasing prices. However, the most pressing constraint on fossil fuel use is that we cannot afford to continue to utilize existing reserves as we have historically because of the associated greenhouse emissions.

Climate Change

It currently appears that we will run out of an atmosphere amenable to life as we know it before we run out of fossil fuels. Figure 2.5 shows how 25% of global greenhouse current gas emissions are from the electricity sector, primarily coal, then gas, and then oil. It may be possible to continue to use fossil fuels while restricting emissions through carbon capture and storage technologies but these are proving to be more difficult to develop into a viable working technology than many proponents originally argued [18]. Climate change is increasingly the driver of transformation, involving a range of fundamental changes that move significantly beyond traditional concerns with economics and reliability.

Ultimately, climate science suggests a requirement for a near complete decarbonization of energy supply by about mid-century, with aggregate emissions starting a steady decline from about the mid-point of this decade. As a result the traditional emphases upon resolving the issues of demand growth and energy security are being both bolstered and augmented through further critical attention to the industry and issues such as its structure, organization, and governance. Given that many sources of emissions such as agriculture have a very high value but limited low-emission options it is conceivable that a complete decarbonization of the electricity industry will be required within a generation. Clearly, such a fundamental transformation cannot be achieved by exchanging supply sources without addressing the fundamental drivers of energy demand. This underscores a requirement to address the systemic stimulation of demand characteristic of the history of the industry through, among other things, empowering end-users.

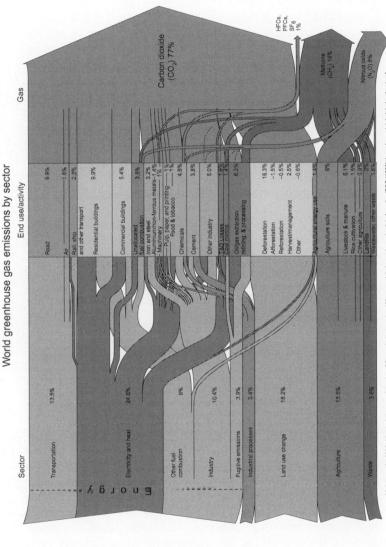

FIGURE 2.5 World greenhouse gas emissions by sector. *(Source: http://maps.grida.no/go/graphic/world-greenhouse-gas-emissions-by-sector. Designer: Emmanuelle Bournay, UNEP/GRID-Arendal.)*

45

PUTTING END-USERS BACK AT THE HEART OF ENERGY DECISION-MAKING: EMERGENT SMART GRID AND DISTRIBUTED ENERGY OPTIONS

The current emphasis on means such as smart meters and dynamic pricing (see Faruqui, Chapter 3 of this volume) conflates consumer choice with the collective response of aggregate demand to real-time prices. However, the emergence of more localized smart grids and distributed energy options, often owned and operated by consumers, effects a return to the situation operating before the consolidation of large-scale integrated grid supply in terms facilitating an effective voice for consumers. As end-users more closely engage with the provision of energy, and in some cases become providers themselves, they are then not only better placed to provision the suite of energy services they have inherited as efficiently and cheaply as possible but also to reflect upon the sustainability of these services and whether or not they might better be either substituted for or, more simply, ceased altogether. This can, and does, result in the prioritization of more sustainable alternatives such as, for example, passive cooling rather than energy intensive air conditioning. In something of a paradigm shift signaling the fundamental significance of these changes the U.S. Federal Energy Regulatory Commission (FERC) has started to echo these ideas. FERC Chairman Jon Wellinghoff identified consumers as "active parts of the grid, providing energy via their own solar panels or wind turbines, a system called distributed generation; stabilizing the grid by adjusting demand through intelligent appliances or behavior modification, known as demand response; and storing energy for various grid tasks" [19].

New Distributed Options

There are a promising range of existing and emerging options. Some of these fit well within current centralized supply style arrangements, such as large-scale wind (Hanser et al., Chapter 10 of this volume) and, perhaps, at some point in the future carbon capture and storage fitted to conventional power plants. Many of the most promising options that might help us address these challenges are, however, distributed energy ones (Hesser and Succar, Chapter 9 of this volume). These include distributed renewable generation such as PV, biomass, small-scale wind, hydro, and highly efficient small-scale fossil fuel plant for cogeneration and trigeneration applications. Energy storage is also important (Wang and Wang, Chapter 5 of this volume) and includes novel possibilities such as electric vehicles (Hindsberger et al. and Armstrong et al., Chapters 18 and 19, respectively, of this volume). Most importantly there are demand-side options including energy efficiency and controllable loads. These all help deliver reduced demand growth, improve energy security, and help reduce emissions.

Energy efficiency (EE) is particularly promising. For example, IEA scenarios involving significant emissions reductions by 2050 see EE playing the major role (with EE in the electricity industry playing a particularly important role, [18]). EE options include not only efficient equipment but also smarter use of equipment, underlining the value of smart grids. Controllable loads can contribute to EE but might instead focus on moving the timing of peak demand, which can involve tradeoffs where control for such objectives actually reduces energy efficiency.

Distributed energy options do, however, present a profound technical, economic, and social challenge to current industry arrangements. Complex new technologies, such as PV, in new locations within the electricity industry at new, small scales require high levels of coordination in order to deliver secure and reliable electricity flows. Smart grids, commonly owned and operated by new industrial players, are an essential element of this coordination of new distributed technologies. Distributed options are often sited on end-user premises, and some, such as cogeneration and controllable loads, require close integration with the range of other activities that end-users undertake. Most EE decisions are investment and operational ones taken by end-users themselves, underlining how the necessary decisions involve a whole new set of stakeholders—the end-users themselves [20].

End-User Engagement

We can distinguish a spectrum of views on smart grids and what they might deliver. On the one hand there are, largely utility-driven and directed perspectives emphasizing more efficient and reliable supply, primarily via means such as smart meters and dynamic pricing, which also share an emphasis upon greater renewable energy deployment. This perspective emphasizes technologies and pricing with users largely engaged through means such as in-house displays, time-of-use tariffs, or utility load control, rather than through the facilitation of more substantive end-user decision-making. Although this approach may improve the reliability and economics of supply while also making the grid more robust to inevitable climate change impacts, it vastly underplays the potential of smart grids in our view.

This potential centers upon returning end-users to the heart of decision-making within an electricity industry set to rapidly and radically transition towards a more sustainable low-carbon future and away from its supply-driven past. So while on the one hand engagement is largely viewed as a matter of mediation by technological and economic means, on the other it is rather seen as a matter of facilitating substantive end-user decision-making beyond pricing and technological interventions.

It's important, however, to acknowledge that this spectrum of views is complex and far from clear-cut. For example, the White Paper "Energy Retailers'

Perspective on the Deployment of Smart Grids in Europe," written by members of the Working Group Demand and Metering of the SmartGrids European Technology Platform for Electricity Networks of the Future,[6] is subtitled "The Smart Grid is only a platform for a Smart Energy Ecosystem, in which Customers play the first violin." However, the emphasis remains upon technology and regulation, with the only major overlap with the arguments articulated in this chapter being at the level of institutional changes, such as those involving the promotion of energy service providers (see the following section). So whereas for EU energy retailers the concept of "customers playing the first violin" remains primarily a matter of technology and pricing we believe that such an approach significantly underplays the potential for more substantive end-user engagement.

Conventional views of community engagement generally assume an economistic model of human behavior centered upon self-interested individuals straightforwardly amenable to economic incentivization. This is the primarily taken-for-granted view underlying the common smart grid emphasis upon smart meters and pricing. Interestingly, however, conventional views also tend to assume, somewhat antithetically, that an end-user's sense of civic responsibility, or "green" values, can be appealed to, via information, education, and/or marketing, so as to facilitate "smart use" understood in terms of frugality, energy efficiency, and a responsiveness to changing conditions (e.g., to only run AC when required, when the sun shines, etc.).[7] However, not only are conventional views commonly ambiguous, privileging both economic self interest and altruistic collective interests, they fail to account for the complexity of real-world consumer behavior.

Today, for example, people wash clothes more frequently than in the past but in a far less energy-intensive manner as a result of not only more energy-efficient washing machines but also improvements to detergent technology that facilitate both shorter and colder washes [21]. This complex intersection of not only economic but also a range of cultural and technical factors is characteristic of consumption more generally but tends to be opaque to the economistic mindset currently predominant.[8] As a result we believe that bringing end-users back into energy decision-making needs to be substantive rather than framed in narrow economic terms. Among the best examples of the potential of such a move to date lies in the recent history of the wind industry.

---

[6]Downloaded from: <http://www.smartgrids.eu/documents/newforum/8thmeeting/ETPSmartGrids_EnergyRetailers_WhitePaper_Final_ExecutiveSummary.pdf>, 31 March 2011.
[7]For a recent review of these kinds of approaches see "Motivation and Behavioural Change" in *Zero Carbon Britain 2030: A New Energy Strategy*, Centre for Alternative Technology, 2010: 147–188 (available from http://www.zerocarbonbritain.com/).
[8]For a recent elaboration of these arguments detailing the broader literature from which it is derived see [22].

Denmark is, perhaps, the leading example. A tax exemption for generating their own electricity meant that by 2001 over 100,000 families belonged to wind turbine cooperatives accounting for 86% of all the turbines installed in the country. This, in no small part, accounts for Denmark's leading role as an innovator and manufacturer of wind turbines today.

Similar cooperative models are in operation today in Germany, the Netherlands, and Australia, with Germany and the Netherlands having particularly active cooperative wind power sectors. The Hepburn Wind Project near Daylesford in Victoria, Australia, is currently catalyzing further related developments such as the recent proposal to establish a community-owned wind park in Mount Alexander Shire, Victoria.[9] Other community-based models are in operation in other jurisdictions, particularly in the United States, most commonly involving the developer or manager of a project sharing ownership with landowners and other community members. Similar models to these should be equally as applicable to other distributed energy options.

While some of these projects are commercial in orientation, many, if not most, of them reflect a corresponding surge in recent years in small, typically local community-led initiatives on climate change, energy use, and other socio-environmental issues. Good examples include Transition Towns initiatives and local CRAGs (carbon reduction action groups). The former centers on the development and implementation of an "Energy Descent Plan" for a particular local area focused upon substantially reducing local dependence on external energy resources, building community, and increasing the availability of local resources (of all kinds). Starting at Totnes in Devon (UK) in 2006 there are now many hundreds of such initiatives globally.[10] Initiatives of these kinds should give many in the electricity industry pause for thought in that they flag a strong, broadly based community disillusionment with the opportunities to address current challenges such as climate change presented by existing, primarily market-based, structures and an enthusiastic willingness to embrace innovate community alternatives.

This disillusionment and the contrasting levels of trust invested in more community- oriented organizations are vividly illustrated in Figure 2.6 (note how retailers fare even worse than utilities). Sharing and mutual trust are some of the primary characteristics defining communities, and, as a result, community initiatives tend to be inherently durable and well placed to effect fundamental transitions such as that required by the requirement to rapidly decarbonize our energy system. One way of envisaging this transition from the supply-side-dominated ESI systems of the past is through the notion of "co-provision" [24],

---

[9]See http://masg.org.au/projects/renewables/community-owned-wind-park-in-the-mount-alexander-shire/. This paragraph primarily drew from http://www.wind-works.org/articles/Euro96TripReport.html and http://e.wikipedia.org/wiki/Community_wind_energy.
[10]See http://www.transitionnetwork.org/.

What organizations do you trust to inform you about actions you can take to optimize your electricity consumption?

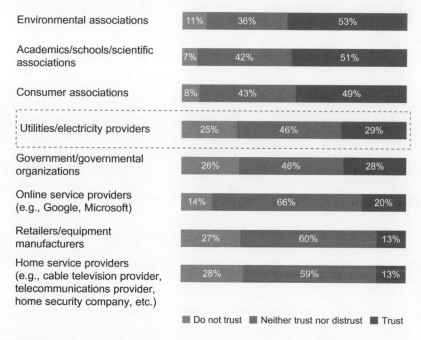

FIGURE 2.6   The levels of trust energy users place in different organizations. *(Source: [23].)*

which envisages a future system that puts consumers on a par with more traditional electricity providers. This is further examined below.

MARKET AND REGULATORY INNOVATIONS FOR SMART ENERGY USE

Smart energy use can only be effective in a market and regulatory environment favorable to the effective exercise of consumer sovereignty, which involves a move away from the historically conditioned supply-side oriented economistic perspectives described above. However, initiatives that currently do this, such as the more community-oriented and inspired ones outlined above will remain, as they do at present, marginal to more mainstream market-oriented initiatives without further regulatory and broader structural support. This is a significant challenge because the emergence of large-scale integrated grids and centralized generation resulted in the disempowerment of consumers as supply-side imperatives came to frame and shape demand-side choice in numerous ways as the above sections have made clear. This section explores some of the market and regulatory innovations favorable to the facilitation of smart energy use conceived as that involving a substantive role for end-user decision-making.

Current Retail Markets

The primary role of traditional industry arrangements has been to achieve cost recovery for the overall cost of supply and to manage the delivery of electricity, viewed as an essential public good shaped by, among other things, available metering technologies. The common result has been flat tariffs across time and location for particular customer classes involving significant cross-subsidies. This is not only economically inefficient but subject to a variety of uncertainties and contingencies that are sometimes reflected in tariffs adjusted to cater to certain groups of users and other systemic contingencies (seasonal, peak, TOU, etc.).

Even in restructured industries such as Australia, retail market arrangements are often an unfinished business, as shown in Figure 2.7. In Australia, for example, all energy users are able to participate directly in the wholesale national electricity market (NEM) subject to metering and NEM membership requirements, but all but a few very large energy consumers currently interface with the NEM via a retailer. The tariff arrangements offered by retailers for large industrial and commercial end-users typically involve a 2- or 3-period TOU energy tariff. Network charges will typically incorporate a "peak load" component or may just be bundled together with the energy tariff [25], and few end-users have direct exposure to NEM spot prices. Falling energy prices due to the success of supply-side electricity industry restructuring have

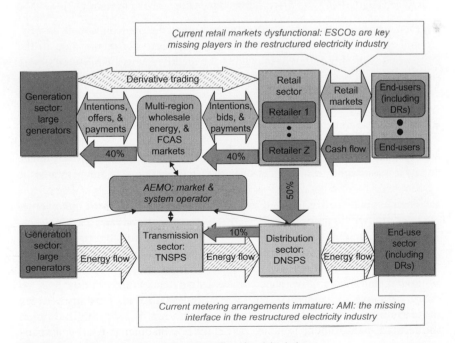

FIGURE 2.7   Schematic of the current Australian electricity industry.

bolstered the historically conditioned end-user complacency about electricity costs, characteristic of ESIs globally, further discouraging effective initiatives aimed at the management of electricity demand [26,27]. Although many retailers offer energy services consulting to larger customers, current market arrangements mean that it is not generally in a retailer's financial interests to promote energy savings [28]. A number of consulting firms offer energy procurement services but they have little influence on distributed resource outcomes.

All but two Australian states in the NEM now have so-called full retail competition where all small businesses and households can choose a retailer. However, most customers continue to receive supply from their default (franchise) provider because of both limited end-user awareness of this option and limited retail competition. At present, most of these customers have only accumulation meters' and profiling is used to estimate varying usage over a typical day. Network pricing generally involves some considerable spreading out of augmentation and maintenance spending over all network customers, and the potential for competition is significantly limited because of the regulated monopoly status of the wires businesses.

A New Approach

This requires an integrative approach embracing the technical, commercial, policy and regulatory, and governance dimensions of the system [17].

- First, a technical framework allowing end-users, distributed resource owners, and energy service companies (ESCOs) to actively participate in the processes of appropriately managing the quality and availability of supply is required. It is critical that this doesn't discriminate against particular technologies (particularly new entrant technologies) and demand-side options such as EE.
- Second, a commercial framework (market design and structure) that facilitates appropriate investment and operation of distributed resource options organized, where appropriate, through ESCOs and that doesn't discriminate against particular technologies or participants (particularly new entrants) is required.
- Third, a policy and associated regulatory framework is required that internalizes climate change and other current electricity industry externalities, while facilitating socially desirable innovation in distributed resources and end-use decision-making and that appropriately assigns risks and accountabilities to end-users, industry participants, regulators, and system operators.
- Fourth, and finally, governance processes are required that will facilitate all of the above. It's essential that these operationalize relationships between consumers and providers, at all scales, moving beyond the traditional demand management mindset that reduces consumers to self-interested actors motivated by little more than price and information.

So, in this approach the system of interest is not conceived as one connecting providers with consumers, understood to be subject to manipulation via, primarily, economic incentives and technological intervention. Rather the system is understood in terms of "co-provision" [24], a notion blurring conventional distinctions between providers and consumers by recognizing the many ways that they might now intersect. Such future systems, facilitative of smart use, involve a recognition of many more players, some of whom both provide and consume for example, and envisate the function of the system as not simply one of transmission but also storage and the smart management of widely distributed, and commonly, intermittent sources of supply.

Frameworks for Engagement

The notion of "co-provision" [24] involves recognizing that the emergence of distributed resource options, and the various innovative arrangements such as cooperatives and mini grids these can involve, requires a rethinking of provider/utility-user relations. While it might suit some customers to retain a fairly conventional relationship with a provider, there will be other consumers, who both draw from the grid and contribute to it, who require different and more flexible arrangements. In their case engagement needs to be conceived as a two-way process. Overall this indicates that ESIs of the future will have to be more complex entities than current ones, involving different kinds of consumers, different kinds of providers, some of whom are also consumers, and a variety of frameworks and arrangements to cater to the interdependencies between them all. Such new frameworks and arrangements will evolve over time to embrace all aspects of the system from the technological hardware, the institutional "orgware" involved in its governance and regulation, to the "software" encompassing the ideas, attitudes, practices, and behaviors of all those participating in it. Figure 2.8 indicates schematically key features of such a system (which should be contrasted with Figure 2.7).

Technical Frameworks

Effective data collection at point of use is imperative. This requires not only half-hour metered data but also measures of power quality and reliability ideally involving data collection at the level of major appliances. Historical data will also be very important for planning purposes. Effective communication, allowing people to act on this data within appropriate time frames, will be critical. Important issues include ones of ownership: do households or those collecting it own the data? Increasingly such data are becoming an important predictive tool for future electricity industry challenges (e.g., historical experience and weather forecasts) and also to facilitate the management of contingencies where a near immediate response can have a particular value. Smart grid technologies both within house, and household to utilities/ESCOs are central elements here. Issues of power and control are central. Traditionally all the power, to intervene and

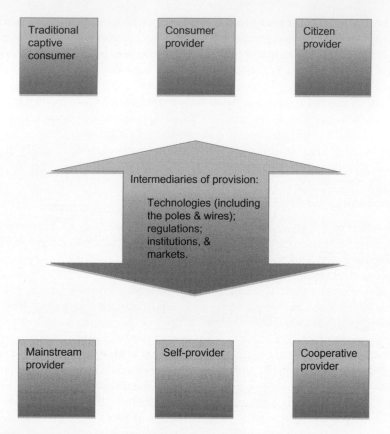

FIGURE 2.8   A schematic of a future smart electricity industry. *(Source: Adapted from van Vliet et al. [24, p. 114])*

control system elements, has remained with system managers, providers, and retailers. As systems move further in the direction of co-provision it will become imperative to find means of sharing this power with the various classes of consumer providers that will emerge.

The presentation of data in clearly understandable and usable formats is an important issue, notwithstanding the automated use of data. Such formats need to be tailored to contexts and to the responses available in different contexts, ideally indicating both the character and degree of suggested responses in a manner easily assimilated by the intended target audience. Technologies for responsive action are critical because if users have no or only limited, options to respond then nothing will be achieved. Above and beyond optimizing information provision, technologies designed to enable the operation of existing equipment to be brought into line with this information are also important. While control decisions might be local, the values informing these decisions are likely to lie in coordinated action requiring communication and decision-making (see the discussion of DataCommsCo in the section "Policy and Governance Frameworks" below) at the system level. Distributed generation options that can provide local "power quality" and power, such as PV, cogeneration, and battery storage, can also be important in these developments.

Commercial Frameworks

Commercially, economically efficient pricing for end-users has significant parallels with that of wholesale markets, involving spot and future market prices in both energy and ancillary services reflecting short to longer term operational and investment aspects of industry economics. However, support allowing small industry participants to actively and effectively participate in the market will also be required and will involve advice and appropriate resources, flagging a significant role for ESCOs or analogous bodies. Traditionally, ESCOs have commonly been seen as primarily suited to delivering energy efficiency to large industrial and commercial customers through performance contracting. However, they have a much wider potential role in terms of end-users (from industry to the commercial and domestic sectors) and in DE options (from EE to micro-generation). ESCOs can play a valuable role in bringing information, knowledge, skills, capabilities, and financing to bear on DE. However, pricing that doesn't include externalities is, by definition, economically inefficient. This will need to address not only carbon and other environmental externalities but also the potential distributional implications of prices, such as the impacts of prices on vulnerable consumers with little capacity to pay (e.g., low income), or limited options to respond (e.g., rental housing) This is, by far, the greatest unfinished business of current commercial arrangements although further innovation will be required to cope with the transition to co-provision.

Policy and Governance Frameworks

Policy must involve, in addition to facilitating traditional foci such as information (helping end-users respond to market signals) and the internalization of important environmental externalities (e.g., carbon pricing), attention to the transition to co-provision illustrated in Figure 2.8. While more traditional command and control regulatory measures such as minimum energy performance standards have a vital role to play, matters such as the financing of nontraditional generators also require attention. Such matters can be critical; for example, the success of Danish wind coops has been vitally dependent on access to competitive finance. This can involve subsidies, low-interest loans or tax breaks, which are a simple way to overcome the upfront cost barrier inherent in many options in order to stimulate and harness demand-side innovation.

Policies may mandate targets for one or more end-user options and use tradable certificates to ensure compliance market-based measures enabling the costs of particular projects to be spread across a range of energy consumers. Policy must also attend to matters such as training and accreditation, testing and standards, and, vitally, ensuring an adequate institutional infrastructure. Smart ESIs of the future will also require policies to guide the centralized decision-making inevitable in an industry requiring high levels of coordination beyond that achievable through conventional price signals and information. For example at the time of writing, the UK experience of rolling out smart meters was suggesting a particularly pivotal role for a market in DataCommsCo's that would take the lead in the facilitation and centralized coordination of communications, data, and security services [29].

Businesses installing EE can also achieve economies of scale where they become specialized at particular actions and have access to cheaper finance. The need for enabling and facilitating regulations and procedures as well as training and accreditation, reflects the broader definition of technology that includes orgware (the associated commercial and governance systems and institutional frameworks and capacity to deploy and integrate the hardware), and software (the knowledge to appropriately apply and use the hardware), in addition to the traditional emphasis upon hardware.

To date governance arrangements for wider policy efforts to support end-user engagement have generally been far less formalized and assured than those for supply-side participants, reflecting, perhaps, their lower priority. Present retail electricity markets and associated regulatory frameworks provide a largely uninformed and unresponsive interface between end-users and wider decision making. Effective end-user advocacy arrangements can be challenging to achieve. Supply-side participants are generally large, focused almost exclusively on the electricity industry, and have considerable shared interests. End-users are far more diverse and generally have only a limited interest in electricity itself.

Formal governance processes must be able to manage these asymmetries between supply- and demand-side stakeholders, particular as these conventional distinctions start to erode. This may require moving beyond straightforward market arrangements so as to further facilitate cooperative and other community-based initiatives that currently are a notable end-user, rather than industry, innovation. Arrangements adequate to both facilitate and interface these with an industry organized on market lines will be required. This may involve governments having to reflect and articulate values not only, as at present, in tune with the rationality of markets but also with the rationality of community values more focused by issues of long-term sustainability.

CONCLUSIONS

The content and character of the existing suite of energy services, along with a narrowing of the dominant understanding of the exercise of consumer choice to, most notably, a matter of simply price and information, reflect the preferences of utility managers, engineers, and regulators, rather than those of consumers themselves. The entrepreneurial exploitation of emerging technological possibilities in the face of aggregate growth in population and wealth has been the primary driver of load growth since the inception of the electricity industry. This dynamic has been central to the development and growth of the industry and is reflected in its supply-side oriented structures and institutional arrangements. However, emerging resource constraints and, most notably, climate change make it imperative to rapidly decarbonize the industry. This chapter has argued that the scale of this challenge requires a substantive re-engagement with end-users so as to grant them a role in energy decision-making not unlike like that obtained in the earliest days of the industry.

Smart grids and distributed energy options have a key role to play in this, but this chapter has argued that the facilitation of such smart energy use requires attending to far more than matters such as technology and pricing, which currently focus thinking about end-user engagement. These emphases assume an economistic model of human behavior that is out of sync not only with the commonly complementary information campaigns focused by collective civic or environmental concerns, but also with the complexity of real-world consumer behavior. Some of the more dynamic and promising decarbonization initiatives have resulted from citizens taking matters into their own hands because of what they regard as the inertia characteristic of more mainstream market-driven approaches. Good examples include Danish wind coops and the Transition Towns movement. Simply mobilizing less carbon-intensive end-use by conventional means, such as via smart meters, therefore risks overlooking a far vaster potential repository of energy savings and is unlikely to be adequate to the scale of change demanded by the imperative to rapidly address climate change.

Facilitating smart energy use is far from straightforward, however, ultimately requiring that the industry revisit its organizational structures so as to turn away from what has become a systemic supply-side bias. It will require attention to technical, commercial, and regulatory matters and a significant re-engineering of governance arrangements. A notable near term opportunity likely exists in retail markets where the facilitation of energy service companies established, resourced, regulated, and governed with end-user interests in mind has the potential to put this agenda into practice. However, in addition to the more common emphasis on load, and, more broadly behavior, control via price, and technology, it will be imperative that these are structured to facilitate end-user choices regarding DE and EE options, in particular, and the synergies they offer. Ultimately, if the ESI does become smart, the notions of supply and demand "sides" will become anachronistic as the system becomes more complex in ways involving both existing categories of consumers and providers and a variety of new ones, many of which transcend these conventional categories.

REFERENCES

[1] R. Fouquet, P.J.G. Pearson, "Seven centuries of energy services: The price and use of light in the United Kingdom (1300-2000)." *Energy J.* 27 (1) (2006) 139–177.

[2] J. McNeill, *Something New Under the Sun: An Environmental History of the Twentieth Century World*, Penguin, London, 2000.

[3] M. Fichman, *Science, Technology and Society: A Historical Perspective*, Kendall/Hunt Publishing, Dubuque (Iowa), 1993.

[4] R.C. Tobey, *Technology as Freedom; The New Deal and the Electrical Modernization of the American Home*, University of California Press, Berkeley, CA and London, 1996.

[5] D.E. Nye, *Electrifying America: Social Meanings of a New Technology, 1880–1940*, MIT Press, Cambridge, 1991.

[6] R.S. Cowan, *More Work for Mother – The Ironies of Household Technology from the Open Hearth to the Microwave*, Basic Books, New York, 1983.

[7] G. Cooper, *Air-Conditioning America: Engineers and the Controlled Environment, 1900–1960*, John Hopkins University Press, Baltimore and London, 1998.

[8] M.E. Ackermann, *Cool Comfort: America's Romance with Air-Conditioning*, Smithsonian Institution Press, Washington, 2002.

[9] R. Arsenault, "The end of the long hot summer: the air conditioner and southern culture." *J. South. Hist.* 50 (4) (1984) 597–628.

[10] EIA (Energy Information Administration), *Trends in Residential Air-Conditioning Usage from 1978 to 1997*, http://www.eia.doe.gov, 2000 (accessed 07.02.08).

[11] EIA (Energy Information Administration) *End-Use Consumption of Electricity 2001*, http://www.eia.doe.gov, 2007 (accessed 07.02.08).

[12] ABS (Australian Bureau of Statistics), 4621.1 – *Domestic Water and Energy Use*, New South Wales, Oct 2006, Canberra, http://www.abs.gove.au, 2007 (accessed 07.02.08).

[13] Integral Energy, Submission to the Legislative Standing Committee on Public Works into Energy Consumption in Residential Buildings, Integral Energy, Sydney, 2003.

[14] G. Prins, "On condis and coolth." *Energy Build.* 18 (1992) 251–258.

[15] G.S. Brager, R.J. de Dear, "*Historical and cultural influences on comfort expectations,*" in: R. Lorch, R. Cole (Eds.), *Buildings, Culture and Environment: Informing Local and Global Practices*, Blackwell, Oxford, 2003, 177–201.

[16] F. Sioshansi, W. Pfaffenberger (Eds.), *Electricity Market Reform: An International Perspective*, Elsevier, Amsterdam, 2006.

[17] H. Outhred, I. McGill, Electricity Industry Restructuring for Efficiency and Sustainability-Lessons from the Australian Experience, ACEEE Summer Study on Energy Efficiency in Buildings, Asilomar Conference Center, Pacific Grove, California, 2006.

[18] *Energy Technology Perspectives 2010: Scenarios &Strategies to 2050*, International Energy Agency, Paris, 2010.

[19] "Reengaging The Disengaged Customers." *EEnergy Informer: The Int. Energy Newsletter* 21 (2) (2011) 3–7.

[20] R. Passey, I. MacGill, *The Economics of Distributed Energy*, Report to CSIRO. Available from: www.ceem.unsw.edu.au, 2009.

[21] E. Shove, "Efficiency and consumption: Technology and practice." *Energy Environ.* 15 (6) (2004) 1053–1065.

[22] *Accenture 2010, Global Consumer Survey.*

[23] B. van Vliet, H. Chappells, E. Shove, *Infrastructures of Consumption: Environmental Innovation in the Utility Industries*, Earthscan, London and Sterling, VA, 2005.

[24] Energetics, *Electricity Pricing Structures*, Report to ESCOSA, 2003.

[25] I. MacGill, H. Outhred, K. Nolles, "Some design lessons from market-based greenhouse gas regulation in the restructured Australian electricity industry," *Energy Policy* 34 (1) (2006) 11–25.

[26] W. Parer, D. Agostini, P. Breslin, R. Sims. *Towards a Truly National and Efficient Energy Market*, Energy Market Review Final Report, COAG, 2002.

[27] Victorian Parliament, *Report of the Environmental and Natural Resources Committee on the Inquiry into Sustainable Communities*, Parliamentary Paper No. 140, 2005.

[28] DECC (Department of Energy and Climate Change)/OFgem (the Office of Gas and Electricity Markets), *Smart Metering Programme – Interim Report on Phase 1 and Future Arrangements*, DECC/Ofgem, London, 2011, p. 2.

[29] E. Shove, "Social theory and climate change: Questions often, sometimes and not yet asked." *Theory Cult. Society* 27 (2–3) (2010) 277–288.

The Ethics of Dynamic Pricing

Ahmad Faruqui [1]

INTRODUCTION

The smart grid has the potential for bringing an immense amount of innovation to the consumption and production of electricity. On the consumption side, it can enable efficient use of energy that can lower societal costs. A key enabler of efficiency is the accurate, cost-based pricing of electricity. In this chapter, we focus on dynamic pricing, which conveys the time-varying nature of electricity costs to consumers.

While the idea of time-variable pricing has been widely practiced in many markets for large commercial and industrial customers, its application to residential and small commercial and industrial customers is in the nascent stage. Since the latter group of customers typically have lower load factors than the system average, the ability to modify their load profiles through dynamic pricing can provide substantial benefits to customers, utilities, and society as a whole. However, two conditions have to be met before dynamic pricing can be successfully implemented in this market segment.

- First, the appropriate type of metering and communication technology— called advanced metering infrastructure or AMI—has to be in place. This is further discussed in Chapter 11 by King and Strapp.
- Second, concerns about the equity of dynamic pricing have to be resolved. It is this second condition, which forms the focus of this chapter and also Chapter 4 by Felder.

Concerns about equity issues have always been associated with changes to the status quo when it comes to any form of energy policy. Nowhere is this more evident than in the pricing of electricity. Under prevailing rates, virtually all small customers typically pay the same flat rate per unit of electricity consumed regardless of the quantity or time-of-use. But since load profiles vary by customer, the cost of serving customers varies. It is more expensive to serve those customers who use relatively more energy in the peak period, and relatively less in the off-peak period, than those who use relatively less electricity in the peak period. In other words, the peakier-than-average customers are subsidizing the less peakier-than-average customers, often without knowing it.

Over a period of time, for a utility with a million customers, the amount of the subsidy can run into the hundreds of millions of dollars. Thus, any attempt to introduce more cost-reflective price schemes, such as dynamic pricing, would result in the elimination of these cross-subsidies. The beneficiaries will be delighted but those who are no longer subsidized will be upset. The latter will find a way to their local regulators and file a complaint. It is the nature of the regulatory process that the complainers who show up get a seat at the table while those who benefit but never show up don't. In the United States, and in fact around the world, in places as far afield as Australia and Britain, opponents of dynamic pricing have filed complaints that the practice is unethical and should not be rolled out. This chapter argues the contrary position—that flat-rate pricing is unethical and it should be pulled back.

The chapter is organized as follows. The section "Background" provides some key definitions, section "The Distributional Effects of Dynamic Pricing" introduces the distributional effects of dynamic pricing, section "The Barriers to Dynamic Pricing" discusses the barriers to dynamic pricing, section "The Unfairness of Flat Rate Pricing" discusses the unfairness of flat rate pricing, section "Dynamic Pricing in Other Industries" discusses dynamic pricing in other industries, section "Overcoming the Barriers to Dynamic Pricing" discusses ways of overcoming the barriers to dynamic pricing, section "The Effect of Dynamic Pricing on Low-Income Consumers" discusses the impact of dynamic pricing on low-income customers, section "Accommodating Potential Objections" reviews potential objections to dynamic pricing, and section "Conclusions" provides the conclusions of the chapter.

BACKGROUND

Dynamic pricing is a form of time-of-use (TOU) pricing where prices during the peak period on a limited number of days can vary to reflect market conditions on a day-ahead or day-of basis. One popular variant of dynamic pricing is critical-peak pricing (CPP) in which prices during the top 40–150 hours of the year rise to previously specified levels designed to recover the full capacity and energy cost of power plants that run primarily during those hours. During all other hours of the year, prices are lower than existing rates by an amount sufficient to leave the bill unchanged for a customer whose load shape mirrors that of the rate class.

An example of CPP is provided in Figure 3.1. Other examples are shown for TOU pricing in Figure 3.2 and real-time pricing (RTP) in Figure 3.3. Combinations of dynamic pricing designs can also be envisaged.

Dynamic pricing has garnered much interest in the country during the past decade since it has the potential for lowering customer energy costs by mitigating the need to install expensive peaking capacity. As can be seen by reviewing load duration curves for various markets around the country, the top 1% of the hours of the year can account for 8–12% of annual system peak demand. In some cases, they may account for as much as 14–18%.

Several studies have been published on the benefits of dynamic pricing. A recent example is the one that was conducted by the New York Independent System Operator [2]. The study, conceived as a *gedanken* or thought experiment, quantified the benefits that would flow from universal deployment of

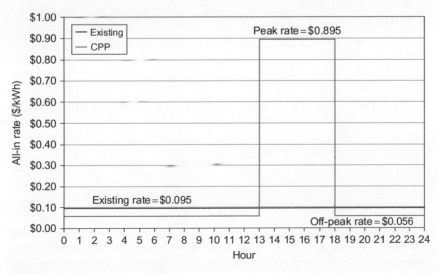

FIGURE 3.1   Illustration of a CPP rate.

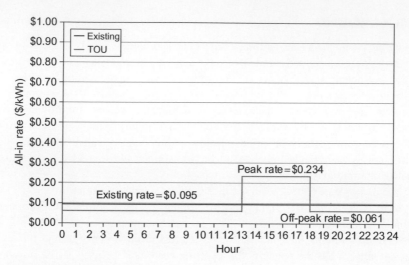

FIGURE 3.2   Illustration of a TOU rate.

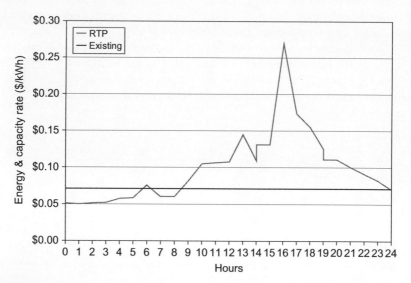

FIGURE 3.3   Illustration of an RTP rate.

real-time pricing in New York State. It used a single year to simulate the benefits [3]. The study found the following benefits:

- *Demand Reduction*: Dynamic pricing would result in system peak demand
  reductions in the 10–14% range, from a projected value of 34,000 mega-
  watts (MW). The reductions would range from 13–16% in New York
  City and 11–14% reduction in Long Island.

- *Cost Reduction:* Total resource costs would decrease by $143–509 million per year, or 3–6%. Market-based customer costs would decrease by $171–579 million per year, or 2–5 %.
- *Economic Efficiency Gain:* Dynamic pricing would improve societal welfare by $141–403 million per year.

Another study by the Demand Response Research Center informed California's decision to deploy advanced metering infrastructure (AMI), a prerequisite for dynamic pricing, to all customers served by the state's investor-owned utilities [4]. The California Public Utilities Commission (CPUC) has ruled that dynamic pricing will become the default rate for all non-residential customers once AMI has been rolled out to them and has suggested that it be extended to residential customers once legal restrictions dating back to the energy crisis on residential tariffs have expired [5].

At the national level, the Federal Energy Regulatory Commission (FERC) filed a staff report with the U.S. Congress in June 2009 that quantified the potential impact of dynamic pricing on a state-by-state level [6]. Several deployment scenarios were presented, ranging from a continuation of current trends to one that included universal deployment. Earlier work has shown that even a 5% drop in demand during critical peak hours can be worth $35 billion [7].

DISTRIBUTIONAL EFFECTS OF DYNAMIC PRICING

For the benefits of dynamic pricing to be realized, not all customers need to respond. In fact, as commonly developed under revenue-neutrality principles, half of the customers whose load factors are better than average will see an immediate reduction in their bills *before* they make any adjustment to their pattern of electricity consumption [8].

To illustrate this point, Figure 3.4 shows the load profiles of three prototypical customers, one whose profile coincides with the class, one whose load profile is peakier than the class average profile, and one whose profile is flatter than the class average.

Figure 3.5 presents the share of peak load in daily load for a representative set of customers who are ordered by their peak shares. The three prototypical customers from Figure 3.4 appear as points along a continuum

Now a prototypical CPP rate is applied to all these customers. The changes in bills brought about by this change in rate design are displayed in Figure 3.6. The cross-subsidies that were inherent in flat rates are removed, and this causes bills to rise for some customers and to fall for others. Since these distributional impacts may vary across utilities, the results are displayed across three utilities in Figure 3.7. Interestingly, there is not much variation across the utilities.

The distributional impacts would also be expected to vary across rate designs, as shown in Figure 3.8.

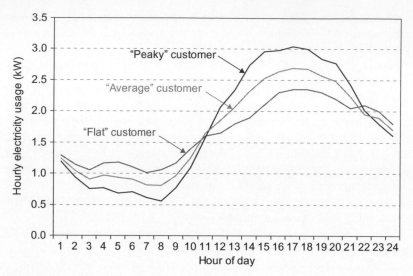

FIGURE 3.4 Average, peaky, and flat usage profiles.

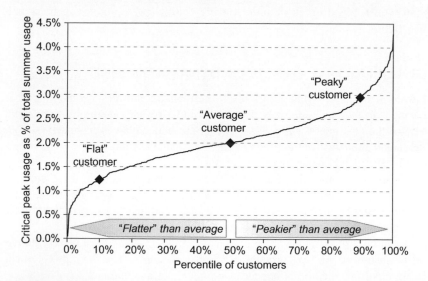

FIGURE 3.5 Distribution of residential customer usage profiles.

So what can be done to offset the adverse impact of moving customers to dynamic pricing rates? Customers who don't see an immediate reduction can lower their bills by reducing their usage during the expensive peak period hours by curtailing some of that use or by shifting some of it to lower-priced hours. As shown later in this chapter, about two-thirds to three-quarters of the customers are likely to see lower bills as a result of dynamic pricing.

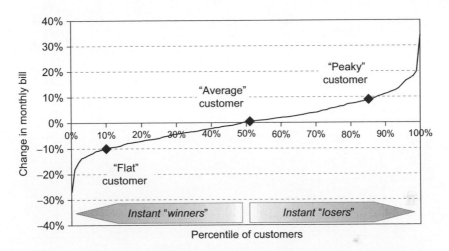

FIGURE 3.6 Distribution of dynamic pricing bill impacts (residential critical-peak pricing).

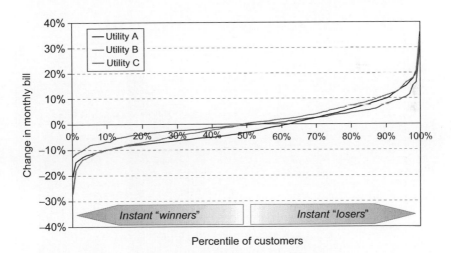

FIGURE 3.7 Comparison of dynamic pricing bill impacts across utilities.

It is important to clear up an important misconception. Under dynamic pricing, customers do *not* have to pull the plug on major end-uses, live in the dark, or eliminate all peak usage in order to benefit. They simply have to reduce peak usage by some discretionary amount that does not compromise their life style, threaten their well-being, or endanger their health. Clearly, the more they reduce, the more they will save. But the choice is up to them.

Over the past several years, 18 pilots have been carried out in North America, Europe, and Australia to assess the magnitude of demand response associated

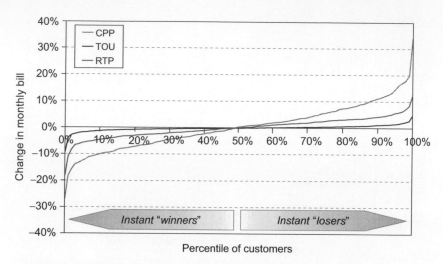

FIGURE 3.8   Comparison of dynamic pricing bill impacts across rate designs.

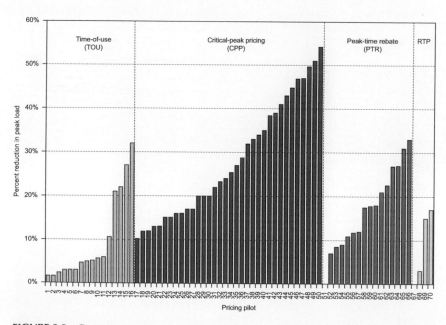

FIGURE 3.9   Customer response in recent pricing pilots.

with dynamic pricing. In just about every case, consumers on average have shown the ability to lower peak usage. Some respond a lot, some respond marginally, and some do no respond at all. The evidence from the 70 most recent tests is presented in Figure 3.9 [9].

Barriers to Dynamic Pricing

Despite the promise of substantial economic gains, the deployment of dynamic pricing has been remarkably tepid, in large measure because of misplaced but recurring concerns about the inequity of dynamic pricing. Approximately 5% of the customers are on AMI today, but less than a tenth of that number is estimated to be on dynamic pricing. The software firm eMeter recently announced that the United States has crossed the 20 million milestone and will add another 50 million smart meters by 2015.[1] If current AMI deployment trends continue, a significant percentage of U.S. customers would have smart meters. However, it is an open question about how many customers would be moved to dynamic pricing in the coming years.

From certain quarters, most notably consumer advocates such as The Utility Reform Network (TURN) in California, concerns have been voiced that dynamic pricing inflicts harm on low-income consumers, seniors and people with disabilities who stay at home a lot, people with medical conditions that require special electrical equipment, people with young children, and small businesses. It is stated that these consumers are unable to curtail peak period usage, in part because they have very little load to begin with.

The underlying premise is that dynamic pricing is unfair. This concern is not confined to the United States. It has shown up recently in the state of Victoria, Australia, where the state government has ordered a review of the smart meter roll out policy after the state's Auditor-General warned that electricity consumers would be worse off [10]. The Essential Services Commission has been asked to conduct a review "to ensure vulnerable Victorians are not disadvantaged." Victoria plans to roll out smart meters to 2.4 million homes and small businesses over the next four years.

The review was triggered by a finding by the state's auditor-general that consumers would have be paying an extra $150 annually under the new metering system. Another study by the University of Melbourne estimated that bills for low-income earners would rise by 30%, or $300 a year.

At the time of this writing, Victoria's Energy and Resources Minister was considering imposing a moratorium on new tariffs until the investigation was completed. The government will establish a consumer working group to consider the impacts of smart meters and help customers access competitive rates. It will also provide $50,000 to the Consumer Utilities Advocacy Centre (CUAC) for a communications campaign to help customers change their usage patterns to maximize the benefits of the new system.

Such concerns are not new. In 1971, Professor William Vickrey of Columbia University wrote a groundbreaking paper on "responsive pricing," his term for what would later be called dynamic pricing [11]. Vickrey, who went on to earn

---

[1]Statement by Chris King of eMeter dated 17 May 2011 and at http://www.emeter.com/smart-grid-watch/2011/us-20-million-smart-meters-now-installed/

the 1996 Nobel Prize in Economic Science, opined, "The main difficulty with responsive pricing is likely to be not just mechanical or economic but political." He felt that people shared the medieval notion of a just price as an ethical norm, and that prices that varied according to the circumstances of the moment were intrinsically evil:

The free market has often enough been condemned as a snare and a delusion, but if indeed prices have to perform their function in the context of modern industrial society, it may be not because the free market will not work, but because it has not been effectively tried.

In 1987, building on many years of work on homeostatic control, Professor Fred Schweppe of MIT co-authored a book that laid out the theory and practice of spot pricing or real-time pricing, the ultimate form of dynamic pricing [12]. Schweppe et al. believed that given the overwhelming efficiency benefits that would flow from dynamic pricing, it was inevitable that deployment of this optimal rate design would soon follow. But it did not.

In 2001, reviewing the slow progress toward dynamic pricing in restructured markets, Eric Hirst of the Oak Ridge National Laboratory lamented, "The greatest barriers are legislative and regulatory, deriving from state efforts to protect retail customers from the vagaries of competitive markets" [13].

It had never been easy to change tariffs in the electricity industry and the problem was not confined to the United States. Back in 1938, the author of a leading British text on costs and tariffs lamented [14]:

There has never been any lack of interest in the subject of electricity tariffs. Like all charges upon the consumer, they are an unfailing source of annoyance to those who pay, and of argument in those who levy them. In fact, so great is the heat aroused whenever they are discussed at institutions or in the technical press, that it has been suggested there should be a "close season" for tariff discussions. Nor does this interest exaggerate their importance. There is general agreement that appropriate tariffs are essential to any rapid development of electricity supply, and there is complete disagreement as to what constitutes an appropriate tariff.

The present tariff position in [Great Britain] is little short of chaos. Even the terminology has not been standardized, and the tariffs themselves appear to be the unbridled whim of the particular undertaking. To quote only one example—taking a single load group (industrial power) and a single type of tariff (the block rate), and considering only the larger undertakings (one quarter of the whole), there were found to be 102 different tariffs! At this rate, the block-rate tariffs alone would muster about 400 different specimens. Kipling might well have said of electricity:

There are nine-and-sixty ways in which the user pays and every single one of them is right.

But change is in the air. In an interview that he gave in December 2008, the former president of the National Association of Regulatory Utility Commissioners (NARUC), Commissioner Fred Butler of New Jersey noted that fundamental changes were coming for energy delivery and pricing. He said that for more than a century "most people have paid for their electricity at the same rate every day of every year, every hour of every day."

"That's going to have to change," Butler noted. "If you're going to have a smart grid, that allows you to measure and have two-way communication between the end-use premises, the utility company, the RTO, and other entities, rates will have to change to be more time-of-use rates or critical peak period rates." With rate changes coming, he added, "We have a massive education campaign that's needed to explain to people why this is happening and why they can adapt their usage of electricity the way they've adapted their telephone usage," waiting for "free nights and weekends" to make calls, Butler says.

While acknowledging that both the FERC-NARUC smart grid collaborative and individual states are working on that massive education campaign and developing programs to effect time-of-use rate changes, "you can only go so fast" to avoid consumer backlash. The process has already begun today in some places, while in other areas the time-of-use changes will take several years. Ultimately, however, Butler concluded, "Pricing five years from now will be very, very different than it is today." As of this writing, very little change had occurred in the industry's pricing practices.

Unfairness of Flat Rate Pricing

The opponents of dynamic pricing, such as Barbara Alexander, use the unfairness argument to present their case [15]. But the presumption of unfairness in dynamic pricing rests on an assumption of fairness in today's tariffs. A flat rate that charges the same price around the clock essentially creates a cross-subsidy between consumers who have flatter-than-average load profiles and those who have peakier-than-average load profiles. This cross-subsidy is invisible to most consumers but over a period of time, it can run into the billions of dollars. An example will suffice to make this point.

Let us divide electricity customers into three groups based on their load profiles: Average Users, whose hourly load profile corresponds to the class peak; Peaky Users, whose load profile has greater than average concentration in the peak period; and Flat Users, whose load profile has less than average concentration in the peak period. Let's set the peak period from noon to 6 PM. Average Users consume electricity in proportion to the ratio of peak to off peak hours so 25% of their consumption occurs during the peak hours. Peaky Users consume 40% during peak hours and Flat Users 10%. Let us also assume that the population is equally divided between the three types of users and that there are a total of 10 million customers in the population of interest. Finally, let us set each customer's average monthly consumption at 500 kWh.

Now we can calculate the total cost of electricity for each of the consumption profiles under two different rates: a flat rate and a TOU rate. A similar approach can be used to estimate costs under dynamic pricing rates, such as CPP.

The flat rate is assumed to be 10 cents/kWh and applies around the clock. The marginal cost of electricity during the peak period is 20 cents/kWh and 6.7 cents/kWh during the off-peak period, and these costs are used to establish the peak and off-peak TOU rates. Table 3.1 summarizes the characteristics of the customer population.

Given these assumptions, we can calculate the total costs incurred by each consumption profile over a 10-year period for both the flat and TOU rates. This is done by multiplying each customer's peak and off-peak consumption by the corresponding rate and summing over both the number of months in the period (120) and the number of customers belonging to each consumption profile (3.3 million). A discount rate of 4% is used to yield a present value. Finally, by subtracting the total costs incurred under the flat rate from the total costs incurred under the TOU rate, we can estimate the cross-subsidy that results from flat rates.

As shown in Table 3.2, while average users do not experience any benefit or loss under the flat rate, flat users are paying $3.92 billion above what they would have paid under a TOU rate and peaky users are benefiting from this subsidy.

TABLE 3.1 Customer Population Characteristics

Consumption Profile	Monthly Consumption (kWh per Customer)			Weight Average Rates (cents/kWh)	
	Peak	Off-Peak	Total	Flat	TOU
Flat	50 (10%)	450 (90%)	500 (100%)	10.00	8.00
Average	125 (25%)	375 (75%)	500 (100%)	10.00	10.00
Peaky	200 (40%)	300 (60%)	500 (100%)	10.00	12.00

TABLE 3.2 Cross-Subsidy Over a 10-Year Period from the Flat Rate

Consumption Profile	Monthly Electricity Cost ($)		Monthly Benefit/ Loss from Flat Rate Cost ($)	Total Benefit/Loss ($ Billions)
	Flat	TOU		
Flat	50.00	40.00	(10.00)	(3.92)
Average	50.00	50.00	0.00	0.00
Peaky	50.00	60.00	10.00	3.92

DYNAMIC PRICING IN OTHER INDUSTRIES

The concept of time-varying rates, while it may be portrayed as being foreign to electricity consumers, is one that those very consumers encounter daily in a variety of applications. Just take the case of a driver looking for a parking space in the downtown of any major metropolitan area. In most cases, the driver expects to pay a sizable parking fee during working hours on weekdays. But he or she knows that parking will be free during evenings and nights on weekdays and typically also free on weekends. In some of the newer parking meters, which have digital technology embedded in them, parking rates vary based on the number of vacant spaces, which will often vary dynamically.

The driver may also find that he or she also has to pay congestion pricing rates in congested areas such as central London in Britain. Another example comes from the San Francisco Bay Area where the Bay Area Toll Authority (BATA) has unanimously approved congestion pricing on the San Francisco-Oakland Bay Bridge [16]. This went into effect in July 2010. Tolls for cars increased from $4 regardless of time to $6 during weekday commute hours, dropping to $4 during off-peak hours on weekdays. On weekends, the auto toll on this bridge became $5. Officials expect the congestion pricing plan to ease commute-period congestion as drivers divert some of their discretionary driving to off-peak hours.

Travelers are likely to encounter dynamic pricing every time they book their flights, hotels, and rental cars. In each of these industries, the fixed costs are very high, and the only way to survive in business is to manage revenues, and therefore yields, by pricing differentially based on demand conditions [17].

Certain cell phone plans also embody time-varying rates. Prices for produce vary seasonally as do movie tickets and sometimes theater prices. The latest industry to introduce dynamic pricing is the sporting industry. This season, the San Francisco Giants plan to introduce dynamic pricing to their fans [18]. This will allow the Giants to offer more price options to patrons since the goal is to have more fans enjoy Giants baseball. Roughly three-quarters of tickets are currently selling for less than they cost last year. Of course, it will cost more to attend popular games. Dynamic pricing will take into account a variety of factors other than seat location. These will include weather, starting pitcher, opponent team, the number of seats already sold, promotion or giveaway day, performance of team, likelihood of making playoffs, day of week, and time of day.

Another team that uses dynamic pricing is the Buffalo Sabres hockey team. For the 41 home games that will be played during the current season, the team will continue with its practice of variable pricing. In this system, each game is designated by a different classification that reflects the capabilities of the opposing team, time of the year, day of the week, rivalries, and games against all-star players [19].

OVERCOMING THE BARRIERS TO DYNAMIC PRICING

Among economists, there are two schools of thoughts when it comes to dynamic pricing. The purist school of thought argues that rates should reflect time-variation in costs, regardless of whether customers respond or not. The pragmatic school of thought argues that rates should reflect time-variation in costs if the societal benefits from so doing exceed the societal costs. Typically, the societal benefits are associated with avoided capacity and energy costs, and the societal costs are associated with implementing AMI.

The challenge is that while net societal benefits might be positive, individual consumer benefits may be positive or negative. A conservative approach associated with the work of Vilfredo Pareto argues that dynamic pricing should only be pursued if at least one consumer is better off and no one is worse off. A more aggressive approach in public policy associated with the work of Hicks and Kaldor would suggest that dynamic pricing is worth pursuing if the gains to the winners exceed the losses to the losers. In other words, if the winners can compensate the losers, go ahead and pursue the policy. Of course, this compensation would not actually be paid because if it were paid, the Hicks-Kaldor solution would collapse to the Pareto solution. Clearly, the Hicks-Kaldor approach would yield much larger societal gains than the Pareto approach.

But that is where the equity argument kicks in and the push back begins. So what can be done to offset the adverse impacts of dynamic pricing? Figure 3.10 shows that by providing an incentive for demand response, dynamic pricing would increase the number of winners from 50% to 75%.

Further gains can be obtained by removing the hedging premium embodied in flat rates [20]. A conservative estimate of the size of the hedging premium is 5%. Once this credit is applied, the share of winners goes up to 92% (Figure 3.11).

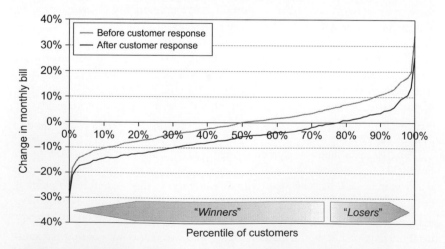

FIGURE 3.10 Distribution of dynamic pricing bill impacts (after customer response).

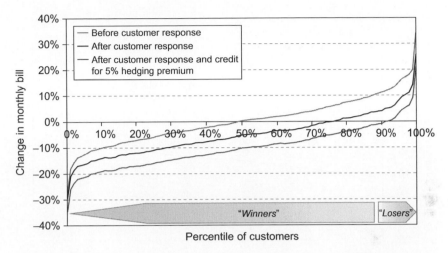

FIGURE 3.11 Distribution of dynamic pricing bill impacts (after customer response and credit for hedging premium).

The Effect of Dynamic Pricing on Low-Income Consumers

How does dynamic pricing affect low-income consumers? More than any other issue, this one crystallizes opposition to dynamic pricing in regulatory proceedings. The contention is that low-income consumers don't use much energy to begin with and therefore are in no position to lower usage during peak period hours. It is also asserted that they lack the know-how and wherewithal with which to curtail peak period usage. Being strapped for cash, they may feel compelled to avoid higher peak period prices and, by reducing energy for essential usage, may cause themselves significant physical harm.

Is this factually correct? There is no documented instance of low-income customers harming themselves through dynamic pricing. In addition, intuition suggests that low-income consumers are likely to have flatter than average load shapes because many of them lack central air conditioning. Thus, one might expect them to come out ahead with dynamic pricing. What are the facts?

New data have recently become available from a large urban utility that shed light on the subject. An analysis of low income customers at this utility is shown in Figures 3.12 and 3.13, which show percentage changes in bills and nominal changes in bills, respectively Figure 3.12 shows that about 80% of low-income customers would gain from dynamic pricing. With a modest amount of demand response, 92% of low-income customers would gain from dynamic pricing.

Then there is the question of whether low income customers are likely to respond to dynamic pricing. The most recent evidence on this topic comes from the experiment with dynamic pricing that was carried out during the summer of 2008 in Washington, D.C. One unique feature of the PowerCentsDC

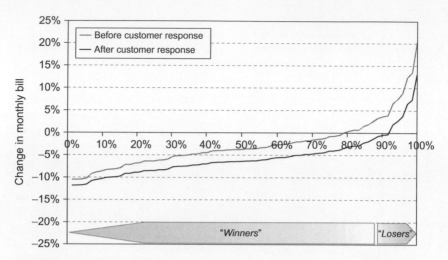

FIGURE 3.12   Bill impacts for low-income customers (expressed as % of monthly bill).

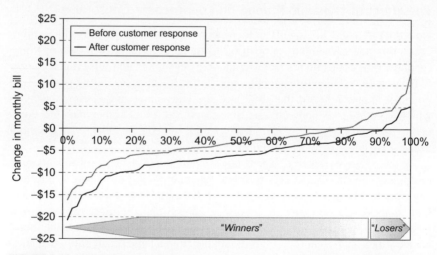

FIGURE 3.13   Bill impacts for low-income customers (expressed as dollars per month).

program is that it actively recruited a group of limited-income customers to understand their responsiveness to dynamic pricing. Of the 857 residential customers in the pilot, 118 were low-income customers. The lead researcher on the project, Frank Wolak of Stanford University, found that the magnitude of demand response, expressed as a percent of their peak load, exhibited by low-income customers to a critical peak pricing rebate program was almost twice as large as that exhibited by non-low-income customers [21].

ACCOMMODATING POTENTIAL OBJECTIONS

Given the potential benefits of dynamic pricing, what practical policies might be contemplated to offset the adverse impact on those customers who might be adversely affected? Several options are available [22].

- Creating customer buy-in. Customers need to be educated on why a century-old practice of ratemaking is being changed. They have to be shown how dynamic pricing can lower energy costs for society as a whole, help them lower their monthly utility bills, prevent blackouts and brownouts, improve system reliability, and lead to a cleaner environment.
- Offering tools. These should allow customers to get the most out of dynamic pricing. At the simplest level, they should be equipped with information on how much of their utility bill comes from various end-uses such as lighting, laundry, and air conditioning and what actions will have the largest response on their bill. At the next level, they could be provided real-time in-home displays that disaggregate their power consumption and tell them how much they are paying by the hour. Finally, they could be provided enabling technologies such as programmable communicating thermostats. Similar examples can be constructed for commercial and industrial customers.
- Designing two-part rates. The first part would allow them to buy a predetermined amount of power at a known rate (analogous to a forward contract), and the second part would give them access to dynamic pricing and allow them to manage their energy costs by modifying the timing of their consumption. They could be allowed to pick their predetermined amount, or it could be based on consumption during a "baseline" period.
- Peak-time rebates. The consumer pays the standard rate but has the opportunity to earn rebates during critical peak periods by reducing consumption relative to an administratively determined baseline.
- Demand subscription service. Each consumer *may* contract for a different "baseline" of demand at a known price and pay for variations in demand from that baseline at real-time prices. A key element of the demand subscription service is that each customer has a choice. For example, the preferred baseline may be zero for a consumer with a flat consumption profile and higher for a consumer with a peaky consumption profile [23].
- Providing bill protection. This would ensure that their utility bill would be no higher than what it would have been on the otherwise applicable tariff but would not preclude it from being lower based on the dynamic pricing tariff. Customers would simply pay the lower of the two amounts. In later years, the bill protection could be phased out. For example, in year one, their bill would be fully protected and would be no higher than it would have been otherwise; in year two, it would be no higher than 5%; in year three, no higher than 10%; in year four, no higher than 15%, and in year five, no higher than 20%. In the sixth year and beyond, there would be no bill protection. Or full bill protection could continue to be offered for a fee.

- Giving customers on dynamic pricing a credit for the hedging premium they no longer need once they move from flat rate pricing to dynamic pricing. Existing fixed price rates are very costly for suppliers to service since they transfer all price and volume risk from the customers to the suppliers. In addition, the supplier takes all the volume risk. In order to stay in business, the supplier has to hedge against the price and volume risk embodied in such open-ended fixed price contracts. It does so by estimating the magnitude of the risk and charging customers for it through an insurance premium. The risk depends on the volatility of wholesale prices, the volatility of customer loads, and the correlation between the two. Theoretical simulations and empirical work suggest that this risk premium ranges between 5 and 30% of the cost of a fixed rate, being higher when the existing rate is fixed and time-invariant and being smaller when the existing rate is time-varying or partly dynamic. For example, a flat and fixed and non-time varying rate may bear a premium of 30% when compared to a real-time pricing rate or a premium on 10% when compared to a critical peak pricing rate.
- Giving customers a choice of rate designs. Dynamic pricing rates, even with all the items mentioned above, may still be too risky for some customers. Thus, they should have the option of migrating to other time-varying rates, perhaps with varying lengths of the peak period and with varying numbers of pricing periods. If the CPP rate (combined with a TOU rate) becomes the default rate, risk-averse customers should have the opportunity to migrate to a fixed time-of-use rate, and risk-taking customers should have the opportunity to migrate to a one-part or two-part real-time pricing rate.

CONCLUSIONS

As a matter of principle, ethical pricing should be cost based and not create subsidies between customers. Flat rate pricing, which has been in place for the past century, creates an enormous subsidy between customers with varying load shapes. It is unethical and needs to be replaced by dynamic pricing. Not only will this be more ethical, it will also improve the economics of the power system and lower costs for all customers.

However, as with any significant change in rate design, it has to be phased in gradually. Several methods for making this gradual change have been discussed in this chapter.

APPENDIX: QUANTIFYING THE HEDGING COST PREMIUM

In defining the benefits of price response, recent analysts have suggested that those who engage in such behaviors realize savings from paying a lower hedge premium. In other words, they get rid of the middlemen (the utility or competitive retailer) and buy directly from the factory, paying wholesale market spot prices or utility RTP prices for their energy consumption. This raises an intriguing question; how large are risk premiums, and are they identical under competitively determined retail prices and regulated rates? However, it's not apparent that the concept of a risk premium as an element of price produced by a regulated, vertically integrated utility is an oxymoron. Traditional rate making bundles costs associated with investment recovery and cost associated with the difference between rates and dispatch costs that might be construed as risk premiums.

Centralized wholesale markets produce transparent spot market prices that provide insight into the risk premiums that competitive retailers build into their prices. If utilities use these prices to establish marginal-cost rates, then price response will improve resource efficiency, and the notion of a risk premium savings is moot.

Traditional Cost of Service

Under conventional embedded cost ratemaking, there is no explicit risk premium added to the energy rate. Overall, the rate includes a provision for the recovery of fixed costs at a rate of return (ROR) that reflects the market's perspective on the enterprise risks a utility undertakes, which largely are associated with generation investments. That ROR premium is folded into the revenue requirement, which is then allocated to classes based on relative load levels and patterns and then incorporated into a bundled rate. There is no way to isolate the risk element; it is inextricably bundled into the rate. Thus, one does not think of traditional rates as having risk premiums. But, implicitly they do, and that is revealed by examining how prices are set in competitive markets.

Competitive Market Pricing

Competitive retailers set prices based on their cost of supply and what customers are willing to pay, the latter determined in part by what their competitors charge. Some competitive retailers are selling generation owned by the same company, while others have to acquire energy to serve their customers' requirements. The integrated generation/retail entity must explicitly consider which is more profitable: to commit capacity to serving customers under fixed retail rates or to sell energy in the wholesale spot market. The specialized retailer faces generation prices that already have taken that opportunity cost into account. So, retail prices implicitly or explicitly embody spot market price expectations, and that includes a provision for risks.

It follows then that in setting prices, a retailer first considers the cost of serving its retail load obligation through spot market transactions. If retail prices are linked directly to wholesale prices, which change every hour, then the retailer passes the cost it incurs in supplying its retail customers directly to the consumer, and there is little or no risk. This works only to the extent that customers are willing to pay prices that change hourly. What about if customers who want to pay a uniform price that changes only periodically (for example every few months or once a year) or to buy from a time-of-use schedule? To accommodate these pricing plans, the retailer must define the risks inherent in committing to serving load under fixed prices. Those risks include the following:

Load risk due to episodic variations in customers' load shapes and levels, due to weather, economic circumstances, and changes in individual customer circumstances (e.g., the need to increase or decrease business or plant output, accommodating a house full of relatives for a week in the summer).

Market load risk—retailers that contract with a utility to serve its default service customers face scale and load shape risk from customers switching to and from utility default service. A larger or different load pattern can result in marginal supply costs that are above the pre-set rate.

Price risk—if the load is being served at a fixed rate through purchases from the spot market, then there is explicit risk associated with the inherent volatility of spot market prices. If the load obligation is being supplied from owned generation assets, then the then the opportunity cost of lost spot market sales defines the price risk. Finally, if the retailer is buying supply from a generation supplier, then that opportunity cost is already incorporated in what it pays.

These risks have to be covered in rates for the retailer to ensure an acceptable return on investment. Consequently, customers who buy power other than at wholesale terms (streaming hourly prices) are paying a risk premium. The higher the degree of temporal aggregation used to price usage, the higher the premium. TOU rates have a higher premium than RTP, and a uniform, fixed rate has a hedging premium that is even higher.

Traders in many commodity markets devise risk premiums from the mean and variance of expected spot market prices, using financial models that rely on predictable market characteristics to determine relative risk. But, is that how competitive electricity retailers set their prices? If that were the case, then the risk premiums in retail prices could be revealed by employing those analytical techniques, in effect reverse-engineering retailers' posted prices. Making the risk premium explicit would aid customers in making usage decision. They could compare the risk premium with buying at spot market prices, first assuming no price response and then factoring in price response behaviors, (and their costs) and deciding which course to take.

Competitive retailers are understandably unwilling to reveal the risk premiums that they add in creating their retail price offerings. Conventional financial models may be employed, but electricity prices do not conform to some of the assumptions these models require, which means that they may not produce consistent and therefore reliable results. The level of hedging premiums therefore remains the subject of speculation.

Comparing the posted prices of competitive retailer products with the cost of paying spot prices for that load is one avenue for establishing risk premiums, albeit a somewhat flawed one. Such a comparison uses already known spot market prices and retail prices that were based on the retailer's price expectations. However, it is at least a rudimentary indicator of implied risk premiums. Applying that reasoning to competitive markets in the Northeast yields implied premiums of 15–40% for a fully hedged service. The difference among retailers' rates for equivalent service reflects their forward market view (each's expectations of prices), along with other transactional considerations, like the cost of operating a retail business (acquiring and servicing customers).

Auctions and RFPs for default service provide another hedge cost indicator. The results of the auction for default service in Illinois caused some to conclude that the implied risk premium was 20–40%. Recent studies of price response by ISO-NE utilized risk premiums that are graduated in the degree of risk of the pricing plan; RTP has the lowest (3–5%), TOU even higher (8%), and the uniform rate had the highest (15%). Under these risk premiums, the analysis concluded that the majority of benefits of price response redound to those that adopt that behavior.

Estimating the Hedging Cost Premium in Flat Electricity Rates

How can the hedging cost premium be quantified? In one approach, the hedging premium is considered to be exponentially proportional to the volatility of loads, the volatility of spot prices, and the correlation between loads and spot prices. This can be represented as follows:

$$\pi = \exp(\sigma_L \cdot \sigma_P \cdot \rho_{L,P})$$

where:

π = Risk Premium
σ_L = Load Volatility
σ_P = Spot Price Volatility
$\rho_{L,P}$ = Correlation Between Load and Spot Price

For example, if price volatility was assumed to be 0.6, load volatility was 0.2, and the correlation between load and the spot price was 0.4, the resulting estimate of the hedging premium would be 5%. In other words, on average, customers are paying 5% more than they would if they were simply exposed to spot prices.

With an assumption about the distribution of these three variables, a Monte Carlo simulation can be used to approximate a distribution around this premium. Assuming that the variables are all triangularly distributed with a minimum of 0 and a maximum of 1, a Monte Carlo simulation of 1,000 iterations produces the hedging premium distribution shown in the following figure.

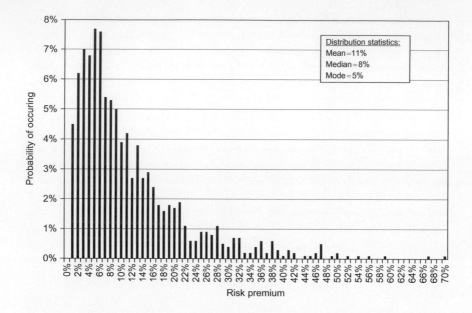

Simulated Distribution of Hedging Cost Premium

The mean, median, and mode of the premium are 11%, 8%, and 5%, respectively. The standard deviation is 10%.

REFERENCES

[1] The views in this chapter are those of the author and not necessarily those of the Brattle Group. The chapter is adapted from an article of the same title that appeared originally in the July 2010 issue of *The Electricity Journal* with permission.

[2] http://documents.dps.state.ny.us/public/Common/ViewDoc.aspx?DocRefId={4FA9C260-CE37-4FA983CA1D79044093A5}.

[3] Of course, benefits over multiple years would be much higher. In a full cost-benefit analysis, all relevant costs would also need to be factored in.

[4] A. Faruqui, R. Hledik, J. Tsoukalis, "The power of dynamic pricing," *The Electricity Journal* April 2009.

[5] Its applicability to residential customers is prevented by state legislation that has frozen portions of residential rates in order to recover the costs of the energy crisis of 2000–2001 from the unfrozen portions.

[6] FERC Staff, *A National Assessment of Demand Response Potential*, Washington, DC, June 2009.

[7] A. Faruqui, R. Hledik, S. Newell, J. Pfeifenberger, "The power of five percent," *Electr. J.*, Vol. 20, October 2007.

[8] Revenue neutrality means that the revenue collected from the class to which the new rate is being applied would not change from the revenue collected under the old rates. In the case of dynamic pricing, this means that the customer who has a load factor equal to the class

average would see no change in her or his bill. Load factor is the ratio of a customer's average demand to her or his peak demand.

[9] Several of the test results are discussed in A. Faruqui, R. Hledik, S. Sergici, "Rethinking pricing: The changing architecture of demand response," *Public Utilities Fortnightly*, January 2010.

[10] http://news.smh.com.au/breaking-news-national/vic-govt-to-review-smart-meters-20100203-nd88.html.

[11] W. Vickrey, "Responsive pricing of public utility services," *Bell J. Econ. Manag. Sci.* 2 (1971) 337–346.

[12] F.C. Schweppe, M.C. Caramanis, R.D. Tabors, R.E. Bohn, *Spot Pricing of Electricity*, Kluwer Academic Publishers, 1987.

[13] E. Hirst, "Price-responsive demand in wholesale markets: Why is so little happening?" ISBN: 0-89838-260-2 *Electr. J.*, Norwell, MA, May 2001.

[14] D.J. Bolton, *Costs and Tariffs in Electricity Supply*, Chapman & Hall, London, 1938.

[15] B. Alexander, "Dynamic pricing? not so fast! a residential consumer perspective," *Electr. J.*, Vol. 23, 39–49. July 2010 and S. Brand, "Dynamic pricing for residential electric customers: A ratepayer Advocate's Perspective." *Electr. J.*, Vol. 23, 50–55. July 2010.

[16] http://www.mtc.ca.gov/news/info/toll_increase.htm.

[17] Robert G. Cross, *Revenue Management*, Broadway Books, New York, NY, 1998.

[18] J. Upton, "Giants make a dynamic move: Team to implement pricing strategy for nonseason tickets." *The Examiner*, February 9, 2010.

[19] http://sabres.nhl.com/club/page.htm?id=39501.

[20] See the appendix at the end of this chapter for a discussion of the hedging premium.

[21] F.A. Wolak, *An Experimental Comparison of Critical Peak and Hourly Pricing: The PowerCentsDC program*, Prepared for The 15th Annual Power Conference, The Haas School of Business, U.C. Berkeley, March 13, 2010.

[22] Additional details are available in Ahmad Faruqui and Ryan Hledik, "Transition to Dynamic Pricing." *The Public Utilities Fortnightly*, March 2009.

[23] H. Chao, "Price-responsive demand management for a smart grid world." *Electr. J.*, Vol. 23, February 2010.

Chapter 4

The Equity Implications of Smart Grid: Questioning the Size and Distribution of Smart Grid Costs and Benefits

Frank Felder

INTRODUCTION

The many and various combinations of technologies encompassed by the term *smart grid* are large and uncertain investments. A major recent report, but by no means the only one, places the cost of smart grid for the United States alone between $338 and $476 billion over a twenty-year period with net benefits estimated in the range of $1,294 to $2,028 billion as shown in Table 4.1 [1]. Stated differently, the cost estimates vary by almost 41% and benefit estimates vary by almost 57%. There is no guarantee that the potential net societal benefits will occur, either for society as a whole or for particular segments of society, particularly low-income families. The above cited study, for example, excludes the costs of generation and transmission expansion needed to support additional renewable resources, which would provide much of the environmental benefit associated with smart grid and also excludes the customer costs for

TABLE 4.1 Summary of Estimated Costs and Benefits of the Smart Grid

	20-Year Total ($ billion)
Net Investment Required (Cost)	338–476
Net Benefit	1,294–2028
Benefit-to-Cost Ratio	2.8–6.0

Source: EPRI [1].

smart-grid-ready appliances needed to capture fully the benefits of smart grid for consumers [1]. Chapter 7 of this text provides rather rough "guesstimates" of the benefits of smart grid, on a global scale and suggest that EPRI may be underestimating the benefits.

Besides questions of the relative size of costs and benefits, a host of equity questions surround smart grid. For the most part, because these smart grid investments would be recovered through mandatory charges assessed in utility rates, they thus raise substantial equity issues that must be considered and analyzed. Whether future utility bills would decrease or not increase as much without smart grid is much less certain. Smart grid advocates are making an analogy with smart phones, computers, and other electronic devices that have taken off over the last twenty years but fail to acknowledge the fundamental difference between smart grid and consumer electronic devices. With smart grid, consumers are being asked to guarantee much if not most of these investments. There is little discussion of the issue and importance of choice. Instead, arguments are made, for example in the case of advanced meters, that one size must fit all. The meter purchase and selection are not left up to the consumer but to the utility and the regulators who must approve the investments. The power of the technological revolution in computers, communications, and electronic devices is that it gives consumers a wide range of choices, including the choice not to purchase a particular device or use a service.

The compulsory nature of cost recovery is critical to any such equity analysis. Even if an individual consumer would be better off as a result of contributing to the recovery of smart grid investments, the requirement to do so may be considered inequitable. There are also other fundamental and critical questions regarding the different equity concerns with smart grid and the conditions under which smart grid would result in equitable outcomes. That being said, the fact that equity issues can be raised should not be a reason to halt debate. Instead, equity considerations must be defined precisely, articulated clearly, and analyzed thoroughly as applied to smart grid. A lot more analysis and discussion are needed to answer key questions, both related to the equity of smart grid investments and to their costs and benefits, before regulators should proceed.

In addition to the compulsory nature of cost recovery is the fact that lower income consumers spend a greater percentage of their income on energy, including electricity, than higher income consumers: the lowest income category in the United States spends about 15% of its income on electricity, whereas the highest income bracket spends approximately 1.5%, ten times less on a percentage basis [2]. Thus, raising the cost of electricity is regressive: it harms lower income consumers more than higher income ones (see Table 4.2). Given how vital electricity is to the health and safety of families, it is not considered to be a commodity but a merit good [3,4]. This means, if one accepts this definition, that its availability and pricing should be based upon need for some, not ability to pay. These three factors—compulsory recovery of costs, energy costs disproportionately affecting the poor, and the fact that electricity is a merit good—combine to raise important equity issues with respect to smart grid.

This chapter begins by describing and characterizing smart grid, an elusive term that perhaps defies precise definition. Some have approached this definitional challenge not by describing the technologies that smart grid encompasses, but by describing the desired attributes, as Hauser and Crandall do in Chapter 1 of this volume. If smart grid means a generation, transmission, and distribution system that is more intelligent than today's because it takes advantage of the leapfrog improvements in informational and computational technologies, it is a useful term. If the term is used to set up a false choice between

TABLE 4.2 U.S. 2005 Electricity Consumption and Costs by Household Income

Household Income	U.S. Households		Annual Electricity		
	Number (million)	%	kWh	Cost	% of Income
Less than $10,000	9.9	8.9%	7,854	$785	15.7%
$10,000 to $14,999	8.5	7.7%	8,710	$871	7.0%
$15,000 to $19,999	8.4	7.6%	9,506	$951	5.4%
$20,000 to $29,999	15.1	13.6%	10,040	$1,004	4.0%
$30,000 to $39,999	13.6	12.2%	11,431	$1,143	3.3%
$40,000 to $49,999	11.0	9.9%	11,650	$1,166	2.6%
$50,000 to $74,999	19.8	17.8%	12,440	$1,244	2.0%
$75,000 to $99,999	10.6	9.5%	13,559	$1,356	1.5%
$100,000 or More	14.2	12.8%	15,382	$1,538	1.5%
Total	111.1	100%			

Source: 2005 Residential Energy Consumption Survey, Energy Information Agency.

"smart" and "dumb," rather than to be informative, it is pejorative and brings discussion to a halt.

Next, important questions related to smart grid underscore that smart grid technologies and policies raise serious and legitimate concerns. Some have argued that dynamic pricing, which smart meters could enable, is more equitable than current flat rates because low-income customers would have lower electricity bills (e.g., as argued by Faruqui in Chapter 3). Others question whether residential ratepayers would benefit from advance metering infrastructure [4–6]. To what extent, if at all, equity analysis of dynamic pricing bears on smart grid more generally is open for discussion. There is nothing close to a majority view, let alone a near consensus, that these technologies should be adopted. As one report concluded: "The mindset of state regulators regarding what constitutes prudent and value-adding smart grid investments is currently in flux" [7]. Even EPRI [1, p. 4–1] acknowledges that estimates of the benefits of smart grid are questionable:

There have been a number of studies which have estimated some of the benefits of a Smartz Grid. Each varies somewhat in their approach and the attributes of the Smart Grid they include. None provides a comprehensive and rigorous analysis of the possible benefits of a fully functional Smart Grid. EPRI intends to conduct such a study, but it is outside the scope of the effort presented in this report.

Note that EPRI's report, which does not contain a "comprehensive and rigorous analysis" of smart grid, is a report on the costs and benefits of smart grid. The claim that smart grid would provide a net positive value to society is far from certain, and society needs to honestly assess these costs and benefits and not be "wowed" by the prospects of smart grid [6]. Even if correct, this claim does not mean smart grid would be equitable under any definitions of equity.

The chapter then turns to discussing different conceptions of equity in the context of smart grid investments and policies. There are multiple definitions of equity, and no one definition is acceptable to all. A complete analysis of smart-grid-related equity concerns, therefore, requires analyzing different equity definitions and considerations. Understanding the conditions under which multiple definitions may lead to the same conclusion is extremely useful because it provides an opening to move forward [2]. The chapter concludes with a discussion of what types of questions and analyses regulators should ask and require in order to evaluate the many issues that smart grid raises, particularly those with equity implications.

SMART GRID IS NOT CONSISTENTLY DEFINED, WHICH LEADS TO A MISMATCH OF ITS COSTS AND BENEFITS

The most revealing quip about smart grid is this: "Smart grid is the best thing ever, although we do not know what it is." Others have noted, perhaps wryly, that smart grid is "variously defined" [8]. It seems as if every report has its own definition. Beginning with the more specific and moving to the more general, here are a few:

1. "The smart grid is a digital network that unites electrical providers, power-delivery systems and customers, and allows two-way communication between the utility and its customers" [9].
2. "A smart grid is the electricity delivery system (from point of generation to point of consumption) integrated with communications and information technology for enhanced grid operations, customer services, and environmental benefits" (U.S. Department of Energy, undated).
3. "Smart grids are electricity networks that can intelligently integrate the behavior and actions of all users connected to it—generators, consumers, and those that do both—in order to efficiently deliver sustainable, economic and secure electricity supplies" [10].
4. "Smart grid is an enhanced electric transmission or distribution network that extensively utilizes internet-like communications network technology, distributed computing and associated sensors and software (including equipment installed on the premises of an electric customer) to provide
   i. smart metering;
   ii. demand response;
   iii. distributed generation management;
   iv. electrical storage management;
   v. thermal storage management;
   vi. transmission management;
   vii. power outage and restoration detection;
   viii. power quality management;
   ix. preventive maintenance improves the reliability, security and efficiency of the distribution grid;
   x. distribution automation; or
   xi. other facilities, equipment, or systems that operate in conjunction with such communications network, or that directly interface with the electric utility transmission or distribution network, to provide the capabilities described in clauses (i) through (x) in paragraph (A)" [11].

The first definition above includes only the digital communication network; the second adds in the electric delivery network including generation; the third adds user behaviors; and the fourth includes everything but the proverbial kitchen sink. Figure 4.1 illustrates these ever-expanding definitions.

It is beyond the scope of this chapter to resolve the smart grid definitional problem, and there are many more variations of its definition not provided. The point here is that if one is not extremely careful about what smart grid does and does not include, it is impossible to assess its benefits, costs, and equity implications. The real danger is that broader definitions are used when discussing benefits, and narrower ones are used when discussing costs. For instance, much is made of smart grid's environmental and sustainability benefits (see

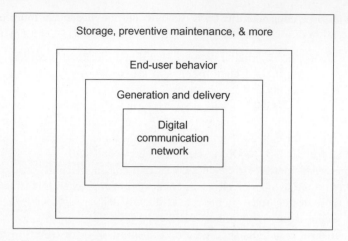

FIGURE 4.1 Diagram of ever-broadening definitions of smart grid.

Chapters 5, 9, 10, and 18 in this text), but much of the costs of achieving those benefits are not included in smart grid cost-benefit analyses [1]. But at best, smart grid enables, and in reality may only facilitate, the use of renewable energy resources. Assuming that smart grid is an enabler, the costs of renewable resources, electric vehicles, and thermal and electric storage, which are very expensive, must be considered in the cost-benefit analysis of smart grid. If advocates of smart grid are going to claim the benefits of smart grid from these additional technologies, they must explicitly include their additional costs, which as noted, as noted above, is not always done.

At worst, smart grid may not be necessary to increase the penetration of cleaner energy technologies. For instance, large-scale wind farms and large numbers of solar facilities exist today without smart grid. A major expansion of nuclear power or carbon capture and sequestration facilities does not require smart grid at all. If this is the case, an analysis is needed to determine the extent to which smart grid facilitates cleaner energy technologies. This could be accomplished by comparing the deployment of renewable resources, electric vehicles, and grid storage with and without compulsory smart grid so that the incremental benefits and costs of smart grid technologies can be isolated.

This "but for" analysis must also incorporate the fact that smart grid technologies will be naturally adopted by utilities over time as existing equipment wears out and the costs of smart grid technologies decline. For instance, the compulsory purchase of smart meters by customers should be compared to what would happen if smart meters were not made compulsory. At some point in the future, smart meters would be installed as replacement meters for existing meters that fail or as new meters for new customers, and if and when the cost savings to utilities of smart meters outweigh their installed cost. In presenting the cost-benefit analysis of smart grid, multiple deployment options must be considered, from naturally occurring smart grid to accelerated deployment in order to obtain a critical mass. EPRI raises the latter issue but does not provide a comparison of costs and benefits under multiple deployment scenarios [1, pp. 2–6]. In fact, it acknowledges that the role of critical mass, tipping point, and penetration of implementation in affecting smart grid costs and benefits is unclear [1, pp. 2–6]. Assuming, which has yet to be demonstrated, that a rapid and compulsory deployment is preferable from a cost-benefit perspective to a naturally occurring one, it is not clear that the additional savings of the former should outweigh the non-compulsory advantages of the latter. In fact, several authors have noted the importance of obtaining customer buy-in, particularly with smart meters, if smart grid is to be successful [6,30]. Rushing deployment may undercut public support.

Furthermore, policy changes may also be required to achieve many of the benefits of smart grid investments. For instance, if regulators do not permit or restrict dynamic pricing or customers do not respond as anticipated to dynamic pricing, then many of the benefits of smart meters will not materialize [12]. Some regulators are reluctant to approve dynamic pricing because of equity concerns based upon positions taken by consumer advocates [4,6]. If transmission expansion to support large-scale wind does not occur, either due to problems with siting or determination of how costs are to allocated "both perennial problems in the United States" then many of the renewable energy benefits that smart grid is supposed to enable may not occur.

There are other fundamental questions related to smart grid that must be addressed. The U.S. Department of Energy, for example, has issued three requests for information listing dozens of fundamental questions about smart grid for stakeholders to address [13]. These questions cover substantive and major issues such as whether it is preferable to wait before adopting major smart grid investments to learn more about the technologies and costs, how to calculate costs and benefits, how to determine who pays what, what should be done if there are cost overruns or benefits do not materialize, how data privacy and cyber security will be ensured (see [14]), and the list goes on. The point here is not to answer these questions, but to suggest that regulators should not consider the compulsory adoption of smart grid paid for by ratepayers until these questions have been coherently addressed.

Take the case of smart meters, a maturing technology, but perhaps one of the most misunderstood in the suite of smart grid technologies. With both costs and need uncertain, and with the prospect of further cost reductions and technological improvements, a compulsory approach to the adoption of smart meters becomes questionable. Only about 6% of electric meters are smart, so our experience is still limited. As with most relatively new technologies, further cost reductions are expected [7]. Before mandating the compulsory purchase of smart meters by ratepayers, regulators should require utilities to evaluate when, if at all, in the future would the utility would be willing to pay for meter replacement because smart meters save more money for the utility than they cost. Recall that many of the benefits of smart meters—for example, reduced maintenance, billing, and customer service costs—benefit the utility. Compulsion has an equity cost. Granted, while hard to quanify it is nonetheless a cost; but it is assumed to be zero when analyzing smart meter and other smart grid deployments. With approximately 60 million smart meters planning to be installed in the United States [15], regulators need to start addressing these questions before it is too late.

But there is even a more fundamental question: some analysts question whether the smart meter should be the gateway to the home for smart grid technologies [7,13]. Perhaps the meter can be directly bypassed and communication with the grid can be done directly with smart appliances. Perhaps a component can be added to existing meters to make them functionally smart. If so, why (subject to obvious safety issues) should that component be owned by the utility? Also, why must a customer use a utility-owned communication network versus the internet?

The above analysis also applies more generally to smart grid investments. Assume that the above definitional and uncertainty issues can be worked through—along with data privacy and security issues, no small challenges—but advocates of smart grid are misapplying cost-benefit analysis. The decision rule employed by advocates of compulsory smart grid investments is that so long as the benefit-cost ratio is greater than 1, or the net present value is greater than zero, then compulsory investments should proceed. This is incorrect because it implicitly assumes that the status quo is static and smart grid would not occur on its own in the future. It may be the case that the natural penetration of smart grid technologies, albeit at a slow pace, has a smaller benefit-cost ratio than a compulsory and accelerated approach. Unless this comparison is made, regulators will not be completely informed. But even if the compulsory approach yields a greater benefit-cost ratio than the naturally occurring approach by saving on rollout costs, regulators may reasonably conclude that the additional expected benefits of the compulsory approach do not outweigh the equity costs of compulsion, which are not incorporated into a benefit-cost analysis. Or regulators may want to consider policies that protect consumers, particularly low-income families, when authorizing smart grid investments. Where exactly to draw that line is not clear, but that should not mean that utilities and regulators ignore the issue.

But there is still another major issue: even if a cost-benefit analysis is performed and used correctly, it may still be factually wrong, that is, in reality the investment may not turn out as expected. At this point, some history is in order. The U.S. electric utility industry does not have a perfect track record with respect to cost effectiveness and completion of long-term investments. Two examples come immediately to mind: nuclear power and long-term power contracts under the Public Utility Regulatory Policies Act of 1978 (PURPA). In both cases, ratepayers ended up being stuck paying for much, if not most, of the cost overruns and out-of-market or stranded costs. Estimates of stranded costs vary, but most range from $100 to $200 billion, in late 1990s dollars [16].

Of course, just because past investments did not always work out does not automatically mean smart grid investments will not do so. But the replacement of existing, working meters with smart meters raises questions about stranded costs. In fact, one utility concedes that its plan to install meters will result in $42 million of stranded costs if 600,000 smart meters are deployed [17]. It is also interesting to compare the cost estimates of smart grid over time. In 2004, EPRI estimated the cost of smart grid at $165 billion (pp. 1–2); seven years later, the cost is between $338 to $476 billion [1, p. 1–4]. Similarly, ERPI estimates of smart grid benefits increased as well. These dramatic increases in costs *and* benefits suggest that these estimates may not be reliable and much more work is needed before they are ready to be used as the basis for ratemaking. EPRI's revised figures are due in part to advances in technologies. The costs associated with smart grid if it addresses privacy or cyber security problems must also be considered [14,18].

THE COMPULSORY NATURE OF SMART GRID RAISES FUNDAMENTAL EQUITY CONCERNS

As stated in the introduction, however one defines smart grid, it is clear that its advocates are drawing an analogy with smart phones, computers, and other electronic devices that have taken off over the last twenty years. What distinguishes smart grid from smart phones, though, are compulsion and choices. Consumers voluntarily use computers, smart phones, and innumerable other electronic devices, along with their innumerable applications. The market was driven by companies anticipating consumer needs, taking the risk on products and services that consumers may want, and investing capital to satisfy consumers without the guarantee of return.

The case with smart grid is completely different. Here, consumers are being asked to guarantee much if not most of these investments with, perhaps, some expectation of lower future utility bills. There is little discussion of the issue and importance of choice. Instead, arguments are made, for example in the case of advanced meters, that one size must fit all. The meter purchase, selection, and timing are not left up to the consumer but to the utility. The power of the technological revolution in computers, communications, and electronic devices is that it gives consumers choices, including the choice not to purchase a particular device or use a service. There seems to be an unwritten premise that utilities must go with a one-size-fits-all strategy; this assumption needs to be demonstrated by smart grid advocates, not just assumed as a given. Much, if not most, of the costs associated with smart grid are assumed to be recovered via compulsory initial increases in electricity rates, which presumably, if advocates of smart grid are correct, would result in lower future rates or less of an increase than without smart grid.

This has profound equity implications, and this issue of compulsion has not been emphasized enough. Compulsion changes everything. One can have legitimate and serious equity concerns with respect to products and services that are purchased voluntarily. Is this used car of the quality claimed by the dealer? Would the average consumer understand the terms and conditions of a mortgage? In these and other cases, the consumer is a voluntary participant, even if misinformed or incompletely informed. The consumer has the choice to walk away from the transaction and pursue other alternatives. Paying for smart grid, though, is mandatory, not voluntary. Forcing customers, particularly poor customers, to pay for smart grid, even if it would make them collectively and individually better off, can be viewed as inequitable. Why should low-income families be required to purchase a smart meter when they cannot afford smart appliances, do not have an internet connection to use the information provided by the meter, do not have a plug-in electric vehicle, and the benefits of smart meters accrue disproportionately to large users of electricity? Granted, society compels its residents to fund all sorts of investments that individuals may or may not desire or benefit from, but society also makes provisions to protect low-income families. Nonetheless, the question remains: is it equitable to require an individual to purchase a particular set of goods and services even though that individual would be better off doing so?

The regulator answering this question does not know whether customers would be better off than doing something else with their money. It may be that the benefit-cost ratio for smart grid is 2:1, but customers may be able to use their money for something with an even higher benefit-cost ratio. Of course, neither the benefits nor costs of smart grid are known with certainty. Take the example of smart meters, which have been the most extensively studied subsystem of smart grid. The ratio of utility operational benefits to broader social benefits of smart metering ranges from approximately 30 to 100% [19]. In the United States, costs of smart meters range from $123 to $532 per meter, with an average of $221. Note that the range is −50 to +100% of the average [20]. As noted above, the cost estimates for smart grid have are large and increasing.

In addition, any benefit-cost analysis looks at ratepayers in the aggregate and not as individuals [21]. Two possibilities need to be considered. It may be the case that the benefits are expected to exceed the costs for the average customer, but not for all customers or subgroups of customers, such as low-income customers. Compelling some ratepayers to pay for smart grid when they receive less benefits than costs can be considered as inequitable. But it is also inequitable if some ratepayers receive a net benefit from smart grid (benefits exceed their costs) but they receive less benefits than others. This may be the case with low-income consumers using smart meters if the cost of smart meters is allocated by household instead of by annual usage. For instance, assume that higher-income households receive $400 in benefits from smart meters but lower-income households only receive $200, yet both pay $150 for the advanced meter. Higher-income customers receive more net benefits than lower-income consumers, an outcome that some may consider inequitable.

Further exacerbating equity concerns is the asymmetric information between utilities (and their vendors) and the regulators and ratepayers. The former have a lot more information about the costs and capabilities of smart grid than the latter. This puts the utilities and vendors in a stronger position than the regulators and customers who lack access to privately held utility/vendor data or information.

The obvious solution to this compulsion problem is to provide customer education, choice to customers, and additional consumer protection [4,6,21,22]. The smart grid discussion has devoted too little effort to figure out how to provide customers with choice. To be sure, there are some difficulties; for example, equipment and deployment costs are lower if uniform decisions are made for a service territory, but that assumes that the initial selection of equipment is correct. If that decision turns out to be incorrect, now a whole service territory has the wrong equipment. There are open issues that need to be analyzed and vetted and not simply assumed to be the case. But it is incumbent upon advocates of smart grid to provide alternatives and explain their implications so regulators and customers can determine the additional costs and benefits of allowing individuals greater choice. In some cases, smart grid proposals have been rejected by regulators due to inadequate justification or inadequately rigorous analysis [23].

SMART GRID RAISES OTHER EQUITY CONCERNS BESIDES COMPULSION

There is no single definition of equity, and any attempt to use only one definition is unlikely to result in clarity or consensus. That being said, it is important to lay out various equity concerns against which to evaluate proposed smart grid plans. As noted above, whether something is compulsory is a fundamental consideration, and the response that compulsion is necessary to save on costs is not sufficient.

Maximizing social welfare using the tool of cost-benefit analysis could be considered an equity framework in the tradition of the utilitarian philosopher Jeremy Bentham [24]—the greatest good for the greatest number—or only an efficiency consideration with no equity content. It would be a mistake to accept implicitly the assumption that a social cost-benefit analysis is the only equity framework and therefore to assume that if smart grid passes such a test, it should be adopted for both efficiency and equity reasons. Proponents of smart grid may, in effect, be making such an assumption by offering a social cost-benefit analysis as the only criterion for evaluation. Any other considerations, under this erroneous assumption, are implicitly irrelevant. It is probably not a coincidence that smart grid advocates have not explicitly stated their belief that equity considerations can be reduced to a social cost-benefit test.

But there are many other major schools of thought that address equity outside of the utilitarian paradigm and should be considered. For instance, some argue that the least well off in society should be made as well off as possible [25]. The focus is not on the average ratepayer, as assumed under a cost-benefit approach, but on those that are worst off, such as low-income families. In contrast, Aristotle argues that goods should be divided in proportion to each claimant's contribution [26]. Under this ethical framework, the costs of smart grid would be assigned based upon the proportion of benefits received. For example, since low-income families use less electricity than most other families, they would pay, if at all, less for smart meters, for instance, than others. Other ethical inquiries start with the premise that ecological and environmental sustainability should be our organizing principles. Under this view, perhaps less resources would be devoted to smart grid and more to renewable resources and energy efficiency, or at least those elements of smart grid that enable cleaner resources would receive more attention. Still others stress a minimal state that should not use coercion to redistribute resources or prohibit activities by people for their own benefit [27]. This approach emphasizes that ratepayers should be given choices and in particular should not be required to purchase functionalities that they do not desire, even if it is for their own good or for society's benefit.

Even within the cost-benefit framework that smart grid advocates have adopted, there are numerous equity issues that have not been brought to light. Both absolute and relative equity concerns must be addressed. For instance, it may be that smart grid technologies do provide net benefits to consumers, an absolute measure, but that consumers' share of the benefits compared to that garnered by the utility is not equitable. This may be the case with monetary benefits or with the allocation of risk between the parties.

For instance, if it were the case that customers received only 10% of smart grid benefits, even if those benefits more than covered their costs, and utilities received 90%, that would not be viewed as equitable. Perhaps many observers assume that virtually all cost savings are passed on to consumers sooner or later and that these cost savings will, including the time value of money, exceed the upfront costs. If this assumption is correct, it needs to be made a lot more explicit, and regulators should consider conditioning smart grid investments on its acceptance by utilities. This relative inequity is compounded by the asymmetry of information between utilities and customers and the fact that utilities have a lot more say in the adoption of smart grid and its particulars than consumers. Similarly, if utilities bore little or no risk for the performance of smart grid whereas customers did, even if such risks were expected to be rewarded, that also is inequitable. In summary, equity concerns include both absolute and relative costs and benefits, risks, and returns.

Another important class of equity concerns pertains to low-income customers [4,21]. Again, issues of absolute and relative equity issues arise. For example, even if low-income customers are better off with smart meters than without, customers who consume more electricity may be receiving a higher share of the benefits for the same cost. If the costs of smart meters are paid by each customer and the meter cost is the same for all residential customers, low-income people who use much less electricity than larger residential customers are paying the same but receiving less benefit. Another example relates to the claimed reliability benefits of smart grid. If costs are allocated by customer rather than by usage, then once again the relative allocation of costs and benefits would not be equitable.

Another limitation of the cost-benefit approach adopted by smart grid advocates is that it reduces benefits to dollars. But an additional dollar to higher-income consumers has less additional benefit (in the language of economists, utility) than it does to lower-income consumers. Presumably, utilities and their vendors, shareholders, and executives are wealthier than low-income consumers. So, if the objective is to provide the greatest good or utility to the greatest number, the types of smart grid technologies that benefit lower-income consumers should be considered and rolled out first because a small improvement in their financial situation is a large benefit or utility. There may be many smart grid technologies and applications that do benefit lower-income and other disadvantaged groups, for instance by improving reliability, reducing electricity consumption and utility bills, and improving monitoring of vital health equipment that some depend upon, among others.

Another common definition of equity within the cost-benefit framework is that costs and benefits should be aligned. That is, customers who add costs to the system—for instance, due to high peak usage—should pay the additional costs due to generating and delivering electricity during peak hours. It may be, as some have argued, that in the case of smart meters combined with some form of dynamic pricing, lower income customers would be better off, at least in aggregate, because they use less on-peak energy than larger customers [28]. Under existing uniform rates, therefore, these lower income customers are paying for some of the costs associated with the peak consumption by higher income customers. If this turns out to be the case when the costs of smart meters are included, it would be fortuitous because it aligns with another, and very different, equity definition under which some customers— low-income, elderly, handicapped, and so on—should have their electricity rates subsidized [2].

This issue of aligning costs with benefits also arises with plug-in electric vehicles. It is not clear why anyone, particularly low-income ratepayers, should have to buy a meter that is capable of supporting an electric plug in vehicle if they do not have such a vehicle. Presumably, families with low income own fewer newer cars than higher income families and therefore are not likely to have a car, let alone a new plug-in electric vehicle. Requiring all ratepayers, particularly low-income ones, to purchase a meter in the unlikely event that sometime in the future they will have a plug-in electric vehicle does not make sense, particularly given the high cost of plug-in electric vehicles and the lack of high market penetration of these vehicles anytime soon.

CONCLUSIONS

The argument here is not that smart grid should be opposed by regulators because it is necessarily inequitable, but that there are important equity concerns that need to be addressed. Each particular smart grid strategy and plan proposed by a utility must be carefully assessed not only for the plan's asserted benefits and costs but also for the distribution of these benefits and costs under different equity lenses. Regulators should not rely upon generic cost-benefit studies of smart grid, particularly when those studies concede that they are not rigorous or complete and do not account for all of the associated costs. This is not to suggest a wide-ranging and never-ending investigation into fundamental philo-sophical questions about social justice. In fact, such an approach would not be successful because it cannot tell regulators how to solve concrete problems [26].

Instead, a much more circumscribed and focused approach is in order. Society compartmentalizes equity issues, such as those with smart grid, because they would otherwise be impractical and overwhelming [26]. The result is that it is perfectly acceptable to address these questions in the context of the norms and precedents that have evolved to deal with specific types of situations [26]. Not surprisingly, laws and regulations regarding the determina-tion of electric rates provide important guidance in evaluating equity concerns with respect to smart grid investments. The process that regulators must go through to make a determination regarding a specific and complete proposal should be thorough and fair, provide access to stakeholders, and require parties to demonstrate their claims based on evidence and reason. The numerous conferences, papers, and presentations heralding smart grid should not replace the ratemaking process.

Case law and ratemaking precedents also provide regulators substantive guidance on equity decision-making. Although considerations of efficiency are important, they are not dispositive. Regulatory rulemaking commonly appeals to other values such as providing consumers information so they are better informed about decisions that affect them and they are better able to respond. Ratemaking policy also considers environmental issues, monetary and other support for low-income households, and assigning costs to those that cause them. Each of these considerations suggest individually and collectively that larger customers who consume more electricity than smaller customers should pay more for smart grid, that additional costs imposed on low-income consumers should be offset, at least partially, and that the elements of smart grid that directly and materially improve their lives should be prioritized over those elements that do not.

Finally, regulators need to push utilities to provide consumers with choice, perhaps the prime benefit that smart grid technologies can provide consumers. Utilities and their vendors must acknowledge that they are asking ratepayers to accept, pretty much on faith, the notion that smart grid, however defined, will benefit society as a whole and that is why they are asking regulators to compel ratepayers to purchase smart grid products and services. Contrast this with advances in computers and communications, to which advocates of smart grid constantly refer as they make their case: consumers voluntarily bought computers and cell phones because the value proposition was clear.

REFERENCES

[1] C. Gellings, EPRI, *Estimating the Costs and Benefits of the Smart Grid: A Preliminary Estimate of the Investment Requirements and the Resultant Benefits of a Fully Functioning Smart Grid*, EPRI, 2011.

[2] F.A. Felder, "The practical equity implications of advanced metering infrastructure." *Electricity J.* 23 (6) (2010) 56–64.

[3] A. Dilnot, D. Helm, "Energy policy, merit goods and social security." *Fisc. Stud.* 8 (1987) 29–48.

[4] B.R. Alexander, "Dynamic pricing? Not so fast! A residential consumer perspective." *Electricity J.* 23 (6) (2010) 39–49.

[5] Synapse Energy Economics, Inc., *Advance Metering Infrastructure: Implications for Residential Customers in New Jersey* (July 8, 2008).

[6] S.A. Brand, "Dynamic pricing for residential electric customers: A ratepayer advocate's perspective." *Electricity J.* 23 (6) (2010) 50–55.

[7] D.J. Leeds, *U.S. Smart Grid Market Forecast 2010–2015*, Executive Summary (Sept. 2010) p. 7.

[8] M. Willrich, *Electricity Transmission Policy for America: Enabling a Smart Grid, End-to-End*, Industrial Performance Center, MIT, MIT-IPC-Energy Innovation Working Paper 09-003, http://web.mit.edu/ipc/publications/papers.html, 2009 (accessed 09.08.11).

[9] Mother Nature Network, *Eco-glossary.* http://www.mnn.com/eco-glossary/smart-grid, 2011 (accessed 22.05.11).

[10] *European Technology Platform for the Electricity Networks of the Future*, http://www.smartgrids.eu/?q=node/163 (accessed 22.05.11).

[11] E. Gunther, *Smart Metering South Dakota PUC Workshop.* http://puc.sd.gov/commission/dockets/electric/2006/el06-018/Guntherpresentation.pdf, 2007 (accessed 22.05.11).

[12] EEnergy Informer, "Dynamic pricing: What if customers don't like it and regulators won't make it mandatory?" *EEnergy Informer*, May 17–18, 2011.

[13] US Department of Energy (U.S. DOE), U.S. Department of Energy, Office of Electricity Delivery and Energy Reliability, *Funding for Smart Grid Activities*, Presentation, undated, http://www.oe.energy.gov/Smart%20Grid%20Request%20for%20Information%20and%20Public%20Comments.htm, 2010 (accessed 09.08.11).

[14] T. Baumeister, *Literature Review on Smart Grid Cyber Security*, University of Hawai'i, 2010.

[15] Pike Research, *57.6 Million smart meters currently planned for installation in the United States.* http://www.pikeresearch.com/newsroom/57-9-million-smart-meters-currently-planned-for-installation-in-the-united-states, 2011 (accessed 2.06.11).

[16] Congressional Budget Office (CBO), *Electric Utilities: Deregulation and Stranded Costs*, http://www.cbo.gov/doc.cfm?index=976&type=0#N_13_, 1998.

[17] A.P. Kelly, *Costs of PECO's Smart Meter Plan, Direct Testimony, Before the Pennsylvania Public Utility Commission*, Aug. 14, 2009 http://www.peco.com/NR/rdonlyres/9D8E9E91-7D41-487F-B884-7A30213812D2/7659/PECOStatementNo4.pdf, 2009.

[18] C. Balough, "Privacy implications of smart meters." *Chi.-Kent L. Rev.* 86 (1) (2011).

[19] B. Neenan, *Smart Policies for a SmartGrid (or, the other way around), Harvard Electricity Policy Group*, La Jolla, CA, 2009.

[20] Smart Grid Watch, http://www.emeter.com/2010/how-much-do-smart-meters-cost/, 2011 (accessed 09.08.11).

[21] N. Brockway, *Smart Grid, Smart Meters: What's Good for Consumers?* Presentation, Colorado PUC, June 7, 2009.

[22] P. Carmody, *Smarting from Resistance to Smart Grid*, Harvard Electricity Policy Group, 2010.

[23] EEnergy Informer, "How smart is your smart grid project?" *EEnergy Informer* (September 19–21, 2010).

[24] J. Bentham, *The Principles of Morals and Legislation*, 1789.

[25] J. Rawls, *A Theory of Justice*, Harvard University Press, Cambridge, MA, 1971.

[26] P. Young, *Equity: In Theory and Practice*, Princeton University Press, Princeton, NJ, 1994.

[27] R. Nozick, *Anarchy, State, and Utopia*, Basic Books, New York, 1974.

[28] A. Faruqui, L. Wood, *Appendix F: Impact of Dynamic Pricing on Low-income Customers: Quantifying the Benefits of Dynamic Pricing in the Mass Market*, Prepared for the Edison Electric Institute, January 2008.

[29] NASUA, Reply Comments by the National Association of State Utility Advocates, August 9, 2010.

[30] P.C. Honebein, R.F. Cammarano, and C. Boise, "Building a social roadmap for the smart grid." *The Electricity Journal*, 24 (4) (2011) 78–85.

Part II

Smart Supply: Integrating Renewable & Distributed Generation

Prospects for Renewable Energy: Meeting the Challenges of Integration with Storage

W. Maria Wang, Jianhui Wang, and Dan Ton

INTRODUCTION

Rapid growth in renewable energy generation has been spurred by concerns such as energy security, fuel diversity, and climate change. Most major economies have government policies supporting renewable electricity. Seventeen countries currently have feed-in tariffs, ten countries have quota obligation systems with tradable green certificates, and four countries have tender systems [1]. In the United States, 36 states and the District of Columbia have set specific standards or goals for a certain percentage of electric power generation and sales to come from renewable sources [2], and utilities in 48 states now offer their customers the option to purchase green power [3].

    Lower costs associated with wind turbines and solar cells due to technology advancements and economies of scale have also contributed to the accelerated growth of these energy markets, notably during the past five years. In 2009, a record 10 GW of new wind power was installed in the United States, resulting in cumulative wind installations of 35 GW [4]. Following the same trend, total U.S. solar electric capacity from photovoltaic (PV) and concentrating solar power (CSP) technologies exceeded 2 GW in 2009, with 1.65 GW being grid-tied. The residential PV market doubled and three new CSP plants were built, resulting in a 37% increase in annual installations over 2008 from 351 MW to 481 MW [5].

As renewable energy technologies mature, and with continued financial subsidies, they are expected to provide a growing share of the world's electricity requirements. Figure 5.1 shows the global growth of renewables compared to other types of generation projected to 2035. Currently, solar PV is the fastest growing renewable technology worldwide at an average of 60% per year, followed by wind power at 27% and biofuels at 18% [1]. However, concerns about potential impacts of high penetration of renewables on the stability and operation of the electric grid may create barriers to their future expansion. The intermittent and variable nature of renewable sources, particularly wind and solar, poses reliability concerns that must be addressed at higher penetration levels. As other chapters in this volume explain, our existing grid is not designed to deal with these types of renewable resources. Current industry practices need to be altered and a smart grid needs to develop for the successful integration of renewables.

Energy storage can serve as an enabling technology for renewables integration by allowing for output firming and dispatchability, as well as other benefits such as load shifting and peak shaving. Renewable energy technologies such as CSP systems have built-in thermal energy storage to extend the generation period beyond the peak solar incidence. A study by the National Renewable Energy Laboratory (NREL) found that even for low penetration levels, adding

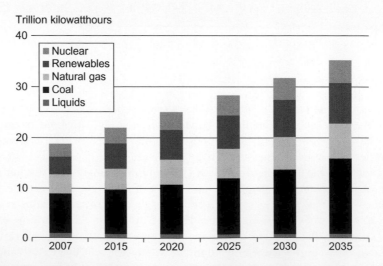

FIGURE 5.1 World electricity generation by fuel, 2007–2035. *(Source: US Energy Information Administration, "International Energy Outlook 2010," http://eia.gov/oiaf/ieo/)*

thermal energy storage can significantly increase the value of CSP through generation shifting, in some cases outweighing the costs of storage [6]. Other technologies such as PV and wind will require a variety of energy storage technologies to help offset their intermittent generation.

This chapter focuses on integration issues surrounding solar power since wind integration is covered in several other chapters. A discussion follows on the possible solutions to these issues using energy storage, which is applicable to the integration of both wind and solar resources. This chapter focuses on storage technologies since Chapters 9 and 10 cover demand response and direct load control for providing capacity firming and ancillary services, and Chapters 18 and 19 cover electric vehicles for reducing wind integration costs in different markets.

The chapter is organized into four sections. Section "High Penetration of Renewables" covers the benefits and issues associated with the high penetration of renewables, focusing on PV, into the electric grid. Section "Energy Storage for Integration" discusses the role of energy storage in mitigating these issues, including how different storage technologies are suited for specific applications. Section "Federally Funded Energy Storage Efforts" highlights research, development, and demonstrations of grid-scale energy storage supported through federal grants and the Department of Energy (DOE). The chapter's main insights are in the concluding section.

HIGH PENETRATION OF RENEWABLES

Benefits

The environmental, economic, and energy security benefits from renewable generation are magnified by increasing its penetration level, that is, the capacity of renewable generation as a percent of peak or total load. To identify these benefits and facilitate more extensive adoption of renewable distributed electric generation, the DOE launched the Renewable Systems Interconnection (RSI) study in 2007. The 15 study reports address a variety of issues related to utility planning tools and business models, new grid architectures and PV systems configurations, and models to assess market penetration and the effects of high-penetration PV systems. As a result of this effort, the Solar Energy Grid Integration Systems (SEGIS) Program was initiated in early 2008. SEGIS is an industry-led effort to develop new PV inverters, controllers, and energy management systems that will greatly enhance the benefits of distributed PV systems.

According to the RSI report on "Photovoltaics Value Analysis," the largest benefits are in cost savings from avoided central power generation and capacity, deferred or avoided transmission and distribution (T&D) investment, and lower greenhouse gas and pollutant emissions [7]. Chapter 7 of this text presents a modeling study that also arrives at similar conclusions: benefits from increased renewable integration arise from reductions in capital expenditure, fuel costs, operation and maintenance costs, and carbon costs. During most hours, with the exception of peak hours, less than 50% of the electricity system capacity is utilized. Thus, a significant portion of the network assets have been built to meet only a few hundred hours of peak demand each year. Consequently, a PV system that produces a high share of its output during on-peak hours and displaces a peaking plant will have a higher benefit.

The RSI study on "Production Cost Modeling for High Levels of Photovoltaics Penetration" found that in the western United States, PV displaces natural gas at low penetration and begins to displace coal at higher penetration [8]. Various strategies to increase production during peak demand periods and increase the benefit from this value include integrating energy storage into the PV system and integrating load management applications with the PV system controls, as schematically illustrated in Figure 5.2.

In addition to cost savings from avoided central generation, there are T&D benefits. Since PV systems can be installed on rooftops and on undesirable real estate, such as brown fields, they can reduce a utility's need to acquire land for construction of new, large-scale generating facilities. Furthermore, locations with congested transmission and/or distribution systems that typically require expensive upgrades could defer these upgrades when PV systems are installed to reduce congestion. The value of deferred T&D upgrades is estimated to be 0.1 to 10 cents/kWh, depending on factors such as location, temperature, and load growth [7].

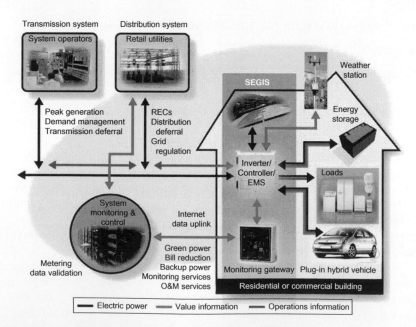

FIGURE 5.2 SEGIS diagram showing the integration of PV with the smart grid. *(Source: US DOE, http://www1.eere.energy.gov/solar/images/segis.jpg)*

Outlook

Government support for renewables has driven their growth across the world in the past decade. In 2010, China outpaced Europe and North America in wind installations by adding approximately 17 GW, becoming the global leader in terms of installed capacity. However, there has been a delay of several months in connecting this capacity to the grid. China is also the leading hydropower producer, followed by the United States, Brazil, Canada, and Russia. For solar PV capacity, Germany remains the leader and is followed by Spain and Japan. The most geothermal power is produced by the United States, followed by the Philippines, Indonesia, Mexico, and Italy. Figure 5.3 shows the current and projected mix of renewable generation (excluding hydropower) for the United States and the world, assuming a business-as-usual scenario in which

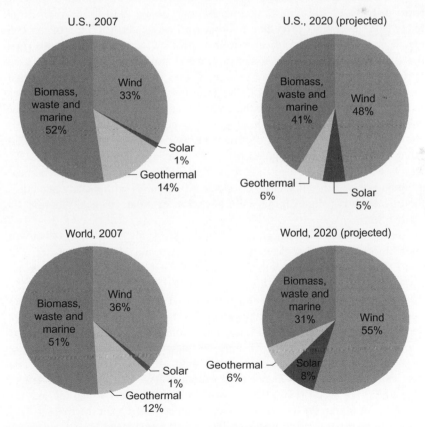

FIGURE 5.3 Renewable generation mix, excluding hydropower, for the United States and the world. *(Data for charts from US Energy Information Administration, International Energy Outlook 2010, Reference Case Projections for Electricity Capacity and Generation by Fuel, DOE/EIA-0484 (2010))*

current regulations and technological trends are maintained. For both the United States and the world, wind and solar are projected to become the majority share of renewable generation as geothermal, biomass, waste, and marine generation lose their current dominance by 2020.

There is considerable room for growth as the ultimate resource potential of wind and solar has barely been tapped to date. Land-based wind, the most readily available for development, totals more than 8,000 GW of potential capacity in the United States alone. The capacity of CSP is nearly 7,000 GW in seven southwestern states, and the generation potential of PV is limited only by the land area devoted to it, which is 100–250 GW/100 km^2 in the United States [9]. However, cost is an issue with all renewable generation. Most solar resources are in the Southwest, and wind resources are most abundant in remote locations with sparse transmission lines.

The DOE goal is to obtain 20% of U.S. electricity capacity, around 200 GW, from distributed and renewable energy sources by 2030 [10]. As of 2009, renewable generation, excluding hydropower, accounted for 3.6% of the U.S. electricity supply, with 51% of that share from wind, 10% from geothermal, 0.6% from solar, and the remainder from wood and biomass [11]. Policy developments at both the federal and state level, coupled with technology improvements funded by the DOE's SunShot Initiative,[1] are helping to create a more receptive marketplace for PV in the United States. The DOE SunShot Initiative aims to make solar energy technologies cost-competitive with other forms of energy by reducing the cost of solar energy systems by about 75% before 2020. By lowering the installed price of utility-scale solar energy to $1/W, which would correspond to roughly 6 cents/kWh, solar energy will be cost-competitive with fossil-fuel-based electricity sources without any subsidies, thereby enabling rapid, large-scale adoption of solar electricity across the United States.

Indeed, scenarios developed as part of the RSI study on "Rooftop Photovoltaics Market Penetration Scenarios" indicate that annual installations of grid-tied PV in the United States could reach 1.4–7.1 GW by 2015, resulting in a cumulative installed base of 7.5–24 GW by 2015 [12]. This study found that the variables with the largest impact on market penetration of rooftop PV were system pricing, net metering policy, extending the commercial and residential federal tax credits to 2015, and interconnection policy (Figure 5.4). Lifting net metering caps and establishing net metering had significant effects on projected PV market penetration in some states. In fact, the projected cumulative installed PV in 2015 increased by about 4 GW. Extension of the federal investment tax credit (ITC) had a critical effect on the PV market and was found to be a prerequisite for the overall success of PV in the marketplace. The projected cumulative installed PV in 2015 increased by 5 GW from a partial to full extension of the ITC.

[1]http://www1.eere.energy.gov/solar/sunshot/

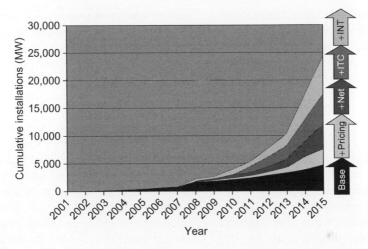

FIGURE 5.4 Influence of system pricing, net metering policy, federal tax credits, and interconnection policy on cumulative rooftop PV installations. *(Source: Paidipati et al. [12] (NREL))*

To address pricing and technology issues, the Solar Energy Technologies Program (SETP), within the DOE Office of Energy Efficiency and Renewable Energy (EERE), conducts research, development, demonstration, and deployment activities to accelerate widespread commercialization of clean solar energy technologies. The goals of the SETP are to make PV cost-competitive across the United States by 2015 and to directly contribute to private sector development of more than 70 GW of solar electricity supplied to the grid to reduce carbon emissions by 40 million metric tons by 2030. The SEGIS awards under this program engage industry/university teams in developing advanced inverters/controllers that integrate a broad range of PV system capacities from <1 kW to >100 kW with the electric grid to meet varying residential, commercial, and utility application needs.

On an international scale, most countries with significant solar installations have national solar missions or programs that set targets and an integrated policy. Several countries have PV feed-in tariffs, which actually had to be reduced in the Czech Republic, Spain, France, Italy, and Germany during 2010 and early 2011 due to unexpected rapid growth in PV deployment that increased policy cost. Capacity expansion was even suspended in some cases. Therefore, more sustainable policies need to be designed that can accommodate the decreasing cost of solar technology.

Besides cost and policy issues, codes, standards, and regulatory implementation are also major barriers to high penetration of grid-tied PV. In the United States, the electric grid safety and reliability infrastructure is governed by linked installation codes, product standards, and regulatory functions such as inspection and operation principles. The National Electric Code, IEEE standards, American National Standards, building codes, and state and federal regulatory inspection and compliance mandates must be consistent to result in a safe and reliable electric grid. Effectively interconnecting distributed renewable energy systems requires compatibility with the existing grid and future smart grid. Uniform requirements for power quality, islanding protection, and passive to active system participation could facilitate the high penetration of PV. National requirements for power quality and active participation of such renewable generation in power system operation must be developed.

As PV technology advances and becomes more competitive, it is expected to supply more residential and commercial loads at the customer's side of the meter. Therefore, PV is being developed in accordance with codes and standards that govern distributed generation, such as IEEE 1547 and UL 1741. These standards, however, are being developed on the important assumption of the low penetration of distributed generation and are focused on simplifying installations for passive system participation. They result in an electric grid that is not designed for a two-way flow of power, especially at the distribution level. The traditional planning process does not consider variable generation such as PV; therefore, the initial response of the electric industry was to exclude it from capacity planning. Current industry practices need to be altered to accommodate the high penetration of renewable generation.

As discussed in the RSI report on "Power System Planning: Emerging Practices Suitable for Evaluating the Impact of High-Penetration Photovoltaics," the emerging practice is to include renewable energy supply early in the planning process and consider it during energy growth forecasts [13]. This practice treats variable renewable generation as a part of the load and thus allows for its full integration into the planning process. In order to forecast effectively, smart grid tools must be able to accurately estimate resource data on wind and solar availability for a given location and time. Dynamic models should be able to include the impacts of resource variability such as cloud cover and wind gusts.

The operational flexibility of the balance of generation portfolio is strategically important so as not to curtail renewable generation. Planning for generation flexibility deals with two aspects of frequency control: economic redispatch of units every five minutes (load following) and automatic generation control (regulation). Both aspects should be evaluated relative to the net load. Understanding the load-following and regulation capabilities of the system is important in determining the system's response to load changes and in evaluating its ability to maintain the frequency within the desired control range. Having spatially diverse renewable resources and energy storage at high penetrations can reduce net load variability at the time scale of load following.

Integration Issues

Resource Intermittency

As mentioned earlier, one important challenge associated with intermittent renewable energy generation is that the generation's power output can change rapidly over short periods of time. Wind and solar generation intermittency can be of short duration or diurnal. The most common causes of short-duration intermittency are gusty conditions and clouds. As a cloud passes over solar collectors, power output from the affected solar generation system drops. Location-specific shading caused by trees and buildings can also cause relatively short-duration intermittency. During these events, the rate of change of output from the solar generation can be quite rapid. These changes in solar irradiance at a point can be more than 60% of the peak irradiance in just a few seconds. However, the time it takes for a passing cloud to shade an entire PV system depends on its speed and height, as well as the PV system size. For PV systems around 100 MW, it will take minutes rather than seconds to shade the system [14]. The resulting ramping increases the need for highly dispatchable and fast-responding generation such as peaker plants or alternatively energy storage to fill in during the decrease in output.

Diurnal intermittency is more predictable, being mostly related to the change of insolation throughout the day as the sun rises in the morning and descends in the evening. During the day, the efficiency of some solar cells may drop as the equipment's temperature increases, reducing PV output. Wind output also tends to be lower during the day and peaks at night when load is the lowest. This attribute favors the use of energy storage to increase the capacity factor of wind turbines. Figure 5.5 illustrates the mismatch between load and renewable generation due to diurnal intermittency. The average daily profiles of wind and solar were modeled assuming 23% wind and solar penetration in the Western Electricity Coordinating Council (WECC). Since wind and solar ramps are usually inversely correlated in the morning and evening, integrating both wind and solar power may reduce load-following and regulation requirements during some hours of the day.

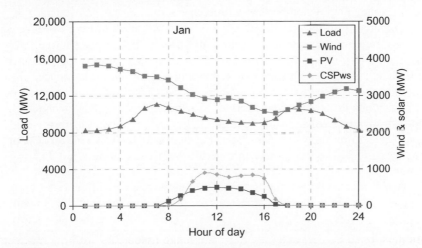

FIGURE 5.5 Arizona load, wind, and solar average daily profiles for January. *(Source: Western Wind and Solar Integration Study, May 2010 (NREL/SR-550-47434))*

Voltage Regulation Problems

The sources of intermittency discussed above that lead to variable PV and wind output can cause potential problems in the reliability and stability of the electric power system, such as in frequency and voltage regulation, load profile following, and broader power balancing. Entities such as the California ISO (CAISO) and the New York State Energy Research and Development Authority (NYSERDA) have conducted studies on issues surrounding the integration of renewable energy. For a 20% wind penetration scenario, CAISO found that all wind generation units should meet the WECC requirements of ±0.95 power factor. This dynamic reactive capability is necessary for voltage control to ensure system stability [15]. NYSERDA also found that wind power needed to meet Low-Voltage Ride Through standards and voltage regulation criteria even at only 10% penetration levels [16]. Both studies found that accurate day-ahead and hour-ahead forecasts of wind and solar generation are essential for reliable operation of the power grid and scheduling of other generation resources and unit commitment, with solar playing a larger role as California moves toward a 33% renewables portfolio standard (RPS) by 2020.[2]

When transients are high, area regulation will be necessary to ensure that adequate voltage and power quality are maintained. Advanced PV system technologies, including inverters, controllers, and balance-of-system and energy management components, are necessary to address voltage regulation issues. At high PV penetration levels, the RSI report on "Distributed Photovoltaic Systems Design and Technology Requirements" suggests that the problems most likely to be encountered are voltage rise, cloud-induced voltage regulation issues, and transient problems caused by mass tripping of PV during low voltage or frequency events [17]. This report discusses several studies where the maximum PV penetration level was found to be anywhere from 25% to 50% before voltage regulation became a problem.

The smart grid integrated with PV and wind will need to have two-layered voltage regulating capabilities from a speed-of-response perspective. Slow regulation (for managing distribution system voltage profiles or microgrid operation[3]) and fast regulation (for addressing flicker and cloud-induced fluctuations) will both be needed in high-penetration scenarios. The low-speed system responds as needed over a period of many tens of seconds or minutes to hold steady-state voltage within the ANSI limits. The second layer is a high-speed system on top of the slow-speed system and serves to moderate rapid changes in voltage and power that result from fluctuating wind and solar resources.

A PV inverter or associated energy storage system could provide voltage regulation by sourcing or sinking reactive power. Implementing this feature

---

[2]Chapter 6 of this text also covers these issues.
[3]Chapter 8 of this text covers microgrids in more detail.

would require modifications to the traditional PV inverter hardware design and current interconnection requirements need to evolve.[4] During a workshop held by the DOE on the "High Penetration of PV Systems into the Distribution Grid" in February 2009, energy storage was identified as a possible solution to solar variability and intermittency; development of small- to mid-scale energy storage solutions was identified as a top RD&D activity. As discussed in section "Energy Storage for Integration," energy storage can be used for power management as an intermediary between variable resources and loads.

Capacity Firming

When PV and wind outputs are low, some type of back-up generation will be needed to ensure that customer demand is met. To address the issues of load profile following and power balancing, renewables capacity firming to decrease variable output needs to occur. Capacity firming offsets the need to purchase or build additional dispatchable capacity. Energy storage can be combined with renewable energy generation to produce constant power. Depending on the location, firmed renewable energy output may also offset the need for T&D investment. Renewables capacity firming is especially valuable when peak demand occurs, and energy storage can even be used for peak shaving.

Intermittent renewable generation is currently mitigated by ramping conventional reserves such as thermal plants up or down based on minute-by-minute and hourly forecasts. A CAISO study found that under the 20% RPS, dispatchable generators need to start and stop more frequently. In particular, combined-cycle generators' starts will increase by 35% compared to a reference case that assumes no new renewable capacity additions beyond 2006 levels [18]. Grid-scale energy storage would provide significantly faster response times than conventional generation, on the order of milliseconds versus minutes. Furthermore, a study by the California Energy Storage Alliance found that the levelized cost of generation for energy storage can be less than that for a simple cycle gas-fired peaker [19].

Therefore, as renewable penetration grows, energy storage will likely become more cost effective and necessary. Most studies conclude that traditional planning and operational practices only suffice for up to 10–15% renewable penetration levels. Although small penetrations of renewable generation on the grid can be smoothly integrated, accommodating more than approximately 20–30% electricity generation from these renewable sources will require new approaches in power system planning and operation. Storage can reduce the amount of dispatchable generation capacity needed to offset ramping of renewable energy generation. Therefore, capacity firming is valuable as a way to reduce load-following resources and improve asset utilization.

---

[4]IEEE 1547 Standard for Interconnecting Distributed Resources with Electric Power Systems.

Storage power and discharge duration for renewables capacity firming are application- and resource-specific. At the lower end, it is assumed that one-half to as much as two hours of discharge duration are needed to firm solar generation, assuming that much of PV output coincides with peak demand, whereas to firm wind generation, a somewhat longer discharge duration of two to three hours is needed. Furthermore, the storage technology used for capacity firming should be reliable so as to provide constant power. The estimated 10-year net benefits associated with firming of PV and wind output are $709/kW and $915/kW, respectively [20].

ENERGY STORAGE FOR INTEGRATION

Energy storage for capacity firming can also minimize curtailment of renewable generation through maximizing energy harvest. As mentioned earlier for energy management applications, storage can also offset the need for additional generation or reserve capacity by continuing to supply power during cloudy or nighttime conditions and addressing power demand surges. Other benefits of energy storage that enable the integration of higher levels of renewable generation include peak shaving or price arbitrage, that is, storing energy during low demand and delivering it back to the grid during peak demand. Stored energy and storage capacity would be managed most effectively with a control algorithm that takes into account estimates of future hourly pricing and renewable generation output. The CAISO Integration of Renewable Resources study modeled regulation and load-following requirements under a 20% RPS, which includes approximately 9 GW of wind and solar power in California. The simulations indicated that the maximum regulation-up requirement will increase 35%, from 278 MW in 2006 to 502 MW in 2012. The maximum hourly simulated load-following up requirement in 2012 is 3737 MW compared to 3140 MW in 2006 [18].

Applications and Technologies

The grid applications for energy storage technologies can be loosely divided into power applications and energy management applications, which are differentiated based on storage discharge duration. Energy applications discharge the stored energy relatively slowly and over a long duration (i.e., tens of minutes to hours). Power applications discharge the stored energy quickly (i.e., seconds to minutes) at high rates. Storage technologies for power applications are used for short durations to address power quality issues, such as voltage sags and swells, impulses, and flickers. The use of storage to prevent voltage rise from the export of power from the customer facility to the grid has been demonstrated in Japan's Ota City PV-integrated distribution system.[5] Technologies used for energy management applications store excess electricity during periods of low

---

[5]Ueda Y. et al., "Performance Ratio and Yield Analysis of Grid-Connected Clustered PV Systems in Japan." In Proceedings of the 4th World Conference on Photovoltaic Energy Conversion, pp. 2296–99.

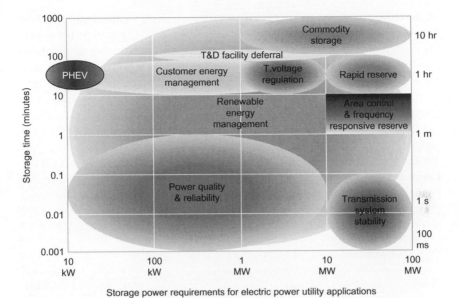

Storage power requirements for electric power utility applications

FIGURE 5.6 Energy storage applications and their associated power and discharge duration requirements. *(Data from Sandia Report 2002–1314)*

demand for use during periods of high demand. These devices are typically used for longer durations to serve functions that include peak shaving, load-leveling, intentional islanding, and renewable energy collection and dispatch. Figure 5.6 illustrates the storage power and discharge duration requirements as a function of application.

Current grid-scale energy storage systems are both electrochemically based (batteries and capacitors) and kinetic energy based (pumped hydropower, compressed-air energy storage [CAES], and high-speed flywheels). For power applications, suitable technologies include flywheels, capacitors, and superconducting magnetic energy storage. For energy applications, suitable technologies include pumped hydropower, CAES, and high-energy sodium-sulfur and flow batteries. The batteries in electric vehicles can also be used for both energy and power applications; they can provide energy management services through controlled charging during off-peak periods, thereby reducing curtailment of wind power, and frequency regulation through vehicle-to-grid capabilities. Such charging schemes would be based on smart grid communication of real-time load, price, and renewable energy generation.[6]

Thermal storage is built into CSP and is also gaining ground on the customer side in the form of heating in sustainable building mass and cooling via

---

[6]More details on linking the charging of electric vehicles to price signals can be found in Chapters 18 and 19.

phase-change systems. Since heating, ventilating and air conditioning are the largest contributors to peak energy demand, thermal energy storage for storing off-peak power and shifting electricity used for air conditioning is becoming popular in commercial buildings. Distributed thermal storage is mainly used for lowering peak demand and shaping load, and needs to be combined with an electric utility's demand response program for maximum benefits. This chapter focuses instead on storage technologies for dispatchable generation.

These technologies for specific applications can be further subdivided according to the scale of storage required, for instance, deployment as a distributed energy resource, or at the distribution feeder, substation, or bulk power system level. Energy storage in a future system will likely be needed in a variety of sizes and configurations to meet needs at all system levels. The wide range of mechanisms, chemistries, and structures of these energy storage technologies enables them to be tailored to meet the power and energy demands of specific applications. Figure 5.7

Technology option	Maturity	Capacity (MWh)	Power (MW)	Duration (hrs)	% Efficiency (total cycles)	Total cost ($/kW)	Cost ($/kW-h)
Bulk energy storage to support system and renewables integration							
Pumped hydro	Mature	1680–5300	280–530	6–10	80–82 (>13,000)	2500–4300	420–430
		5400–14,000	900–1400	6–10		1500–2700	250–270
CT-CAES (underground)	Demo	1440–3600	180	8	See note 1 (>13,000)	960	120
				20		1150	60
CAES (underground)	Commercial	1080	135	8	See note 1 (>13,000)	1000	125
		2700		20		1250	60
Sodium-sulfur	Commercial	300	50	6	75 (4500)	3100–3300	520–550
Advanced Lead-acid	Commercial	200	50	4	85–90 (2200)	1700–1900	425–475
	Commercial	250	20–50	5	85–90 (4500)	4600–4900	920–980
	Demo	400	100	4	85–90 (4500)	2700	675
Vanadium redox	Demo	250	50	5	65–75 (>10000)	3100–3700	620–740
Zn/Br redox	Demo	250	50	5	60 (>10000)	1450–1750	290–350
Fe/Cr redox	R&D	250	50	5	75 (>10000)	1800–1900	360–380
Zn/air redox	R&D	250	50	5	75 (>10000)	1440–1700	290–340
Energy storage for ISO fast frequency regulation and renewables integration							
Flywheel	Demo	5	20	0.25	85–87 (>10000)	1950–2200	7800–8800
Li-ion	Demo	0.25–25	1–100	0.25–1	87–92 (>100,000)	1085–1550	4340–6200
Advanced lead-acid	Demo	0.25–50	1–100	0.25–1	75–90 (>100,000)	950–1590	2770–3800

FIGURE 5.7 Energy storage characteristics by application. (*Source: Electric Energy Storage Technology Options: A White Paper Primer on Applications, Costs, and Benefits. EPRI, Palo Alto, CA, 2010. 1020676. © 2010 Electric Power Research Institute, Inc. All rights reserved*)
Note 1: Refer to the full Electric Power Research Institute (EPRI) report for important key assumptions and explanations behind these estimates.

tabulates the main characteristics including size, performance, and cost of these storage technologies according to their power or energy application for renewables integration.

Centralized storage, such as pumped hydropower and CAES, is most likely to be applied at the supply side, (i.e., transmission or bulk system level), to manage variations in output from solar plants and wind farms via capacity firming. Pumped hydropower uses off-peak electricity to pump water from a low-elevation to a high-elevation reservoir. The stored energy is delivered to the grid by releasing the water through turbines to generate power. The United States has pumped hydropower facilities in 19 states that provide about 23 GW of capacity. Out of all the energy storage options, pumped hydropower is the most established technology; however, it has a higher capital cost compared to CAES. CAES uses off-peak power to pump air into a storage reservoir such as an underground salt cave. The air is released through a turbine to meet power demand. As seen in Figure 5.7, underground CAES is the cheapest bulk energy storage option. However, lack of data and analysis on suitable sites has limited its use. The United States has only one 110-MW CAES plant in Alabama. A barrier to both pumped hydropower and CAES development is assessment of resource availability. While pumped hydropower has achieved widespread deployment, all of the suitable locations currently being used provide only a small fraction of baseload electricity needs.

At the distribution level and the customer-distributed resource location, more compact and short-duration forms of energy storage for power applications (batteries, flywheels, and capacitors) are more likely to be used. Superconducting magnetic energy storage (SMES) is an experimental technology that may also be used for power applications. The different storage technologies are shown in Figure 5.8. The use of batteries, flow batteries, flywheels, and ultracapacitors for power applications has gathered considerable steam in recent years as they are well suited for rapid compensation of power fluctuations from wind and PV. In fact, many recent distribution-scale demonstration projects have successfully established the value of these technologies for frequency regulation, intentional island transitioning, and other such applications. Such distributed storage technologies serve the demand side; these mobile, modular technologies are preferred for microgrids and off-grid communities.

The RSI study on "Enhanced Reliability of Photovoltaic Systems with Energy Storage and Controls" observed a significant improvement in the three reliability indices—critical SAIDI (average duration of critical load interruptions), critical SAIFI (average number of interruptions per customer), and unserved critical load (UCL, annual unserved critical load [kWh] on a circuit)—when PV and battery energy storage were deployed at each home within a community. The presence of more than ~5 kWh of battery capacity per home reduced each index to nearly zero [21]. In order to reap maximum benefits from power management applications, the storage technology needs to have a

Electronic flywheel
systems
(Beacon Power)

Superconducting
magnetic energy
storage (SMES)

Batteries
(lithium ion, lead acid,
sodium-sulfur, etc.)

Ultracapacitor
(ESMA corporation)

Pumped storage hydro Raccoon
Mountain hydropump storage (TVA)

FIGURE 5.8   Various energy storage technologies. *(Source: Whitaker et al. [17] (SAND2008-0944 P))*

roundtrip efficiency of 75–90%, a system lifetime of 10 years with high cycling, a capacity of 1 MW to 20 MW, and a response time of 1 to 2 seconds [22].

For most PV applications, lead-acid technology has been the preferred energy storage technology due to its maturity, low cost, and availability. However, its low energy density, short cycle life, and high maintenance requirements have deterred wide-scale use in the electric grid. A number of lead-acid battery manufacturers, such as East Penn in the United States and Furukawa in Japan, are manufacturing prototype batteries for hybrid electric vehicles to overcome the main disadvantages of valve-regulated lead-acid (VRLA) batteries by using new carbon formulations for the anodes. These formulations promise to reduce sulfation, thereby increasing the cycle life and available energy. Before applying such technologies to the grid, however, a better understanding is needed of how particular applications such as peak shaving will affect the battery life.

Other advanced technologies such as molten salt batteries are currently being developed for utility-scale (> 1MW) applications. For instance, sodium-sulfur batteries have high energy density and are low cost; however, high operating temperatures between 300 to 350°C limit their use. Other molten salt batteries such as sodium/nickel-chloride, or ZEBRA batteries, have been developed for transportation applications and are currently being considered for some grid-scale applications, such as peak shaving. Figure 5.9 groups the major energy storage technologies according to their suitability for certain applications.

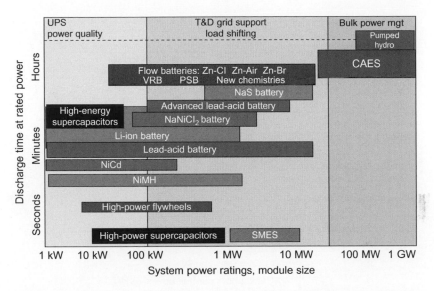

FIGURE 5.9 Energy storage technologies according to specific application power and discharge duration requirements. *(Source: Electric Energy Storage Technology Options: A White Paper Primer on Applications, Costs, and Benefits. EPRI, Palo Alto, CA, 2010. 1020676. © 2010 Electric Power Research Institute, Inc. All rights reserved)*

Cost-Benefit Analysis

The main issues preventing widespread deployment of these energy storage technologies are the current high capital cost (Figure 5.10) and market structure that make it difficult to quantify and capture all of their value streams across the electric grid. Aggregated benefits are not accounted for, cost recovery is complicated by the regulatory vacuum in terms of how to categorize energy storage as an asset, (i.e., as transmission, distribution, generation, or load), and there is a lack of communication to electric utilities that energy storage is more economical than gas-fired peakers. These barriers result in underinvestment in energy storage despite its social and economic benefits [15].

With the exception of pumped hydropower and perhaps CAES, the other energy storage technologies are expensive options. As a result, they are not widely used on a large-scale commercial basis for long-duration applications, which require several hours of power output at the storage device's rated power capacity. For arbitrage and load-following applications, the target capital cost for commercialization is $1,500 per kW or $500 per kWh, with an operations and maintenance cost of $250–$500 per MWh for a discharge duration of 2 to 6 hours [22]. These requirements mean that costs need to be lowered for technologies such as lithium-ion batteries, electrochemical capacitors, and advanced flywheels for grid-scale applications. Placement flexibility could be important for the economics of energy storage given

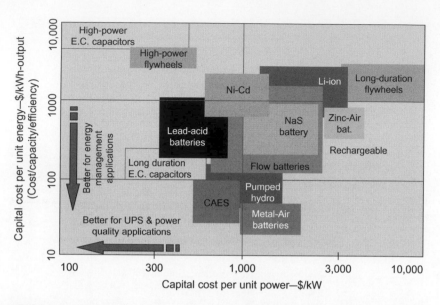

FIGURE 5.10 Capital cost of various energy storage technologies. *(Source: Electricity Storage Association, http://www.electricitystorage.org/ESA/technologies/)*

that electrochemical storage devices are not constrained to a specific geographic topology and hydrological system, unlike CAES and pumped hydropower systems.

With the growing contributions of intermittent energy resources across the United States, load-balancing requirements are expected to grow. A Pacific Northwest National Laboratory (PNNL) study has estimated the balancing requirements for the 2019 timeframe under a 14.4 GW wind scenario in the Northwest Power Pool (NWPP). This study examined various scenarios for meeting balancing requirements using an array of technologies, including sodium-sulfur and lithium-ion batteries, combustion turbines, demand response, and pumped hydropower. The main insights were that sodium-sulfur was the least costly option whereas pumped hydropower was the most costly option, and that storage should be able to accommodate ~25% of projected 2019 wind generation for the NWPP.

These results indicate that energy storage, and particularly electrochemical storage, technologies can compete with conventional combustion turbines when used to meet specific load-balancing requirements with high ramp rate requirements. This finding has general applicability beyond the investigated NWPP footprint [23].

Energy arbitrage opportunities, however, may not be the key driver for large deployment of energy storage, at least not in the near term, that is, 2010–2019. Results from a Sandia National Laboratories (SNL) analysis of the Pennsylvania, New Jersey, and Maryland (PJM) region indicated that arbitrage benefits for

Benefit type	Avg. gross benefit ($/kW-year)	Benefit capture (% of gross)	Net benefit ($/kW-year)
Electric supply			
Electric energy time-shift	77	50%	39
Electric supply capacity	75	100%	75
Ancillary services			
Load following	112		
Area regulation	195		
Electric supply reserve capacity	20	75%	15
Voltage support	56	50%	28
Grid operations			
Transmission support	27		
Transmission congestion relief	12	75%	9
T&D upgrade deferral, 50th percentile*	584*	100%	584*
T&D upgrade deferral, 70th percentile*	752*	100%	752*
T&D upgrade deferral, 90th percentile*	919*	100%	919*
Reliability (15 min. −1 hour)	93		
Power quality (10 seconds)	93	50%	47

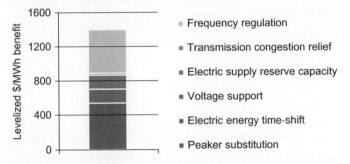

FIGURE 5.11 System benefits of energy storage according to application. *(Source: Eyer et al. [20] (SAND2010-0815))*

10 years of storage operation are on the order of $300/kW, whereas single-year T&D capacity upgrade deferrals are worth as much as $1000/kW of storage installed [24]. These numbers are consistent with those from a more recent SNL study covering the entire United States, summarized in Figure 5.11. These benefits appear to be additive in the case of application synergies, such as storage used for capacity firming, voltage support, and arbitrage [20]. Aggregating energy storage benefits will make a stronger case for their widespread deployment by increasing the benefit-to-cost ratio.

Figure 5.12 provides a perspective of the level of maturity based on installed capacity of grid-tied storage globally. These numbers suggest that there is significant room for cost and performance improvements of the less mature technologies such as compressed air and batteries, while pumped hydropower, due its maturity, is not likely achieve cost reduction—at least at the same rate as the nascent battery technologies. Research and development on energy storage systems, specifically batteries, are expected to lower their costs.

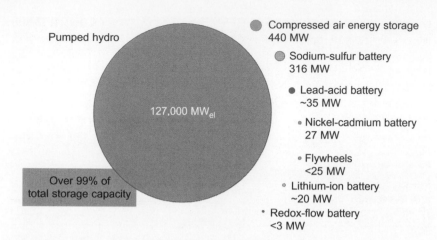

FIGURE 5.12   Worldwide installed storage capacity for electrical energy. *(Source: Electric Energy Storage Technology Options: A White Paper Primer on Applications, Costs, and Benefits. EPRI, Palo Alto, CA, 2010. 1020676. © 2010 Electric Power Research Institute, Inc. All rights reserved)*

R&D Directions

Successful development of renewable energy storage systems will require comprehensive systems analysis, including economic and operational benefits and system reliability modeling. Systems should be analyzed based on the requirements of the application. The analysis should include an investigation of all of the possible storage technologies suitable for the application and the operational/cost/benefits tradeoffs of each. R&D is needed to quantify the value of energy storage to the grid depending on the application, and economic viability needs to be assessed by comparing value to the lifecycle cost. Models for analysis of energy storage value as a function of time, location, market, solar profile, etc, need to be developed. Software-based modeling and simulation tools represent a key component of successful systems analysis. For instance, the Regional Energy Deployment System (ReEDS) model[7] developed at NREL was used to quantify the value that storage can add to wind under the 20% penetration scenario in the "20% Wind by 2030" report.[8] ReEDS integrates storage technologies such as CAES, pumped hydropower, and batteries with renewable generation such as wind and solar.

Besides systems analysis, control algorithms to optimize application of energy storage and enable real-time dispatch need to be developed. For electrochemical storage, advanced battery management systems can be developed to address some of the charge/discharge issues. The U.S. Coast Guard is sponsoring an effort to develop the Symons Advanced Battery Management System

---

[7]http://www.nrel.gov/analysis/reeds/.
[8]http://www.20percentwind.org/.

(ABMAS) for off-grid, PV-storage-generator hybrid systems. Initial results using the ABMAS system show a 25% reduction in fuel use and improved battery charging and discharging profiles, thus promising increased battery lifetime.[9] Similar management systems are needed for grid-connected PV-storage systems and applications.

By themselves, energy storage devices (batteries, flywheels, etc.) do not discharge power with a 60-Hz AC waveform, nor can they be charged with 60-Hz AC power. Instead, a power conditioning system is necessary to convert the output. Under the SEGIS initiative, the DOE Solar Energy Program is currently developing integrated power conditioning systems for PV systems. These systems include inverters, energy management systems, control systems, and provisions for including energy storage. It is anticipated that charging and discharging control algorithms for different battery technologies will be included in the SEGIS control package.

The main R&D needs for battery technologies address the following aspects of their use:

- Increasing power and energy densities;
- Extending calendar- and cycle-life;
- Increasing efficiency;
- Increasing reliability;
- Ensuring safe operation; and
- Reducing costs.

FEDERALLY FUNDED ENERGY STORAGE EFFORTS

Several programs at the Department of Energy are funding energy storage research for both grid-scale and transportation applications. Federal support for stationary energy storage mainly stems from the Office of Electricity (OE), which funds projects to improve basic materials for battery, electrolytic capacitor, and flywheel systems to reduce cost and enhance capabilities, improve the modeling capabilities of compressed-air energy storage, and develop advanced components and field-test storage systems in diverse applications.

The Renewable and Distributed Systems Integration (RDSI) program within the OE focuses on integrating renewable energy, distributed generation, energy storage, thermally activated technologies, and demand response into the electric distribution and transmission system. This integration is aimed toward managing peak loads, offering new value-added services such as differentiated power quality to meet individual user needs, and enhancing asset use. The program goal is to demonstrate a 20% reduction in peak load demand by 2015, through increased use of both utility- and customer-owned assets.

---

[9]Corey, G. "Optimizing Off-grid Hybrid Generation Systems." *EESAT 2005*, Conference Proceedings.

The American Recovery and Reinvestment Act (ARRA) allocated $185 M for deploying and demonstrating the effectiveness of utility-scale grid storage systems. The goal is provide a ten-fold increase in energy storage capacity to improve grid reliability and facilitate the adoption of variable and renewable generation resources. Three projects on large battery systems (total 53 MW) will address the variable nature of wind energy and aid in the integration of wind generation into the electric supply. The additional projects include two CAES (450 MW), one frequency regulation (20 MW), five distributed projects (9 MW), and five technology development projects.

Other ARRA-funded energy storage projects have been awarded by the Advanced Research Projects Agency–Energy (ARPA-E), which funds high-risk, translational research driven by the potential for significant commercial impact in the near-term. The funded projects under the "Grid-Scale, Rampable, Intermittent Dispatchable Storage (GRIDS)" topic area include nine battery (e.g., sodium-beta, liquid metal, flow batteries, metal-air), one SMES, two fly-wheel, one CAES, and one fuel cell.

The EERE also funds stationary energy storage projects through the SETP and Wind and Water Power Program, as discussed earlier in this chapter. SEGIS projects range from optimizing interconnections across the full range of emerging PV module technologies to lowering manufacturing costs through integrated controls for energy storage and development of new inverter designs. In the area of sustainable pumped storage hydropower, the DOE intends to provide $11.8 million in funding toward projects that begin construction by 2014 and integrate wind and/or solar.

The Office of Science also supports energy storage R&D through its Basic Energy Sciences (BES) program. The core program conducts fundamental research to understand the underlying science of materials and chemistry issues related to electrical energy storage. BES will be initiating a Batteries and Energy Storage Hub in FY 2011 with a planned funding amount of $35 million. This particular Energy Innovation Hub will address specific areas of research that were identified in the BES workshop report titled "Basic Research Needs for Electrical Energy Storage" that include efficacy of materials architectures and structure in energy storage, charge transfer and transport, electrolytes, multi-scale modeling, and probes of energy storage chemistry and physics at all time and length scales. Fundamental research on electrochemical storage technologies is also funded through Energy Frontier Research Centers across the United States.

CONCLUSIONS

As the fastest growing renewable energy sources worldwide, solar PV and wind power are gaining a stronghold in the electric grids of the United States and the world. The issues surrounding their intermittency need to be addressed so that this growth can be sustained, especially in the context of integration with smart grids that are being planned and deployed. Variations in energy output at 20–30% penetration levels of renewables may cause reliability problems in the electric grid, such as voltage fluctuations, and require a significant amount of reserves for capacity firming. The challenge of renewable resource intermittency can be met using a variety of energy storage technologies in lieu of conventional generators. The energy storage capacity for renewable generation varies from kW to hundreds of MW, depending on whether it will be used for power or energy management, (e.g., frequency regulation or capacity firming).

The optimal technologies for addressing renewable energy integration will be application-specific and will scale with the size of variable generation, ranging from pumped hydropower and CAES for centralized, bulk storage at PV plants and wind farms to batteries and electric vehicles for distributed storage near rooftop PV installations. The increasing amount of distributed renewables has triggered an evolution from centralized to distributed storage. Smart grid deployment is facilitating this transition since integration of distributed storage requires more intelligent control, advanced power electronics, and two-way communication. Both central and distributed energy storage are required for source-load matching in a smart grid with high levels of renewable penetration. A cost-benefit analysis needs to be done to determine whether certain storage applications should address the supply or demand side.

Key benefits of energy storage include providing balancing services (e.g., regulation and load following), which enable the widespread integration of renewable energy; supplying power during brief disturbances to reduce outages and the financial losses that accompany them; and serving as substitutes for transmission and distribution upgrades to defer or eliminate them. A smart grid is needed to maximize benefits from load shifting and ancillary services. To maximize these benefits and minimize costs, each energy storage technology needs to be optimized for certain applications. In particular, battery technologies are promising due to their wide range of chemistries and operating conditions for providing services that cover several applications. Over \$380 million in federal funds are supporting energy storage R&D in the United States to lower their cost and improve performance.

REFERENCES

[1] International Energy Agency, *Clean Energy Progress Report*, 2011. http://www.iea.org/papers/2011/CEM_Progress_Report.pdf (accessed 04.04.11).

[2] US DOE, Office of Energy Efficiency and Renewable Energy, *Database for State Incentives for renewables & Efficiency (DSIRE)*, http://www.dsireusa.org/documents/summarymaps/RPS_map.pptx, May 2011 (accessed 15.04.11).

[3] US DOE, Office of Energy Efficiency and Renewable Energy, *State-Specific Utility Green Pricing Programs*, http://apps3.eere.energy.gov/greenpower/markets/pricing.shtml?page=1, August 2010.

[4] American Wind Energy Association, *US wind industry annual market report: Year ending 2009*, 2010.

[5] Solar Energy Industries Association, *US Solar Industry: Year in Review 2009*, April 15, 2010.

[6] R. Sioshansi, P. Denholm, The value of concentrating solar power and thermal energy storage, *IEEE Trans. Sustain. Energy* 1 (2010) 173–183.

[7] J.L. Contreras, L. Frantzis, S. Blazewicz, D. Pinault, H. Sawyer, *Photovoltaics Value Analysis*, Prepared by Navigant Consulting Inc. for the National Renewable Energy Laboratory, NREL/SR-581-42303, February 2008.

[8] P. Denholm, R. Margolis, J. Milford, *Production cost modeling for high levels of photovoltaics penetration*, National Renewable Energy Laboratory, NREL/TP-581-42305, February 2008.

[9] American Physical Society, *Integrating renewable electricity on the grid*, 2010 (accessed 15.04.11). http://www.aps.org/policy/reports/popa-reports/upload/integratingelec.pdf

[10] US DOE, Office of Electricity Delivery & Energy Reliability, *Smart Grid Research & Development Multi-Year Program Plan (MYPP)*: 2010–2014, http://www.oe.energy.gov/Documentsand Media/SG_MYPP.pdf, 2010 (accessed 15.04.11).

[11] US Energy Information Administration, *Electric Power Annual 2009*, DOE/EIA-0348, April 2011.

[12] J. Paidipati, L. Frantzis, H. Sawyer, A. Kurrasch, *Rooftop Photovoltaics Market Penetration Scenarios*, Prepared by Navigant Consulting Inc. for the National Renewable Energy Laboratory, NREL/SR-581-42306, February 2008.

[13] J. Bebic, *Power System Planning: Emerging Practices Suitable for Evaluating the Impact of High-Penetration Photovoltaics*, Prepared by GE Global Research for the National Renewable Energy Laboratory, NREL/SR-581-42297, February 2008.

[14] A. Mills, M. Ahlstrom, M. Brower, A. Ellis, R. George, T. Hoff, et al., Dark shadows: understanding variability and uncertainty of photovoltaics for integration with the electric power system, *IEEE Power & Energy Mag.* 9 (3) (2011) 33–41.

[15] California ISO, *Integration of renewable resources*, November 2007.

[16] New York State Energy Research and Development Authority, *The Effects of Integrating Wind Power on Transmission System Planning, Reliability, and Operations*, Prepared by GE Energy for NYSERDA, March 2005.

[17] C. Whitaker, J. Newmiller, M. Ropp, B. Norris, *Distributed photovoltaic systems design and technology requirements*, Sandia National Laboratories, SAND2008-0946 P, February 2008.

[18] California ISO, *Integration of renewable resources – Operational requirements and generation fleet capability at 20% RPS*, August 2010.

[19] J. Lin, G. Damato, *Energy Storage – A Cheaper and Cleaner Alternative to Natural Gas-Fired Peaker Plants*, Prepared by Strategen for California Energy Storage Alliance, 2011.

[20] J. Eyer, G. Corey, *Energy storage for the electricity grid: Benefits and market potential assessment guide*, Sandia National Laboratories, SAND2010-0815, February 2010.

[21] D. Manz, O. Schelenz, R. Chandra, S. Bose, M. de Rooij, J. Bebic, *Enhanced Reliability of Photovoltaic Systems with Energy Storage and Controls*, Prepared by GE Global Research for the National Renewable Energy Laboratory, NREL/SR-581-42299, February 2008.

[22] US DOE, Office of Electricity Delivery and Energy Reliability, and Advanced Research Projects Agency—Energy, *Advanced Materials and Devices for Stationary Electrical Energy Storage Applications*, Prepared by Nexight Group, December 2010.

[23] M. Kintner-Meyer, M. Elizondo, P. Balducci, V. Viswanathan, C. Jin, X. Guo, et al., *Energy storage for power systems applications: A regional assessment for the northwest power pool (NWPP)*, Pacific Northwest National Laboratory. PNNL-19300, April 2010.

[24] J. Iannucci, J. Eyer, B. Erdman, *Innovative applications of energy storage in a restructured electricity marketplace, Phase III Final Report*, Sandia National Laboratories, SAND2003-2546, March 2005.

Chapter 6

The Smart Grid Vision and Roadmap for California

Heather Sanders, Lorenzo Kristov, and Mark A. Rothleder

INTRODUCTION

Two parallel forces have been converging in recent years and are now driving major changes in the power industry. The first of these two forces is the growing demand for cleaner sources of energy, now manifest in ambitious renewable and greenhouse gas policy mandates and the increasing adoption of distributed renewable generation. The second force is the accelerated emergence of new technologies that are capable of transforming all aspects of electricity production, transportation, and consumption. A major contributor to this convergence of forces is the broad array of emerging technologies and capabilities referred to as *smart grid*, which promises to be a key enabler of momentous industry advances.

For several years now the California ISO (CAISO) has recognized that the changes underway require a thorough inquiry into all areas of our core responsibilities, including reliable grid operation, efficient spot markets, open-access transmission service, grid planning, new generator interconnection, and integration of large amounts of renewable resources into the supply fleet—to determine how our rules and practices should be revised to align with and support the broader evolution of the industry. In this context the CAISO views its role as twofold.

- First, the CAISO is the entity responsible for operating a reliable grid and efficient spot markets, and as such we must continue performing these functions at the highest level of excellence throughout the coming changes.
- Second, the CAISO is an engaged participant in and facilitator of these industry changes.

By virtue of our unique position as the grid and market operator and the planner of new transmission infrastructure (see boxes below), the CAISO is situated at the nexus between, on the one hand, the panoply of emerging technologies and their wide-ranging potential applications, and on the other hand, the detailed, practical needs of operating the grid, planning grid expansion, connecting new resources, and enabling their participation in the spot markets under an environmental policy program that is dramatically altering the makeup of the supply fleet. Thus, the CAISO is in a central position to identify the most promising linkages between capabilities and needs, and to facilitate their implementation through changes to our rules and practices.

California ISO by the Numbers

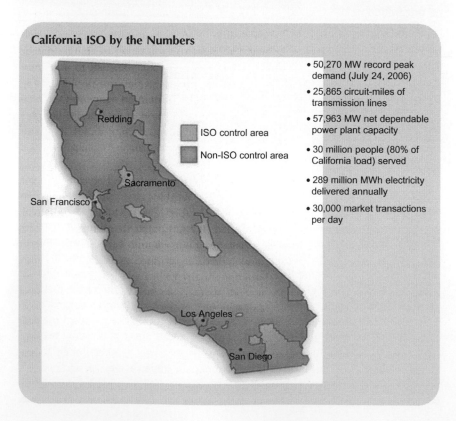

ISO control area

Non-ISO control area

- 50,270 MW record peak demand (July 24, 2006)
- 25,865 circuit-miles of transmission lines
- 57,963 MW net dependable power plant capacity
- 30 million people (80% of California load) served
- 289 million MWh electricity delivered annually
- 30,000 market transactions per day

The California ISO began operation in 1998 in conjunction with California's restructuring of its power sector to enable wholesale and retail competition. The CAISO is responsible for the reliable operation of the transmission grid under its control, and operates day-ahead and real-time spot markets for energy, ancillary services, and open-access transmission service, as well as year-ahead and monthly markets for financial transmission rights. In 2009 the CAISO implemented a comprehensive redesign of its market structure, in which it adopted locational marginal pricing of energy at approximately 3000 pricing nodes where energy transactions occur.

Resource reciprocity

Welcome to the Western Grid

- CA is one of 14 states within Western Electricity Coordinating Council

- Resource sharing enhances reliability, helps achieve renewable targets, and manages cost

- A quarter of all the electricity that keeps the lights on during the summer comes from other parts of the West, including parts of Canada and Mexico

Source: CAISO, 2011.

Although the subject of this chapter is the CAISO's vision and roadmap for smart grid, the reader should view this as part of a much broader strategic framework for navigating and shaping industry evolution over the next decade. In order to best use our unique position to facilitate and influence the coming changes for the benefit of California consumers and power industry participants, for the past few years the CAISO has been engaged in strategic visioning and action planning efforts both within our organization and in collaboration with the state agencies and policy makers that have energy-related responsibilities.

Pursuant to our own 10-year strategic framework, the CAISO has already made major revisions to our transmission planning and new generator interconnection processes to better align with the environmental policy and technology drivers of change.[1] In addition, we have a current initiative with our stakeholders to design spot market changes to address the many operational and commercial challenges related to renewable integration and the participation of new technologies. We have also performed several innovative studies to quantify the impacts on grid operation and market outcomes of the increased participation of variable renewable resources. In 2010 we completed a study based on grid conditions with 20% renewable energy,[2] and are nearing completion of a study focusing on 33% renewable energy by the year 2020, which is now a state legislative mandate. The CAISO is also working with participating transmission owners and the Western Electricity Coordinating Council (WECC)

---

[1] Documentation of the CAISO's revised transmission planning process can be found at http://www.caiso.com/242a/242abe1517440.html. Documentation of the CAISO's 2010 enhancements to its generation interconnection procedures can be found at http://www.caiso.com/275e/275ed48c685e0.html.

[2] The CAISO's 20% integration study is posted at http://www.caiso.com/23bb/23bbc01d7bd0.html.

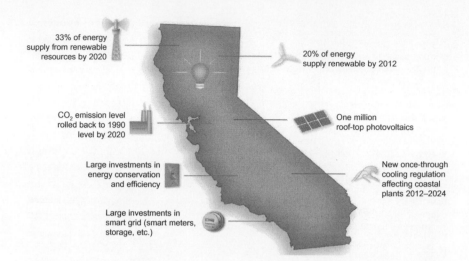

FIGURE 6.1   Key smart grid drivers for CAISO. *(Source: California ISO, Smart Grid Roadmap and Architecture, December 2010)*

on installing advanced monitoring devices called synchrophasors that will greatly improve visibility to the status of the CAISO grid and the entire western interconnection. Most notably, at the end of 2010 the CAISO moved into a new home, custom-built for the CAISO to meet the highest green building standards and housing the country's most advanced control center for monitoring and reliably operating the grid with a renewable-rich supply fleet.

As further described in the CAISO's 2010 *Smart Grid Roadmap and Architecture*,[3] the CAISO envisions California's transmission grid in the year 2020 to be brimming with efficient, clean wind and solar energy that responds to grid operator instructions and dependably contributes to system reliability. The CAISO is working closely with policy makers and industry participants to foster the development of the smart devices, software systems, and market rules and policies needed for grid evolution. This surge of innovation is driven by California's energy and environmental policy goals, as highlighted in Figure 6.1, which include a legislative mandate to procure 33% of the state's retail electricity needs from renewable sources by 2020, promoting energy efficiency, increasing levels of distributed generation, and reducing greenhouse gas emission levels to 1990 levels.

Among its main features, the 2020 grid and spot markets will enable full participation by various types of storage technologies. These devices will store or discharge energy at appropriate times, thereby firming up the variability of renewable resources and shifting renewable energy supply to more useful time periods. Storage could, if developed as hoped, supply ancillary services

---

[3]http://www.caiso.com/2860/2860b3d3db00.pdf.

products such as regulation, which is critical for maintaining system frequency within very narrow limits, perhaps even more effectively than conventional resources. Another feature of the transmission grid and spot markets made possible by smart grid technologies is the everyday use of demand response and broad participation of price-responsive demand, a topic discussed in several other chapters in this volume. Smart technologies are expected to empower residential and commercial consumers by providing them the timely information they need to manage their energy use, for example by shifting it to times when supply conditions and prices are most favorable. The CAISO continues to be a strong advocate of consumer empowerment through smart devices, because as the grid operator we fully understand that delaying the implementation of these technologies will likely result in further reliance on conventional fossil-fuel generation to balance renewable variable generation, which would be contrary to the goal of diversifying our generation fuels and increasing the renewable energy share of consumption.[4]

California's modern grid will leverage existing technologies, such as synchrophasors, to perform at its peak capabilities. Up until now, synchrophasor data had been used for offline analysis, but in the smart grid it will be used for near real-time on-line monitoring and possibly for control. New technologies including smart meters and smart substations will help the local distribution systems, owned and operated by utilities, to match the sophistication of the high-voltage transmission system. These specialized devices could communicate demand levels, output from distributed generation, and system conditions that the CAISO will need to monitor to manage a grid that is more complex than ever before. The result will be a thriving electricity sector that is competitive and efficient—all to the benefit of our wholesale customers and ultimately to retail consumers.

The CAISO is also actively pursuing initiatives that will determine system impacts and needs under different levels of renewable resources and different mixes of resource types, as well as changes to load levels and patterns that will likely occur as Californians purchase substantial numbers of hybrid and all-electric vehicles. A recently published CAISO report on integrating renewable resources provides operational requirements and generation fleet capability under a 20% renewables portfolio mix,[5] while forthcoming studies will characterize system conditions and operating requirements under a 33% renewables energy standard.

There is no question that the rapid pace of new technology development and the increasing impact of environmental policies will change the power industry in ways that cannot be fully predicted at this time. The direction of change is

---

[4]Documentation of several CAISO activities in the area of demand response can be found at http://www.caiso.com/1893/1893e350393b0.html.

[5]California ISO, *Integration of Renewable Resources—Operational Requirements and Generation Fleet Capability at 20% RPS*, August 31, 2010, available at http://www.caiso.com/2804/2804d036401f0.pdf.

quite clear, however, as are many of the major challenges, so the CAISO will continue to develop and adopt the best new applications and devices that will enable our operators to better monitor and manage the real-time grid, both within our own balancing area and in coordination with our neighbors and the entire western interconnection. Smart grid infrastructure will be at the heart of these innovations.

In the following sections, we first describe the operational challenges associated with integrating large quantities of variable renewable generation into the grid, which the CAISO has identified and begun to assess quantitatively through its integration studies. Smart grid technologies will play a significant role in enabling grid operators to maintain reliability in the face of these challenges. We then identify the objectives and the major domains of activity of the CAISO's smart grid program, including advanced forecasting and grid monitoring, demand response, distributed energy resources, storage devices, and cyber security. We then provide more in-depth sections on each of these areas. At the end we provide a key to frequently used smart grid and related acronyms.

OPERATIONAL CHALLENGES AND MARKET IMPACTS OF RENEWABLE INTEGRATION

On April 12, 2011, Governor Jerry Brown signed legislation adopting a target of 33% renewable energy by 2020, the most aggressive renewables portfolio standard in the United States. In doing so Governor Brown clarified that, "While reaching a 33% renewables portfolio standard will be an important milestone, it is really just a starting point—a floor, not a ceiling. With the amount of renewable resources coming on-line, and prices dropping, I think 40%, at reasonable cost, is well within our grasp in the near future." Well before this new legislation was adopted, California had legislation mandating a 20% renewables standard, and an executive order by former Governor Arnold Schwarzenegger mandating 33% renewables by 2020. As a result, the CAISO had begun studying the implications of and preparing for these targets for several years before the new legislation was signed.

The operational challenge with variable renewable resources such as wind and solar resources is to maintain constant balance between supply and demand, given the inherent variability and unpredictability of wind and solar output.[6] The expected expansion of these types of capacity on the grid requires the industry to review and revise its operational tools and practices for maintaining

---

[6]Variability in this context refers to the propensity of the output of variable renewable resources to change dramatically within intervals of minutes or even seconds, whereas uncertainty refers to the distribution of forecast errors associated with predicting average renewable output over a specific future time interval, even when such forecasting is done as little as five to ten minutes ahead of the target interval. From the operational perspective, variability affects the need for regulation service, whereas uncertainty requires the CAISO to ensure that sufficient dispatchable capacity is available to respond to five-minute dispatch instructions.

system energy balance, and to develop new market products and market rules for procuring, compensating, and allocating the costs of the required balancing services. This section describes some of the technical challenges the CAISO has been assessing through its renewable integration studies, and identifies the promising approaches for meeting these challenges, focusing particularly on the potential benefits of smart grid technologies.

In addition to operating its own balancing authority areas, the CAISO is also electrically interconnected and synchronized to other balancing authority areas in the Western Electricity Coordinating Council or WECC. Therefore the CAISO, as a balancing authority area that relies heavily on import and export flows, must maintain its net interchange with each of its neighbors as part of maintaining energy balance within its own system.

The net interchange is the amount of scheduled net imports from and exports to neighboring balancing authority areas.

- When the CAISO is a net importer, it is generating less energy from resources within its area than its demand requires and making up the difference from other balancing authority areas.
- When the CAISO is a net exporter, it is generating more than its demand requires, and its extra energy is meeting the demand of other balancing authority areas that are importing from the CAISO.

If the CAISO is not able to balance its system due to the variability of its supply resources, it will cause the frequency of the interconnection to fall or rise and will result in inadvertent transfers from other balancing authority areas in the interconnection as others attempt to return to balance and maintain frequency. Leaning on one's neighbors in this way also creates energy accounting issues between balancing authority areas.

Not maintaining balance under normal conditions causes devices that are synchronized to the frequency of the system to operate slower or faster, depending on whether there is excess demand or excess supply, respectively. In more extreme situations, if the system slows down too much, there is a risk of load shedding, whereas if the system frequency is too fast there is risk of damage to generation equipment and tripping of generation. Figure 6.2 illustrates using a simple water flow analogy the balancing of generation and load.

The mechanisms for maintaining system balance are best understood in terms of three time frames for the system to respond to disturbances or imbalances. The first and most critical time frame is what occurs in the first 30 seconds after a disturbance in which a large amount of generation is removed. In this time period inertia and governor response are the primary frequency control mechanisms deployed to arrest frequency deviation. The second time period is what happens between 30 seconds to ten minutes. In this time frame secondary frequency controls such as regulation are deployed to control frequency. For the third period, starting at five minutes, economic dispatch of resources through the five-minute real-time market provides the tertiary control mechanism

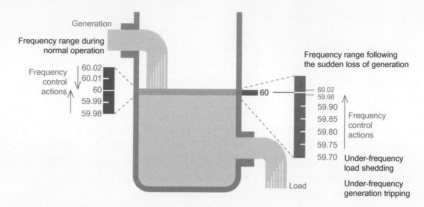

FIGURE 6.2   Ranges of power system frequency during normal operations and following the sudden loss of generation. *(Source: Use of Frequency Response Metrics to Assess the Planning and Operating Requirements for Reliable Integration of Variable Renewable Generation, December 2010, Ernest Orlando Lawrence Berkeley National Laboratory, LBNL-4142E)*

deployed to maintain balance between supply and demand based on forecasted conditions.

While maintaining balance under normal conditions is a challenge, maintaining balance after a disturbance or a contingency in which a large resource or group of resources trip off-line simultaneously can have more far reaching reliability impacts. Figure 6.3 illustrates how primary, secondary, and tertiary frequency controls act to restore frequency to its normal operating level after a disturbance. After a disturbance, primary frequency response measures are designed to deploy and automatically arrest frequency. While frequency is decaying, demand actually increases as motor load increases as an inverse function of frequency. Primary frequency controls such as inertia and governor control will automatically activate to first arrest frequency decay and then recover frequency. Inertia is the result of large spinning masses continuing to spin, creating a resistance to the system slowing down due to the imbalance. Thus inertia becomes a shock absorber or resistance to change in frequency after a disturbance.

Governor control on the other hand is an independent control mechanism designed into generators that automatically increases the output of the generators as the frequency begins to drop. The more frequency drops, the more the generator increases its output up to the resource's maximum output capability or fuel supply availability. For example, a steam turbine whose governor responds due to a frequency disturbance will open up the steam valves and release steam into the turbines, increasing the generator's output. If the steam pressure in the turbine is not built up sufficiently, however, the steam output may only be able to increase the generator's output for a limited amount of energy or for a limited time. Moreover, not all generators have governors, and in some cases generators with governors may have their governors blocked and will not respond to a

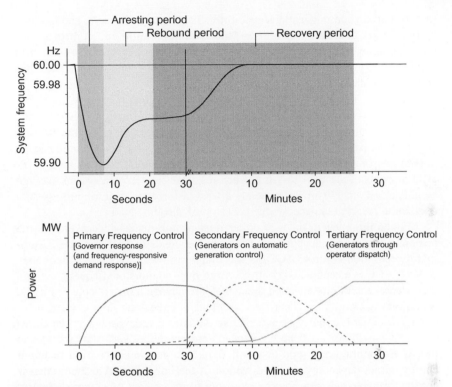

FIGURE 6.3 The sequential actions of primary, secondary, and tertiary frequency controls following the loss of generation, and their impacts on system frequency. *(Source: Use of Frequency Response Metrics to Assess the Planning and Operating Requirements for Reliable Integration of Variable Renewable Generation, December 2010, Ernest Orlando Lawrence Berkeley National Laboratory, LBNL-4142E)*

frequency drop. This is where smart grid technologies such as synchrophasors offer tremendous promise for improving the grid operator's visibility to over-stressed operating conditions and thereby enabling the operators to take timely corrective actions.

The challenge of arresting frequency decay after a disturbance event increases with high penetration of renewable resources such as wind and solar photovoltaic, as these resource types are unable to provide frequency control services. In the case of wind, there is generally little inertia response resulting from the spinning generators and no governor response to increase output in response to a disturbance event. In the case of solar photovoltaic there is no spinning mass providing inertia and no governor response. Any response would be from the inverter converting the DC power from the solar panels to the AC current of the grid, and currently inverters are not programmed to provide any response. As technologies develop it may be possible for inverters to assist with frequency response, but this will require implementation of smart grid monitoring and communication capabilities. To the extent that solar photovoltaic resources are distributed resources, the communication and control issues are even more complicated. In the case of solar thermal there is some inertia and potentially some governor response as well, since a solar thermal plant has converted solar energy into steam that turns a steam turbine.

The above discussion explains a primary operational concern with higher penetration levels of variable energy resources such as wind and solar photovoltaic. To the extent these resources displace conventional resources on the transmission system, there can be a significant loss of the inertia and governor response needed to recover from frequency disturbance events. If the system does not have sufficient amounts of these primary frequency response services, the system may not be able to adequately arrest frequency after a disturbance, requiring the system operator to resort to the undesirable backstop of shedding load.

Maintaining system balance under normal non-disturbance events is the responsibility of the secondary and tertiary frequency control measures. Automatic generation control (AGC) is the first form of the secondary frequency control. AGC is a centralized control system that monitors the frequency within the balancing authority area and the actual interchange transfers between it and its adjoining neighbors. This measurement is called the area control error (ACE). If either frequency or interchange deviates from expected or scheduled levels ACE will deviate from a balanced zero condition, and AGC will detect the ACE deviation and send dispatch signals to generators to raise or lower energy output depending on the direction of the imbalance. A resource that is certified capable and agrees to be available to respond to AGC signals is said to provide regulation service. The CAISO procures regulation service from certified resources that offer to provide the service on an hourly basis through its day-ahead and real-time market processes. Every four seconds, the AGC system monitors for ACE deviations and sends dispatch signals to those resources providing regulation service.

Tertiary control for balancing the system is performed by committing and dispatching resources over a longer interval to meet the expected forecast of demand, taking into account any forecast deviations of supply. The CAISO performs this balancing function through its real-time economic dispatch market, which can commit short-start and quick-start resources every 15 minutes and issue dispatch instructions to resources every 5 minutes. This tertiary control is also sometimes referred to as real-time balancing or load following. The latter term, though used traditionally to refer to real-time balancing, is no longer really accurate in the context of large amounts of variable renewable resources, however, as the real-time dispatch will actually follow load net of variable resource production.

Figure 6.4 illustrates the relationship between the secondary and tertiary mechanisms—regulation and load following, respectively—in balancing the system. Load following entails the use of five-minute dispatch instructions to make up the difference between the hourly schedule and the average net load forecast for each five-minute interval. Regulation provides the more granular

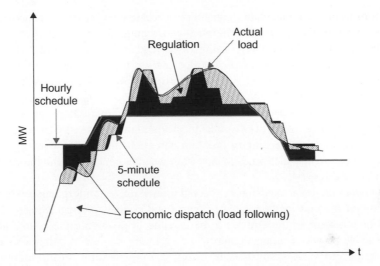

FIGURE 6.4 Regulation and load following. *(Source: California ISO, Integration of Renewable Resources: Operational Requirements and Generation Fleet Capability at 20% RPS, August 31, 2010)*

adjustments necessary to meet the difference between the response of the dispatched resources to the CAISO's instructions and the actual net load as it varies within each five-minute interval.

Increased levels of variable renewable generation raise yet another complication for system balancing. The process described above for arresting the frequency decay when a contingency event occurs—when a generation resource is lost, for example—is supplemented by the dispatch of contingency reserves and the use of emergency ratings of grid facilities, which help the system respond to disturbances. The potentially large swings in the output of intermittent renewable resources are not classified as contingencies, however, so contingency reserves and emergency ratings are not available to the grid operator, even though such swings may be as large as the loss of a generation unit. This means that the system must be able to respond to these changes without using resources designated as contingency reserves or any emergency line ratings, and may even have to respond to a contingency event while it is responding to the intermittency of the renewable generation.

Describing the primary, secondary, and tertiary balancing mechanisms this way helps to convey how the variability and uncertainty of renewable resources can challenge grid operations. Before the variability and uncertainty associated with renewable resources became a concern, supply resources were all highly controllable and predictable, and therefore the largest contributor to real-time imbalance was the variability and uncertainty of the load. With increasing penetration of renewable resources, the supply side of the operational balance is also variable and uncertain, with supply variation offsetting load variation in some instances and adding to it in others in a manner that is not easy to predict.

For example, during the morning when load is naturally increasing, wind production in California is typically decreasing, thereby increasing the need for capacity that the CAISO can dispatch up to follow net load. At the same time, solar production will typically increase as the morning sun rises and can somewhat offset decreasing wind production, although there can be a temporal gap between the fall-off of wind production and the upswing of solar. Obviously, the complementary problem can occur during the evening ramp, the exact nature of which tends to vary with weather conditions and the daylight savings time regime.

Other operational challenges expected to increase with more renewable generation on the grid include voltage control and congestion management. Conventional resources are required to be capable of providing a certain amount of reactive power or voltage control, but currently no such requirements exist for wind and solar resources. In order to impose such requirements the CAISO will need to establish the operational need, which the CAISO is currently assessing through its integration studies. Until such requirements are imposed on interconnecting variable renewable generation, the CAISO expects the operational challenges of maintaining voltage and avoiding cascading voltage collapse conditions to increase.

Congestion occurs when there is insufficient transmission capacity to support total transfer of energy from areas with supply to areas with demand. Although congestion is not a new phenomenon created by renewable resources, it is expected to become more challenging with the increase in variable renewable generation.

- First, the frequency and magnitude of congestion will increase in areas high in wind or solar development potential to the extent renewable resources are allowed to interconnect to the grid prior to the completion of sufficient transmission upgrades to transfer all their energy output to demand.[7]
- Second, the high level of wind output during off-peak hours means that there will often be insufficient dispatchable conventional resources on line to manage congestion.
- Finally, the potential for large rapid swings in output from wind and solar resources in response to changing weather conditions can create situations

---

[7]This highlights a further complication that the CAISO is addressing through its initiatives on transmission planning and generation interconnection. The variable nature of renewable generation means that these resources will have lower transmission utilization rates than conventional generators. A conventional generator with a capacity of 500 MW will likely use a 500 MW capacity transmission line often, whereas a wind or solar generator with a 500 MW capacity will likely use the full 500 MW infrequently. The planning and interconnection processes need different approaches for determining the most cost-effective way to build transmission to wind and solar regions than have been used traditionally for meeting the transmission needs of conventional resources.

where congested areas quickly switch from having too much generation to too little generation on-line. In these cases the grid operators may have to take manual or out-of-market actions to resolve congestion, including curtailment of the renewable resources.

While operationally necessary at times, curtailment of renewable resources is counter to the objective of the renewables portfolio standard to maximize the amount of energy supplied by renewable resources. As described in a number of chapters in this volume, smart grid technologies, particularly in conjunction with storage facilities participating in the CAISO market, will enable more efficient utilization of the grid in achieving the state's renewable energy goals; for example, by storing renewable energy when transmission is congested or there is excess supply and delivering it at times when the energy can be utilized.

These real-time operational challenges have a planning aspect as well, as the CAISO must consider whether the generation fleet will be capable of managing the variability of the new renewable resources in each year of the planning horizon, given the current uncertainty about future retirement, repowering, and construction of new dispatchable resources. State policy has introduced another issue here. California has determined that power plants can no longer rely on a cooling process known as "once-through cooling," which takes water in for cooling purposes and then returns it to the environment. Concern about the environmental damage from this process has led the State Water Resources Control Board to determine that all such power plants must either retire or repower over the next decade; yet many of these resources provide much of the inertia and dispatchable capacity the system relies on for real-time balancing.

The uncertainty about the future availability of existing conventional resources is further complicated by the changing economics of conventional resources. As larger amounts of the state's energy is provided by renewable resources, conventional dispatchable generation will likely see their spot market energy revenues decrease while they experience greater numbers of startups and shutdowns, increased ramping, increased hours of operation at low loading levels, and generally lower capacity factors. Thus the CAISO's integration studies are examining both the magnitudes of operational services such as regulation and load following that the system will need with different levels of renewable penetration, as well as the potential changes in revenue patterns that conventional resources will face due to the spot energy price impacts of large amounts of wind and solar energy.

This information will be used on the one hand to inform investment decisions by developers and procurement decisions by load-serving entities, and on the other hand by the CAISO to develop changes to its market structure to identify and procure sufficient generation services—possibly including new ones—to maintain reliable operation and to compensate the providers of those services in a manner that reflects their value and contributes to their commercial viability. In developing these changes, the CAISO intends to define its market procurement and compensation provisions in a way that is technologically neutral; that is, to define market products and their compensation based on the services needed by the grid, independent of the technology that provides the services, so that the ISO is not in the role of influencing which types of technologies will be successful. This approach will allow competitive forces to provide California with the most efficient generation fleet to meet its needs for both renewable energy and a reliable electricity grid.

To understand the extent of these impacts at increased levels of renewable resources, the CAISO has conducted several analyses, both collaboratively and independently, over the past several years, including a study released in 2007 that focused on the operational and transmission requirements of wind integration.[8] The CAISO's revised 20% integration study released in August 2010 and the 33% integration study currently in progress build on those prior efforts.[9] The purpose of the revised 20% study was to assess the operational impacts of an updated renewable resources portfolio that includes 2,246 MW of solar, and to evaluate in more detail the operational capabilities of the existing generation fleet, as well as changes to their energy market revenues. The study utilized several analytical methods, including a statistical model to evaluate operational requirements, empirical analysis of historical market results and operational capabilities, and production simulation of the full CAISO generation fleet.

The results presented in the 2010 study have significant operational and market implications. From an operational perspective, the CAISO is concerned with the extremes of potential impacts, particularly large, fast ramps that are difficult to forecast and likely to occur more frequently with larger amounts of wind and solar resources on the grid. Thus, key objectives of the simulations we conducted were to estimate the capabilities of the fleet to meet these operational needs and clarify possible changes to market and operational practices to ensure that the system can perform as needed under these extreme conditions. The study identified the maximum values of simulated operating requirements, such as load-following and regulation, by operating hour and by season. In addition, to clarify how more typical daily operations may change, the 2010 study report provided distribution statistics for most of the simulated requirements and capabilities to facilitate both operational and market preparedness.

---

[8]California ISO, *Integration of Renewable Resources—Transmission and Operating Issues and Recommendations for Integrating Renewable Resources on the ISO-Controlled Grid* (Nov. 2007), available at http://www.caiso.com/1ca5/1ca5a7a026270.pdf.
[9]California ISO, *Integration of Renewable Resources—Operational Requirements and Generation Fleet Capability at 20% RPS*, August 31, 2010, available at http://www.caiso.com/2804/2804d036401f0.pdf.

The following graphs illustrate overgeneration conditions during morning hours of a simulated May 28, 2012, date. Figure 6.5 shows the makeup of the supply output compared to system load, and indicates an oversupply condition between 5:00 AM and 8:00 AM. Figure 6.6 shows the exhaustion of downward ramping capability during the same time period. Figure 6.7 then shows the impact of these conditions on the area control error (ACE).

Turning to some preliminary results from the 33% renewables integration study, Figures 6.8 and 6.9 quantify the amounts of upward and downward

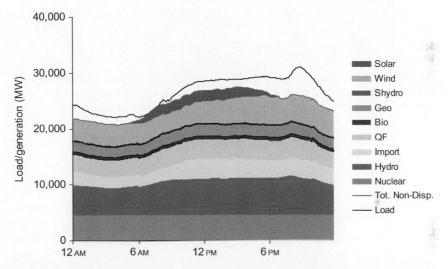

FIGURE 6.5   Simulation of supply output compared to load for May 28, 2012.

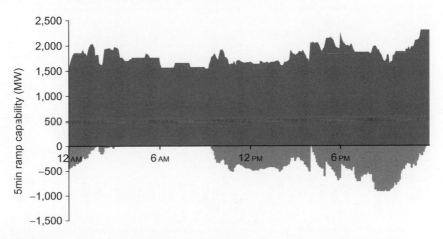

FIGURE 6.6   Simulation of upward and downward ramping capability for May 28, 2012.

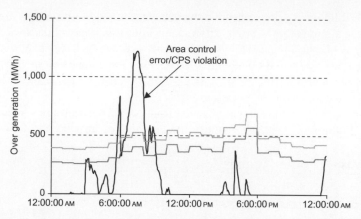

FIGURE 6.7   Detailed over-generation analysis for May 28, 2012. *(Source for Figures 6.5–6.7: California ISO, Integration of Renewable Resources—Operational Requirements and Generation Fleet Capability at 20% RPS, August 31, 2010)*

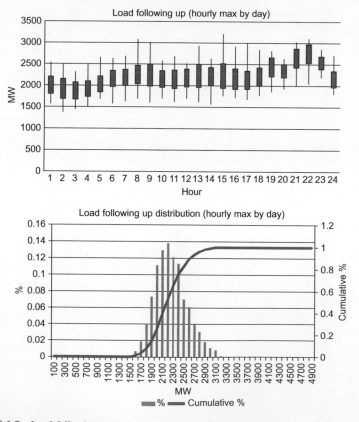

FIGURE 6.8   Load following up requirements at 33% renewable supply.

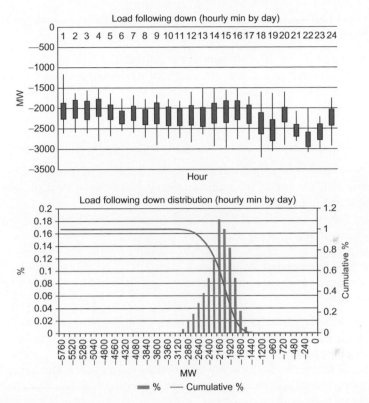

FIGURE 6.9   Load following down requirements at 33% renewable supply. *(Sources for Figures 6.8 and 6.9: California ISO, preliminary results of 33% renewable integration study, not yet published.)*

load-following capability that we expect to need to maintain system balance, based on a projection to the 33% level of the solar and wind resource mix observed today in the renewable energy bilateral procurement patterns of the load-serving entities inside the CAISO balancing authority area.

THE CAISO SMART GRID OBJECTIVES

As further described in other chapters of this volume, the *smart grid* is the application of technologies to all aspects of the energy transmission and delivery system that provide better monitoring and control and more efficient use of the system. The CAISO's goal is to enable and integrate all applicable smart technologies while operating the grid reliably, securely, and efficiently, and to facilitate efficient competitive markets that engage and empower consumers while meeting state environmental and energy policies.

To this end, the CAISO will research, pilot, implement, and integrate smart grid technologies that:

- Increase grid visibility, efficiency, and reliability;
- Enable diverse generation including utility-scale renewable resources, demand response, storage, and smaller-scale solar PV technologies to fully participate in the wholesale market; and
- Provide enhanced physical and cyber security.

The expected benefits from smart grid technology deployments include:

- Ability to recognize grid problems sooner and resolve them proactively;
- More efficient use of the transmission system to defer or displace costly transmission investments;
- Consumers' capability to react to grid conditions, making them active participants in their energy use; and
- Leveraging conventional generation and emerging technologies when possible, including distributed energy resources, price-responsive demand, and energy storage, to address the challenges introduced by variable renewable resources.

The research, pilots, and implementation efforts to modernize the grid will provide the basis for evaluating and understanding new technologies as well as verifying the economics and workforce requirements for deploying them. These efforts will require working closely with CAISO stakeholders. The research and pilot efforts should accomplish a number of important objectives that contribute to smart infrastructure development:

- Provide real-world experience with a new technology;
- Help characterize the technology's benefits;
- Identify what is needed to integrate the technology; and
- Provide the basis for conducting a cost assessment of the technology.

If the industry is to benefit from emerging technologies and the capabilities they support, the efforts must extend beyond the research and pilot stage. It will be important for stakeholders to take information from the research and pilot work to develop business models and policies that bring the technology forward to full commercial implementation.

The California Smart Grid Roadmap[10] is divided into five capability domains, which will guide CAISO activities over the next ten years. Other ISOs will likely be undertaking initiatives in the same domains, though their relative priorities may differ based on their specific policy objectives and the operational challenges they foresee for their own systems. Load-serving utilities will also undertake smart grid activities in domains that include more customer-facing

---

[10]California ISO, *Smart Grid Roadmap and Architecture*, December 2010, available at http://www .caiso.com/2860/2860b3d3db00.pdf.

devices. While each domain area on its own would significantly transform the grid, when combined these capabilities will fundamentally change how the grid will be managed and operated to reliably provide energy where and when it is needed under the smart grid context. The five domain areas listed are discussed in more detail below.

- Advanced Forecasting
- Synchrophasors
- Advanced Applications
- Enabling Demand Response, Storage, and Distributed Energy Resources
- Cyber Security

Advanced Forecasting

Today, regional load forecasting sets the stage for determining what resources are likely to be called upon to supply the necessary energy and energy reserves the CAISO requires to maintain a reliable grid. These forecasts are largely based on what is called a conforming forecast; that is, based on a specific region and actual load pattern, the forecast algorithm will compare actual load history with current weather and geographic data to create a new load forecast used to procure and manage energy supply the day-ahead, hour-ahead, and near real-time market processes.

As the grid evolves, forecasting capabilities will need to improve to address a number of significant grid balancing challenges that will emerge. Energy intermittencies brought about by renewable resources (illustrated in Figure 6.10), incentive-based demand response programs, significant distributed generation, and the proliferation of plug-in electric vehicles will all introduce non-conforming forecasting elements to our current forecasting models and algorithms.

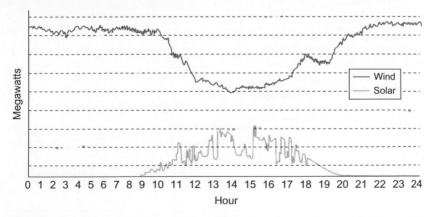

FIGURE 6.10   Illustration of wind and solar generation intermittency. *(Source: CAISO Smart Grid Technology Center)*

The CAISO will need to employ advanced forecasting techniques to produce the most accurate prediction of load, grid conditions, and generation resource status, thereby determining the most reliable and cost-effective scheduling and unit commitment plans.

As more variable energy resources are added to the energy supply mix, the potential negative impact to the grid increases. For example, cloud formation and movement of clouds over solar farms can cause sudden drops and increases in energy output. Changing wind patterns and wind gusts can create conditions of significant, often intra-hour, changes in wind energy output. New weather tracking systems and weather prediction technologies using light waves and sound waves are being studied and deployed in the field to monitor and relay current weather conditions, which can be utilized to update the latest forecasts. New intra-hour prediction tools are being developed to assist grid operations to determine ramping requirements, which can dynamically change based on the percentage of variable resources in use and availability of conventional resources that can be called upon or curtailed when variable resources do not perform as forecasted. Figure 6.11 illustrates how expectations of future forecasts can fall in a broad range. This range—the forecasting confidence band—needs to be as narrow as possible to effectively and efficiently commit resources.

In addition to improving its own forecasting of the output and variability of renewable generation, the CAISO will work to encourage resources to improve their own forecasting. For example, the CAISO's market design will encourage variable resources to improve their forecasting capability in order to best manage their exposure to real-time prices. These individual market participants should generally be in the best position to make accurate forecasts about their own resources.

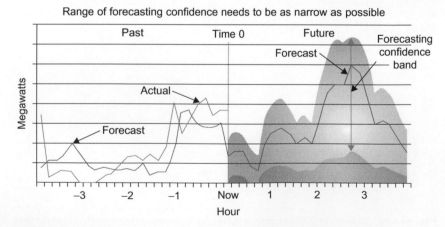

FIGURE 6.11   Forecasting confidence band. *(Source: CAISO Advanced Grid Technology Center)*

The ability to manage energy supply through demand response programs and eventually through price-responsive demand will play a meaningful role in managing peak loads and maintaining system balance by providing consumers with the ability to have greater control over their energy consumption and associated costs.[11] How exactly consumers will interact in the future with demand response programs, dynamic pricing, and emerging home energy management technologies is still uncertain. However, the ability to forecast changes in load due to consumer response to time-of-use pricing or dynamic price signals based on grid conditions will be important for maintaining a reliable just-in-time energy supply.

The amount of rooftop solar photovoltaic and other distributed generation is expected to grow significantly as electricity consumers take advantage of cost reductions and incentives that subsidize the cost of installation. Emphasizing the potential benefits of using distributed resources to efficiently reduce environmental impacts, Governor Brown recently articulated a target of 12,000 MW of distributed generation capacity in California. As these distributed resources supply a larger percentage of the total energy demand, their variability can significantly affect load and CAISO load forecasts unless the CAISO has visibility into the potential generation capability of these resources and can forecast their behavior similar to larger scale wind and solar farms.

The proliferation of plug-in electric vehicles (PEV) has the potential to significantly change energy consumption patterns on the distribution level. Predictions can be made in terms of where and when PEV charging will occur, taking account of the various rate programs adopted by load-serving entities to help control the timing of the charging, and these consumption patterns will need to be incorporated into CAISO forecasting and ramp prediction tools in order to maintain reliability and resource adequacy requirements.

The CAISO smart grid roadmap includes research and implementation of advanced load and generation forecasting technologies and techniques combined with intra-hour ramp prediction tools expected to reduce the associated risks of adding variable and distributed generation resources by reducing forecasting error. Accurate forecasts will lead to more optimal unit commitment that will help account for forecast uncertainties and better use of renewable resources.

The CAISO is collaborating with researchers to investigate the use of sky tracker technology to track cloud movement, which can result in improved solar forecasting in the two-hour to five-minute ahead forecasting interval. Additionally, the use of LiDAR (light detection and ranging) technology that reflects light waves off dust and rain particles in the atmosphere will improve our ability to reduce forecasting errors for solar resources.

---

[11]Documentation of several CAISO activities in the area of demand response can be found at http://www.caiso.com/1893/1893e350393b0.html.

Synchrophasors

Having the ability to monitor grid conditions and receive automated alerts in real time is essential for ensuring reliability. System-wide and synchronized phasor measurement units (PMUs) take sub-second readings that provide an accurate picture of grid conditions. The CAISO's work in this area focuses on obtaining, displaying, and storing synchrophasor data.

Deployment of synchrophasor technology is accelerating under recent U.S. Department of Energy initiatives. Most relevant to the CAISO, the Western Electricity Coordinating Council's *Western Interconnection Synchrophasor Project* (WISP) will almost triple the number of deployed PMUs to over 300. Figure 6.12 shows the currently installed and desired future PMUs in the WECC, many of which are covered in the WISP effort. The project will also develop common software suites that improve situation awareness, system-wide modeling, performance analysis, and wide-area monitoring and controls. Among the challenges related to using synchrophasor technology are that the communications infrastructure lacks the bandwidth to handle the data traffic

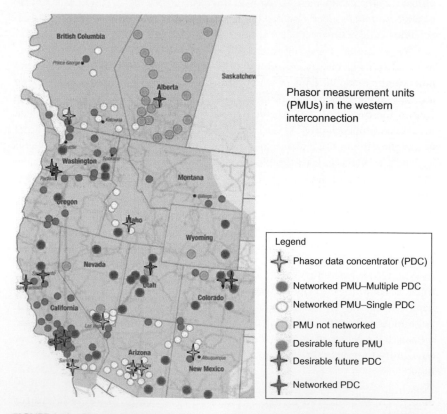

Phasor measurement units
(PMUs) in the western
interconnection

Legend

✈ Phasor data concentrator (PDC)

● Networked PMU–Multiple PDC

○ Networked PMU–Single PDC

◔ PMU not networked

◕ Desirable future PMU

✚ Desirable future PDC

✚ Networked PDC

FIGURE 6.12   Phasor measurement units in the western interconnection. *(Source: www.NAPSI.org)*

produced by the smart devices, needs enhanced security, and must maintain a high degree of reliability if the data are to be used for control decisions. Another major challenge is the lack of available software applications that assimilate and provide grid operators meaningful, understandable visual displays of the extensive data produced by the smart devices.

Phasor units measure voltage and electric current physical characteristics. This data can be used to assess and maintain system stability following a destabilizing event within and outside the CAISO footprint, which includes alerting system operators to take action within seconds of a system event. This capability reduces the likelihood of an event causing widespread grid instability. Moreover, having detailed monitoring data will allow the CAISO and other balancing authorities to identify potential issues before they become actual issues and take steps to proactively resolve them.

Phasor data are also useful in calibrating the models of generation resources, energy storage resources, and system loads for use in transmission planning programs and operations analysis, such as dynamic stability and voltage stability assessment. The technology may have a role in determining dynamic system ratings and allow for more reliable deliveries of energy, especially from remote renewable generation locations to load centers.

The CAISO currently uses phasor data on a real-time basis for basic monitoring and on a post-mortem basis to understand the cause and impact of system disturbances. Data from 57 phasor devices stream at a rate of 30 scans per second collecting more than three gigabytes of data per day. The CAISO has already begun to receive real-time phasor data from some of its neighboring balancing authorities, and by the end of 2011 will be receiving data from additional phasor locations in the Western Electricity Coordinating Council area that will further enhance visibility to grid conditions. Critical to the synchrophasor roadmap is implementing a robust, standards-based communication infrastructure with monitoring and alert capabilities.

Advanced Grid Applications

The CAISO relies on advanced grid applications to monitor grid conditions, recognize possible sources of instability and provide prices and control signals to system resources. This information is used in tandem with economic models to solve reliability problems in the most cost-effective way. These applications need to evolve into more forward-looking and pro-active systems, rather than only reacting to real-time conditions, in order to truly enhance grid operations.

Integrating phasor data as well as other measurements made possible by smart grid technology can enhance a number of applications used today for managing the grid. Advanced applications for monitoring, dynamic (on the fly) assessments of grid conditions, and automated controls are slowly emerging. Because the technology and communication infrastructure for synchrophasors is only now being implemented, developing applications to use this data is lagging. Also, inserting more inputs into modeling algorithms adds significant complexity to an already complicated system.

Increased variable generation on the grid is expected to bring challenges in terms of decreased system inertia, which reduces the margins to maintain stability. Phasor data availability may lead to algorithms to measure this effect in real time and provide needed feedback that can be used to take preventive measures, such as scheduling additional conventional generation or sending signals to flywheels or demand response applications.

For example, if phasor data analysis detects that oscillations in frequency are beginning to develop in an area that produces high amounts of variable renewable generation, the CAISO could step in and dampen those oscillations by quickly curtailing the variable generation and replacing it with generation of higher frequency stability before the oscillations grow to the point of risking collapse. Table 6.1 lists several potential applications for the data collected from synchrophasors.

Increased use of price-responsive demand and distributed resources to manage the grid will require the development of feedback loops providing continuous and automatic adjustments based on the updated measurements. For example, if the CAISO anticipates a potential supply shortage and the prices in the real-time market rise to bring on increased supply and to decrease demand, the CAISO must monitor the responses of the suppliers and the load to ensure that the system is adjusting as expected and, if not, to take further action. In the past, to control the grid the operator or the market dispatch algorithm issued instructions whose results were highly reliable, because they directed dispatchable generators to increase their output or directed load-serving entities to curtail load. In contrast, the controls envisioned with the smart grid are indirect controls: instead of directly curtailing load the CAISO market raises the price of energy, which will signal price-responsive consumers to modify their behavior. Feedback loops must be developed so that the operator is informed of the responses to these indirect control mechanisms and can adjust the controls if the results turn out not to be as expected.

The CAISO has a suite of market and power flow systems and tools that determine the best use of available resources based on economics and reliability. The tools include an energy management system, a modeling system that estimates the status of the statewide grid, system event analysis, voltage assessment, automatic economic unit commitment and dispatch for the real-time and day-ahead markets, a load forecasting tool, and plant outage scheduler (as shown in Figure 6.13). Under development is a voltage stability analysis application that calculates voltages at different locations on the system to determine those near limits and sends alerts to grid operators. Integrating this functionality into the market systems will enable the CAISO to commit units based on the voltage information.

The applications roadmap includes activities to advance monitoring capabilities, the systems and algorithms to determine the best use of the grid, including dynamic thermal line ratings, and automated adaptive generation control that uses response forecasts of demand, storage, and other system

TABLE 6.1 Potential Advanced Applications Utilizing Synchrophasor Data

Application	Data input	What it does	Expected Benefits
Small signal analysis (SSA)	Synchrophasor	Performs oscillation detection, damping computation, and mode identification.	By detecting and identifying low-damping operating conditions, operators can take preventive control actions to increase the system's damping.
Dynamic model validation	Synchrophasor	PMU sub-second resolution data allow operators to obtain the dynamic response of components (gens, loads, renewable resources).	By validating current dynamic models with PMU data, planning and operating engineers will obtain more accurate results when performing dynamic stability and voltage stability studies.
Voltage sensitivity analysis (VSA)	Synchrophasor	Assess the current operating point and power-to-voltage sensitivities at a sub-second resolution.	Incorporated with model-based VSA application it provides operators visibility of current operating point vs. collapse point (unstable conditions).
Phase angle difference dynamic limits (PADDL)	Synchrophasor	Dynamically computes the angle difference limits across pre-defined transmission paths.	Monitor stress across the transmission system.
Event playback	Synchrophasor	Provides the ability to play back events at a sub-second resolution.	Automatically saves event files and allows the user to perform post-disturbance analysis.
State estimator (SE)	Synchrophasor & SCADA, CIM/XML	Estimates the state (voltage magnitudes and angles) and provides results on network topology and flows. These results are used in grid operations and markets.	Provides redundancy of measurements for improved bad data detection and allows for cross-validation between PMU measurements and SE results.
Nomogram validation	Synchrophasor & SCADA, CIM/XML	Better assess the system operating conditions with respect to stability limits, and consequently validate or improve existing nomograms.	Synchrophasor data can provide for less conservative nomograms (operational boundaries).

Source: CAISO Five-year Synchrophasor Vision; not yet published.

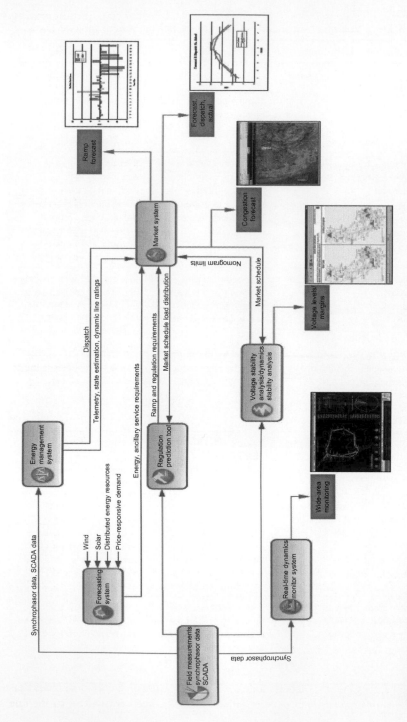

FIGURE 6.13 Advanced applications in use or in development at the CAISO. *(Source: CAISO Advanced Grid Technology Center)*

resources. The roadmap also calls for investigating and implementing automated decision-making and control systems. Of course, unforeseen problems may prevent or delay some technologies coming to market, while technology advancements may bring about new systems and applications that are not even contemplated at this time. These uncertainties contribute to the complexity in upgrading the grid while, at a minimum, maintaining current levels of reliability.

Enabling Demand Response, Storage, and Distributed Energy Resources

Among the highest priorities for the CAISO is to identify the viable smart grid technologies that will aid in understanding what is happening on the grid and will support active participation in California's wholesale energy market. The need to expand demand response, both existing programs and future price-responsive demand, is driving infrastructure needs, which include smart devices and control systems that can collect data, present it to the power users, and then relay their decisions back to the load-serving utilities or third-party aggregators (also called curtailment service providers). The enabling technologies include but are not limited to:

- Building automation systems—the software and hardware needed to monitor and control the mechanical, heat and cooling, and lighting systems in buildings that can also interface with smart grid technologies; and
- Smart homes—similar to smart building technologies, except designed for the home where devices communicate with the smart grid to receive and display energy use and costs, as well as enable energy users to reduce or shift their use and communicate those decisions to the load-serving entities. These technologies are also known as home automation networks (HAN). Figure 6.14 identifies devices that may be part of the future smart home.

If the technologies develop as hoped, power users will also be able to receive real-time prices or indicators of grid conditions that aid their decision-making processes. For instance, if the grid is under stress, consumers could elect to configure devices that automatically respond to these indicators to shift or curtail use even before wholesale prices rise or system events occur. This is one reason, along with price responsiveness, why the CAISO needs to better understand how consumers use demand response capabilities so that we can predict responsive behaviors that will affect forecasts and energy resource unit commitments.

Among the challenges to overcome:

- Enhancing current market models, which are based on operational characteristics of conventional generation (natural gas, nuclear, hydro), to include

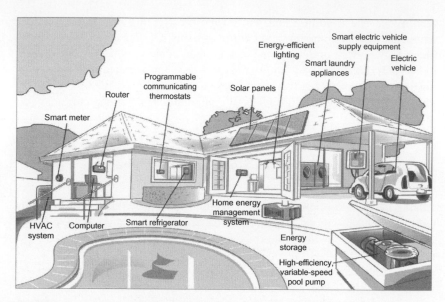

FIGURE 6.14   Devices enabling the future smart home. *(Source: CAISO Advanced Grid Technology Center)*

models of distributed generation and the full participation of demand-side resources, including eventually price-responsive demand;

- Determining minimum monitoring and telemetry requirements to enable more cost-effective participation for many small aggregated demand resources; and
- Maturing standards such as OpenADR[12] to enable demand response.

Besides conducting the research and analysis to form the market theories that aid industry understanding of how demand response and price-responsive demand should work under real conditions, the CAISO will pursue pilots and demonstration projects that help prove or disprove expectations. Figure 6.15 illustrates the data flow resulting from demand participating in the CAISO market.

Smart grid technologies focused on consumers holds the promise of providing visibility of their real-time use, the current condition of the grid, and their energy costs. With this information, consumers can make choices about how to adjust their energy usage manually, for example by turning down or off the air conditioner, or automatically by setting thresholds managed by smart grid technologies. Direct consumer grid interaction and impact are possible, but only if a host of other challenges are overcome, including closing the gap between the wholesale market and retail prices, specification of communication standards

---

[12]OpenADR, developed by Lawrence Berkeley National Laboratory, is a set of rules that specify how building and facility managers can implement automated demand response in energy management systems.

Demand response is enabled through a communications network that can relay instructions between the ISO,
energy control systems at the facility, and energy-consuming devices. The ISO market provides signals for
demand response, which gives buyers of electricity the opportunity to respond to grid conditions and
sell power curtailments. In the future, consumers may also respond to price signals.

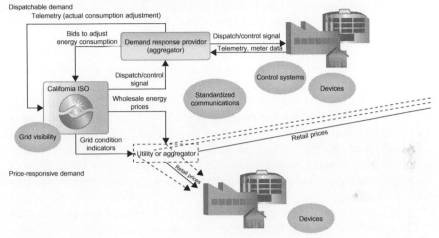

FIGURE 6.15   Demand participation data flow in the CAISO market. *(Source: CAISO Advanced Grid Technology Center)*

for exchanging this information between end-users, distribution companies and the ISO, and improving data confidentiality and network security.

The ISO is stepping up its activities to understand and demonstrate how storage technologies will play a role in the advancement of renewable integration in conjunction with the smart grid, including:

- How different types of storage behave (flywheels, batteries, etc.);
- How they fit into grid operations and can participate in CAISO energy and ancillary services markets;
- How they can efficiently and effectively provide regulation service and operating reserves;
- How they can efficiently and effectively shift energy deliveries from off-peak periods to peak load periods; and
- How they can co-locate with renewable resources to assist in more efficient use of transmission capacity.

Identifying and creating standards that technologies must meet become increasingly important and difficult as the ramping capabilities of renewable resources expand, increasing the need for capacity to be available that can follow net demand up and down. As it becomes more feasible to use different types of demand-side resources during high-renewable production to maintain reliable grid conditions and mitigate unfavorable conditions, this should reduce the need for and the associated costs of building new dispatchable generation and in some cases, new transmission lines.

Currently, the CAISO has market mechanisms and products, such as proxy demand resources that allow aggregators access to the wholesale market, supporting the increased participation of storage, demand response, and distributed energy resources and enabling these resources to enjoy comparable treatment as generating resources. As yet, however, no model exists that allows these resources to participate fully. Meanwhile, Western Electricity Coordinating Council rules are evolving, albeit slowly, to allow participation in spinning reserve and regulation markets.

The CAISO is actively participating in wholesale smart grid standards development efforts led by the National Institute of Standards and Technology (NIST) through the North American Energy Standards Board (NAESB) and the ISO/RTO Council (IRC). The CAISO is also closely involved with demand response policies being considered at the California Energy Commission and smart grid proceedings at the California Public Utilities Commission.[13]

The enabling demand response, storage, and distributed energy resources roadmap includes pilots to better understand technology capabilities, expectations for continued participation in national standards development efforts, and developing and piloting approaches for reflecting grid conditions that can be directly sent to smart grid devices.

Cyber Security

Cyber security becomes a priority concern as additional technologies connect to grid systems and provide more real-time data as well as two-way communications. The need exists to assess risks and vulnerabilities all along the communications chain from data sources to consumers, much of which is outside CAISO control. There is little doubt that situations will emerge that require new security controls and monitoring to ensure that grid monitoring, operations, and control systems are not compromised. At the same time it is clear that much of the potential benefit from smart grid will be due to the significantly increased information available to the CAISO, participating transmission owners, generators, and consumers. For this benefit to be realized, the cyber security rules must not be overly burdensome or so slow to be developed that they impede the progress of the new technologies.

A number of national forums are addressing security concerns. One is the National Institute of Standards and Technology that recently released *NISTIR 7628, Guidelines for Smart Grid Cyber Security*. This is a three-part document covering smart grid from a high-level functional requirements standpoint.

---

[13]Information on California Energy Commission programs can be found at http://www.energy.ca .gov. Information on California Public Utilities Commission proceedings can be found at http:// www.cpuc.ca.gov/puc/.

Among the challenges associated with cyber security is to tailor policies for power system monitoring and control applications, which are complex and industry and application specific. Implementing, maintaining, monitoring and improving information security so it is consistent with the organizational requirements and process are also issues to address.

The roadmap for cyber security addresses the evaluation and implementation of secure and standard protocols where applicable. It also calls for creating centralized security management and auditing as well as a situational awareness dashboard.

CONCLUSIONS

By now the entire electricity industry must recognize that the traditional ways of performing nearly all core activities are becoming obsolete, and that the reforms needed for the twenty-first century will involve unprecedented flexibility to adapt to change and new ways of thinking about supply and demand. Traditionally, operating the grid involved dispatching large thermal and hydroelectric resources to meet demand that was virtually entirely exogenous and could respond to grid conditions only through utility-administered programs. Transmission planning and generator interconnection procedures only had to deal with a modest, predictable rate of annual growth in load and the occasional addition of new supply resources. Even in regions where new spot market systems were implemented and have evolved over the past two decades under independent system operators and regional transmission organizations, these new markets were largely designed around traditional assumptions about the nature of supply and demand.

Today it is obvious that we can no longer operate electric power systems under the traditional assumptions. Over the next few years the industry will undergo tremendous changes, many of which will be the direct result of the technologies that comprise the smart grid, while others are being driven by environmental policy mandates that the smart grid will facilitate. These advances in technology and public policy are abandoning the traditional nature of electricity supply and demand, and are empowering consumers to choose their sources and manage their uses of energy in ways that were not possible just a few years ago but are now becoming possible with the emergence of smart grid capabilities.

In short, the future electricity industry with the smart grid will turn the traditional structure on its head. It will be a future where power flows from the distribution grid to the transmission grid as well as the other way, and where demand, rather than just being a passive consumer of energy, will quickly adjust its behavior in (often automated) response to price signals, which in turn reflect system conditions. Further, the distinction between load and generation will break down as more and more distributed resources such as solar photovoltaics are installed. These changes will require more information to be transmitted to energy end-users, and will also provide an opportunity for the

CAISO and distribution operators to receive significantly more data about the status of the grid and the consumption and production of all the supply and demand resources.

For the CAISO, one of the main implications will be the tremendous increase in data that must be received, transmitted, processed, understood, and responded to. The markets will need to accommodate many new types of entrants, most of whom will be smaller than the existing market participants. Further, unlike today where the grid is controlled directly through instructions to a small number of generation resources which can be counted on to respond, in the future the control of grid resources will be more indirect, in the form of adjusting prices and allowing the market participants to respond as they choose to the new price signals. This will require new feedback loops that sense how these indirect control systems are functioning and continually make adjustments to achieve the desired results.

The potential for efficiently controlling the grid to ensure power is delivered when and where needed will be greatly enhanced by the new technologies of the smart grid, but there is a huge amount of work to be done to realize the potential. For those of us working at the CAISO these are exciting times. The authors of this chapter span the whole range of CAISO core functions, from smart grid strategy and implementation (Sanders), to grid operation and spot market performance (Rothleder), to market redesign and infrastructure planning policies (Kristov). We see state environmental policy as the main driver of the transformation of the supply fleet, while smart grid and other new technologies such as energy storage provide the means to achieve the environmental goals. At the CAISO this means undertaking several parallel initiatives to facilitate and prepare for the new world, while maintaining through cross-functional collaboration a view of the big picture that reveals how all the changes interact and all the pieces fit together.

ACRONYMS

AGC	Automatic Generator Control
CMRI (ISO Application)	CAISO Market Results Interface
DNP	Distributed Network Protocol
DR	Demand Response
DRS	Demand Response System
DSA	Decision Support Applications
EMMS	Enterprise Model Management System
EMS	Energy Management System
EPDC	Enterprise Phasor Data Concentrator
GIS	Geographic Information System
HTTPS	Hypertext Transfer Protocol Secure
IEC 61850	International Electrotechnical Commission
LMP	Locational Marginal Pricing
NAESB	North American Energy Standards Board
NASPI	North American SynchroPhasor Initiative

OASIS	Organization for the Advancement of Structured Information Society
OASIS (ISO Application)	(California ISO) Open Access Same-Time Information System
OpenADR	Open Automated Demand Response
PDC	(Synchro) Phasor Data Concentrators
PEV	Plug-In Electric Vehicles
PMU	(Synchro) Phasor Measurement Unit
PV	Photovoltaic
RIG	Remote Intelligent Gateway
RTDMS	Real-Time Dynamics Monitoring System
SCADA	Supervisory Control and Data Acquisition
SIBR (CAISO Application)	Scheduling Infrastructure Business Rules
SOAP	Simple Object Access Protocol
VSA	Voltage Stability Analysis
WECC	Western Electricity Coordinating Council
WISP	Western Interconnection Synchrophasor Project

Chapter 7

Realizing the Potential of Renewable and Distributed Generation

William Lilley, Jennifer Hayward, and Luke Reedman

INTRODUCTION

In a traditional network such as the one shown in Figure 7.1, electricity is produced by large centralized plant located remote to the user. These plant typically convert energy contained in a fuel (e.g., coal, gas, or nuclear material) into electricity via some form of spinning machine, typically a turbine. The output from these prime movers is fed to a generator, which develops electricity at low voltage. This electricity is then converted to a high voltage for efficient transport through the use of a step-up transformer. The electricity travels through the transmission network toward the end-user at high voltage to reduce losses. When the electricity nears major load centers (e.g., a town), it enters the more widely spread distribution network for transport to numerous end-users. When entering the distribution network the voltage is brought to a lower voltage level by a step-down transformer. This step might typically occur a number of times before reaching the final consumer.

Typically the amount of power produced by a given plant is determined by a central control authority or market operator. In Australia's eastern states, for example, this is the Australian Energy Market Operator (AEMO). In the

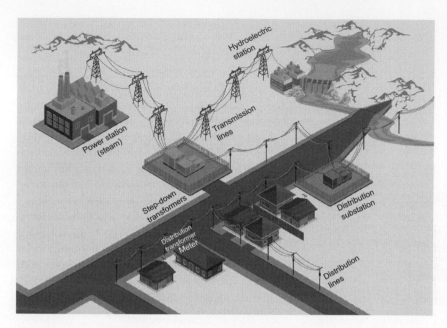

FIGURE 7.1   A simplified view of electricity generation and transfer. *(Source: CSIRO)*

United States, market operators are called independent system operators (ISOs) or regional transmission organizations (RTOs).[1] These organizations control the dispatch of power to meet system-wide demand. Dispatch takes into account issues such as scheduled outages, power flows including losses, the price offered by each generator for supplying electricity, and a prediction of aggregated demand. The system is then balanced through small changes to dispatch and ancillary services, which control frequency and voltage.

Because these large centralized plant are being fed a consistent source of fuel, their output is readily controlled and predictable. In response to concerns about climate change as well as fuel diversity, energy security, and a host of other reasons, there is a movement toward bringing large renewable generators into the supply system. These systems are typically connected where there is a good natural resource and where there is access to the high voltage transmission system or higher voltage sections of the distribution network. A number of these renewable generators operate by capturing a source of energy, which is variable by nature, for instance the wind or sun. As a consequence their output is less controllable and less predictable; hence these plants are referred to as intermittent renewable generators. Because their output can vary, their use can be problematic for the finely tuned electricity system, which must balance supply and demand within quite stringent limits.

---

[1]Chapter 6 describes CAISO, and Chapter 17 describes PJM.

The rise in electricity prices in many developed countries has been driven by expenditure on distribution networks to meet growing demand from large consumer devices such as air conditioners. The use of this equipment can lead to large demands on supply at certain times of the year, in this case on very hot days. The network must be rated to meet this large demand that typically occurs on only a small number of days per year. In response to rising prices to deal with this demand, there has been a trend toward the introduction of measures to better understand and control demand and to provide local supply to avoid transmission and distribution losses. This local generation is referred to as distributed generation (DG), also often referred to as embedded generation.

The introduction of DG into a distribution network poses potential problems to a system essentially designed to cater to one-way flow from large centralized plant located in remote locations to the end-user far away. These new two-way flows need to be measurable and controllable to ensure that issues around safety and performance are not unduly affected by the use of DG.

Adapting the way in which energy is used and supplied is a major challenge facing the world's economies as they attempt to reduce emissions in response to climate change and to reduce large expenditures in the supply and transfer of energy. Ensuring that new sources of energy supply and management can be integrated with existing technical and economic frameworks is a challenge being addressed through the emergence of smart grid infrastructure and control techniques, including intermediary steps such as minigrid architecture, further described in Chapter 8. In this chapter the results of a modeling analysis are provided that considers the value that smart grids may provide by enabling the increased use of intermittent renewable and distributed generation.

As smart grids represent a new and evolving way in which energy is generated and delivered, the cost and benefits are yet to be well characterized. Programs such as Smart Grid, Smart City in New South Wales, Australia (http://www.ret.gov.au/energy/energy_programs/smartgrid/Pages/default.aspx), and SmartGridCity in Boulder, Colorado (http://smartgridcity.xcelenergy.com/), have been developed to explore these issues and report outcomes to industry and the wider community. In other studies such as a recent report by EPRI [1], efforts have been made to quantify at a high level the potential cost and benefits associated with smart grids.

In the case of EPRI's report, the major benefits considered are:

- Allowing direct participation by consumers;
- Accommodating all generation and storage options;
- Enabling new products, services, and markets;
- Providing power quality for the digital economy;
- Optimizing asset utilization and operation efficiently;
- Anticipation and response to system disturbances;
- Resilience to attack and natural disaster.

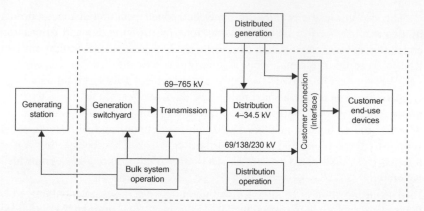

FIGURE 7.2 Modeling scope for EPRI cost/benefit analysis of smart grid in the United States; the dashed line represents the components of the energy sector in the scope of the EPRI study [1]. *(Source: EPRI [1])*

However, as can be seen in Figure 7.2, the benefits associated with changes to the development and use of centralized and distributed generation were outside the scope of their study.

Potential changes that may be required by the smart grid to deal with intermittency include:

- Better forecasting techniques for grid-connected wind and solar generation (e.g., the Australian Wind Energy Forecast System [AWEFS] in Australia and AEMO [2]) to allow more accurate dispatch of supply to match demand;
- Better control of the output of intermittent renewable generators to constrain plant ramp rates, that is, the rate in which output varies (e.g., the semi-scheduled rules within the Australian National Energy Market; AEMO [3]);
- The use of storage including electric vehicles (see Chapters 5, 18, and 19) to increase revenue earned by renewable generators;
- The adoption of new architectures such as mini grids (see Chapter 8) that can provide local areas with high penetration of intermittent generation through a combination of sophisticated control of generation devices and demand.

This chapter attempts to put a value on the benefits of a smart grid on a global scale. The analysis posits that greater amounts of renewable and distributed generation can be facilitated by a smart grid. Previous studies such as EPRI [1] do not estimate the benefits of a smart grid on the integration of renewable and distributed generation. This chapter presents modeling of the global electricity sector to examine the impact of intermittency constraints on renewable generation. Varying this constraint in the model is a means to estimate the potential benefits of a smart grid in facilitating greater deployment of renewable generation.

Section "Modeling Approach" presents the methodology of the economic modeling. Section "Results and Discussion" presents results and discussion of the modeling. Finally, section "Conclusions" provides conclusions resulting from the analysis.

MODELING APPROACH

The modeling in this chapter complements the approach used by EPRI [1] in their U.S. study by examining—using simple assumptions—the economic benefits derived from increasing levels of intermittent distributed and renewable generation in the grid over a long time frame. In the context of increasing global electricity demand, there are numerous supply-side options that may become economically feasible over time. The model assumes that smart grids will allow increasing levels of intermittent renewable and distributed generation into electricity networks. Prior to discussion of the scenarios, a brief overview of the modeling framework is presented below.

Modeling Framework

This chapter provides an estimate of the potential economic benefits from the integration of smart grids through the application of CSIRO's Global and Local Learning Model (GALLM).[2] GALLM is an international and regional electricity sector model that features endogenous technology learning via the use of experience curves. GALLM projects the global uptake of electricity generation technologies under business as usual and alternative policy environments. Currently the model is separated into three zones; developed countries,[3] less developed countries,[4] and Australia [4]. The model estimates the least-cost mix of different electricity generation technologies to meet electricity demand in each region over time, factoring in different initial generation capacities [5], different resources [6–9], and different electricity demand growth rates [8].

In the model, the uptake of a limited set of DG technologies has been considered by modeling generic combined heat and power (CHP), generic fuel cells, and rooftop PV. Large-scale intermittent technologies include wind, solar thermal, large-scale PV, wave, and ocean/tidal current. Non-intermittent large-scale technologies in the model are black and brown coal, pf (pulverized

---

[2]The Commonwealth Scientific and Industrial Research Organisation (CSIRO; http://www.csiro.au) is Australia's national science agency and one of the largest and most diverse research agencies in the world.
[3]Developed countries in the model include Austria, Belgium, Bulgaria, Canada, Croatia, Czech Republic, Denmark, Finland, France, Germany, Greece, Hungary, Iceland, Ireland, Italy, Japan, Latvia, Luxembourg, Netherlands, New Zealand, Norway, Poland, Portugal, Romania, Russia, Slovenia, Spain, Sweden, Switzerland, United Kingdom, Ukraine, and United States.
[4]Less developed countries are countries not included as developed countries.

fuel); black and brown coal combined cycle; black and brown coal with carbon capture and storage (CCS); gas open and combined cycle; gas with CCS; nuclear; biomass; hot fractured rocks; conventional geothermal; and hydroelectric. Table 7.1 lists the main technology cost and performance assumptions of GALLM.

TABLE 7.1 Technology Cost and Performance Assumptions of GALLM

Technology	Capital Cost ($/kW Sent Out)	Efficiency HHV (%)	Capacity Factor (%)	O&M ($/MWh)	Learning Rate (%)
Black coal, pf	1948	35.1	80	5.48	NA
Black coal, IGCC	3004	41.0	80	7.78	2.0
Black coal with CCS	4348	25.2	80	9.83	5.0
Brown coal, pf	2895	28.0	80	7.34	NA
Brown coal, IGCC	3320	41.0	80	8.40	2.0
Brown coal with CCS	7380	17.0	80	10.55	5.0
Gas open cycle	449	20.0	20	19.97	NA
Gas combined cycle	892	49.0	80	7.70	2.0
Gas with CCS	2900	40.0	80	25.69	2.2
Nuclear	3971	34.0	80	6.99	3.0
Hydro	3246	NA	20	21.98	NA
Biomass	2924	26.0	45	15.54	5.0
Solar thermal	5898	NA	25	24.23	14.6
Hot fractured rocks	4633	NA	80	11.99	8.0 (BOP)
Conventional geothermal	2878	NA	80	11.99	8.0 (BOP)
Wave	7000	NA	50	32.13	9.0
Ocean current	5200	NA	35	38.61	9.0
Wind	1518 (DEV); 1742 (AUS); 1389 (LDC)	NA	29	15.33	4.3 (turbine) 11.3 (AUS inst) 19.8 (global inst)

TABLE 7.1 Technology Cost and Performance Assumptions of GALLM—cont'd

Technology	Capital Cost ($/kW Sent Out)	Efficiency HHV (%)	Capacity Factor (%)	O&M ($/MWh)	Learning Rate (%)
PV rooftop	10529 (DEV); 9960 (AUS); 11858 (LDC)	NA	20	2.14	20.0 (module) 17.0 (BOS all regions)
PV large scale	6969 (DEV); 6615 (AUS); 7867 (LDC)	NA	20	12.83	As above
CHP	1600	42.1	41	6.53	NA
Fuel cells	12500	50.0	30	55.5	20.0

Source: CSIRO.
Notes: NA: not applicable; HHV: Higher heating value; IGCC: Integrated gasification combined cycle; O&M: operating and maintenance; BOP: Balance of plant; DEV: Developed countries; AUS: Australia; LDC: Less developed countries; inst: installations.

To encourage the development of CHP, heat credits[5] have been added into the objective function of the model. The CHP heat credit was initially 31.08 AU $/MWh, and this is reduced by 0.025% per year to reflect the fact that there may be an oversupply of heat.[6] European feed-in-tariffs have also been included for rooftop PV, assuming 0.19 AU$/kWh and reduced by 8% per year over 20 years. Table 7.2 provides a list of the fuel cost and CO_2 emission rate assumptions used in the modeling.

In GALLM, most technological development occurs as a result of global technology deployment such that all countries benefit from the spill-over effects of other countries investing in new technologies. Wind and PV are exceptions. Both wind and PV were assigned two experience curves: a global curve for the prime mover; and a local curve for installation and balance of system (BOS), where BOS only applies to PV costs. The twin experience curves for wind turbines and their installation in developed countries are shown in Figure 7.3.

Another feature of GALLM is that capital cost reductions can differ from that expected by the learning curve. For example, before the global economic crisis in 2008, the capital cost of energy technologies was extremely high. In regard to wind energy, the price rise was due to high demand and the resultant

---

[5]This is an allowance for heat production [10].
[6]At the time of writing the Australian dollar (AU$) is roughly at parity with US$.

TABLE 7.2 Fuel Cost and Emissions Assumptions of GALLM

Fuel Type	Cost of Fuel (AU$/GJ)	Emissions (kgCO$_2$/GJ)
Brown coal	0.5	93.6
Black coal	1.0	95.29
Natural gas	0.9–30.0	62.9
Biomass fuel	0.6–10.0	0
Uranium	0.7–30.0	0

Source: CSIRO.

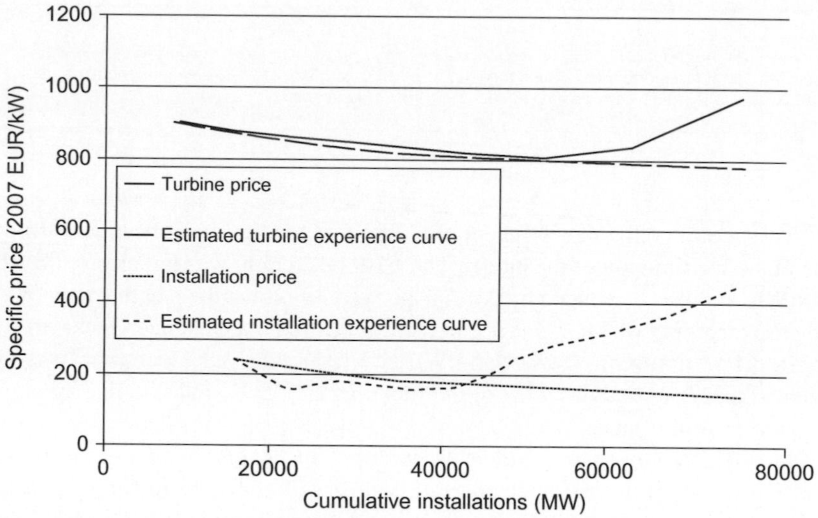

FIGURE 7.3 Experience curves for wind turbines and their installation in developed countries. *(Source: CSIRO, Note: International turbine wind turbine data with the experience curve and international installation data with the experience curve. Each data point represents a year where the first data point for turbines is from 1998 and the first data point for installation is from 2000.)*

increased profit margins and higher materials prices that this allowed [11]. These market forces have been included in GALLM as a "penalty" constraint; if demand for one technology exceeds one third of total required new installed capacity, then a premium is placed on the price of that technology based on historical data for wind turbines [8]. One effect of implementing the penalty constraint in the model is that it creates a disincentive for too rapid an uptake of any single energy technology [4].

Scenario Definition

The purpose of this chapter is to estimate the potential benefits of greater renewable and distributed generation that may be facilitated by the roll out of smart grids on a broad scale. To capture this, a new constraint is implemented in GALLM. It constrains the uptake of technology by imposing a variable constraint or cap on generation from intermittent renewable technologies including wind, solar thermal, PV, wave, and ocean current. Four scenarios that vary this constraint were considered in the current modeling:

- Case A: a base case with a maximum 20% cap on generation of intermittent renewable technologies by region;
- Case B: a maximum 30% cap in 2030 on generation of intermittent renewable technologies by region that increases linearly from 20% in 2020;
- Case C: a maximum 40% cap in 2030 on generation of intermittent renewable technologies by region that increases linearly from 20% in 2020;
- Case D: a maximum 50% cap in 2030 on generation of intermittent renewable technologies by region that increases linearly from 20% in 2020.

In Case A, for example, the model does not force the deployment of intermittent renewable technologies in each region to 20%. Rather, the 20% is a maximum share of electricity demand in each region that can be met by intermittent renewable technologies if it is economic to do so. This 20% cap, while quite arbitrary, represents a value similar to limits often quoted in industry discussion (e.g., UKERC [12] and Myers et al. [13]). The 20% cap also represents an aspirational renewable target in some countries (e.g., Australia ORER [14] and the EU-25 ECCA [15]) and some U.S. states [16], to be reached by the year 2020. Figure 7.4 shows the current technology

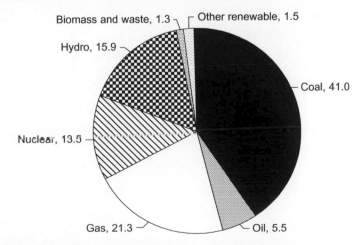

FIGURE 7.4   Percentage of world power generation by fuel type. *(Source: IEA [17])*

mix of world power generation. It shows that intermittent renewables, labeled as "other renewable" in Figure 7.4, represent a small proportion of current global power production.

While some regions may have higher or lower targets, it is worth noting that such targets are generally set at a high level that simply assumes the intermittent sources can be accommodated without problems or constraints. Ultimately the actual amount that can be accommodated will be highly dependent on the type and size of generation technology, the location of the source of generation, the topology of the network, and market trading structures. In the analysis presented here, it is assumed that the same limit applies everywhere. The modeling then considers a future where the percentage of intermittent generation is able to increase on the presumption that smart grid technologies and techniques discussed in this book enable an increase in their use.

For results displayed in this chapter, GALLM was operated under a business-as-usual (BAU) case and two different carbon price scenarios in which the price varied within regions in GALLM as shown in Figure 7.5. The first case, the 550 ppm, represents a lower price path consistent with a target of 550 parts per million (ppm) CO_2 equivalent (CO_2-e) atmospheric concentration corresponding to 2.8–3.2°C average global temperature increase by 2050, consistent with other studies by the International Energy Agency (IEA) and the UN's Intergovernmental Panel on Climate Change or IPCC [18,19]. The second case, the 450 ppm, represents a higher price path with a target of 450 ppm CO_2-e corresponding to 2.0–2.4°C temperature increase. As a point of comparison, the target ranges for the United States, the European Union, and Japan all correspond to entitlements for a global agreement between 450 ppm and 550 ppm atmospheric concentration of CO_2-e [20].

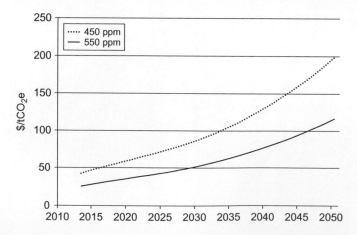

FIGURE 7.5 Carbon price trajectories used in the modeling for the 450 ppm and 550 ppm cases. *(Source: Commonwealth of Australia [21])*

The corresponding carbon prices associated with the 450 ppm and 550 ppm targets are shown in Figure 7.5, taken from the Commonwealth of Australia [21] report examining the impacts of carbon pricing on the Australian economy. This report showed that higher carbon prices facilitated greater uptake of intermittent renewable technologies. For the 550 ppm scenario the prices start at AU \$20 per ton of CO_2-e and increase by 4% each year. For the 450 ppm case the starting price is AU\$43 per ton of CO_2-e rising at 4% per year. For modeling purposes it is assumed that a global emissions trading scheme (ETS), similar in nature to the one adopted in the EU, is adopted by all developed countries on commencement in 2013 and that developing countries begin trading in 2025.

It should be noted that there is currently no national or global commitment to implement any of the modeled schemes. Hence these two carbon price trajectories are only a guide to possible future carbon prices were such schemes to be agreed and implemented some time in the future.

Other groups, such as the International Energy Agency, in their modeling use an emissions reduction target scenario, the "BLUE Map," consistent with reaching 450 ppm of atmospheric concentration of CO_2-e, rather than assigning a carbon price [5,19]. This chapter uses the carbon prices in Figure 7.5, rather than an emissions reductions target used by the IEA. This is because GALLM is a partial equilibrium model and thus only covers electricity generation, not other emitting sectors such as transport and agriculture. Figure 7.6 shows the estimated emissions reduction under the 550 ppm and 450 ppm scenarios and the IEA "BLUE Map" scenario.

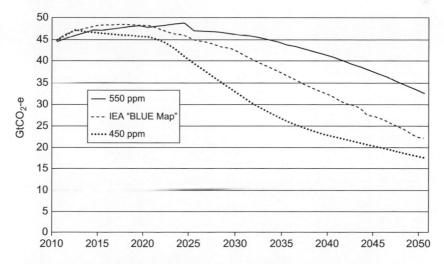

FIGURE 7.6 Projected global greenhouse gas emissions under 550 ppm, 450 ppm, and IEA "BLUE Map" scenarios. The 550 ppm and 450 ppm scenarios include emission reductions from forestry and agriculture, whereas the IEA "BLUE Map" only includes energy-related emission reductions. *(Source: Commonwealth of Australia [21]; IEA [19])*

RESULTS AND DISCUSSION

Modeling Results

In Figure 7.7A a plot is provided for the BAU case with a 20% cap under the Case A scenario previously described. In Figure 7.7B the BAU case for a 50% cap under the Case D scenario is displayed for comparison. Both figures clearly show that global electricity demand is expected to grow significantly, increasing two-and-a-half times by 2050. In these figures coal represents all brown and black coal sources with and without carbon capture and sequestration (CCS). Gas represents both open cycle and combined cycle turbines with and without CCS. Non-intermittent renewable generation, NI RG, includes biomass, hot fractured rocks, conventional geothermal, and hydroelectric. RG represents intermittent renewable generation sources including large-scale PV and solar thermal, wind, wave, and ocean/tidal current. Small-scale rooftop PV is not included here as it is captured within DG, which represents small-scale technologies such as PV, CHP, and fuel cells.

Figures 7.7A and 7.7B show that the absence of a carbon price has not prevented the uptake of solar-based renewable technologies and CHP. All of these technologies receive some form of government support in the model, such as feed-in tariffs or heat credit. Solar thermal has the additional advantage of providing some peaking power. These technologies, with the exception of CHP, have high learning rates as described in Table 7.1; therefore by increasing their cumulative capacity their capital costs fall over time. Black coal pulverized

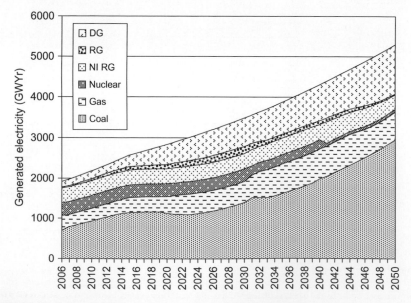

FIGURE 7.7A   Case A: Global technology mix assuming business as usual.

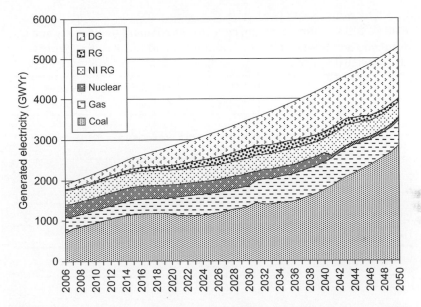

FIGURE 7.7B   Case D: Global technology mix assuming business as usual. *(Source: CSIRO)*

fuel plant continues generating beyond 2050, and black coal combined cycle, with its higher efficiency, absorbs the growing demand for electricity. Under this scenario very little CCS technology is developed. This is also the case for ocean energies, biomass, and geothermal plant.

The modeling results indicate that the total amount of intermittent generation does not necessarily reach the cap in all three regions in the model, but it does reach the cap in at least one. This level of detail is not captured in these figures, which simply present the total worldwide generation. However, this means that even under a BAU case there are financial benefits to be gained by using smart grids, assuming they increase the amount of intermittent generation that can integrated into the grid.

For comparative purposes, results from the IEA [19] projections for the year 2050 for uptake of various electricity generation technologies under their baseline scenario are also presented. The results are shown in Table 7.3. It is not apparent if any intermittent constraints have been made by the IEA in their modeling, and therefore they are compared with the range of outputs in the modeling presented here.

Major differences between the modeling presented here and the IEA results lie in the share of coal to nuclear and the share of renewable generation, both intermittent and non-intermittent, and DG. The differences between coal and nuclear arise because of two issues:

● The first is that this modeling uses a flat price for coal and a cost curve for uranium, so that the more uranium is used the more expensive it becomes.

TABLE 7.3 BAU Cases A, Assuming a 20% Maximum Intermittency and D, Assuming a 50% Maximum Intermittency, and IEA Baseline Scenario Projected Share of Electricity Generation Technologies in the Year 2050

	BAU Case A (% Share)	BAU Case D (% Share)	IEA Baseline (% Share)
Coal	55	54	41
Gas	13	12	14
Nuclear	1	1	10
Non-intermittent renewable	7	7	15
Intermittent renewable	1	2	6
Distributed generation	23	24	14

Source: CSIRO.

- Second, this modeling has a higher capital cost for nuclear than the IEA [19], which makes nuclear a less attractive option in GALLM. The IEA nuclear costs are primarily U.S. based while the figures in this analysis are global, including European costs informed by recent data from manufacturers. Furthermore, these baseline cases place no cost on greenhouse gas emissions; therefore there is no need to build zero-emission technologies unless they are economic.

The IEA results show a bigger share of generation from non-intermittent renewable generation, most likely due to constraints in GALLM, particularly for hydroelectric generation. The IEA's biomass component includes waste, which is modeled here as DG since this can encompass bagasse and other locally produced biomass waste that is used as fuel close to where it has been harvested. GALLM predicts more DG mainly because of rooftop PV, which has a high learning rate compared to other technologies, and its capital cost becomes quite low in time, reaching 1400 $AU/kW by 2050. In this modeling there are no limits on the amount of rooftop PV that can be constructed, aside from the cap on intermittent sources. It is worth noting, however, that an earlier study [22] in Australia showed that PV installation was economically constrained rather than physically constrained by available roof space, assuming slightly less than one quarter of available rooftop space was used.

In Figure 7.8A a plot is provided for the 450 ppm case with a 20% cap (Case A). In Figure 7.8B the 450 ppm case for a 50% cap (Case D) is displayed for contrast. These results show that for a grid constrained by 20% intermittent resources, large centralized low-emission plant such as coal with CCS, shown

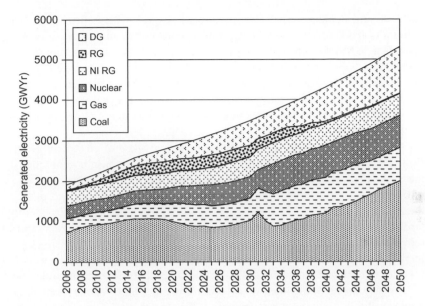

FIGURE 7.8A   Global technology mix under a 450 ppm case with a 20% cap on intermittent sources.

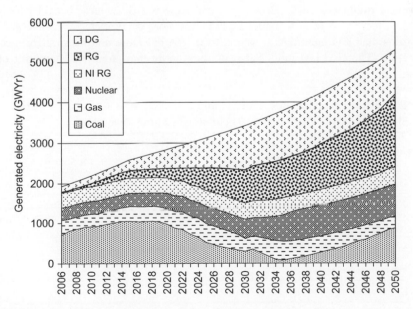

FIGURE 7.8B   Global technology mix under a 450 ppm case with a 50% cap on intermittent sources.

as Coal, and nuclear play a strong role, while gas turbines provide a large amount of the generation mix because of its cheap price and its use in peaking operation. When the constraints on intermittency are relaxed, the strong price of carbon results in large amounts of rooftop PV, in the DG set, and solar thermal and wind, in the RG set, being economically deployed over the period to 2050. In this case nuclear also plays a strong role, while gas and coal with CCS, shown as Coal, are deployed at much lower rates than in the more constrained intermittency case.

For comparison, Table 7.4 shows the projected share of electricity generation technologies in the year 2050 under the 450 ppm case compared with the IEA [19] "Blue Map" scenario, which has also been designed to achieve 450 ppm. In Case A where intermittent sources are limited to a cap of 20%, the modeling shows a 21% share in DG. In this case little large-scale intermittent renewable generation is predicted. When the intermittency constraint is relaxed in Case D, the DG and large-scale intermittent renewable technologies increase to a 54% share. In contrast, the IEA in their predictions shows a 34% total for intermittent renewables. The technologies that have increased the most due to the relaxing of the intermittent constraint are solar PV, solar thermal, and wind. In the IEA's case, the renewable technologies that have increased their share the most compared to the BAU case are wind and solar. Geothermal triples its share but it is starting from a lower BAU generation capacity.

The differences between Case A and the IEA BLUE Map are again in the shares of nuclear vs. coal and of renewable generation, both intermittent and non-intermittent, vs. DG. However, there are fewer differences between Case D and the BLUE Map scenario. Case D has more intermittent renewable generation than non-intermittent, whereas the IEA has more non-intermittent generation. The GALLM model also predicts more than double the amount of

---

TABLE 7.4 450 ppm Cases A and D and IEA BLUE Map Scenario Projected Share of Electricity Generation Technologies in the Year 2050

	450 ppm Case A (% Share)	450 ppm Case D (% Share)	IEA BLUE Map (% Share)
Coal	37	16	12
Gas	16	6	7
Nuclear	15	15	24
Non-intermittent renewable	10	8	23
Intermittent renewable	1	34	19
Distributed generation	21	20	15

Source: CSIRO.

intermittent renewable generation compared to nuclear, whereas the IEA has more nuclear than intermittent renewable generation.

These differences reflect the generous 50% intermittent constraint, which is allowing more large-scale intermittent renewable technologies with high learning rates, such as solar thermal, wave, and wind, into the market. The more these technologies are deployed, the lower their capital costs become due to assumptions on learning rates. The IEA does not state the limits used in their modeling but the results suggest they may use a cap lower than 50% on intermittent technologies, yet not as low as 20% since both intermittent renewable generation and DG contain intermittent technologies and the total of these is 34%.

It is worth reiterating that the modeling simply examines the savings that might be achieved by allowing an increase in intermittent generation—mostly attributed to the presence of smart-grid-enabling infrastructure. It does not specify how this may be achieved, only the savings that might be realized through their use. There are many possible ways in which intermittent generation may be better integrated into electricity networks including the use of storage, better management of demand through increased consumer awareness and appliance automation, and better forecasting of demand and supply for instance as discussed below and covered in more detail in other chapters of this book.

The modeling displayed above shows that in later years, traditional peaking plant such as gas turbines become less prevalent and slower reacting plant such as nuclear begin to dominate large centralized facilities. In simple terms in current electrical networks the balance between supply and demand is provided by these large plant, which receive appropriate signals from a central control to ramp their supply up or down as required. Ongoing developments in transmission and centralized dispatch fit within the wider smart grid paradigm, and it will need to continue to evolve to accommodate the technical performance characteristics of the generation mix as it changes in time.

One example of a smart grid response to intermittency is the semi-scheduled rules for large intermittent plant in Australia [2]. In this case the market operator (AEMO) established rules for the operation of large intermittent plant that constrains the divergence of output of these plants from their nominated dispatch levels. By constraining large-scale fluctuations, the market operator is able to maintain system balance more readily than it could if these sources were left uncontrolled. In a related activity AEMO also commissioned a wind forecasting system to improve efficiency of overall dispatch and pricing, and to permit better network stability and security management [2]. It is expected that solar forecasting will be built in the near future. If non-renewable large-scale centralized plant begin to be dominated by slow reacting facilities such as nuclear, then mechanisms such as those used by AEMO may become even more important to maintain system balance.

Much of the context for current smart grid discussion relates to distribution networks. A major reason for this is that historically they have acted passively, with system balance provided by large centralized plant matching aggregated demand to supply delivered through the transmission network as noted above. These large-scale operations can be considered "smarter" than their distribution counterparts as they already contain aspects of measurement and control. In many ways the emergence of smart grid technologies in distribution networks is driven by a need to bring these systems into line with the automated measurement and control of the large high-voltage networks.

One emerging response in distribution networks is demand management through automated control of individual devices—such as air conditioners, pool pumps, refrigerator compressors, and so on—and consumer response to varying tariff structures, such as dynamic pricing.[7] In the future this is likely to play a significant role in controlling local demand as covered in Chapters 2 and 9; demand management may make it more easy to balance supply, particularly if it is supplied by more variable sources.

The role of storage in buffering the impacts of intermittency is an area of growing interest in both transmission and distribution systems, as further described in Chapter 5. There are many different types of storage devices, such as batteries, thermal, compressed air, flywheels, ultra-capacitors, and superconducting magnetic devices. These devices can store and release energy at significantly different timescales and efficiencies. This variation in performance affects their potential application and arbitrage opportunities in the energy market. In contrast to these statically located devices, the emergence of electric vehicles, covered in Chapters 18 and 19, provides both a temporally and spatially varying storage device that may contribute both to the use and supply of energy, with its effects most pronounced in distribution networks.

A critical aspect of the modeling presented here is how the energy mix changes over time under different scenarios. The results show the emergence and dominance of fundamentally different types of generation technology, which vary in time and by end-use sector. In an earlier analysis [22] of the Australian energy market and the potential impact of DG, the modeling showed there is a large potential for gas-fired co- and tri generation to economically reduce emissions in the near future. In time, this potential is taken over by renewable technologies as their price becomes more favorable from increased learning and higher carbon prices. These findings are repeated here in the global modeling for rooftop PV and to a lesser extent CHP.

In this case there are important implications for distribution network planning due to the manner in which the generation devices are connected into the grid. Taking the Australian case as an example, the increased use of large gas-fired co- and trigeneration plant is more ideally suited in the commercial and industrial sectors where waste heat can be well utilized and it is predicted to be more economically viable in the near term. PV on the other hand is predicted to be predominantly installed in the residential sector in later years after it becomes more economically attractive.

---

[7]Chapters 8 and 12 describe these options in more detail.

Gas-fired CHP units are typically connected to the grid through a generator as a synchronous machine. These machines can produce and sustain large fault currents. A fault current is an abnormal current in a circuit due to a fault (usually a shortcircuit). The maximum (or making) fault current occurs in the first 20 ms while the steady-state fault current follows after approximately 40–60 ms. To protect a circuit, the fault current must be high enough to operate a protective device as quickly as possible, and the protective device must be able to withstand the fault current. A calculation of fault currents in a system determines the maximum current at a particular location, and this value determines the appropriate rating of breakers and fuses. Changes to fault currents from this type of installation are already posing challenges to some network operators particularly in central business districts (CBDs). Considerable effort will be required to manage their potential impact by appropriately locating control equipment such as superconducting fault current limiters, one example of an emerging smart grid technology.

Intermittent renewable generators on the other hand are connected to the grid via inverters that change DC output from the generating device (e.g., solar panel) to an AC waveform for export to the grid. These devices do not sustain large fault currents but can add harmonic distortions to the network, change voltage profiles on feeders, and be disconnected if voltage levels exceed preset bounds. In this case the installation of an integrated voltage control system including software and hardware will be required as part of a smart grid solution to ensure the successful integration of DG. Potential equipment includes automatic load tap changers, switched capacitors, medium voltage sensors, customer meters, and inverters able to operate at lagging or leading power factor.

Detecting and isolating faults, restoring operation, controlling voltage, and controlling real and reactive power flows will be some of the significant issues for smart grids to address if these generators are to reach their full economic and environmental potential. Since the type of technology and location in which it is installed will vary in time, the development of the smart grid will need to be well planned and flexible accommodate this complex evolution in supply and demand. This has implications both for technological development and associated policy and regulation. In countries such as Australia where the once vertically integrated system has been separated for economic efficiency reasons, this coordination may be potentially more difficult. Furthermore the disaggregation may make it harder to measure and attribute the value of individual actions to the system as a whole, which could inhibit the uptake of the most efficient solutions [22].

As noted above, the present analysis does not attempt to specify how the smart grid should evolve to meet the potential change in energy supply. Instead the focus is on the savings that might be achieved by allowing intermittent renewable technologies to reduce greenhouse gas emissions. Below the discussion considers how the savings may compare to costs of smart grids already identified by organizations such as EPRI.

Value of Potential Benefits

While the previous figures show how the mix in generation technologies could vary in time, Table 7.5 provides a summary of the undiscounted savings that may be achieved by increasing the amount of intermittent generation in the grid. This analysis has not specified how the intermittency may be accommodated as many of these technological issues are examined elsewhere within this book. Instead it focuses on the relative savings that could be realized from changes to capital expenditure, operation and maintenance, fuel and carbon prices, and revenues generated from feed-in-tariffs and heat credits for centralized and decentralized generation. From the table it is obvious that by 2050 very significant savings—potentially up to AU$20 trillion—could be achieved worldwide if the amount of intermittent supply can be increased in response to the world's reduction of greenhouse gas emissions.

The value of AU$20 trillion does not include the costs actually spent on upgrading the network to a smart grid to enable intermittent sources to reach their potential. To estimate this cost, estimates from EPRI [1] for the United States have been extrapolated to a global scale. EPRI [1] estimates that upgrading the U.S. network to a fully functioning smart grid will cost between US$338 billion and 476 billion by 2030. Assuming that this level of expenditure allows the modeled levels of intermittency to be achieved, and extrapolating the upper end cost ($476 billion) using IEA data on U.S. and world demand, by 2050 roughly AU$6.4 trillion will be needed to upgrade the global electricity networks to a smart grid, all else being equal. When this is combined with the savings in generation from utilizing more distributed generation and large-scale intermittent technologies, it means that the total global cumulative savings from installing a smart grid, assuming it allows 50% intermittent generation onto networks, could be as high as AU$14 trillion by 2050 if the world acts to reduce atmospheric concentration of CO_2e to 450 ppm.

It is important to note that the modeling indicates most of these savings come in later years, from 2040 onwards, after current long-lived stock with sunk costs have retired and large amounts of new generation come online

TABLE 7.5 Undiscounted Cumulative Savings AU$ Billion from Increasing Intermittency Relative to the 20% Base Case (Case A) by 2050

	Case B (Max 30%)	Case C (Max 40%)	Case D (Max 50%)
BAU	2680	2678	2695
550 ppm	9577	8688	10522
450 ppm	12024	18997	20763

Source: CSIRO.

and are able to become profitable; that is, run for sufficient time to recover initial investment. It is also important to note that the result is formed through comparison to a base case where there is an assumed intermittency limit of 20%. This limit is quite arbitrary for the purpose of this high-level study and in reality will vary by location due to the types and size of generation assets, installation location of generation within the grid, topologies of the electrical network, and market mechanisms.

While this examination is a simple high-level assessment, the outcomes show that it is an important consideration as the full benefits of smart grids will only be captured by examining the long-term changes that can occur through advanced operation of the electricity network, which takes into account all costs and benefits. The study also highlights the complexity in designing a smart grid that takes into account all potential changes in an evolving electricity system as consumers react to rising prices and the challenge of reducing emissions.

This study has shown that substantial savings could be obtained by allowing an increase in the use of centralized and distributed intermittent renewable generation. These savings are above and beyond those typically noted in cost-benefit assessments of smart grids in which network-specific aspects such as increased reliability, customer engagement, and asset performance are considered. While these factors are exceptionally important, the high-level analysis presented here shows that even further benefit can be captured by taking into account the role of smart grids in providing an increase in the amount of intermittent renewable generation able to participate in electricity markets.

CONCLUSIONS

This chapter presents results focused on the potential net benefits that may occur by allowing a greater proportion of global energy supply to be met by intermittent renewable and DG resources out to 2050. It is assumed that the introduction of smart grids will enable a larger proportion of generation to be provided by intermittent and local generation devices. These savings come from reductions in capital expenditure, fuel costs, operation and maintenance costs, and carbon costs, and revenue from feed-in-tariffs and heat credits for DG.

Modeling presented here clearly shows that savings from allowing an increased proportion of intermittent renewable and distributed generation can be very significant when considering how the world may meet the dual challenge of reducing emissions of greenhouse gases while accommodating the ongoing growth in demand. These savings are only realized by considering the long-term change to energy supply because of the lifetimes of the assets involved. This has important implications for smart grid use, planning, and development, which will be needed to ensure these renewable technologies reach their full potential.

When coupled with more traditionally noted benefits such as increased reliability, security, and consumer awareness, the development of a smart grid appears to be a very favorable mechanism to help the world reduce its greenhouse gas emissions while maintaining current levels of supply enjoyed in many of the world's developed countries.

REFERENCES

[1] EPRI, *Estimating the Costs and Benefits of the Smart Grid. A Preliminary Estimate of the Investment Requirements and the Resultant Benefits of a Fully Functioning Smart Grid.* 1022519 Final Report, March 2011.

[2] AEMO, *Australian Wind Energy Forecasting System (AWEFS).* http://www.aemo.com.au/electricityops/awefs.html, 2011 (accessed 30.03.11).

[3] AEMO, *Semi-Dispatch of Significant Intermittent Generation: Proposed Market Arrangements.* http://www.aemo.com.au/electricityops/140-0091.html, 2010 (accessed 20.03.11).

[4] J.A. Hayward, P.W. Graham, P.K. Campbell, *Projections of the Future Costs of Electricity Generation Technologies: An Application of CSIRO's Global and Local Learning Model (GALLM).* CSIRO Report EP104982, http://www.csiro.au/resources/GALLM-report.html, 2011.

[5] IEA, *Energy Technology Perspectives 2008*, IEA, Paris, 2008.

[6] M.Z. Jacobson, Review of solutions to global warming, air pollution, and energy security, *Energy Environ. Sci.* 2 (2009) 148–173.

[7] Carnegie Wave Energy, *Wave Energy as a Global Resource.* http://www.carnegiecorp.com.au, 2010 (accessed 07.01.11).

[8] IEA, *IEA Wind Energy, Annual Report 2007*, IEA, Paris, http://www.ieawind.org, 2008.

[9] I.B. Fridleifsson, R. Bertani, E. Heunges, J.W. Lund, A. Ragnarsson, L. Rybach, The possible role and contribution of geothermal energy to the mitigation of climate change, in: O. Hohmeyer, T. Trittin (Eds.), *IPCC Scoping meeting on Renewable Energy Sources*, Lübeck, Germany, 2008.

[10] IEA, *Projected Costs of Generating Electricity: 2010 edition*, IEA, Paris, 2010.

[11] D. Milborrow, Dissecting wind turbine costs, *WindStats Newsletter* 21 (2008) 2.

[12] UKERC, *The Costs and Impacts of Intermittency: An Assessment of the Evidence on the Costs and Impacts of Intermittent Generation on the British Electricity Network*, Imperial College London, 2010, ISBN 1 90314 404 3.

[13] K. Myers, S. Klein, D. Reindl, Assessment of high penetration of solar photovoltaics in Wisconsin, *Energy Policy* 38 (11) (2010) 7338–7345.

[14] ORER, LRET/SRES – *The Basics.* http://www.orer.gov.au/publications/lret-sres-basics.html, 2011 (accessed 15.05.11).

[15] ECCA, *The EU Climate and Energy Package.* http://ec.europa.eu/clima/policies/package/index_en.htm, 2011 (accessed 30.03.11).

[16] P.M. Jansson, R.A. Michelfelder, Integrating renewables into the U.S. grid: is it sustainable? *Electricity J.* 21 (6) (2008) 9–21.

[17] IEA, *World Energy Outlook 2010*, IEA, Paris, 2010.

[18] IPCC, Climate Change 2007: *Mitigation of Climate Change.* Contribution of Working Group III to the Fourth Assessment Report of the Intergovernmental Panel on Climate Change, B. Metz, O.R. Davidson, P.R. Bosch, R. Dave, L.A. Meyer (Eds.), Cambridge University Press, Cambridge, 2007.

[19] IEA, *Energy Technology Perspectives 2010*, IEA, Paris, 2010.

[20] R. Garnaut, *The Garnaut Review 2011: Australia in the Global Response to Climate Change.* http://www.garnautreview.org.au/update-2011/garnaut-review-2011.html, 2011.

[21] Commonwealth of Australia, *Australia's Low Pollution Future: The Economics of Climate Change Mitigation.* http://www.treasury.gov.au/lowpollutionfuture/report/default.asp, 2008.

[22] CSIRO, *Intelligent Grid – A Value Proposition for Distributed Energy in Australia.* CSIRO Report ET/IR 1152, http://www.csiro.au/resources/IG-report.html, 2009.

Chapter 8

What Role for Microgrids?

Glenn Platt, Adam Berry, and David Cornforth

INTRODUCTION

While many of the technologies described in this book, from low-emissions generation sources to new voltage control or energy storage devices, have the potential to bring great benefit to electricity systems, they are not without their challenges. Fundamentally, the complexity associated with integrating such plant into an already labyrinthine electricity distribution system is limiting enthusiasm for many of these technologies. The key to addressing this issue is to minimize the changes felt at the distribution level by simplifying the interface to these new resources. Microgrids represent just such a simplification.

In essence, a microgrid is a collective of geographically proximate, electrically connected loads and generators. While microgrids were traditionally viewed as a technology used in remote area power supplies, they can operate either connected or disconnected, or "islanded," from the wider utility grid. When operating as a member of the wider utility grid, microgrids effectively introduce a new level of hierarchy to the utility grid, where network assets and microgrids themselves can operate at the level of the utility grid; alternatively they may operate a level below this, internal to the microgrid, but separated from the wider utility grid operation. In a microgrid the challenge of controlling large numbers of distributed resources is reduced to an internal process, operating solely within the microgrid. By using microgrids to abstract the challenges of coordinating and controlling multitudes of distributed resources away from the wider utility grid, the interoperability challenges facing today's smart grids are eased significantly—grids can use conventional command and control techniques, ignorant of the machinations at lower layers, where the microgrid control system manages things.

While microgrids are bounded in scope, the challenges of ensuring stable and reliable operation of these systems should not be underestimated—particularly in systems with a high penetration of intermittent renewable supply. Challenges in this case range from interfacing the microgrid to the wider electricity system, through to implementing microgrid control systems without requiring expensive communications and control infrastructure.

This chapter explores the microgrid concept in the wider context of the smart grid. Sections "The Microgrid Concept" and "Key Technologies" detail the microgrid concept and its benefits, before a discussion of the challenges facing microgrid operation. The final section of the chapter provides examples of deployed microgrids that are exploring these issues in practical detail.

BACKGROUND

Traditionally, power networks have been based on a radial topology, where one generator is attached to many consumers in a tree-like structure. As shown in Figure 8.1, large, high-voltage transmission networks act like the trunk of the tree, carrying electricity long distances away from the generator and towards the electrical loads. Closer to these loads, distribution networks act like the branches of the tree, interconnecting loads and the long-distance transmission network.

The radial nature of modern power distribution networks is one of the core challenges in modernizing the control and operation of contemporary power systems. For example, adding automated protection devices to existing centralized control schemes places a heavy load on data communication networks. Similarly, as the network becomes more complex the processing power of

Traditional
grid:

Microgrids:

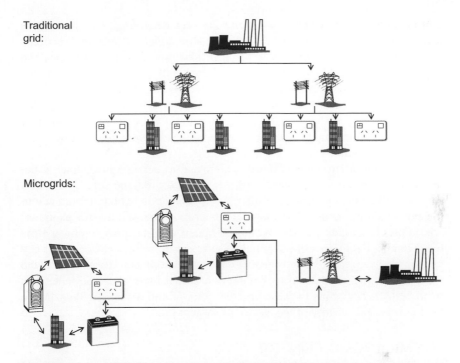

FIGURE 8.1 Conceptual differences between the traditional grid, which is hierarchical, and a grid including microgrids.

the central control system becomes a challenging factor [1]. In addition to these challenges of computing power, such centralized control schemes are challenged by "point-of-attack" issues where system-wide failures can occur if the central controller fails.

Another challenge to the reliable operation of current power systems comes from the growing prevalence of distributed generation—electricity generation sources that are typically much smaller than conventional power stations, and located close to electrical loads in the network. As discussed in Chapter 7, these distributed generators have a number of benefits that are driving their uptake, with advocates claiming benefits such as curtailment of transmission and distribution losses, greater robustness in the face of extreme weather events or attack, improved reactive power support, and decreased deployment time.

Importantly, distributed generators are often based on low or zero-emission generation sources—from highly efficient gas turbines through to renewable energy systems. Moreover, their close proximity to loads means that the efficiency of distributed generators can be further improved by their waste heat—for example to heat or cool nearby buildings. Though the range and depth of distributed generation technologies are great, their use is not without significant challenges. Fundamentally, it is expected that localized generators will be able to supply power back into the grid when local supply exceeds local demand. This challenges the traditional assumption of a unidirectional power distribution network, as power can now flow in both directions, and it may be quite common for power to flow locally, in varying directions, between distributed generators and nearby consumers [2]. With the growth in such devices, system planning and operation studies that assumed unidirectional power flow are no longer accurate, and the reliable operation of power systems incorporating large amounts of distributed generation becomes increasingly problematic [2].

While their environmental benefits are clear, as detailed in Chapter 6, the introduction of renewable energy sources such as wind and solar generation can exacerbate the challenges of distributed generation. The intermittency in the power supply available from such generation (caused by, for example, wind gusts or clouds passing overhead) means that not only is power flow bidirectional, but the power being fed into the system from such sources at any one time can vary randomly, depending on local environmental conditions. In the European Union, the increasing penetration of distributed generation such as wind farms is beginning to cause problems that, if unmitigated, will challenge the integrity and security of the electricity system [2].

THE MICROGRID CONCEPT

Fundamentally, many of the challenges discussed in the previous section are due to the inflexible way power systems are traditionally operated. In particular, their centralized, radial design limits the dynamic reconfiguration that is possible in the system. Though some reconfiguration can occur at the edges of the "tree," the core structure remains in place, is difficult to change in-situ, and is at risk of failure.

Considering this issue, contemporary power systems research has been increasingly focused on how best to integrate distributed energy resources such as renewable generation devices into the larger electricity grid. Of the various methodologies available, grouping distinct distributed resources so they represent a single generator or load to the wider electricity system is the technique that least affects existing infrastructure. When such loads and generators are located within close geographical proximity of each other, such a system is often referred to as a *microgrid*. More specifically, in this work we define a microgrid as a collection of controllable and physically proximate distributed generator and load resources, where there are multiple sources of AC power and at least one of these is based on a renewable energy technology such as wind or solar energy.[1]

---

[1]In some literature the term "minigrid" is also used in reference to groups of distributed generators and loads. In this work "minigrid" is considered a synonym of "microgrid."

A microgrid may or may not be connected to the wider electricity grid. Here, an *isolated microgrid* is defined as a microgrid that is not connected to the utility grid in any way, shape, or form; it is a distinct island for which no point of common coupling (PCC) exists. A *connected microgrid* is defined as a microgrid that *may* be connected to the utility grid; it may operate as a distinct island, but features a point of common coupling (PCC) that allows interaction with the utility grid, most typically to facilitate import/export of power.

There are a variety of reasons why microgrids are gaining more interest from the wider research and deployment community. On the one hand, microgrids represent a way of coordinating the growing number of sites with local on-site generation. For example, a university campus with roof-mounted solar cells and a diesel-powered backup generator can be transformed into a microgrid by adding intelligent control systems to the generators and linking these to load controllers to form a dynamic self-contained energy system. On the other hand, microgrids represent an entirely new way of powering remote or rural communities—rather than one centralized, often diesel-powered generating station, these communities can be powered by a large number of low-emissions generators, linked with appropriate load control.

The spectrum of possibilities between these two examples is quite wide, and, across the range of references at the end of this chapter, some ambiguity certainly lies in the wider community regarding this term. To assist, Table 8.1 gives our

TABLE 8.1 A Vision for Microgrids—Key Characteristics

Quality	Explanation
Intelligent	Capable of sensing system conditions and reconfiguring device operation and system topology given goals such as reliability of supply, cost of operation, and emissions minimization.
Efficient	Based primarily on clean renewable generation sources and intelligent energy-efficient loads.
Resilient	Able to reconfigure to withstand device failure and avoid total system collapse. Able to maintain quality of supply despite issues on the wider electricity grid.
Dynamic	Constantly changing in order to meet the intelligent and resilient goals above.
Load Integrated	Includes intelligent integration of local loads with generation, matching supply and demand and maximizing infrastructure utilization.
Flexible	Facilitates easy introduction of new loads and generators and the adoption of new communications technologies, without requiring significant infrastructure change.

particular vision for what features a microgrid should have, and later sections provide examples of deployments with these characteristics, while Table 8.2 provides some examples of what, in this chapter, *is not* considered to be a microgrid.

Figure 8.2 shows an example of a connected microgrid system deployed at an industrial campus. Power is obtained from the grid to supply loads, in this case the offices, as on a conventional site, but can also be obtained from the embedded generation, including wind turbines, solar photovoltaic panels, and a microturbine. All of these feed power to the site via inverters. The battery

TABLE 8.2 What a Microgrid Is Not

Commonly Used Name	Typical Example	Why This Is Not a Microgrid
Remote area power supply	A remote community supplied by a small number of relatively large (100 kW+) generators	There are only a small number of generators, with a centralized control system linking them. Any dynamic control is typically limited to throttling the generator, and no load control is incorporated. Such systems do not exhibit the underlying challenges found in multiple small-generator microgrids.
Local renewable energy system	A single house or building with a local renewable energy system (for instance, a wind or solar generator). May be grid connected	Typically such systems operate using a DC bus and single inverter, so do not fit into our definition of a microgrid having multiple AC sources. Further, the loads and generators in such a system rarely exhibit any form of intelligent, dynamic control.
Local backup power supply	A single house or building with a local backup power supply (such as batteries or a fossil-fuel generator)	Typically such systems operate using a single generation source or battery supply, so do not fit into our definition of a microgrid having multiple AC sources.
Grid-connected peaking plant	Relatively large (>1 MW) generators interspersed throughout a distribution network for meeting peak demand	Connected to an inter-meshed distribution system, these have multiple points of connection to the wider grid, violating our definition that a microgrid has one single point of connection. Such a topology demonstrates different electrical characteristics from a microgrid with only a small number of electrical busses.

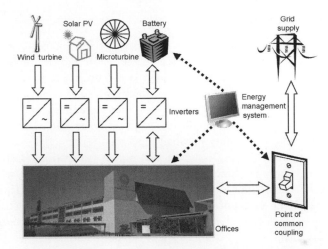

FIGURE 8.2   Simplified diagram of a commercial microgrid.

is equipped with a two-way power inverter that can absorb power when excess is available or supply power when there is a shortage. If embedded generation is greater than the required load, excess power can be exported to the grid via the two-way point of common coupling (PCC).

The site detailed in Figure 8.2 can be also isolated from the grid at the PCC, in which case the loads are supplied from the embedded generation only. In this case, the microgrid control system must carefully control which loads are allowed to draw power, in order to balance supply with demand. Fluctuations in the renewable generation due to changing weather conditions are balanced by judicious charging and discharging of the batteries, under the control of the microgrid control system. Some of the key features of the microgrid shown in Figure 8.2 that set it, and microgrids in general, apart from more traditional power distribution systems are:

- The use of distributed generation sources, including renewable and combined heat and power (CHP) generators;
- The introduction of dynamic load control systems;
- The use of multiple grid-connected inverters;
- The ability to island the system; and
- A point of common connection (for connected microgrids).

These features, while they may occur in isolation in traditional electricity systems, combine to form a unique set of opportunities and hurdles when deployed as a microgrid.

KEY TECHNOLOGIES

To fully realize the benefits possible from a sophisticated microgrid, as opposed to the relatively simple systems described in Table 8.2, a number of key technologies are required. On the generation side, a diverse range of generation types is particularly useful, allowing the microgrid to take advantage of the different availability characteristics of particular generation sources such as solar, wind, or fuel generators. On the demand side, controllable loads, typically with some discretionary ability in when they need to be activated, allow the microgrid to flexibly schedule demand to match generation. Lastly, some form of energy storage, either electrical, such as batteries or flywheels, or thermal, such as heat retained in an air-conditioned building's walls, is useful to aid the balancing of supply and demand. Finally, a sophisticated sensing, communications, and control system is needed to link all these microgrid assets together. Key characteristics of this system should include:

- Advanced high-speed control methods are needed to maintain microgrid stability and provide intelligent and dynamic operation, despite having no central point of grid "inertia" [3,4].
- Advanced sensing, diagnostics, forecasting, and adaptation technologies are needed to provide the microgrid control system with detailed and up-to-date information on the status of the microgrid [4,5]. Care must be taken to ensure that a minimum number of sensors are used and that the total capital outlay is also minimized.
- Integrated communications systems link the various resources in the microgrid to ensure reliable operation, even during typical communication outages [5].
- Advanced materials to provide economic, high-performance energy storage devices, such as batteries or flywheel technologies, will facilitate microgrids of growing complexity and size.

THE ADVANTAGES OF MICROGRIDS

While the previous sections suggest microgrids as a natural path in the evolution of the wider electricity system, there are also a number of discrete benefits for owners and operators of microgrids.

Autonomy of Supply

By operating their own network of local generators and coordinated loads, a microgrid owner/operator need no longer be reliant on the wider electricity grid. In maintaining the ability to island from the wider grid, a microgrid can ensure robust and reliable supply for its enclosed loads, isolated from faults on the wider electricity system. The microgrid can also provide enhanced power quality compared to the wider grid, useful in critical applications such as semiconductor manufacture. The Palmdale Water Treatment Plant microgrid in the United States [6] and CSIRO Energy Technology microgrid in Australia [7] are examples of microgrids operating with a grid connection, but in being able to operate in islanded mode, they are able to ensure the maintenance of local supply during time of macro grid failure.

Coordination of Distributed Generation

As the penetration of distributed generators in traditional (non-microgrid) electricity systems increases, the types of challenges described earlier grow in concern. Microgrids present a way of addressing these challenges—by coordinating geographically proximate generators, a microgrid can avoid the challenging interactions that conventional distributed generators display when attached directly to the wider grid. Microgrids do this in a way that does not require a large, capital-intensive, and technically challenging centralized control system, facilitating greater penetration of distributed generation without many of the limiting challenges. Many of the microgrids referred to in section "Existing Installations with Microgrid Features" of this chapter include a plurality of generation types, and, particularly in the research microgrids, there is a great deal of work investigating the benefits available from better coordination of these various generators.

Aggregation of Resources

Connected microgrids, with their single point of connection to the wider electricity system, implicitly aggregate all the resources within the microgrid to appear as one large, single controllable resource. This resource may appear to shed load, where this outcome could occur by increasing local generation within the microgrid; or by shedding controllable load, it could absorb controllable amounts of power by ramping up loads/reducing local generation, or it could provide system support through reactive power control. This single large resource provided by the microgrid has greater flexibility, reliability, sophistication of control, and less infrastructure requirements than significantly less sophisticated mechanisms such as direct load control systems, which shed load through (for example) air-conditioner management. The Kythnos [8] and CSIRO [7] microgrids are real-world examples of where multiple generators, as well as loads, are being aggregated and managed through local microgrid control schemes, thus removing the complexity of controlling large numbers of small generators from the macro grid operator.

High Penetration of Renewable Supply

The traditional view of electrifying remote areas was that while some renewable sources could be included, a large centralized fossil-fueled generator was needed to provide "spinning reserve" and act as inertia for the generation system. Microgrids challenge this theory, being designed around multiple small generators, without any single large source of inertia. With appropriate control and management mechanisms, microgrids can allow high penetration of renewable resources in powering remote communities, without needing the fossil-fueled generators. Many of the microgrids described in Table 8.3 are operating with greater than 50% of the electrical load being met by renewable generation supply.

THE CHALLENGES FOR MICROGRID ADOPTION

Though the microgrid concept is powerful and carries with it numerous advantages from both an economic and environmental perspective, it is unrealistic to assume that the construction of a full-scale system will be without significant challenges. These challenges come from the intersection of new technologies, emerging standards, dated regulations, and practical goals, and are introduced in the following sections.

Control of Microgrids

Control of microgrids is one of the most frequently cited issues facing this technology. Control in this context refers to the scheme and apparatus for ensuring that various parameters of the microgrid lie within limits that will guarantee the desired behavior of the system, including the effects of the microgrid upon the wider grid and the safety of personnel. Taken as a whole, the task of controlling multiple distributed sources and loads across a microgrid to achieve stability is highly complex. While traditional centralized control approaches are generally used in a microgrid, the complexity of algorithms required for centralized control of a growing and more dynamic distribution system, and the communications bandwidth required to support that control, has led to suggestions that a measure of decentralization could ameliorate this burden. With this in mind, newer methods have been suggested based, for example, on artificial neural networks and fuzzy controllers [3,9].

Planning and Design of Microgrids

The placement of distributed resources into microgrid topologies is a multi-faceted problem of greater complexity than may first be evident. Issues of installation cost, environmental impact, line loss, grid connectivity, reliability, resource longevity, reuse of waste heat, capacity for intentional islanding, and physical constraints all affect the decision-making process. The interaction between these distinct objectives, the complexity of power flows, and the relatively small number of existing practical microgrid installations mean that heuristics or rules-of-thumb are inappropriate if an optimal or near-optimal

TABLE 8.3 Some Microgrid Installations from Around the World

Name	Location	Research/Features	Capacity	Storage	Generation Types						
					Fuel Cells	Reciprocating engines	Micro-turbines	Other Non-Renewables	Wind	Solar	Other Renewables
Cape Verde wind-diesel systems	Cape Verde Islands, Africa	A collection of remote islands with wind–diesel systems; wind penetration levels between 14% and 6%.	Between 4 MW and 12 MW reciprocating engine and 600 kW and 900 kW wind	Unknown		x			x		
CICLOPS microgrid	Soria, Spain	Designed for experimentation; not strictly a microgrid as the three sources are coupled to a single common DC bus.	20 kVA reciprocating engine 595 Ah battery 5 kW solar 7.5 kW wind	Battery		x			x	x	
Coyaique Power System	Coyaique, Chile	Isolated community; manually operated control.	16.9 MW reciprocating engine 4.6 MW hydro 2 MW wind	Unknown		x			x		x
CSIRO Energy Technology Centre microgrid	Newcastle, New South Wales, Australia	Real-world system designed to also operate as an experimental facility; allows islanding, import and export of power.	110 kW PV 60 kW wind 150 kW turbines	Three kinds of battery; fourth type planned.			x		x	x	

(Continued)

195

TABLE 8.3 Some Microgrid Installations from Around the World—cont'd

Name	Location	Research/Features	Capacity	Storage	Generation Types						
					Fuel Cells	Reciprocating engines	Micro-turbines	Other Non-Renewables	Wind	Solar	Other Renewables
Fernando de Noronha island	Fernando de Noronha, Brazil	Isolated microgrid servicing approximately 2,500 people; aiming for 25% wind penetration.	2 MW reciprocating engine 225 kW wind	Unknown		x			x		
Flores Island microgrid system	Flores Island, Azores, Portugal	Studies have investigated impact of penetration levels on system stability; achieved >50% wind penetration.	600 kW reciprocating engine 1.48 MW hydro 600 kW wind	Unknown		x			x		x
Herbec Plastics	Ontario, New York, USA	Provides 100% of energy needs for the site.	750 kW turbine 250 kW wind	Unknown			x		x		
Kings Canyon microgrid	Kings Canyon, Northern Territory, Australia	Supplies a tourist park; features computer-controlled energy management.	225 kW solar 650 kW reciprocating engine (approx)	Unknown		x				x	
Kythnos microgrid	Kythnos Island, Greece	One of the first microgrids in the world (operational since 2003);	12 kW PV 85 kWh battery 5 kW reciprocating engine	Battery		x			x	x	

TABLE 8.3 Some Microgrid Installations from Around the World—cont'd

| | | | | | Generation Types | | | | | | |
Name	Location	Research/Features	Capacity	Storage	Fuel Cells	Reciprocating engines	Micro-turbines	Other Non-Renewables	Wind	Solar	Other Renewables
		services a remote community; no utility grid connection.									
Lolland microgrid	Island of Lolland Denmark	Using CHP; services 10 homes; uses small methane-burning turbine.	11 MW wind 15 kW turbine	Hydrogen	x		x		x		
Miquelon Wind-Diesel Project	Miquelon, Canada	Between 20% and 35% wind penetration; integrates wind into a pre-existing diesel grid.	5.2 MW diesel 600 kW wind	Battery		x			x		
NTT Facilities' Tohoku Fukushi University microgrid	Tohoku Fukushi University, Sendai, Japan	Services five university buildings, an aged care facility, high school, and water treatment plant; uses AC and high-voltage DC; connected to utility grid.	50 kW PV 250 kW fuel cell 700 kW reciprocating engine 800 kVA battery	Battery	x	x				x	

(Continued)

197

TABLE 8.3 Some Microgrid Installations from Around the World—cont'd

Name	Location	Research/Features	Capacity	Storage	Fuel Cells	Reciprocating engines	Micro-turbines	Other Non-Renewables	Wind	Solar	Other Renewables
Palmdale Water Treatment Plant	California, USA	Plans to allow island mode; storage will be used for blackout support; will offer ancillary services.	950 kW wind 1350 kW reciprocating engine 250 kW hydro	Ultra-capacitor		x					x
Sandia National Laboratories (SNL) research and the Distributed Energy Technology Lab (DETL)	Albuquerque, New Mexico, USA	Storage, islanding testing, and scheduling of PV and wind studies.		Battery, flywheel, and superconductors	x	x		x	x	x	
Subax residential microgrid	Subax, Xinjiang, China	Services 60 homes; appears to be an isolated microgrid.	30 kW reciprocating engine 4 kW solar 16 kW wind	Battery		x			x	x	
Utsira microgrid	Island of Utsira, Norway	Research and development project; fully autonomous; linked to mainland by sea cable; services 10 homes.	1.2 MW wind 50 kWh battery 5 kWh flywheel 48 kWh electrolyser 10 kW fuel cell 55 kW reciprocating engine	Battery, flywheel, and hydrogen	x	x			x		

system configuration is required. In a bid to move away from heuristics, contemporary research [10–14] has examined automating this design process through a range of computational intelligence techniques.

Irrespective of the planning approach chosen, the need for reliable and comprehensive data further complicates the task of planning a microgrid development. Any quality microgrid plan will be built on knowledge of the types of load profile that are to be expected, the seasonal characteristics that will affect renewables, and the specifics of technologies that may be used, for instance. Where possible, it is likely that much of this data will need to be generated through models, which raises concerns both with respect to cost and accuracy. Moreover, for some microgrids, the exact nature of loads may not be known, which introduces significant noise into the planning process.

Cost of Microgrids

A barrier to the uptake of microgrids is often the perceived higher financial cost of distributed energy resources compared to the available generation from centralized power stations. In particular, the higher per-kilowatt price of distributed generators is often cited as a drawback, while doubts remain as to the ongoing maintenance and operation costs associated with newer distributed technologies. Recent studies, however, have found that the cost of generating power in a microgrid is comparable with present electricity supply, as long as support for photovoltaics is available [15]. At this early stage of the technology's development, it is clear that the most economic use of microgrids remains in developing countries without infrastructure, since the installation cost of the microgrid must be compared with the cost of installing high-voltage transmission lines [16].

Financial cost features regularly in optimization studies of microgrids, typically offset against other desired properties, such as CO_2 reduction. A great deal of effort is being given to developing microgrid models that will help calculate this offset. A case study of the hypothetical installation of distributed generation in a hotel in the United States reveals a 10% cost saving and 8% CO_2 saving [17].

Integrating Renewables into the Microgrid

A central assumption of the microgrid concept is the inclusion of renewable generation sources. Though such resources afford the microgrid many advantages, not least in the areas of environmental impact and fuel costs, they also complicate the planning and operation of the microgrid significantly due to their intermittent production and their need for power inverters. While Chapters 6, 10, and 11 of this book discuss broader issues around the integration of renewable generation, the following sections consider this challenge in the particular context of microgrids.

Inverters

The use of renewables will almost certainly necessitate the inclusion of inverters, both to convert from direct to alternating currents where necessary and to provide some level of frequency control. The key to the successful integration of inverters into the microgrid is to facilitate parallel operation of inverters serving potentially heterogeneous sources without loss of synchronization, propagation of harmonics, or loss of stability in general. In the case where a small number of inverters are used, this presents a challenging, though well-studied, dilemma. If the inverters use centralized or master-slave control, current sharing is enabled but high-bandwidth links are required to, at the least, facilitate the distribution of error signals [18]. In contrast, distributed, on-board control reduces the bandwidth but at the cost of synchronization difficulties [18].

For larger microgrids with numerous inverters, the available literature is less comprehensive, and doubts remain as to the scalability of techniques proposed for small-scale systems. In particular, given the complex interactions that may occur between inverters [18], there is an increasing potential for problematic emergent phenomena to appear as the size grows. Predicting and controlling such behavior are difficult and have not been sufficiently explored in preceding works.

Intermittency of Renewable Generation and the Need for Storage

Given the intermittency of power supply from wind and solar generators, the obvious response is to add energy storage to intermittent sources, allowing for stored reserves to be accessed when environmental conditions are unfavorable. The fundamental challenge with such an approach lies in selecting a suitable *amount* of storage and in optimally *using* that storage.

For an isolated microgrid, the storage must be sufficient to satisfy gaps between generation and load across both small and large time-scales. Thus, in order to correctly size the storage devices in advance, accurate and reliable behavioral models for intermittent sources are required. The use of such models inevitably raises concerns about quality, and the fidelity of intermittent source models is seldom high. Moreover, even if the models are accurate, selecting storage that minimizes cost while satisfying both short- and long-term commitments is a complex optimization task.

Dispatch of stored energy is complicated by a similar need to serve both short- and long-term goals. In order to assess which stores should be accessed and at what rate they should be discharged, it will be necessary to implement intelligent control systems that are capable of handling noisy and dynamic data. Such systems must also correctly assess when energy storage devices should charge and how. Beyond the complexity of building such intelligent controls, it is also likely that the controllers will require high-speed communications between microgrid devices in order to capture system state information.

Islanding

A central benefit of the connected microgrid is the capacity to ride-through failures that occur on the utility grid with limited loss of localized service. By rapidly disconnecting from a faulting utility grid system (intentional islanding), adjusting local generation, and shedding non-priority loads, particularly high levels of reliability can be guaranteed for priority resources residing on the microgrid. Achieving this goal though is not as straight forward as the high-level description may imply.

- First, the fault condition must be detected, which is quite a challenge given the significantly different fault levels between the utility grid and microgrid, and the two-way power flows across the PCC.
- Second, the entire microgrid must be coordinated to take collective action, including disconnecting from the utility grid, and either black-starting (shutting down and re-starting in island mode, the simplest path of action), or maintaining supply (which requires very fast response times from loads and generators on the microgrid and sophisticated interfacing hardware between the microgrid and utility grid).
- Third, all controllers must change their logic to support operating in this different mode, which may require new operating parameters for inverters, and fault detection states for protection equipment.
- Fourth, the loads in the microgrid must be adjusted to match the available generation capacity, which is almost certainly reduced. In its simplest form, this will require a prioritized table of loads to drop when generation capacity is limited; however, ideally an intelligent load management scheme would be used to dynamically and continuously match load with available supply.
- Fifth, when the fault is cleared, the microgrid must be able to make the transition back to normal service, which will require re-synchronization of all components in the microgrid with the operating state of the wider utility grid.

Microgrid Modeling

Microgrids are expensive to install and set up for the purposes of research, and the first approach will usually be to derive some kind of model. Comprehensive models exist for traditional rotating generators, but models for distributed generation are much more problematic, as these devices rely on power electronics interfaces such as chargers and inverters. Models of the operation of microgrids and their component devices are required to analyze issues such as microgrid stability, fault behavior, interaction with the utility grid, and the transition between grid-connected and island modes of operation. As with all models, there are important issues relating to system calibration, scope of assumptions, and result validation. Though it is a practical reality that all microgrid models will only ever be an approximation of real-world behavior, accuracy is paramount to success (as discussed by Jayawarna et al. [19]), and care must be taken to identify and consider the impact of simplifications on model fidelity. Currently, there is a distinct scarcity of commonly accepted comprehensive models for microgrid analysis.

Energy Management

To achieve goals such as the minimization of fuel consumption, an energy management system (EMS) is expected to be responsible for generator dispatch and for adjusting generator set-points and parameters. However, this adjustment has implications for the stability of the microgrid, and therefore the stable limits of parameters must be well known for a range of operating conditions [20,21].

The importance of the EMS is underlined by the fact that it may be responsible for bidding on a deregulated market, the performance of which may determine whether or not the microgrid is able to meet its economic goals. The EMS may also be responsible for reconfiguration of the microgrid during transition to and from island mode [22,23].

This is a complex problem, and it is clear that traditional industrial automation techniques are unlikely to meet these needs. There is therefore a need for the development of a new generation of management systems [24]. In response, a redundant hierarchical tree structure was proposed in Ishida et al. [25], while [26] develops a full-scale small-signal model for the entire microgrid and obtains settings via dynamic analysis. A less traditional approach to an intelligent EMS is using an artificial neural network [9]. Though such approaches are promising, it is clear that more investigation is needed.

Finally, a related issue is the need for prediction of generation required. Though a host of works exist in this area, including [27] who use a neural network approach, the fidelity of the prediction will invariably be drawn into question, while inevitable short- and long-term inaccuracies leave any intelligent system prone to sub-optimal performance.

Policy, Regulation, and Standards

The microgrid concept is relatively new, and it is not surprising therefore that the regulatory framework for integration of microgrids into the wider grid is still developing [28]. However, getting this framework right is essential as it deeply affects the economic benefits of microgrids [29]. It is clear that the current framework poses some barriers for the uptake of the microgrid [15]. For example, IEEE Standard 1547 requires that grid-connected power inverters can detect a grid fault and shut down in that event. Consequently, commercial inverters have been designed to do that, and there is no incentive for them to offer island transition and support of uninterruptible supply in microgrids. This standard is driving research into the static switch, which can disconnect and reconnect in sub-cycle times [30].

At the operational level, there is no agreed policy relating to how microgrids will behave under different operating conditions, such as with light or heavy loads, with or without grid connection, or after communication failure (see, for example, Phillips [4]). Research is being conducted into standards for the operation of microgrids [5], and a new generation of energy management systems are being designed to meet the challenges, in particular to meet standard EEC61970 [24]. The work, however, is only in its infancy.

EXISTING INSTALLATIONS WITH MICROGRID FEATURES

Although the microgrid concept has only been formalized in recent years, there are numerous real-world networks that satisfy at least some of the requirements of a functional microgrid. These practical installations provide valuable insight into both the operational issues that must be addressed and the advantages that are offered within this new power paradigm. To understand the general trends in deployed microgrids, we reviewed the information available in this area. While the area is so immature that it is difficult to obtain direct references detailing particular installations, our research has uncovered a variety of information on today's existing microgrids, uncovering approximately 40 significant microgrid installations worldwide. A sample of these is listed in Table 8.3.

Currently, many of the reviewed systems are relatively simple—based on bi-generation setups (most often wind-diesel), with a non-renewable source used to offer baseline support to an intermittent renewable technology. These generation sources are seldom co-located with loads and instead mimic the prevailing centralized generation and distribution approach, albeit with a more interesting mix of outputs, a smaller range, and lower capacity. Since most such systems are used to service small remote communities, such as those on Cape Verde Island and Flores Island, to name but a few, it also seems unlikely that they offer any level of control over loads, be it through demand management or otherwise.

In reviewing the general trends across existing microgrid installations, we analyzed the types of generation most commonly used in existing deployments worldwide. Figure 8.3 shows that wind turbines are by far the most popular renewable generation technology used in today's microgrids. Somewhat unexpected is the prevalence of fuel cells, which are used in significant numbers of the microgrids we have seen, despite its fledgling nature as a technology and relative expense. Indeed, fuel cells are used more frequently than the much more mature microturbine technology. It is possible that the value of fuel cells in CHP systems, where they can provide impressive thermal capacity, and their flexibility in size make them a viable alternative generation source in this context.

Following our generation analysis, we then considered the total peak capacity of today's installed microgrids, detailed in Figure 8.4. Considering this information, there is a reasonably even distribution of microgrid sizes across the reviewed systems, covering microgrids that generate less than 20kW to those that produce more than 60MW.

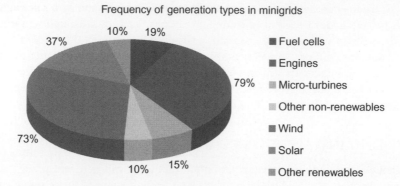

FIGURE 8.3   Characteristics of existing microgrid installations.

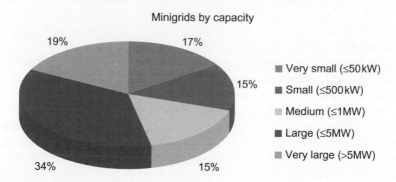

FIGURE 8.4   Size of existing microgrid installations.

FUTURE EVOLUTION OF MICROGRIDS

Considering the deployments detailed in the previous section, there is ample evidence that microgrids are being constructed across a range of sizes, with a diversity of generation, integration with loads, and interactivity with the grid. That said, microgrids are not *yet* the smart-grid stepping stone that their proponents would suggest.

While the size range of currently installed microgrids suggests that such systems have a place in a variety of grid deployments—from remote islanded communities to major grid-connected urban developments—the reality is that very few of these systems are operating in any particularly sophisticated manner.

Very few of the microgrids in Table 8.3 are able to operate in both islanded and grid-connected modes, likely due to the complexities (and expense) associated with switching between these two operating states. As discussed earlier, managing control of the microgrid when grid connected is quite different from managing control of the microgrid when operating in island mode. Further, at this stage, the fast switches required to allow smooth transfer between islanded and grid-connected mode are relatively uncommon and quite expensive.

Additionally, very few of the microgrids in Table 8.3 have any particularly sophisticated integration of load with supply in the microgrid. Where such integration does exist, it is often relatively simple, with simple load prioritization tables used to schedule a fixed priority list of loads so that total demand is matched to total supply. As discussed earlier in this chapter, for more optimal performance, such systems need to operate dynamically, respecting the changing priorities of loads. They will need to forecast future demand and also the future generation available to meet that demand, and with this information at hand, proactively *plan* how to match generation and supply.

While currently built microgrids have much potential to still be realized, we can remain positive that the issues raised in the preceding paragraphs are the subject of a growing amount of research. That said, while quite sophisticated control techniques are now being demonstrated, further research is needed to reduce costs and improve the simplicity of installation and operation. Ultimately, the sophisticated microgrid control schemes being demonstrated in the research space must move from the "bleeding edge" to reliable and dependable commercial products.

Lastly, aside from improvements internal to microgrid operation, at this stage there are few examples of electricity utilities themselves using microgrids as the beneficial asset they might be. While a number of the microgrid deployments in Table 8.3 have had utility involvement, often these are islanded systems, and the microgrid is not seen by the utility as a wider macro-grid asset. Although the Utsira microgrid in Norway [31], and the Ausgrid Smart Grid Smart City project in Australia [32] are early exemplars of utilities considering how microgrid could be an asset to wider operations, we need further detailed results data, and more deployment examples, before utilities will genuinely see this technology as a true stepping stone to the smart grid.

CONCLUSIONS

While microgrids were once considered a technology limited to remote area power supplies, they now offer a transitionary path from the utility grid operations of today to the intelligent, dynamicm and flexible smart grids of tomorrow. By aggregating large numbers of generators, loads, and storage devices together so they appear to the utility grid as a single controllable asset, microgrids ease the burden of control for the utility grid, simplifying issues such as the integration of large numbers of renewable energy devices, or managing system operation when given a complex multitude of control devices and constraints.

Microgrids also offer significant benefits to the end-user, helping to isolate them from utility grid issues, manage multiple on-site loads and generators, or improve local power quality. Microgrids are not without their challenges however, with a variety of issues challenging the early adopters of this technology, and limiting the number of deployments of this technology in developed electricity networks.

These challenges are being addressed by a growing number of researchers and system developers worldwide. While many of today's microgrid installations remain relatively simple, there are a growing number of microgrids being built with an increasing variety of generation types, integration of load and generators, and control systems that interact dynamically with the wider utility grid. These systems are a shining exemplar of the very near future, where microgrid technologies will be relied on as our first major stepping-stone towards the grid of the future.

ACKNOWLEDGMENT

This work was supported by funding from the Australian government as part of the Asia Pacific Partnership on Clean Development and Climate.

REFERENCES

[1] M. Amin, Toward self–healing infrastructure systems, *Computer* (2000).

[2] J. Schmid, P. Strauss, N. Hatziargyriou, H. Akkermans, B. Buchholz, F. Van Oostvoorn, et al., *Towards Smart Power Networks: Lessons Learned from European Research FP5 Projects*, European Commission Directorate–General for Research Information and Communication Unit, Brussels, 2005.

[3] S.M.T. Bathaee, M.H. Abdollahi, *Fuzzy-Neural Controller Design for Stability Enhancement of Microgrids*, 42nd International Universities Power Engineering Conference, 2007.

[4] L.R. Phillips, *Tasking and Policy for Distributed Microgrid Management*, IEEE–Power–Engineering–Society General Meeting, 2007.

[5] B. Kroposki, C. Pink, T. Basso, R. DeBlasio, *Microgrid Standards and Technology Development*, IEEE Power Engineering Society General Meeting, 2007.

[6] Hoffman, *Integration of Renewable and Distributed Energy Resources – U.S. Progress and Perspectives*, International Conference on Integration of Renewable and Distributed Energy Resources, 2006.

[7] The CSIRO Energy Centre, http://www.csiro.au/places/Newcastle.html, 2011 (accessed 11.05.11).

[8] The Kythnos Microgrid, http://www.microgrids.eu/index.php?page=kythnos&id=2, 2011 (accessed 11.05.11).

[9] G. Celli, F. Pilo, G. Pisano, G.G. Soma, *Optimal Participation of a Microgrid to the Energy Market with an Intelligent EMS*, International Power Engineering Conference, 2005.

[10] M.S. Rios, S.M. Rubio, Sequential optimization for siting and sizing distributed generation (DG) in medium voltage (MV) distribution networks, *IEEE Power Tech*, 2007.

[11] M.R. Vallem, J. Mitra, S.B. Patra, *Distributed Generation Placement for Optimal Microgrid Architecture*, IEEE Transmission and Distribution Conference and Exhibition, 2006.

[12] L.F. Ochoa, A. Padilha-Feltrin, G. Harrison, Evaluating distributed generation impacts with a multiobjective index, *IEEE Trans. Power Deliv.* 21 (3) (2006) 1452–1458.

[13] E. Haesen, M. Espinoza, B. Pluymers, Optimal placement and sizing of distributed generator units using genetic optimization algorithms, *Electrical Power Qual. Util. J.* 11 (2005) 97–104.

[14] E. Haesen, J. Driesen, R. Belmans, *A Long-Term Multi–Objective Planning Tool for Distributed Energy Resources*, IEEE Power Systems Conference and Exposition. PSCE '06, 2006.

[15] S. Abu-Sharkh, R.J. Arnold, J. Kohler, R. Li, T. Markvart, J.N. Ross, et al., Can microgrids make a major contribution to UK energy supply? *Renew. Sust. Energ. Rev.* 10 (2) (2006) 78–127.

[16] B.K. Blyden, W.J. Lee, *Modified Microgrid Concept for Rural Electrification in Africa*, General Meeting of the Power-Engineering-Society, 2006.

[17] C. Marnay, G. Venkataramanan, M. Stadler, A.S. Siddiqui, R. Firestone, B. Chandran, Optimal technology selection and operation of commercial–building microgrids, *IEEE Trans. Power Syst.* 23 (3) (2008) 975–982.

[18] S. Kong, D. Cornforth, A. Berry, *A New Approach to the Design of Multiple Inverter Systems Using Evolutionary Optimization*, IEEE Power Engineering Society General Meeting (PESGM), 2009.

[19] N. Jayawarna, C. Jones, *Operating Microgrid Energy Storage Control During Network Faults*, IEEE International Conference on System of Systems Engineering, 2007.

[20] E. Barklund, N. Pogaku, M. Prodanovic, C. Hernandez-Aramburo, T.C. Green, *Energy Management System with Stability Constraints for Stand–Alone Autonomous Microgrid*, IEEE International Conference on System of Systems Engineering, 2007.

[21] N. Barklund, M. Pogaku, M. Prodanovic, C. Hernandez-Aramburo, T.C. Green, Energy management in autonomous microgrid using stability-constrained droop control of inverters, *IEEE Trans. Power Electron* 23 (5) (2008) 2346–2352.

[22] H. Gaztanaga, I. Etxeberria-Otadui, S. Bacha, D. Roye, *Real–Time Analysis of the Control Structure and Management Functions of a Hybrid Microgrid System*, 32nd Annual Conference of the IEEE Industrial Electronics Society, 2006.

[23] S. Conti, A.M. Greco, N. Messina, U. Vagliasindi, *Intentional Islanding of MV Microgrids: Discussion of a Case Study and Analysis of Simulation Results*, International Symposium on Power Electronics, Electrical Drives, Automation and Motion, 2008.

[24] Y.Y. Zhang, M.Q. Mao, M. Ding, L.C. Chang, *Study of Energy Management System for Distributed Generation Systems*, 3rd International Conference on Electric Utility Deregulation, Restructuring and Power Technologies, 2008.

[25] S. Ishida, C. Roesener, H. Nishi, J. Ichimura, *Discussion of Aspects in Energy Management with Demand Response System*, 11th IEEE International Conference on Emerging Technologies and Factory Automation, 2006.

[26] F. Pilo, G. Pisano, G.G. Soma, *Neural Implementation of Microgrid Central Controllers*, 5th IEEE International Conference on Industrial Informatics, 2007.

[27] S. Chakraborty, M.D. Weiss, M.G. Simoes, Distributed intelligent energy management system for a single–phase high-frequency AC microgrid, *IEEE Trans. Ind. Electron* 54 (1) (2007) 97–109.

[28] M. Costa, M.A. Matos, J.A. Lopes, Regulation of microgeneration and microgrids, *Energy Policy* 36 (10) (2008) 3893–3904.

[29] P.M. Costa, M.A. Matos, *Economic Analysis of Microgrids Including Reliability Aspects*, 9th International Conference on Probabilistic Methods Applied to Power Systems, 2006.

[30] D. Klapp, H.T. Vollkommer, *Application of an Intelligent Static Switch to the Point of Common Coupling to Satisfy IEEE 1547 Compliance*, IEEE-Power-Engineering-Society General Meeting, 2007.

[31] The Utsira Project, http://www.eahia.org/pdfs/EFH_Industry_Pres_Utsira_ExCo_Mtg_5–05.pdf, 2011 (accessed 11.05.11).

[32] Smart Grid Smart City, http://www.ret.gov.au/energy/energy_programs/smartgrid/Pages/default.aspx, 2011 (accessed 11.05.11).

Chapter 9

Renewables Integration Through Direct Load Control and Demand Response

Theodore Hesser and Samir Succar

INTRODUCTION

As described in other chapters of this volume, the smart grid is a combination of enabling technologies that span the structural topology of the power grid. Its mandate includes a broad range of societal benefits that broadly encompass the notion of grid reliability, security, versatility, and resilience. While this points to a large class of technologies that can be leveraged toward a broad range of applications, from the point of view of carbon emission reductions, renewables integration is of particular interest. The carbon reductions attributed to smart grid deployment can be attributed to four broad categories:

- Increased deployment of renewable energy through increased grid flexibility.
- Decreased carbon intensity of energy goods and services through energy efficiency.

- Reduced electric consumption in response to market signals (conservation).
- Displacement of petroleum consumption through vehicle electrification.

The impact of the smart grid on the cost of integrating renewable energy is often described in qualitative terms, but rarely quantified in a rigorous way. Despite the fact that enabling variable energy resources (VERs) integration is often described as the largest source of carbon emission abatement from the smart grid (Figure 9.1), this quantity is usually estimated in terms of an arbitrary increase in renewables deployment [1] or by estimating incremental revenue diverted to other sources of abatement as a proxy [2]. However, these approaches do not address the mechanisms by which smart grid technology will facilitate increased deployment of renewables.

The purpose of this chapter is to look at the role of load as a source of flexibility that can facilitate increased deployment of variable generation from renewable sources. How do VERs affect reliability and how can those effects be quantified in ancillary services markets? What types of DR exist, and which are better suited to address the needs of renewables integration? Finally, is the DR resource big enough to make a dent in the demands that wind and solar energy will place on the system?

In Chapter 10 Hanser et al. focus on the related issue of how to determine the incremental demand for ancillary services stemming from VER deployment and propose a methodology for estimating the optimal sizing of critical peak pricing programs to serve that need.

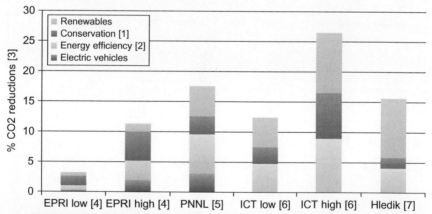

[1] In home display direct feedback and consumption impacts of load shifting
[2] Reduced line losses, conservation voltage reduction, advanced voltage control, measurement and verification of efficiency programs and accelerated efficiency deployment
[3] Reductions in power sector emissions measured relative to DOE/EIA 2030 Reference Case except ICT study where reductions are based on 2020 Reference Case
[4] Electric Power Research Institute (EPRI), 2008 *The Green Grid: Energy Saving and Carbon Emissions Reductions Enabled by a Smart Grid* EPRI 1016905
[5] Pacific Northwest National Lab (PNNL). 2010. *The Smart Grid: An Estimation of the Energy and C02 Benefits*, PNNL-19112 Pacific Northwest National Lab, Richland, WA
[6] The Climate Group, 2008 *Smart2020 Enabling the Low Carbon Economy in the Information Age*
[7] Hledik R. 2009 "How Green is the Smart Grid?" *The Electricity Journal* 22(3):29–41

FIGURE 9.1 Power sector greenhouse gas (GHG) emissions reductions from smart grid deployment. *(Source: Pratt et al. [2])*

In this chapter we focus on the supply side of renewables integration, the full portfolio of DR resources, and the size of the resource as it relates to specific requirements of the grid. We begin with a description of the variable resource integration challenge and a short description of grid support services at various characteristic timescales. We then identify various classes of VER integration costs. In the following section, we develop a topology of DR resources that provides the basis for analyzing the ancillary service requirements relative to DR resource potential. Finally, we discuss DR deployment potential, identify the resources best suited for renewables integration, and quantify their potential relative to the demands of integrating large quantities of variable generation onto the grid.

INTEGRATING TEMPORALLY VARIABLE AND GEOGRAPHICALLY HETEROGENEOUS RESOURCES

Implicit within high renewable generation scenarios is the heterogeneous geographic distribution of variable energy resource (VERs) such as wind and solar. Economics drive deployment of utility-scale renewable technologies to regions with the highest quality wind and solar resources. The geographic constraints of resource distribution and the temporal constraints of resources variability pose unique challenges to reliable system operation as the share of generation from renewables grows. The path to the grid's decarbonization invariably flows through the confluence of grid reliability and efficient capital outlay. In that context, a flexible yet resilient power grid will be a key enabler of cost-effective renewables integration and rapid carbon reductions.

The Eastern Wind Integration and Transmission Study (EWITS) exemplifies this type of regional heterogeneity in its analysis of 20% and 30% wind penetration scenarios for the Eastern United States [3]. Achieving these penetration milestones by the year 2024 represents an aggressive trajectory for renewable energy deployment and requires significant new infrastructure upgrades. A comparison of regional peak wind capacities with regional peak loads is shown in Figure 9.2. It is notable that certain regions such as Southwest Power Pool (SPP) are projected to install new wind capacity that exceeds total current system peak demand in order to reach a penetration of 20% across the East.

Because a large fraction of the generation that wind capacity produces will necessarily be consumed outside the region, such high-penetration scenarios imply a strong role for transmission and other flexibility resources to manage resource variability. The take-away is that in scenarios aimed at large-scale deployment of renewables, VERs are likely to develop in regionally clustered pockets in order to access the highest quality, most economic resources.

An analysis of wind capacity from the North American Electric Reliability Corporation (NERC) Long Term Reliability Assessment (LTRA) illustrates a similar trend [4], with relative wind capacities highest in those areas with the strongest wind resource (Figure 9.3). This wind capacity represents actual projects that are

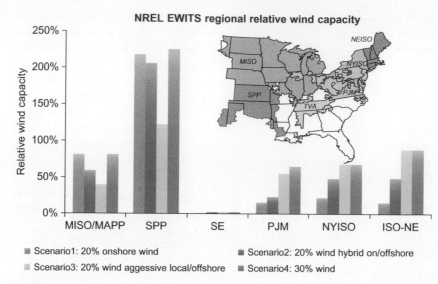

FIGURE 9.2   Regional wind capacities in EWITS scenarios relative to 2006 peak load by region. *(Source: NREL [3])*

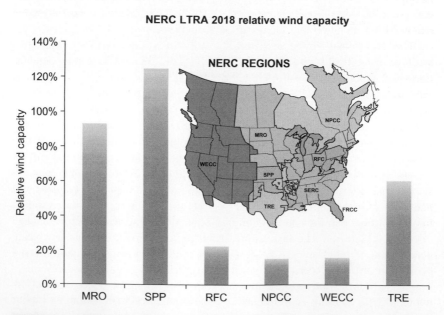

FIGURE 9.3   NERC region 2018 wind capacities in 2009 LTRA relative to 2018 peak load by region. *(Source: NERC [4])*

in the interconnection queue, as opposed to the hypothetical projections developed in EWITS. The LTRA data translate to 26% wind relative wind capacity by 2018. Figures 9.2 and 9.3 demonstrate how resource heterogeneity drives uneven development of the resources. In high wind penetration scenarios, this type of development trend can amplify the operation impacts of integrating variable generation.

The challenges associated with resource distribution and grid integration described above for the Eastern United States are mirrored elsewhere, including the Western United States, Europe, and China [5–7]. Therefore, while the focus of this analysis will be on the demand-side flexibility resources in the United States, the operational issues addressed here will have strong relevance for all the regions of the world currently engaged in large-scale VER deployment.

Although the output of wind and solar generators cannot be controlled directly, the system's ability to accommodate the variability inherent in load is often sufficient to address the impact of VERs at modest penetration levels. However, as the quantity of variable generation grows, the operational impact of VERs will increase. Combining weakly correlated resources can help smooth out the fluctuations, but when the deployment of VERs is highly clustered as indicated above, it will be more difficult to leverage this type of geographic diversity. In this case the variations in wind output will instead be highly correlated in regions with the greatest density of high-quality resources, making resource integration that much more challenging.

Reliably integrating large quantities of variable generation will ultimately require a portfolio of solutions that facilitate resource sharing across regions and enhance system flexibility. In the long term, building high-capacity transmission lines that allow for increased power flow will ameliorate the supply and demand imbalances associated with the geographic heterogeneity of relative wind capacities. In the short and long term, a portfolio of reliability-focused tools will be needed to firm wind capacity on a variety of temporal scales. Aggregating balancing authorities, high-fidelity wind forecasting, utility-scale storage facilities, new transmission, flexible natural gas and hydroelectric generation, and demand response (DR) all augment the grid's flexibility. Quantifying the relative value of each flexibility resource in a flexibility supply curve [8] would enable an economically efficient path to significant VER integration.

Direct load control (DLC) DR can be an important component of an integration portfolio. The dispatchable demand-side flexibility that DLC provides can increase the grid's reliability and ensure meaningful carbon reductions through the accelerated low-cost integration of VERs.

DEMAND RESPONSE: PAST, PRESENT, AND FUTURE

A historical perspective of DR is necessary to properly understand its future potential as an important component in the portfolio of VER integration solutions. Historically, system operators used manual load shedding to maintain grid reliability during emergency situations. Although such blunt DR measures are necessary to prevent catastrophic grid failure, they are not the best way to protect the integrity of the system. Manual load shedding does not represent Pareto optimality. Automatic load shedding caused by under-frequency and under-voltage switches are in some ways a brute-force form of DR as well.

Today, more sophisticated DR programs differ from historical demand response in that utility customers are given the choice to shut down during times of peak demand and are incentivized to do so accordingly. The incentive structure, driven by the lucrative economics of peaking generation, provides participating customers with payments exceeding their operational value for short durations a few times a year. Current DR programs represent a Pareto improvement because both utilities and customers profit from the exchange.

For example, Baltimore Gas and Electric Company estimates that the capital cost of DR, at US$165/kW, is three to four times cheaper than the cost of installing new peaking generation, which is around US$600–800/kW [9]. This insight is critical to the assessment of the relative value of flexibility resources. In this example, DR significantly trumps peaking units as the flexibility resource of choice. In addition, DR is more efficient than peaking plants or batteries in that it involves no energy conversion and therefore sidesteps any losses associated with thermodynamic or electrochemical processes. Covino et al. discuss the relative economics of DR and peaking capacity in the PJM context in Chapter 17 of this volume.

The historical focus on DR as a last line of defense against cascading system failure has recently been superseded with an offensive strategy of incentivizing reliability-focused peak reduction. Incentive-based peak load reduction DR programs have experienced tremendous growth in recent years. From 2006 to 2008 the number of entities offering DR programs rose from 126 to 271, a 117% increase. Also, the size of peak load reduction from existing DR resources, relative to the national peak, rose from 5% in 2006 to 5.8% in 2008, a 16% increase.

The New England Independent System Operator's (ISO-NE) DR programs increased from 0.4% of forecasted system peak demand in 2002 to 3.9% in 2007: a 57% levelized compound annual growth rate (CAGR) [10]. Enrollment in PJM's DR programs sustained a 17% levelized CAGR, growing from 2100 MW in 2002 to 4600 MW in 2007. Figure 9.4 illustrates the DR's growth in the largest U.S. market, PJM, post-2007. Of note is the depreciating clearing price that results from DR's growth relative to peak capacity. Thus, a curtailment service provider's business model suffers from revenue cannibalization as a result of successful market growth.

Future DR programs are anticipated to expand from the commercial and industrial (C&I) sectors into the residential with dynamic pricing programs and advanced metering infrastructure (AMI). Advanced metering infrastructure and the IT overlay of smart grid technology will enable low-latency communication between loads and system operators (or third-party aggregators) that will facilitate the real time operational control of loads.

The Federal Energy Regulatory Commission's (FERC) biennial staff report on AMI deployment states that AMI penetration increased by more than a factor of 6

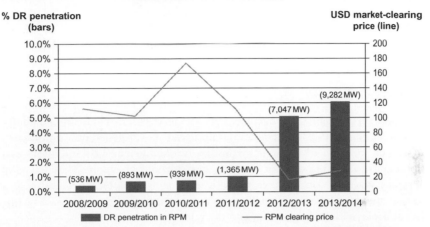

FIGURE 9.4   Growth of demand response penetration in the PJM interconnection. *(Source: Bloomberg New Energy Finance [36])*

in just two years, growing from 0.7% in 2006 to 4.7% in 2008 [11]. The American Reinvestment and Recovery Act provided $3.4 billion for AMI manufacturing, implementation, and innovation.[1] Bloomberg New Energy Finance projects that 58 million homes in the United States will be equipped with smart meters by 2014, with 50 million homes equipped by 2012. Whether AMI rollout will simply enable faster data collection or facilitate vast DLC distribution is yet to be determined.

In addition, Order 745 (RM10-17) issued by the Federal Energy Regulatory Commission in March 2011 will allow regional transmission organizations and independent system operators to pay DR resources the full market price for energy under certain conditions. This increased compensation for demand-side resources, if widely implemented in the organized markets, could fuel increased growth in DR, beyond the levels previously contemplated.

Future Pareto improvements will occur when the externalities of climate change are properly incorporated into the macro economic analysis. When the goal of minimizing consumer costs is better informed through valuation of environmental externalities DR's function will transition from a role limited to peak demand reduction to cover a broader gamut of applications including VER firming.

Variable generation increases the need for ancillary services that DR can provide, and this will open the door to continued smart grid deployment—as the key enabler of the grid's decarbonization. Recent studies have shown that supporting enhanced penetrations of wind and solar generation is among the greatest potential reductions in carbon emissions attributable to smart grid

---

[1]DOE (2009). October Press Release. http://energy.gov/news/8216.htm.

technology [1,2,12]. The use of DR for these applications will bring about a dramatic change in the frequency of DR utilization. Currently DR programs are called upon a handful of times a year during extreme events. Under a DR-VER-firming paradigm, DR would be utilized more frequently in normal grid operations, as opposed to contingency operations, but for very short durations to avoid sacrificing the original intent of the energy service.

ANCILLARY SERVICES

Ancillary service markets in the electricity sector are paramount to grid reliability and minimizing the consequences of capricious power generation. In order to maintain voltage and frequency stability on the system, grid operators call upon units to provide standby increases or reductions in supply. These standby services, known as ancillary services, were established after deregulation as a means toward valuation of the services that vertically integrated utilities traditional deployed in-house to maintain the grid's stability.

Ancillary service markets are categorized based on the timescale of the response to a request. Figure 9.5 illustrates the temporal and operational topology of existing ancillary service markets. Spinning reserves, for instance, must

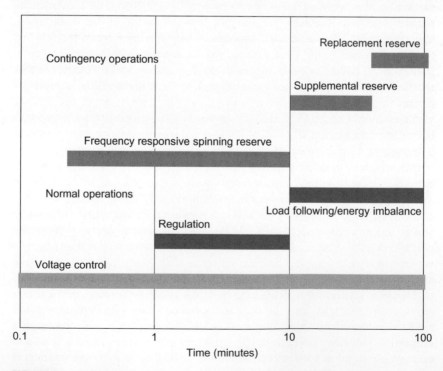

FIGURE 9.5 Temporal and operational classifications in the ancillary service topology. *(Source: Kirby [38])*

be provided within roughly 10 minutes of the grid operator's request. Regulation services, in comparison, are intended for minute-to-minute fine-tuning of the grid frequency. Typically these services are the most difficult for conventional generators to provide, and typically command the highest price. In the absence of ancillary services, a violation of the grids narrow frequency thresholds (+ or − 1%) will result in cascading failure, as synchronous generators disconnect to protect their systems [13].

Regulation Services

Both regulation and load-following services are utilized on a daily basis as part of the grid's normal operations. Generators providing regulation typically incur costs for efficiency loss due to throttling and ramping, environmental costs due to increased emissions, increased O&M costs due to equipment wear and tear, and lost opportunity costs in the energy market [14]. Consequently, a price premium is placed on regulation services based on these costs and the difficulty of closely following regulation requests with conventional thermal units [15].

Historically, regulation markets have not reflected differences in speed or accuracy of delivery. This is indeed changing as new ancillary service market products have appeared in ISO-NE that pay premiums for fast-acting regulation services. These "mileage-payment" market products are also under review in CAISO and PJM [37]. In Chapter 6, Sanders et al. discuss the management of system frequency in the context of the California market.

Procuring higher quality regulation control may reduce the system's regulation requirements and reduce the cost of regulation services. A regulation market structure that discerns the quality of regulation services will augment the grid's reliability and support the development of alternative regulation providers. Dispatchable loads are able to provide high-quality regulation service. The response times of low-latency smart grid protocols are generally less than 500 ms compared to fossil-fuel generation, which sometimes requires 1–10 seconds to provide regulation services [16].

Solid-state controls are expected to perfectly follow system operator regulation commands, making DLC the regulation provider of choice. Brendan Kirby notes that loads with large adjustable speed drives, or solid-state power supplies, are ideal candidates [15]. A body of literature has outlined ways in which electric vehicles could provide regulation services in strategies collectively referred to as vehicle-to-grid technology [17–19]. Hindsenberger et al. present an analysis of this approach in the context of New Zealand's system in Chapter 19. Using loads to provide regulation services, as opposed to wind turbines, has the added reliability advantage that the power system will not lose both its regulation and energy supply resource if the wind dies down. Also, additional regulation services will be crucial to the deployment of utility-scale solar PV. There is the preliminary indication that solar PV presents a greater need for regulation services than wind energy.

Load Following

Load following and energy imbalance ancillary service markets are best suited to managing the ramping events triggered by high-capacity wind integration. Dispatchable DR can be utilized as a tool for smoothing wind ramping events.

Large-scale wind ramp events, like conventional contingencies, occur infrequently and also are typically slower than conventional load ramps [20]. Michael Milligan emphasizes that the vast majority of wind ramp events observed in the Eastern Wind Integration and Transmission Study have ramp rates of less than 4,000 MW/hr, which, in the context of a footprint-wide load of 300,000 MW, are relatively easy to accommodate with moderate system flexibility and sufficient transfer capacity in the transmission system [3].

The asymmetry of extreme wind and load ramps (Figure 9.6) illustrates the greater frequency of economically constrained conditions, as opposed to reliability-constrained conditions. Essentially meteorology and the physical relation of a turbine's power output being proportional to the wind's velocity cubed, result in wind up-ramps that are less "peaky" than turbine down-ramps [20]. Compounding this asymmetry is the fact that load down-ramps, which occur when people turn things off, are less "peaky" than load up-ramps. In other words, people come home and turn things on all at the same time, but turn things off over longer periods of time with a greater degree of randomness. Thus wind turbine operators are left with a greater likelihood of needing to "spill energy" (an economic constraint) than procure additional peak capacity from a battery or

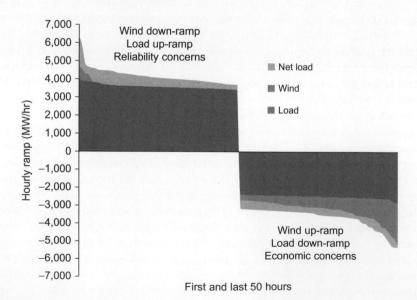

FIGURE 9.6   First and last 50 hours of a ramp duration curve indicates the asymmetric nature of wind and load ramps. *(Source: Kirby and Milligan [20])*

natural gas turbine (a reliability constraint). Loads well suited to profit from spilled energy during extreme wind up-ramps include water pumping, irrigation, municipal treatment facilities, thermal storage in large buildings, industrial electrolysis, aluminum smelting, electric vehicle charging, and shale oil extraction [20]. Incentivizing these loads to take advantage of excess power during extreme wind up-ramps parallels market structures that incentivize industrial loads to turn off during extreme reliability events.

Energy markets are generally able to obtain a great deal of load-following response from intermediate and peaking generators without explicitly paying for them [14]. There is often no premium placed on maneuverability in load-following ancillary service markets. Adequate incentive for load-following DR participation will most likely occur during events that exceed the maximum ramp rate of plants already in operation during the event. The economics of peaking plant operation will typically set the price in load-following markets.

The same market forces that drive reliability resources will govern load-following DR. Utilizing DR as a resource for extreme wind ramp events is analogous to utilizing DR as a resource for extreme heat waves. Both events, after all, stem from high- and low-pressure systems driven by synoptic-scale Rossby waves. One key difference, however, is that economic load-following applications will likely demand higher dispatch frequency than reliability modes typically require. As discussed below, this will have important implications both for the economics that drive those applications and the resources topology necessary to serve them.

Contingency Reserves

Spinning, non-spinning, and supplemental operating reserves all classify as contingency reserves on different temporal scales. These resources are called into action for reliability purposes, typically when a generator unexpectedly shuts down. Immediate response is required from reserves to provide frequency and voltage support for a short duration (minutes for spinning and hours for non-spinning). Milligan notes that because wind plant outages typically only occur at the level of individual turbines and because short-duration fluctuations are averaged out over the array, wind output typically changes only by small fractions of its nameplate capacity. By contrast, a conventional generator can trip during peak load hours and therefore presents greater reliability constraints than wind and solar [14].

Advanced metering infrastructure and low-latency communication networks allow certain responsive loads to respond faster, and at a lower cost, to contingency events than traditional generators. Quite simply, it is more efficient to turn loads off for short durations than to leave generators "spinning" for long durations. Residential and commercial air conditioning, refrigeration, pumping loads, hot water heating and industrial aluminum smelting all qualify as attractive spinning reserve loads [15]. These loads are ideal because their use generally coincides with peak spinning reserve pricing. Steady revenue streams and minimized disruptions encourage the use of DLC participation. Spinning reserves are called up relatively infrequently (every few days), and for short durations (11 minutes for spin and non-spin) [20]. The aforementioned loads would be able to collect additional revenue streams without sacrificing the intent of the original energy service.

WIND INTEGRATION COST

Wind integration costs are typically defined as those incremental costs incurred in the operational time frames that can be attributed to the variability and uncertainty introduced by wind generation. These integration costs can be minimized by thoroughly utilizing ancillary services provisioned by DLC. The increased operational costs, brought onto the bulk power system, can be temporally disaggregated to include regulation, load following, and unit commitment. The integration costs associated with the merit order effect are sometimes treated separately and sometimes bundled into unit-commitment costs. Studies find that the cost of integrating wind rises with greater wind penetrations [5]. DLC DR is capable of decreasing the regulation and load-following portions of the integration costs. Nine of the sixteen wind integration studies surveyed by Wiser disaggregate integration costs by temporal classification [21]. With the caveat that each study represents a different relative wind capacity, ranging from 11% to 48%, Figure 9.7 illustrates the average share of each temporal classification's integration cost.

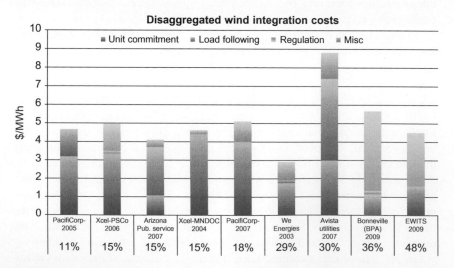

FIGURE 9.7   Wind Integration costs, temporally disaggregated. *(Source: Wiser and Bolinger [21])*

Unit-commitment integration costs are typically greater than regulation or load-following integration costs. Unit-commitment cost-mitigation strategies for wind integration include, but are not limited to, aggregating wind plant output over large geographic regions [22], consolidating balancing authorities [23], increasing the fidelity of wind forecasting [24] intra-hour wind scheduling [25], making better use of physically (as opposed to contractually) available transmission lines, dynamic thermal line rating [26], and improved unit-commitment algorithms that incorporate adaptive load management [27]. Economic DR programs that provide dynamic pricing signals to participants can be utilized to mitigate the unit-commitment costs of wind integration.

Minimizing regulation and load-following integration costs is, of course, also essential to maximizing wind capacity integration. It is worth noting that Figure 9.7 is a representation of integration costs, not capacity. Therefore, since the price of regulation is several times that of other services, the cumulative MWs of required regulation are an even smaller fraction of the ancillary MWs required to integrate wind. The annual average price ($/MWh) of regulation up and down is typically two to ten times that of spinning reserves and twenty to thirty that of non-spinning reserves [28].

The clear trend among the studies surveyed is that unit-commitment costs account for the majority of operational costs associated with VER integration. While there are exceptions, these are typically attributable to variations in accounting methodology employed by the study authors. For example, the Avista study, which focused on wind integration in the Pacific Northwest region, attributes the largest fraction of integration costs to load-following services [29]. This anomaly is driven by the study's mingling of wind forecasting error with load-following requirements. Wind forecasting errors generally fall into the unit-commitment bin, although some amount of intra-hour forecasting errors falls into load following as well.

Strategies for decreasing regulation and load-following integration costs are less extensively documented than those of unit commitment. Utilizing DR to firm VERs through ancillary services provides such a strategy.

DEMAND RESPONSE RESOURCE TOPOLOGY

To examine the relative merits of various DR resources classes and their suitability for VER integration-driven applications, it is helpful to deconstruct the panoply of DR resources into a finite framework disaggregated by relevant metrics. Figure 9.8 presents a topology of DR resources that is framed in terms of resource dispatchability and ancillary service timescale. Because VER integration is dissected into the timescales necessary to attribute incremental operational costs into discrete ancillary markets, it is instructive to frame the DR resource topology in the same way.

Because of typical latencies, tariff-based DR programs are not suitable for dispatch on timescales relevant to daily reliability concerns. The price structure

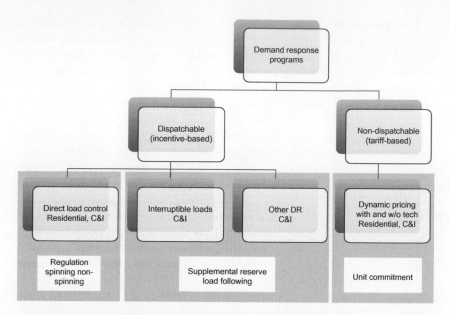

FIGURE 9.8   Demand response resource topology.

of these programs can be tweaked to optimize the unit-commitment process with VERs. In a simplified sense, this amounts to charging more for electricity when demand is high and the system is strained. Dynamic pricing with enabling technology refers to rate structures that offer customers time-varying electricity prices on a day-ahead or real-time basis. Dynamic pricing without enabling technology refers to the same program structures, but without the appropriate technology to automate price-response behavior.

One important drawback of price-responsive DR is that the marginal contribution to system reserves declines with penetration [30]. In other words, the effective load-carrying capability (ELCC) of economic DR is subject to the law of diminishing returns as a reliability asset. Conversely, truly dispatchable DR can be controlled directly by the system operator and can therefore theoretically achieve an effective load-carrying capacity of up to 100%. This means that, in reliability terms, dispatchable DR resources could potentially be valued at the same level as thermal generating capacity.

The next level of the DR topological tree distinguishes the five classes of DR type outlined in FERC's National Assessment of Demand Response Potential [31]. Interruptible tariffs refer to DR programs where customers agree to reduce consumption during system reliability problems in return for incentive payments. These programs are typically only available for large commercial and industrial customers. Interruptible tariff DR resources are dispatchable with time spans ranging from seconds to hours, implying the ability to bid into both regulation, spinning, and load-following ancillary service markets [10].

Other DR programs refer to capacity bidding, demand bidding, and other aggregator offerings to medium and large commercial and industrial customers. Some of these programs are price triggered, while others are triggered by reliability events. For the purposes of this analysis other DR programs are considered to be dispatchable with time spans ranging from hours to days.

Direct load control (DLC) refers to customer end uses that are directly controlled by the utility or third-party aggregators and are shut down or moved to a lower consumption level as the market dictates. A wide range of DR resources exist that could be utilized by DLC programs. Residential DLC resources include water heating, pumping loads, refrigerators defrost cycles, battery charges for consumer electronics, washing machines, lights, and stoves. Commercial DLC resources include air conditioners, heat pumps, refrigerated warehouses, electric water heaters, dual-fuel boilers, HVAC systems with thermal storage, and lighting. Industrial DLC resources include induction and ladle metallurgy furnaces, air liquefaction facilities, gas and water pumping, agricultural irrigation, aluminum smelting, and various electrolysis facilities such as chlor-alkali, potassium hydroxide, magnesium, sodium chlorate, and copper [15].

In the commercial and residential sectors, DLC loads are often modeled as what a building owner can submit to DR markets, such as air conditioning for spinning reserves. However, DLC potential from the tenant's perspective is often overlooked. Tenants of large commercial buildings represent a vast potential DR resource that could be accessed by means of competitive DR bidding processes. During the Empire State Building retrofit, Johnson Controls Inc. discovered that it was able to submit 5% of a 10 MW peak load to DR programs from centrally controlled devices. When tenants were allowed to collectively bid into DR programs, facilitated by the building's management, DR capacity grew to 25% of building peak load.[2] During extreme reliability events, when incentive payments were at their peak, tenants were literally "unplugging copy machines." This five fold increase is just one anecdotal example of the tremendous savings inherent in opening up DR facilitated through tenant bidding as opposed to solely building manager response.

The final level of the DR topological tree in Figure 9.8 shows the ancillary services relevant to the class of DR on the basis of the response latency. Since VER integration costs are binned into the same market structure, this facilitates a comparison of DR resource potential and integration requirements. By treating the structure of ancillary services markets as the interface between DR resource potential and VER integration costs, one can estimate the extent to which DR can enable high penetration of wind and solar. Building on the central metaphor of a flexibility supply curve, the framework provided by this type of topological categorization of DR resources elucidates the width, and potentially the height, of the DR portion of that curve.

---

[2]Based on conversations with Paul Rode, Senior Project Manager of the Empire State Building energy retrofit.

It is important to note that interruptible tariff DR programs are well suited to manage reliability-constrained load-following ramps of wind-down and load-up. These scenarios are less frequent than the economically constrained ramps of wind-up, load-down that are well suited to forward capacity DR programs (other DR).

This final level of the resource topology reveals that DLC is well positioned to meet the incremental ancillary service requirements imposed by large-scale deployments of variable renewables. The ability of DLC to be dispatched within short timescales allows it to provide non-spinning, spinning, and even regulation services that account for the lion's share of integration requirements (Figure 9.7). Hindsberger et al. describe how DLC can provide frequency regulation in Chapter 19.

DEMAND RESPONSE POTENTIAL ASSESSMENT

Because DR is used primarily for reliability applications in current markets, resource potential is typically expressed in units of capacity. Peak load reduction is especially relevant for reliability and peak shaving applications; thus a metric of % MW peak reduction is appropriate. A relevant metric for economic DR might be ELCC. The ELCC metric is defined as the amount of new load that can be added to a system after a new unit is added without increasing the system's loss of load expectation above its initial value [30]. The adoption of this method is encumbered by the fact that the calculation of ELCC is more computationally intensive than alternative methods. In addition, because ELCC is a highly non-linear quantity that embodies a broad set of system characteristics, the results are very location dependent, with limited comparability across systems. Nevertheless, capturing the true capacity value of DR resources will be increasingly important as applications for load flexibility evolve.

A comparison of several DR potential studies surveyed indicates that DR could reduce system peak load by roughly between 5 and 15% (Figure 9.9). Although the range of values is relatively large, even in the most conservative case, DR has the potential to offset a large fraction of load. Notably however, the most detailed, bottom-up estimates of resource potential provide the largest estimates, which suggests DR's true potential lies closer to the top of this range.

EPRI's "Assessment of Achievable Potential from Energy Efficiency and Demand Response Programs in the United States" represents the most conservative estimate of DR deployment potential. EPRI's model considers the impact of energy efficiency on peak savings before evaluating DR. This arbitrary loading order creates an energy efficiency peak shaving bias. The residential and commercial sectors were modeled bottom-up with GEP's loadMAP model, while the industrial sector was modeled top-down with data from the Energy Information Administration's 2008 Annual Energy Outlook. Direct load control measures modeled include personal electronics, refrigerators, air conditioners

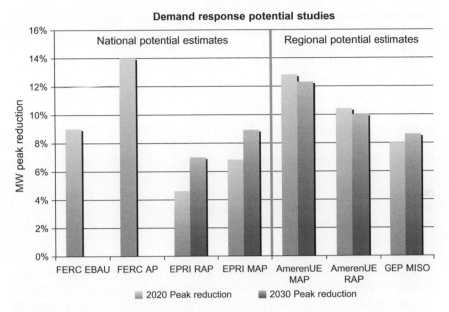

FIGURE 9.9   Peak shaving estimates from existing demand response potential studies. *(Authors compiled from FERC [31], GEP [33,34].)*

(C&I and residential), residential water heating, commercial lighting, and other, unspecified industrial DLC processes [32].

"The AmerenUE DSM Market Potential Study" provides a Midwestern region focus, as opposed to the national EPRI estimate. The study models the residential and C&I sectors from the bottom up, as opposed to EPRI's top-down econometric method for the industrial sector [33], and, like the EPRI study, utilizes the Global Energy Partners LoadMAP model. EPRI's Midwest economic potential in 2030 is 12.3% while AmerenUE's economic potential is 16.6%, reflecting a more extensive list of energy-efficiency measures. However, EPRI's Midwest realistic achievable potential is 7.5% in 2030, compared with 7.3% for AmerenUE. The dramatic drop in AmerenUE's achievable potential is attributed to stark market acceptance rates. In particular, direct load control DR potential estimates include only residential water heating and air conditioning.

"The Global Energy Partners Assessment of Demand Response and Energy Efficiency Potential for the Midwest ISO" provides another bottom-up assessment of demand-side management potential [34]. Data were provided by local utilities where available, but did not incorporate state energy efficiency (EE) portfolio standards (Illinois, Indiana, Michigan, Ohio). This report assumes stagnant EE and DR growth by 2020, corresponding to the untenable assumption of when, if ever, market saturation will occur. The AmerenUE report assumes a similar inflection point for EE and DR potential. The largest DLC DR potential corresponds to residential air conditioning and hot water heating, representing 93% of DLC potential in 2030. Commercial DLC programs make up the remaining 7%, although industrial DLC potential was not modeled.

FERC's "National Assessment of Demand Response Potential" is the most comprehensive state-by-state database of DR potential yet developed. The report estimates the amount of DR that could be deployed based on the number of participating customers, availability of dynamic pricing, advanced metering infrastructure, the use of enabling technologies, and varying responses of different customer classes [31]. These scenarios are Business-as-Usual, Expanded Business-as-Usual, Achievable Participation, and Full Participation.

This analysis focuses on the Expanded Business as Usual (EBAU) and Achievable Potential (AP) scenarios of the report as a credible range for DR deployment. The Business as Usual scenario assumed no additional growth beyond current participation levels, which is clearly not credible in light of current market growth. Likewise, the Full Participation scenario assumes a level of dynamic pricing program adoption not likely in the current market environment.

FERC's EBAU scenario departs from the BAU scenario by assuming "best practice" participation levels for the current mix of DR (Figure 9.10). Partial deployment of AMI and a small number of customers (5%) choosing dynamic pricing DR programs are also assumed. Of the panoply of DLC loads described in this report, only three were considered in FERC's EBAU DLC modeling: residential and commercial air conditioning and irrigation loads in a few states. Consequently, FERC's EBAU DLC potential is a relatively conservative estimate of available DLC potential.

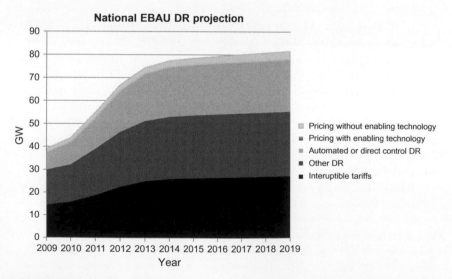

FIGURE 9.10 FERC's EBAU scenario disaggregated by program type with increasing participation levels in current DR programs. (Source: FERC [31])

FERC's Achievable Participation scenario assumes that AMI infrastructure is universally deployed by 2019, that a dynamic pricing tariff becomes the default, and that 60–75% of customers stay on dynamic pricing rates. The key difference between the EBAU and AP scenario is the potential peak reduction unlocked by dynamic pricing DR programs enabled through AMI. The initial growth of IT, DLC, and other DR programs closely tracks the EBAU scenario for the first few years until pricing programs take off with AMI deployment.

The diminishing capacity of DLC after the year 2013 is an odd characteristic of Figure 9.11 that deserves attention. Figure 9.12 highlights the observed decrease in DLC MWs by region. The FERC study is not only assuming that DLC participation peaks by the year 2013, but also that formerly subscribed DLC resources will be displaced by dynamic pricing capacity.

This assumption convolves two discrete categories of DR as outlined in the topology of this report. Switching DLC capacity to dynamic pricing transfers DR resources from the dispatchable (maximum ELCC) branch to the non-dispatchable (minimum ELCC) branch. Such a shift is suboptimal from the perspective of the grid's reliability, efficiency, and flexibility in VER integration.

Furthermore, to the extent that increased VER penetration drives greater demand for ancillary services applicable to DLC operation (regulation, spin, and non-spin), it will be unlikely that dynamic pricing displaces DLC capacity in this way. From this perspective, the assessment of DLC potential in the FERC report can be regarded as a conservative estimate, especially in the context of continued growth in the deployment of variable energy resources.

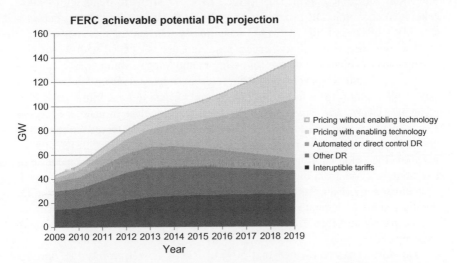

FIGURE 9.11   FERC's AP scenario disaggregated by DR program type with 100% AMI rollout and 60–75% of customers engaged with dynamic pricing rates. *(Source: FERC [31])*

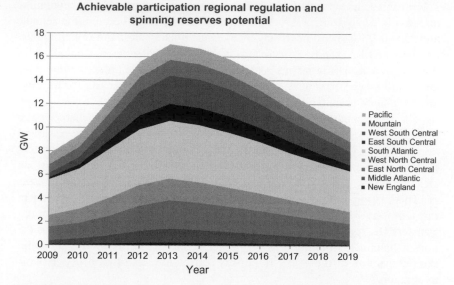

FIGURE 9.12 DLC potential from FERC's AP scenario disaggregated by region. *(Source: FERC [31])*

WIND INTEGRATION AND DEMAND RESPONSE

Illustrating DR's ability to decrease wind integration costs requires contrasting DR's evolution with the evolution of ancillary service requirements that originate from wind integration. Due to the FERC AP scenario treatment of DLC relative to economic DR programs, as well as the overall diminishing cumulative MWs of DLC, FERC's EBAU scenario was chosen for this analysis.

The evolution of ancillary service requirements associated with wind integration was established with the following methodology. Smith et al. estimated the capacity required for total operating reserves to increase from 5% to 7% in a 25% RPS scenario for a four-utility combined balancing area with a peak load of about 21 GW [35]. This corresponds to a 40% increase in ancillary reserve requirements. Upper and lower bounds of total operating reserves (contingency and normal operations) in power systems range from 5% to 10% [28]. Combining these facts lead to an upper and lower bound on the incremental increase in operating reserves from high wind penetration scenarios of 2% and 4% of peak system capacity. Data from Heffner's international survey of loads providing ancillary services indicates that PJM obtained operating reserves at 4.75% of peak capacity and that ERCOT obtained operating reserves at 7.5% of peak capacity [28].

The dataset also indicated that 50% of PJM's operating reserves derived from regulation and instantaneous spinning reserves and that 75% of ERCOT's operating reserves derived from regulation and spinning reserves.

The majority of this capacity corresponds to spinning rather than regulation reserves (typically ~0.7% of load) [2]. Smith et al. estimated that the regulation requirement increases from 0.65% in the base case to 0.75% with a 25% RPS [35].

Putting these disparate studies and percentages together yields upper and lower bounds on regulation, spinning, and non-spinning reserve requirements of 1% to 3% of peak capacity. Upper and lower bounds on load following and supplemental reserve requirements were established to be 0.5% to 2% of peak capacity. Peak capacity in 2019, not including the growth in DR, was back calculated from FERCs EBAU scenario. A relative understanding of DR's ability to mitigate wind integration costs is achieved by combining the cumulative incremental increase in DR from FERCS EBAU scenario with the upper and lower bound cumulative increase in reserve capacities (Figure 9.13). It must be emphasized that this method of comparison provides insights into the orders of magnitude involved but does not offer a finely tuned projection of the future.

The variations between regions in the incremental ancillary services needed to accommodate large VER penetrations reflect differences in balancing authority size, transmission system topology, available transfer capacity, market design, load shape, resource diversity and temporal correlation, diurnal and

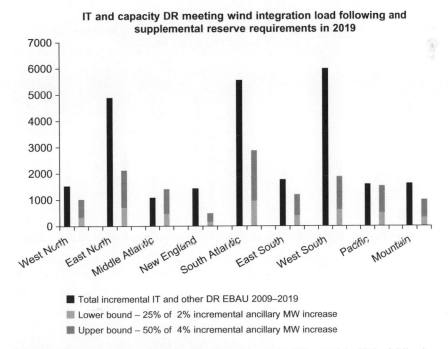

FIGURE 9.13 Comparison of interruptible tariff and capacity DR potential with load following and supplemental reserve requirements for wind integration at 25% penetration.

seasonal wind profile, system scheduling frequency, and a host of other variables. The largest, most diverse system with the most flexible generation mix and stiffest transmission system will have the lowest need for incremental firming capacity and vice versa.

Total incremental DLC potential from FERC's EBAU scenario is of the same order as incremental ancillary firming capacity required for significant wind penetrations (Figure 9.14). Inspection of the preceding figures reveals that load following and supplemental reserve requirements are satisfied by interruptible tariff and capacity-based DR but that regulation, spin, and non-spin requirements are not always met by DLC DR.

The upper bound of regulation, spin, and non-spin requirements exceeds the DLC EBAU MWs in all regions. It is important to note that some of the interruptible tariff DR is actionable on spinning and non-spinning reserve time-scales. Although the topology described in this study draws a line between interruptible tariff DR programs and DLC programs, some interruptible tariff DR resources are capable of providing instantaneous response. LBNL's 2009 empirical review of existing DR resources revealed that 15% of IT DR is capable of response from 1–30 minutes in SPP and 35% in MISO. The EBAU FERC scenario assumes that each mutually exclusive DR topology grows under aggressive scenarios without transferring MWs from other DR types.

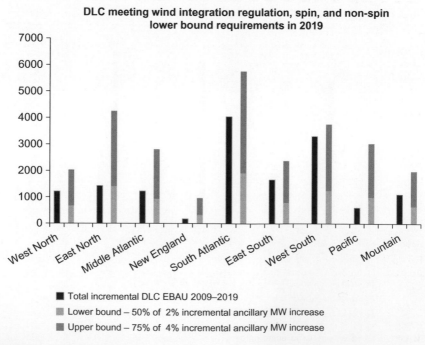

FIGURE 9.14 DLC compared to regulation, spin, and non-spin wind integration requirements.

In reality, commercial and industrial loads that are currently signed up for interruptible tariff DR could switch to DLC if it was better suited to provide valuable fast-acting response. In addition, given the limited scope of resources included in FERC's EBAU DLC potential assessment, this represents a lower bound of available DLC potential.

CONCLUSIONS

DR is a rapidly growing resource in the electricity sectors portfolio that can help facilitate large penetrations of renewables on the grid. Instantaneously dispatchable DLC offers the greatest value on the regulation and spinning reserves timescales. Existing capacity and tariff based DR programs are well suited to provide load following and contingency reserves during economic and reliability-constrained wind ramps.

Even under relatively conservative assumptions, total incremental DLC potential is of the same order as the incremental ancillary firming capacity required for significant wind penetrations. Moving forward, this resource potential could be further augmented through building automation and software innovations that enable industrial, commercial, and residential loads to be dispatched into DR markets without requisite oversight by a sophisticated energy user.

This suggests DR has the potential to take a prominent place in the portfolio of wind integration strategies. If these resources have access to clear price signals and market products that facilitate their transparent participation, DR could provide an important source of distributed flexibility for the grid as a whole. Dispatchable DR resources will be of particular importance when it comes to integrating renewable resources with variable output and meeting the increased demand for ancillary services that VERs will drive.

ACRONYMS

AP	Achievable Potential
AMI	advanced metering infrastructure
BAU	Business-As-Usual
C&I	commercial and industrial
CAGR	compound annual growth rate
DR	demand response
DOE	Department of Energy
DLC	direct load control
ELCC	effective-load carrying capacity
EWITS	Eastern Wind Integration and Transmission Study
FERC	Federal Energy Regulatory Commission
GHG	greenhouse gas
IT	interruptible tariffs
LOLE	loss of load expectation
LTRA	Long Term Reliability Assessment
MAP	Maximum Achievable Potential
NERC	North American Electric Reliability Corporation

NREL   National Renewable Energy Laboratory
PNNL   Pacific Northwest National Laboratory
RAP   realistic achievable potential
VER   variable energy resource

REFERENCES

[1] R. Hledik, How green is the smart grid? *Electricity J.* 22 (3) (2009) 29–41.

[2] R. Pratt, P. Balducci, C. Gerkensmeyer, S. Katipamula, M.C.W. Kintener-Meyer, T.F. Sanquist, *The Smart Grid: An Estimation of the Energy and CO_2 Benefits*, Pacific Northwest National Laboratory, Richland, WA, 2010.

[3] NREL, *Eastern Wind Integration and Transmission Study*, Prepared by Enernex Corporation for the National Renewable Energy Laboratory, Knoxville, TN, 2010.

[4] NERC, *Long–Term Reliability Assessment*, North American Electric Reliability Corporation, Princeton, NJ, 2009.

[5] H. Holttinen, P. Meibom, B. Parsons, E. Ela, A. Orths, M. O'Malley, *Impacts of Large Amounts of Wind Power on Design and Operation of Power Systems, Results of IEA Collaboration*, 8th International Workshop on LargeScale Integration of Wind Power into Power Systems as well as on Transmission Networks of Offshore Wind Farms, Bremen, Germany, 2009.

[6] M.B. McElroy, X. Lu, et al., Potential for wind–generated electricity in China, *Science* 325 (5946) (2009) 1378–1380.

[7] NREL, *Western Wind and Solar Integration Study*, Prepared by GE Energy for the National Renewable Energy Laboratory, Schenectady, New York, 2010.

[8] P. Denholm, E. Ela, B. Kirby, M. Milligan, *The Role of Energy Storage with Renewable Electricity Generation*, National Renewable Energy Laboratory, Golden, CO, 2010.

[9] A. Vojdani, Smart integration, *IEEE Power Energy Mag.* 6 (6) (2008) 71–79.

[10] P. Cappers, C. Goldman, D. Kathan, Demand response in U.S. electricity markets: empirical evidence, *Energy* 35 (4) (2010) 1526–1535.

[11] FERC, *Assessment of Demand Response and Advanced Metering*, Federal Energy Regulatory Commission, Washington, DC, 2008.

[12] EPRI, *The Green Grid: Energy Savings and Carbon Emissions Reductions Enabled by a Smart Grid*, Palo Alto, CA, 2008.

[13] A. von Meier, *Electric Power Systems: A Conceptual Introduction*, John Wiley & Sons, Inc., Hoboken, New Jersey, 2006.

[14] B. Kirby, M. Milligan, *Capacity Requirements to Support Inter–Balancing Area Wind Delivery*, National Renewable Energy Laboratory, Golden, CO, 2009.

[15] B.J. Kirby, *Load Response Fundamentally Matches Power System Reliability Requirements*, IEEE Power Engineering Society General Meeting, June 2007.

[16] A. Brooks, E. Lu, D. Reicher, C. Spirakis, B. Weihl, Demand dispatch, *IEEE Power Energy Mag.* 8 (3) (2010) 20–29.

[17] W. Kempton, J. Tomic, Vehicle–to–grid power implementation: From stabilizing the grid to supporting large–scale renewable energy, *J. Power Sources* 144 (1) (2005) 280–294.

[18] J. Tomic, W. Kempton, Using fleets of electric–drive vehicles for grid support, *J. Power Sources* 168 (2) (2007) 459–468.

[19] H. Lund, W. Kempton, Integration of renewable energy into the transport and electricity sectors through V2G, *Energy Policy* 36 (9) (2008) 3578–3587.

[20] B. Kirby, M. Milligan, *Utilizing Load Response for Wind and Solar Integration and Power System Reliability*, Windpower 2010, Dallas, TX, 2010.

[21] R. Wiser, M. Bolinger, 2009 *Wind Technologies Market Report*, U.S. Department of Energy, 2010.

[22] C.L. Archer, M.Z. Jacobson, Supplying baseload power and reducing transmission requirements by interconnecting wind farms, *J. Appl. Meteorol. Clim.* 46 (2007) 1701–1717.

[23] B. Kirby, M. Milligan, *Combining Balancing Areas' Variability: Impacts on Wind Integration in the Western Interconnection*, Windpower 2010, Dallas, TX, 2010.

[24] E. Natenberg, J. Zack, S. Young, J. Manobianco, C. Kamath, *A New Approach Using Targeted Observations to Improve Short–Term Wind Power Forecasts in the Tehachapi Pass of California*, Windpower 2010, Dallas, TX, 2010.

[25] M. Milligan, K. Porter, E. DeMeo, P. Denholm, H. Holttinen, B. Kirby, Wind power myths debunked, *IEEE Power Energy Mag.* 7 (6) (2009) 89–99.

[26] K. Hur, M. Boddeti, et al., High–wire act: ERCOT balances transmission flows for Texas–size savings using its dynamic thermal ratings application, *IEEE Power Energy Mag.* 8 (1) (2010) 37–45.

[27] M. Ilic, *Unit Commitment for Sustainable Integration of Large-Scale Wind Power and Responsive Demand*, FERC Technical Conference on Unit Commitment Software, Docket AD10-12, Washington, DC, 2010.

[28] G. Heffner, C. Goldman, B. Kirby, M. Kintner-Meyer, *Loads Providing Ancillary Services: Review of International Experience*, Ernest Orlando Lawrence Berkeley National Laboratory, Orlando, FL, 2007.

[29] Enernex, *Avista Corporation Wind Integration Study*, Avista Corporation, Knoxville, TN, Prepared by EnerNex Corporation, 2007.

[30] R. Earle, E.P. Kahn, E. Macan, Measuring the capacity impacts of demand response, *Electricity J.* 22 (6) (2009) 47–58.

[31] FERC, *A National Assessment of Demand Response Potential*, Federal Energy Regulatory Commission, Washington, DC, 2009.

[32] EPRI, *Assessment of Achievable Potential from Energy Efficiency and Demand Response Programs in the US*, Electric Power Research Institute, Palo Alto, CA, 2009.

[33] GEP, *AmerenUE Demand Side Management (DSM) Market Potential Study*, AmerenUE, Walnut Creek, CA, Prepared by Global Energy Partners, LLC, 2010.

[34] GEP, *Assessment of Demand Response and Energy Efficiency Potential for Midwest ISO*, Midwest ISO, Walnut Creek, CA, Prepared by Global Energy Partners, LLC, 2010.

[35] J.C. Smith, M.R. Milligan, E. DeMeo, B. Parsons, Utility wind integration and operating impact state of the art, *IEEE Trans. Power Syst.* 22 (3) (2007) 900–908.

[36] A. Cheung, Demand response in the US organized markets: A snapshot, *Bloomberg New Energy Finance*, 2010.

[37] T. Hesser, Smart grid North America: Moving past stimulus, *Bloomberg New Energy Finance*, 2011.

[38] B. Kirby, *Frequency Regulation Basics and Trends*, Oak Ridge National Laboratory, Oak Ridge, TN, ORNL/TM-2004/291, December 2004.

Chapter 10

Riding the Wave: Using Demand Response for Integrating Intermittent Resources

Philip Hanser, Kamen Madjarov, Warren Katzenstein, and Judy Chang[1]

Chapter Outline

INTRODUCTION

Renewable power's share in total generation is expected to grow significantly over the next decade and beyond. As of early 2011, 29 U.S. states have adopted renewable portfolio standards (RPS) and 7 states have established goals.[2] The U.S. Energy Information Administration projects that the share of generation coming from renewable fuels, including conventional hydro, will grow from 11% in

---

[1]The views expressed in this chapter are solely those of the authors and should not be construed to be either those of *The Brattle Group* or those of *The Brattle Group's* clients.
[2]Source: Database for State Incentives for Renewables & Efficiency (DSIRE) [1].

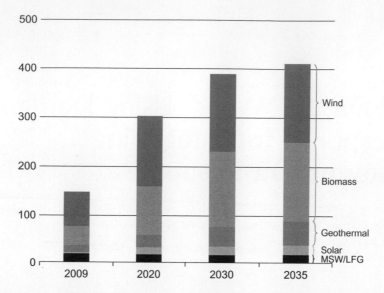

FIGURE 10.1   2009–2035 renewable electricity generation by energy source (billion kWh). *(Source: Annual Energy Outlook 2011, U.S. DOE, EIA)*

2009 to 14% in 2035 (see Figure 10.1).[3] Furthermore, states like California have established ambitious targets for the level of renewable penetration for their electricity grids—California's target stands at 33% renewable generation by 2020.

System operators and planners have increasingly focused their attention on the challenge of integrating these growing quantities of renewable generation, primarily wind and solar. Unlike conventional generation, wind and solar plants cannot be dispatched on a predetermined schedule, and their output is dependent on variable and largely unpredictable fluctuations in weather conditions—wind, cloud cover, air temperature, etc. This variability and unpredictability are the two principal characteristics of wind and solar generation that affect power system operations and reliability. To further complicate matters, system operators are required to accommodate renewable generation as a "must-take" generation resource and allow it to deliver its variable and unpredictable output on the system without significant curtailment. With growing renewable generation penetration levels, therefore, the issues of variability and unpredictability must be addressed in a comprehensive manner.

The range and depth of issues associated with the integration of significant levels of intermittent renewable generation have become the focus of much attention, discussion, and research for utilities, regional transmission organizations,[4] public utility commissions, and other entities tasked with the reliable operation of

---

[3]Annual Energy Outlook 2011 [2].

[4]Chapter 6 by Sanders et al, for example, discusses the issues confronting CAISO with 33% RPS by 2020.

the grid. Over the past few years, a number of studies by these entities have been completed. The North American Electric Reliability Corporation (NERC) examined the emerging importance of the operational and reliability impacts that renewable resources may have on the grid in the 2008 report titled "Accommodating High Levels of Variable Generation."[5,6] Among the ISO/RTO reports are the 2007 CAISO Integration of Renewable Resources studies [3], and that of 2007–2011,[7,8] the New England Wind Integration Study (2010),[9] NYISO 2010 Wind Generation Study,[10] and the Analysis of Wind Generation Impact on ERCOT Ancillary Services Requirements (2008).[11] The utility studies include the Minnesota Wind Integration Study (2006),[12] the Wind Integration Study for Public Service Company of Colorado (2006, 2008),[13] and the study, Operational Impacts of Integrating Wind Generation into Idaho Power's Existing Resource Portfolio (2007).[14] These studies, and many others, have provided a range of tools and methodologies designed to establish the needed changes in operational practices, conventional resource procurement, and ancillary service deployments.

While the reports differ in some technical aspects, each recognizes that a large penetration of solar and wind generation will result in a greater need for ancillary services such as regulation and load-following. In addition, flexible resources providing those services will be expected to exhibit faster ramp rates and will be called on more frequently to provide such services. Some studies have suggested that a new type of regulation service might need to be defined in order to recognize the faster response and faster ramping capabilities that will be needed in the future. The Federal Regulatory Energy Commission (FERC) has responded to this last issue by issuing a proposal to create a new market for regulation that rewards storage technologies for providing such service.[15]

---

[5]http://www.nerc.com/docs/pc/ivgtf/IVGTF_Outline_Report_040708.pdf.

[6]One of the seemingly mundane operational questions faced is simply creating a uniform approach to the reporting of renewable resource outages and deratings that is comparable to that for conventional generation. See http://collaborate.nist.gov/twiki-sggrid/pub/SmartGrid/PAP16Objective1/NERC_GADS_Wind_Turbine_Generation_DRI_100709_FINAL.pdf.

[7]http://www.caiso.com/Documents/Integration-RenewableResourcesReport.pdf.

[8]http://www.caiso.com/Documents/IntegrationRenewableResourcesOperationalRequirementsand GenerationFleetCapabilityAt20PercRPS.pdf.

[9]http://www.iso-ne.com/committees/comm_wkgrps/prtcpnts_comm/pac/reports/2010/newis_es.pdf.

[10]http://www.nyiso.com/public/webdocs/newsroom/press_releases/2010/GROWING_WIND_Final_Report_of_the_NYISO_2010_Wind_Generation_Study.pdf.

[11]http://www.uwig.org/AttchB-ERCOT_A-S_Study_Final_Report.pdf.

[12]http://www.state.mn.us/portal/mn/jsp/content.do?contentid=536904447&contenttype=EDITORIAL &hpage=true&agency=Commerce.

[13]http://www.nrel.gov/wind/systemsintegration/pdfs/colorado_public_service_windintegstudy.pdf http://www.uwig.org/CRPWindIntegrationStudy.pdf.

[14]http://www.idahopower.com/AboutUs/PlanningForFuture/WindStudy/default.cfm.

[15]Frequency Regulation Compensation in the Organized Wholesale Power Markets, Notice of Proposed Rulemaking, 18 CFR 35 (Feb. 17, 2011), FERC Stats. & Regs. 61,124 (2011) (NOPR). http://www.ferc.gov/whats-new/comm-meet/2011/021711/E-4.pdf.

Fewer attempts have been made, however, to study in greater detail the various roles that *load* can play in providing ancillary services and thus actively assist in the integration of renewable generation. This chapter provides an overview of the nature of sub-hourly flexible operating reserves given their growing importance in balancing intermittent generation on the system grid. This chapter is organized as follows: "Overview of System Operation Flexibility Issues" discusses flexible operating reserves in the context of increasing penetration of renewable generation; "Wind Integration Studies and Ancillary Service Requirements" reviews essential methodologies and tools deployed by wind integration studies; "Additional System Challenges" and Load as a Flexible Resource – System Programs and Experience" point to some additional system challenges in accommodating large penetration levels of renewable generation; "Structuring Load Control Products for Intermittent Resources" reviews recent experience in deploying load as a flexible operation reserve; and "Conclusion: Benefits of Substituting Load for Generation" discusses the use of optimal inventory control to design direct load control programs for providing ancillary services such as regulation and load-following, as well as the potential benefits from such an approach in integrating intermittent resources.

OVERVIEW OF SYSTEM OPERATION FLEXIBILITY ISSUES

The existing power grid system has historically been designed to serve the continuously fluctuating levels of load from one instance to the next. The grid, however, was not designed to simultaneously serve both fluctuating loads and relatively unpredictable and highly variable generation such as wind and solar generation. Although conventional load also exhibits pronounced variability, modern forecasting techniques have been quite successful in using real-time and historical electricity usage and weather data to predict conventional electricity demand. This significant level of load predictability has given system operators the ability to update and develop monthly, weekly, and day-ahead resources schedules and, thus, serve load with the optimal size and type of resources through a daily optimized unit-commitment process.

In a traditional system, where virtually all generation resources are controllable and dispatchable, with the exception of some hydro generation, and the majority of electricity demand is served by "baseload" generation such as nuclear and coal, the variations in load are met by generation that can rapidly respond to instructions to adjust their output to the oscillating portions of electricity demand above the "baseload" levels. Meeting the constant variations in electricity demand is more complex than this and actually requires several levels of generation scheduling and generation unit types.

First, on the highest resolution timescales such as second-to-second and minute-to-minute, certain generators are interconnected to the grid in a way that allows them to automatically respond to the varying demand levels. Those generators provide what are known as "frequency control" and "regulation" services. *"Primary frequency control* involves the autonomous, automatic, and rapid action, that is, within seconds, of a generator to change its output to compensate for large changes in frequency. Primary frequency control actions are especially important during the period following the sudden loss of generation, because the actions required to prevent the interruption of electric service to customers must be initiated immediately (i.e., within seconds)."[16]

In addition to primary frequency control, the grid operator must have the capability to provide secondary frequency control. *"Secondary frequency control* involves slower, centrally (i.e., externally) directed actions that affect frequency more slowly than primary control (i.e., in tens of seconds to minutes). Secondary frequency control actions can be initiated automatically or in response to manual dispatch commands. *Automatic generation control* (AGC) is an automatic form of secondary frequency control that is used continuously to compensate for small deviations in system frequency around the scheduled value."[17] Secondary frequency control actions can be initiated automatically or in response to manual dispatch commands.[18]

While AGC is usually referred to as *regulation* service, its definition in the sub-hourly time domain varies among control areas. Some studies of renewable integration have modeled regulation service on the 10-minute time interval due to high-resolution wind data limitations.[19] As Figure 10.2 shows, on the other hand, the California Independent System Operator (CAISO) considers regulation to be secondary frequency control that opposes deviations within the 5-minute time scale.[20]

Load-following is another type of ancillary service that has not yet been formally defined in most energy markets. *Load-following* refers to the relatively fast and dynamic adjustment of generation to meet sub-hourly blocks of demand (i.e., 10 or 15 minutes). Load-following provides a similar function to regulation but its time scale is often longer since the time it encompasses allows for larger and steeper changes in real-time demand. For example, as Figure 10.3 illustrates, CAISO considers load-following to be the service that follows the deviations between the five-minute and hourly block schedules.

As discussed above, the reliable moment-to-moment operation of the power grid requires the continuous dispatch of energy blocks to meet the up and down variability of demand across several time frames—from instantaneous to sub-hourly and hourly. In addition, once large levels of variable and unpredictable solar and wind generation are added to the grid, even more resources will have to be devoted to serving the increased variability of net load—that is, load net of

---

[16]Eto, Joseph H. et al., *Use of Frequency Response Metrics to Assess the Planning and Operating Requirements for Reliable Integration of Variable Renewable Generation*, Lawrence Berkeley National Laboratory, LBNL-4121 at p. 9. (http://certs.lbl.gov/pdf/lbnl-4142e.pdf).

[17]Ibid. at p. 9.

[18]Ibid. at p. 9.

[19]NREL [4].

[20]Further discussion is provided in Chapter 6 by Sanders et al.

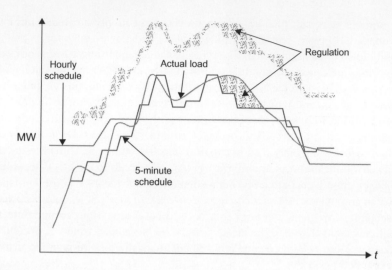

FIGURE 10.2   Regulation service in the CAISO. *(Source: CAISO Integration of Renewable Resources Report, Fig. 5–19[21])*

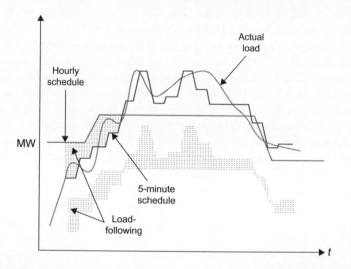

FIGURE 10.3   Load-following service in the CAISO. *(Source: CAISO Integration of Renewable Resources Report, Fig. 5–15[22])*

renewable generation. In simple terms, this need for constant redispatch and readjustment of flexible generation levels is amplified and complicated significantly by increasing levels of renewable real-time generation on the grid. The estimation of future needs for regulation and load-following services, as well

---

[21]http://www.caiso.com/1ca5/1ca5a7a026270.pdf.
[22]Ibid. A similar diagram appears in Chapter 6.

as the characteristics of the generation units that would provide them, has been the main focus of renewable integration studies.

Collectively, the dispatch of generation to compensate for the constant fluctuations in net load needed to maintain system equilibrium and reliability can be referred to as "balancing services." Depending on the specific magnitude and characteristics of the future balancing service needs, they can be provided by already existing generation resources and by a range of properly designed demand-side products (that is, contracts with demand-side resources designed to address the relevant balancing resource needs). The rate at which those flexible generation units' output can be varied becomes the dominant and most essential technical feature because of wind-generating variability and the steep up and down ramps needed to regulate and follow load net of wind ("net load"). It is essential, therefore, that future resource planning and system design are founded on a critical analysis of the optimal generation technology mix as renewable wind and solar generation becomes significant and even potentially dominant resources on the U.S. power grid.

Deregulated markets and large RTOs already largely provide a centralized medium for procuring regulation via the day-ahead market structure. Load-following, on the other hand, is generally not an explicitly defined market product and is provided by a range of spinning units. These resources are dispatched to provide spinning reserves of various time-span definitions (10-minute, 30-minute spinning reserves, etc.). Traditionally, regulation and load-following have been provided by conventional generation units, which meet the technical requirements associated with those services—ramp rates, ability to follow rapid instructions, and so on. The majority of renewable integration studies, however, have identified that a large penetration of wind generation on the grid will inevitably bring a significant increase in regulation and, especially, load-following needs both in terms of capacity, but also ramping capability and frequency of dispatch. This has prompted the focused interest in deploying battery and other storage technologies as highlighted in the recent proposal issued by the FERC with regards to frequency regulation.[23]

WIND INTEGRATION STUDIES AND ANCILLARY SERVICE REQUIREMENTS

System owners and operators undertake wind, and in some cases solar, integration studies to (1) determine the capability of their systems to integrate large penetrations of variable renewable energy, (2) estimate the growth in flexible services required for a reliable system, and (3) estimate the increased system costs of integrating variable renewable energy sources.

Wind integration studies are complex system studies that typically involve challenging modeling tasks such as developing a highly resolved time series of

[23]Frequency Regulation Compensation in the Organized Wholesale Power Markets, Notice of Proposed Rulemaking, 18 CFR 35 (Feb. 17, 2011), FERC Stats. & Regs. 61,124 (2011) (NOPR).

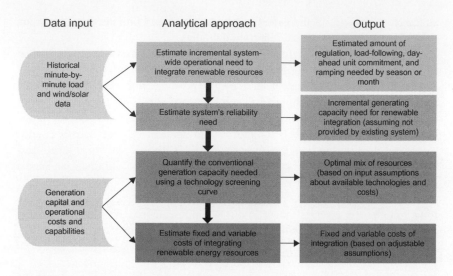

FIGURE 10.4 Renewable integration model (RIM) structure and output. *(Source: Chang et al. [5])*

very large penetrations of wind power (>15%) or the commitment and dispatch of a large electricity market. Most integration studies use proprietary methods and models to estimate the growth in ancillary services required to integrate future wind plant additions. The proprietary nature of these studies protects against the disclosure of sensitive or confidential market information, but it also precludes critical analyses of individual model components and hinders identifying best practices. The Brattle Group and Pacific Gas and Electric collaborated to create an open source Renewable Integration Model (RIM) to enhance the discussion of identifying the appropriate methods and assumptions to use for integration studies.[24]

Figure 10.4 outlines the important inputs, components, and outputs of the RIM. Important inputs include highly time-resolved historical wind and solar datasets that are used to characterize a region's current wind and solar power variability. This characterization forms a basis from which the variability of larger penetrations of wind power can be estimated. Estimating larger penetrations of wind power involves key assumptions about the degree that wind power balances itself through the smoothing effect of diversifying the geographic locations of installed wind plants. The same is true for solar power. Additionally, the other primary set of inputs is a characterization of the current generation portfolio of the system of interest as well as associated capital costs and operational costs.

From these inputs, RIM estimates four key components of a system: (1) the incremental system-wide operational needs of a system in order to integrate the energy generated from renewable resources; (2) the incremental system-wide

---

[24]For more information, please see Chang et al. *Renewable Integration Model and Analysis.* IEEE Transmission and Distribution Conference and Exposition, 2010.

reliability needs of a system to maintain a stable system with larger penetrations of variable renewable energy; (3) the required generator portfolio to meet the needs identified in 1 and 2; and finally, (3) the fixed and variable costs incurred integrating the renewable energy resources.

ADDITIONAL SYSTEM CHALLENGES

As already noted, here are several important challenges facing balancing systems that are integrating renewable generation. First, the uncertainty and variability of renewable wind and solar generation require that system operators procure increasing amounts of balancing services. Furthermore, the types of generation technologies expected to provide those growing amounts of balancing services will have to be characterized by rapid up- and down-ramping capabilities, as well as minimally short start-up and cool-down times. It is still an open question whether these new operating requirements can be entirely met by existing generation resources. Moreover, cycling and peaking units that would be providing such services might potentially be faced with difficulties in realizing sufficient profits, given the unpredictable patterns of dispatch calls. It is essential that a range of future empirical studies and analyses focus on resolving those issues—the resulting solutions will inevitably be driven by the specific weather, load, and economic conditions across various control areas and system operators.

Second, as wind and solar penetration increases significantly, it is likely that wind and solar generation resources will displace some of the marginal peaking and cycling fossil-fuel units. As a result, wind and solar generation could also force some traditional baseload plants to operate in a way that resembles cycling units. Most baseload plants, however, are not designed to operate in that way. Forcing them to do so will very likely reduce their capacity factors and revenues and increase their heat rates—all of which is, of course, significantly disadvantageous to baseload generators. In addition, the increased wear and tear due to more frequent cycling of units, which are not traditionally designed to cycle, will result in higher operation and maintenance costs, further reducing their profitability. The combined effect of these first two situations is illustrated in Figure 10.5.[25]

Third, system operators are increasingly concerned with an empirical system operation challenge known as "over-generation." In many geographical regions, wind is most likely to be strongest during off-peak periods such as late night and early in the day. This creates a problem—the higher wind generation output tends to come at a time when demand is at its lowest. Since the system must accommodate the renewable generation as must-take, this requires that

---

[25]Source: ERCOT Energy Seminar 2009, Chairman Barry T. Smitherman, Public Utility Commission of Texas, November 12, 2009. http://www.puc.state.tx.us/about/commissioners/smitherman/present/pp/GDF_Suez_111209.pdf.

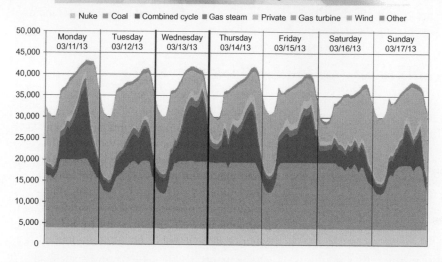

FIGURE 10.5   Increased cycling of conventional generation. *(Source: ERCOT Energy Seminar 2009, Public Utility Commission of Texas, November 12, 2009)*

conventional resources reduce their output to "make room" for the higher wind output. While some conventional generation can be easily dispatched down, the situation is exacerbated when the only generation left to reduce is baseload generation. Most baseload plants have relatively high minimum generation level requirements[26] and long start-up and shut-down times—they cannot be turned off economically and reliably overnight if they will have to be turned back on again the next day. This creates tension because neither baseload generation nor wind generation owners have an economic interest in seeing their generation curtailed during periods of over-generation.

In fact, those generator owners would be willing to receive zero or slightly negative prices for their output before they would agree to reduce their production levels.[27] This situation is particularly challenging because wind generators are

---

[26]For example, the largest coal plants are estimated to have a minimum generation level of 40–50% relative to their peak generation capacity. See Shankar [6].

[27]In many cases, wind plants that qualify for Renewable Energy Credits (RECs) and/or Production Tax Credits (PTC) are willing to receive negative market energy prices because their opportunity cost of not producing would be the foregone values of RECs and PTCs, which in most cases are greater than $20/MWh. In some cases, the generator is paid both for the implied value of RECs and PTCs through their long-term contracts with load-serving utilities even if the power is curtailed. This results in generators willing to curtail even before receiving negative prices. However, under those circumstances, rate payers who ultimately pay for the renewable generation contracts may be left paying twice, once for the environmental attributes that they never received due to the curtailment, and also for the coal generation that could not be backed off.

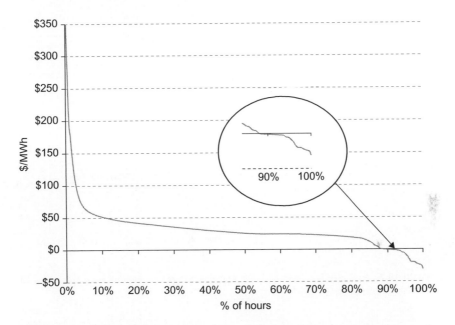

FIGURE 10.6   Real-time price duration curve ERCOT West Zone 2010. *(Source: ERCOT market data from http://www.ercot.com)*

often likely to see higher outputs at night, which places greater pressure on conventional base generation to step in and provide higher levels of output during the day after it has already been asked to dial down at night—as a result, both energy margins and operation and maintenance costs are altered. Negative energy prices have already been observed in Texas and the Midwest. As Figure 10.6 demonstrates, approximately 10% of hours had negative real-time prices in ERCOT's West Zone.[28] Figure 10.7 illustrates similar conditions experienced by MISO,[29] where as many as 5% of the hours were characterized by negative real-time energy prices in 2009.

Finally, given that increasing levels of renewable portfolio standards must be met via a substantial penetration of wind generation, which has near-zero marginal costs, wind generators are typically operated as must-run generation. Operated as must-run, they force the grid operator to reduce the output of existing marginal generation resources, triggering all of the adverse impacts discussed earlier. In deregulated markets, however, generators may have no revenues sources beyond hourly energy sales and capacity market payments. Therefore, the consequent revenue reductions could result in some plants exiting the market, reducing the availability of conventional cycling and peaking

---

[28]Source: ERCOT market data from http://www.ercot.com.
[29]Source: *2009 State of the Market Report Midwest ISO.*

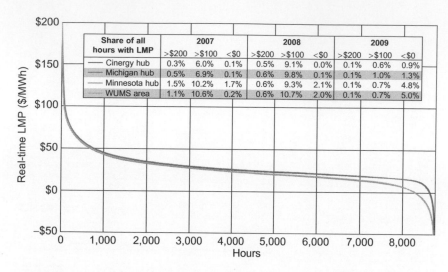

FIGURE 10.7 Real-time price duration curve Midwest ISO: pricing hubs all hours 2009. *(Source: 2009 State of the Market Report Midwest ISO)*

generation at a time when it might be increasingly needed to provide flexible generation services.

Increased operation and maintenance costs and higher heat rates are not the only negative effects of increased cycling of conventional generation. Higher heat rates due to inefficient and more frequent cycling instructions also result in increased emissions—higher emissions are directly translated into adverse environmental impacts and higher air emission compliance costs. In order to provide flexibility services for greater levels of renewable generation penetration, conventional units will have to operate at lower output levels to preserve their ability to respond to increased net load-following and regulation ramps. Those units are most likely to be combined cycle gas turbines.

In general, combined cycle gas turbines (CCGT) have low NOx burners, which reduce their nitrous oxides emissions by lowering the temperature of combustion. Unfortunately, this type of emission control has considerably reduced effectiveness if the generator operates at less than 60% of its nominal rating—this is likely to occur when it is operating as a regulation or load-following service resource for a system with large levels of renewable generation installed. This scenario is also likely in the instance when "must-take" renewable generation forces reductions in CCGTs, generation output level or heat rate efficiency. Furthermore, baseload units that have been forced to cycle but are not designed to do so will also, as outlined above, suffer from increased heat rates—as a result, those generators will be burning significantly greater amounts of fossil fuels to generate the same level of electricity. Again, the increased fuel usage will lead to higher fossil-fuel emissions.

LOAD AS A FLEXIBLE RESOURCE – SYSTEM PROGRAMS AND EXPERIENCE

Given all of the issues discussed above with regard to the use of generation resources as a means of integrating intermittent resources, demand response is rapidly becoming a potential alternative for the provision of flexibility services to the grid. Given a proper telemetric and communication infrastructure, load entities can be just as effective in providing flexibility services, such as regulation and load-following, while demonstrating desirable response times and ramping capabilities. The benefits of relying on demand response to provide regulation and load-following are multiple. On the one hand, as we have already discussed, there is growing empirical evidence that a higher penetration of renewable resources will result in faster ramping rate requirements for units providing regulation and load-following. Conventional generation with such specifications comes at a higher cost; in addition, alternative technologies such as battery storage are still quite expensive and limited in their sustained response time. Load, on the other hand, can effectively provide a high-quality service exhibiting both instantaneous response and fast ramp rates. Furthermore, relying on demand response will avoid the issues of increased cycling, higher emissions, and higher O&M costs that arise with conventional generation.

ISOs and utilities have begun to realize the benefits of relying on demand response for regulation and load-following. The success of enabling demand response customers to bid into capacity markets has led systems to open up their energy and ancillary service markets to entities providing load control. For example, the New York Independent System Operator (NYISO) has four demand response programs: the Emergency Demand Response Program (EDRP), the ICAP Special Case Resources (SCR) program, the Day Ahead Demand Response Program (DADRP), and the Demand Side Ancillary Services Program (DSASP).[30] The EDRP and SCR programs are capacity programs in which load resources are curtailed in energy shortage events in order to maintain a reliable system. The DADRP program allows load resources to bid into NYISO's day-ahead energy market in a method similar to generators. Finally, the DSASP program allows load resources to provide load-following and regulation services.

Capacity programs are still NYISO's most popular product for demand response, followed by energy markets, and then ancillary services. By August 2010, approximately 2,495 MW of demand response enrolled in NYISO's EDRP and SCR programs while only 331 MW of demand response enrolled in NYISO's DADRP and no entity has qualified for the DSASP.[31] While no load entities have yet to qualify to participate in NYISO's DSASP market,

---

[30]Source: NYISO [7].
[31]Ibid.

several customers have expressed interest and are currently going through the registration and qualification process.

The story is similar for demand response programs in the PJM system. Like NYISO, PJM allows load entities to participate in their capacity, energy, and ancillary service markets. And like NYISO, demand response participation is almost an order of magnitude greater in PJM's capacity market (8,683 MW) versus its energy market (1,727 MW).[32] But unlike NYISO, load is an active participant in PJM's ancillary service markets, particularly PJM's synchronized reserve market that provides load-following services. In the first quarter of 2011, demand response provided on average 84,551 MWh of synchronized reserve service in PJM, which translates to an average demand response capacity of 118 MW.

The Midwest Independent System Operator (MISO) demand response participation is also similar to NYISO and PJM. It allows demand response resources to participate in its capacity, energy, and ancillary service markets. But where MISO differs is that it has 17 MW of load from an aluminum smelter providing regulation service, making it the only ISO or RTO to have a load entity providing flexible operating reserve service on timescales shorter than 10 minutes.

As described below, events in ERCOT have shown the importance of load resources in integrating wind and solar power. ERCOT has approximately 2,200 MW of demand response participating in their Load acting as Resource (LaaR) program, and the load resources regularly provide up to half of ERCOT's responsive reserve requirement.[33,34] In addition, ERCOT also has experienced a significant build-out of wind power with a cumulative nameplate capacity of over 10 GWs of wind power installed by the end of 2010 providing 6.4% of the electricity in Texas.[35] The differences in wind generation's operation from conventional fossil-fuel or hydro generators have led ERCOT's system operators to adapt their methods to successfully integrate the increasingly larger amounts of wind energy fed into their system. The updated methods have not been perfect, and there have been events that threatened the stability of the system. One such event occurred in ERCOT on February 26, 2008, and ERCOT's LaaR resources played a pivotal role in maintaining the stability of ERCOT's electrical grid.

On February 25, 2008, ERCOT had forecasted its hourly MWs of load and wind power expected on its system and had arranged for adequate generation plus a reserve of 1,400 MW. ERCOT learned on February 26 that its load forecast did not adequately capture a large and rapid increase in demand of 2,550

---

[32]Source: Monitoring Analytics [8].
[33]ERCOT's LaaR program is a program where load resources with interruptible capabilities provide balancing energy service (load-following ancillary service).
[34]Source: Newell and Hajos [9].
[35]Source: American Wind Energy Association [10].

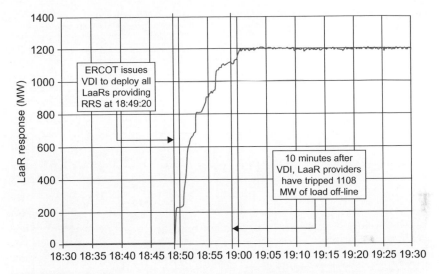

FIGURE 10.8   ERCOT LaaR responsive reserve deployment during February 26, 2008, event. *(Source: ERCOT [11])*

MW that occurred between 6:00 PM and 6:41 PM. Unfortunately, this rapid increase in demand coincided with a drop-off in wind power production that had begun earlier, happened faster, and was larger by 1,000 MW than their wind forecast predicted.[36]

These two events combined to deplete ERCOT's operating reserves over the course of 40 minutes from 6:00 PM to 6:40 PM. At 6:41 PM, ERCOT implemented step 2 of its Emergency Electric Curtailment Plan (EECP) and called upon 1,150 MW of LaaRs to reduce the load. Within 10 minutes of receiving instructions to curtail, 1,108 MW of LaaRs had curtailed their demand (Figure 10.8). According to ERCOT, the deployment of LaaRs halted the decline in system frequency and "restored ERCOT to stable operation."

Although utilities and RTOs have successfully enrolled large quantities of demand response in capacity, energy, and ancillary service markets, these demand response products have been designed to help a system in one direction, namely by reducing load. This has been adequate in the past because historically the common emergency events that a system encountered have been shortages in generation. Yet the significant build-out of wind power combined with RPS obligations and the constraints of operating limits of conventional generators to increase the likelihood of over-generation events where a system produces more energy than is being consumed. As a result, there will be increasing demand for the ability to deploy increased load instead of only shedding wind power or turning a power plant off.

---

[36]Source: ERCOT [11].

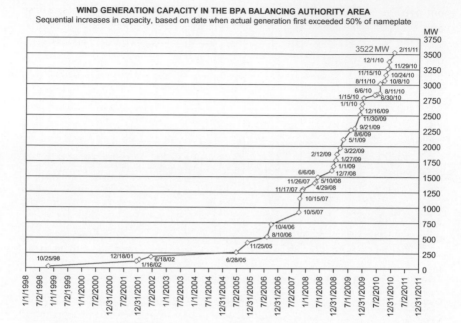

FIGURE 10.9 The growth of wind power capacity in BPA's balancing area. *(Source: BPA [13])*

A recent case in point is Bonneville Power Authority's (BPA) experience in dealing with over-generation events in the first half of 2011. BPA is the transmission owner and operator for 75% of the high-voltage transmission lines in the Pacific Northwest. Approximately 89% of BPA's 18.3 GW peak capacity resources are hydropower (31 federal dams), 6.3% are nuclear (1 nuclear power plant), and the remaining 4.6% are primarily wind power.[37] But because BPA is the regional transmission owner, it is also responsible for the distribution of an additional 27.3 GW of peak capacity, of which 5.9 GW is coal and 8.5 GW is natural gas.

Over the last five years, BPA has seen a rapid growth in wind power within its balancing area (Figure 10.9), and BPA estimates the installed wind power capacity in its balancing area will double in the next few years.[38] The large penetration of wind power in BPA (>30% of installed capacity in BPA's balancing area providing 13% of BPA's energy in 2010) has resulted in its heavy reliance on hydro generators to provide balancing services. Hydropower is an excellent resource to pair with wind generation due to hydropower's fast response times, ramp rates, and lack of CO_2, NO_x, and SO_x emissions. But BPA's hydropower facilities have operating constraints to ensure the health of multiple flora and fauna species within the river systems in Oregon and Washington. These hydropower operating constraints are similar in nature,

[37]Source: BPA [12].
[38]Source: BPA [13].

but more rigid, than emission limitations of fossil-fuel power plants. These operating constraints are flow constraints and river chemistry constraints; thus BPA's dams have minimum and maximum flow rates as well as dissolved oxygen and nitrogen levels they must maintain for the river system downstream of their discharge. As BPA has experienced over the last few years, these environmental constraints have increased the operational challenges of integrating wind power by reducing the capability of BPA's hydropower facilities to provide balancing services and increasing the occurrence of over-generation events.

The increased frequency of over-generation events led BPA to formalize an environmental redispatch policy that provides BPA with a formal plan for curtailing generation in its system. When over-generation conditions are imminent, BPA will first redispatch thermal generators to minimum power levels that generators are capable of or are necessary for reliability reasons. If this redispatch does not alleviate the threat of an over-generation event, BPA will then curtail wind generation on the system in order to allow its hydropower facilities to meet their environmental constraints.[39]

Current demand response programs would not have helped BPA alleviate over-generation events as curtailing load would only have worsened the conditions. Instead, it would have been beneficial if BPA could have dispatched customers to *increase* the load in their balancing area. This capability does not exist except with customers enrolled in real-time pricing programs. RTP responses to reduced prices, however, are relatively small for most customers. Thus system operators in the coming years will look to develop more advanced flexible load control programs that can provide both up and down flexible reserve services.

In order to effectively deploy demand response as an ancillary service resource, a system first needs to determine the nature and size of the flexibility services it would need to integrate significant penetrations of variable renewable generation. From this it can then employ an optimal inventory algorithm to determine the amount of load participation needed and the optimal manner in which it should be deployed. The concepts and design of an optimal inventory algorithm for load control products are discussed in the next section.

STRUCTURING LOAD CONTROL PRODUCTS FOR INTERMITTENT RESOURCES

To deploy demand response products for ancillary services, system operators need to translate the total capacity and energy ancillary service needs into a number of contracts for specified blocks of energy to be delivered by participating load. The implementation of demand response for the provision of

---

[39]Ibid.

regulation and load-following is to some extent similar to programs designed to meet critical peak conditions by enrolling participating loads that are willing to be curtailed.[40] Critical peak programs of such a nature are generally directly linked to the seasonal characteristics of peak demand on a given system. For example, if the control area demand peaks in the summer, curtailable load programs are implemented only for summer months, allowing the system operator to exercise a limited number of curtailment calls over the summer period. In contrast to critical peak programs, ancillary service needs persist across all months of the year. As a result, load control products must allow for monthly differences in the number of participants and amount of energy contracted.

Intermittent generation increases the seasonality and variability of ancillary service needs. The seasonality of wind speed and, by extension, wind generation output can be observed across various time-scales—from intra-daily and weekly cycles, to multi-year phenomena such as El Nino. The variability of wind, as it pertains to the need for regulation and load-following, is largely affected by short-term weather patterns. The results reported by wind integration studies establish the quantity of regulation and load-following that would be needed to maintain reliable system operations while accommodating increasing penetrations of renewable generation. A portion of those ancillary service needs can be contracted for provision via demand response with monthly differences that recognize the seasonal patterns in output and variability of renewable generation.

The system operator will need to contract for a sufficient amount of energy deliveries to ensure that the system will be able to meet any actual outcome of MWh energy of ancillary service needs with a given probability (e.g., 95% likelihood of meeting the need). The total amount of energy deliveries has to be decomposed into interruptible contracts with a three-dimensional property. Each contract should be designed to represent (1) a block of MW quantity; (2) the fixed number of contiguous hours at a time over which the quantity can be called upon; and (3) the maximum annual number of times a multi-hour delivery block can be called. Consequently, the total size of the contracted controllable load energy will be represented by the collection of all contracts (defined by a block of capacity that can be exercised up to a maximum number of sequential hours up to a maximum number of times per year).

For example, a single contract could specify that the customer can be called to provide one kW for no more than five contiguous hours at a time and can be called to do so at most 10 times per year. This contract, therefore, provides at maximum 50 kWh per year. It is possible, however, to design two contracts that allow for the same maximum amount of total energy per year, but in reality differ in how they can be exercised by the system operator. In the preceding

---

[40]An extensive treatment of design and implementation of curtailable load contracts can be found in Oren and Smith [14].

example, the customer agrees to interruptions over no more than 50 hours per year with no single interruption lasting more than 5 hours. Alternatively, a contract may be designed to require a maximum of 50 hours of interruptions *without* a limitation on the length of each interruption. The latter contract provides more value to the system operator and requires fewer unique program participants. This difference arises from two underlying challenges.

First, the system operator faces an inventory problem—enough total energy needs to be contracted from individual program participants to meet a designated portion of overall system ancillary service needs. For example, if the average energy need is 10 MWh with a standard deviation of 0.5 MWh, then a total of 11 MWh would need to be contracted for delivery.[41] If each contract was worth 50 kWh per year, then one potential solution would be to contract with 11 MWh/50 kWh = 220 participants (each agreeing to the 50 kWh contract described above).

There is, however, a second challenge—the system operator is also faced with a "stacking" problem. While the total contracted quantity might satisfy the identified *total* energy need for ancillary services, if the contract limits each program call to a limited number of contiguous hours (e.g., no more than 5 contiguous hours per each of 10 annual calls), then the system operator might run out of energy blocks that can be "stacked" to meet a volatile spike in ancillary service needs because some calls were already exercised at less than the 5-hour limit. When a contract is structured as a "call" contract rather than a "total energy" contract, each time a call is used for less than the maximum allowed amount of hours, the energy from the unused hours is "lost." To address this complication, the initial solution of 220 participants could be relaxed upward using a general rule-of-thumb informed by historical experience.[42]

There are several dimensions to establishing a pricing mechanism for the ancillary service demand response program. Continuing the preceding illustrative example, two principal contract designs can be considered. Program participants can be offered a "total energy contract"—in essence it commits each customer to providing a total of 50 kWh of energy annually, which can be called over 50 hours without a limit on the duration of each program period.[43] On the other hand, a "call contract" might set forth a different condition—each customer can be called no more than 10 times per year, with each call lasting for no more than 5 contiguous hours. Estimating the value provided by participating loads in each of the two contract scenarios has to rely on measures of avoided cost and capacity value.

---

[41]That is, an amount equal to the mean (10 MWh) plus at least two standard deviations (2 × 0.5 MWh).
[42]For example, the system operator might enroll *an additional* 37 participants (~365 days/10 MWh) if the contract is of the "call" type.
[43]This allows for the possibility that the operator might ask a participant to respond to load control interruptions for a number of contiguous hours.

A critical peak pricing (CPP) demand response program developed by Southern California Edison (SCE) provides a useful framework in the present context.[44] Under this CPP program, the utility may call a maximum of 15 critical events per year. Demand response participants receive notification the day before they are likely to be called upon.[45] In addition to the limited number of calls per year, each event can only last a maximum number of hours. Under these conditions, the capacity value of the load resources is estimated as the product of their expected deliverable capacity and the avoided cost of capacity. The avoided cost is based on the capacity value of a combustion turbine, but must be derated to capture the reduction in value due to the lack of availability of the load resource all year round. To estimate the derating factor, the utility can run simulations in which participating load is dispatched to meet critical system needs. Similarly, for demand response providing ancillary services, the simulations will model the percentage of times the load resources were available to meet regulation and load-following needs. Consequently, the derate factor can be decreased, thus increasing the capacity value, by either offering contracts with greater number of calls, longer call durations, or "energy" type stipulations. Given this framework, an "energy" type contract for 50 kWh as defined in the illustrative example earlier will be more valuable than a "call" type contract for the same amount of energy.

CONCLUSIONS

The primary system and social benefits of using load resources instead of generators for ancillary services are increased portfolio diversity, reduced fuel and O&M generator costs, reduced customer costs, and lower emissions. Introducing wind into a system increases a system's variability and leads to reduced utilization of conventional fossil-fuel generators and as a result reduced revenues. Unless other capacity and energy sources are introduced into a market, prices will rise as conventional generators seek to recover their lost revenues. Demand response is a perfect complement to the increased penetration of wind energy because demand response programs can provide capacity and energy in limited quantities.

Fuel costs can be avoided by calling upon load resources to integrate variable renewable resources instead of conventional fossil-fuel generators when the economics to do so are favorable. O&M costs are also reduced because maintenance schedules for fossil-fuel power plants are determined by how much cycling a plant does through its capacity range. By calling on conventional generators less for ancillary services, generators are able to extend the

---

[44]Chapter 23 by Braithwait and Hansen discusses CPP rates and empirical outcomes of those programs in California.
[45]This discussion follows closely [15].

amount of time between periods when they are overhauled and repaired. Customer costs can be reduced because they will receive revenue payments for capacity and energy as participating load resources. Finally, carbon dioxide, nitrogen oxides, and sulfur dioxide emissions are also reduced when load resources are substituted for fossil-fuel generators due to decreased fuel usage as well as avoided efficiency declines that occur when a fossil-fuel power plant produces power below its specified nameplate capacity.[46]

REFERENCES

[1] *Database for State Incentives for Renewables & Efficiency (DSIRE)*, http://www.dsireusa.org/documents/summarymaps/RPS_map.pptx (accessed 17.03.11).

[2] EIA, *Annual Energy Outlook 2011 Early Release Overview*. U.S. Energy Information Administration, 2008, 9.

[3] CAISO, *Integration of Renewable Resources*. California Independent System Operator, November 2007–July 2011, http://www.caiso.com/Documents/Integration-RenewableResourcesReport.pdf.

[4] NYISO, *NYISO Supplement and Errata to NYISO Annual Report on Demand Response Programs*. Federal Energy Regulatory Commission. Filing; Docket No. ER01-3001-000.

[5] NREL, *Eastern Wind Integration and Transmission Study*. National Renewable Energy Laboratory, February 2001, 134–148, http://www.nrel.gov/wind/systemsintegration/pdfs/2010/ewits_final_report.pdf.

[6] J. Chang, K. Madjarov, R. Baldick, A. Alvarez, P. Hanser, *Renewable Integration Model and Analysis*. IEEE Transmission and Distribution Conference and Exposition, 2010.

[7] R. Shankar, *Low Load/Low Air Flow Optimum Control Applications*. EPRI. TR-111541, December 1998.

[8] Monitoring Analytics, *PJM State of the Market Report 2010*. Monitoring Analytics. March 3, 2011.

[9] S. Newell, A. Hajos, *Key Issues for Demand Response*. The Brattle Group. Presentation to the Northeast Energy and Commerce Association's Energy Efficiency and Demand Response Conference. October 21, 2010.

[10] AWEA, *Wind Energy Facts: Texas*. American Wind Energy Association. http://www.awea.org/learnabout/publications/upload/1Q-11-Texas.pdf (accessed 24.05.11).

[11] ERCOT, *ERCOT Operations Report on the EECP Event of Electricity Reliability Council of Texas*, February 26, 2008.

[12] BPA, *2009 BPA Facts*. http://www.bpa.gov/corporate/about_BPA/Facts/FactDocs/BPA_Facts_2009.pdf (accessed 25.05.11).

[13] BPA, *BPA's Interim Environmental Redispatch and Negative Pricing Policies: Administrator's Final Record of Decision*. Bonneville Power Authority, May 2011.

[14] S.S. Oren, S.A. Smith, Design and management of curtailable electricity service to reduce annual peaks, *OR Practice* 40 (2) (1992) 213–228.

[15] California Public Utilities Commission. *Appendix A of Rulemaking 07-01-041 Straw Proposals for Load Impact Estimation and Cost Effectiveness Evaluation of Southern California Edison Company*, San Diego Gas & Electric Company, and Pacific Gas and Electric Company

[46]For more information, see Katzenstein and Apt [16] or Denny and O'Malley [17].

filed on July 16, 2007. CA CPUC, R.07-01-041, http://docs.cpuc.ca.gov/efile/REPORT/ 72728.pdf (accessed 30.03.11).

[16] W. Katzenstein, J. Apt, Air emissions due to wind and solar power, *Environ. Sci. and Technology* 43 (2009) 253–258.

[17] E. Denny, M. O'Malley, Wind generation, power system operation, and emissions reduction, *IEEE Trans. Power Syst.* 21 (2006) 341–347.

Smart Infrastructure, Smart Prices, Smart Devices, Smart Customers, Smart Demand

Chapter 11

Software Infrastructure and the Smart Grid

Chris King and James Strapp

INTRODUCTION

The smart grid promises many benefits to utilities, consumers, and society—as much as $131 billion per year for the United States.[1] As described in other chapters of this book, these benefits include utility operating efficiencies, better use of existing grid assets, enhanced energy efficiency, greater reliability, reduced peak demand, and integration of distributed renewable energy resources. Achieving these benefits requires investing in hardware such as

---

[1]McKinsey & Co., *McKinsey on Smart Grid,* Summer 2010, at 6.

smart meters, communication networks, line sensors and controllers, automated switchgear, capacitor bank controllers, in-home devices, and other equipment. The amount is significant, with estimates from ranging from a "bare bones" approach costing $50 billion to as high as $476 billion for 150 million electric customers in the United States[2]—and an order of magnitude higher to implement smart grid for all 1.7 billion electric meters globally.[3] By way of progress, U.S. installations surpassed 20 million meters in May 2011, with nearly 50 million committed by 2015[4]; global spending on smart grids is expected to reach $46 billion by 2015.[5]

Nowhere are \ the potential benefits greater and the profile higher than with those smart grid applications that have a direct impact on the energy customer. The smart grid technologies associated with the home or business include smart meters, home energy devices connected through a so-called home area network or HAN, and smart appliances. Delivering smart grid application functionality requires a complex integration of these millions of devices in environments that challenge communications technologies. More than the other smart grid technologies, these devices have a direct daily impact on energy consumers' lives and all the associated implications. This chapter is focused on the software platform necessary for successful deployment and operation of smart grid technologies associated with the consumer's premises.

Just as a computer is useless without its operating system software and software applications, smart grid capital investment can deliver little of the desired benefits without the necessary accompanying information technology (IT) systems. And just as with computers, the IT systems include both software platforms—analogous to the PC's operating system—and software applications.

Smart grid software platforms are the underlying elements that support applications (Figure 11.1). For example, a software platform is needed within the utility back office to link individual applications with smart grid communications networks. This allows multiple applications to use the same network, such as having the billing system receive usage data and the outage management system receive outage alarms from the same smart meters and over the same smart meter communications network. Another example is in a home or building, where the software platform underlying the HAN enables communications between devices, further described in Chapter 15.

On the applications side, a plethora of product offerings has emerged. Many of these are in the utility back office, such as meter data management (MDM) and demand response management (DRM). Such applications process data

[2]EPRI, *Estimating the Costs and Benefits of the Smart Grid*, March 2011.
[3]ABS Energy Research, *Electricity Meter Report*, June 2009.
[4]Chris King, *U.S.: 20 million smart meters Now Installed*, May 17, 2011. Available at http://www.emeter.com/smart-grid-watch/2011/us-20-million-smart-meters-now-installed/.
[5]ABI Research, *Smart Grid Applications: Smart Meters, Demand Response, and Distributed Generation*, Research Report, July 2010.

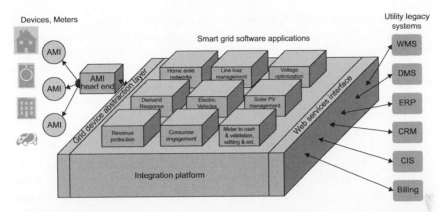

FIGURE 11.1 Software platform links smart grid devices with utility legacy systems and provides a platform on which to develop new software applications. *(Source: eMeter Strategic Consulting, Smart Meter Best Practices Webinar, April 22, 2010.)*

provided by smart meters, sensors, or other devices and produce useful outputs for utility business processes. For example, the MDM performs validation, editing, and estimation (VEE) on usage data received from smart meters, making the results available to the billing system as needed to generate customer bills. Many other applications are consumer-oriented, such as consumer engagement websites where end-users can view their consumption data online or subscribe to weekly or monthly emails to have the usage reports "pushed" to them.

Implementing the IT systems that go along with the smart grid may seem daunting to utilities—and not all of the early implementations have gone smoothly—but there are many cases that demonstrate best practices for success. Building on these successes, exemplary utilities are actively beginning to achieve the desired benefits of smart meters and smart grids. Utility projects highlighted here are in Texas and Ontario, Canada. The authors have worked closely with the project principals in both cases and have detailed knowledge. These two case studies illustrate the approaches of a distribution utility in a deregulated retail market, with a mixture of IT responsibilities between the utility and centralized systems serving multiple utilities. The two cases are analogous to, but more challenging than, cases of integrating systems serving multiple departments within a single vertically integrated utility. In other words, the chosen cases illustrate the same principles and needs of a vertically integrated utility but with additional complexity. In addition, many regulators require that utilities interface with third parties to share customer usage and other data, in cases where customers have authorized the sharing of their data.[6]

This chapter is organized into four sections. "Introduction" reviews the goals of the IT infrastructure included in a smart grid deployment. These

[6]California Public Utilities Commission. D.11-07-056. July 28, 2011.

include delivering utility operating benefits, grid efficiencies, and consumer benefits. Importantly, they also include "future-proofing" to accommodate new technologies and new regulatory policies, both of which continue to evolve more and more rapidly. "Smart Grid IT Challenges" describes the IT challenges faced by utilities implementing smart grid projects. These challenges include the handling of imperfect data and communications—think dropped cell phone calls or corrupt data files on a computer hard drive, which are not that common, but do happen regularly—sharing of the data and communications by multiple applications, and integration of all of the software and hardware elements. "Essential Smart Grid Software Platform" describes the essential software platforms that provide the underlying infrastructure to realize desired smart grid functionality for the premises. This section presents a data integration platform and key interfaces, as well as the standards associated with those interfaces. Specific examples are smart grid Interoperability Panel (SGIP) Priority Action Plan (PAP) 10 for the data model for providing energy information to consumers and Smart Energy Profile, associated with ZigBee, for sending and receiving data and control signals between smart meters and smart appliances, thermostats, and other devices. "Smart Grid Applications" elucidates some of the major software applications that provide the smart grid functions for customers, such as billing of dynamic pricing rates, meter data management, net metering, consumer engagement, outage management, HAN management, and demand response management.

Throughout the chapter, examples from the sophisticated and mature smart grid programs in two jurisdictions, Texas and Ontario, are used to highlight and explain the concepts discussed. Both of these programs distribute IT responsibilities between competitive energy retailers, electricity distributors, and central market players in an architecture that illustrates the IT complexities involved in smart grid deployments. Sidebars provide background information on the history, structure, and scopes of the smart grid in both jurisdictions. The chapter ends with our findings and conclusions.

SMART GRID IT CHALLENGES

Investing in the smart grid may be one of the largest capital investments that a distribution utility makes. The associated IT infrastructure represents a new breed of enterprise information system that requires comprehensive integration and management tools—a smart network application platform. Such a platform supports basic meter data management (MDM) functionality but many more capabilities as well.[7] These additional capabilities are essential in minimizing

---

[7]Meter data management refers to processing of smart meter data for use in billing. This includes receiving the data from the head end of a smart meter network; performing validation, editing, and estimation (VEE); storing the data in a database; and sending it to the billing system according to each customer's billing schedule.

the capital and operating costs of the smart grid, as well as maximizing business benefits realization.

When evaluating solutions as part of a smart grid investment, utilities generally have two overriding and fundamental objectives:

- **Maximize Business Benefits Realization**—most utilities that have deployed advanced metering or smart meter systems have yet to realize benefits beyond reduced meter reading costs. A good application platform enables added benefits through the addition of specific data applications, the ability to apply business rules to raw data such as outage alarms, and a robust integration platform to allow the implementation of new applications and new data collection technologies.
- **Manage and Minimize Risk**—prominent smart grid implementation risks include technology obsolescence, integration risk—as a rule, the more interfaces required, the greater the risk—and changes required to core utility business systems such as billing and the customer information system (CIS), and scalability. The complexities of the IT challenge should not be underestimated. Robust applications and formal IT integration approaches must be applied to the implementation and operation.

Achieving these objectives requires understanding the challenges. As with any large IT project, there are many obstacles that must be overcome to achieve the desired results. The goal is typically a reliable system that generates accurate results—for example, customer bills—and one that enables the use of efficient business processes—for example, processes that are primarily automated and require minimal human intervention.

Standing in the way of achieving this goal are five key challenges to smart grid IT systems:

- **Obtaining data reliably** from smart meters, sensors, controllers, smart appliances, and any other devices included in the overall smart grid project.
- **Ensuring that the data accurately record** or represent what they are supposed to record or represent.
- **Defining data systems of record**—which database is the master, when, as with the smart grid, multiple IT applications are using the same data.
- **Managing a large and diverse communications network** involving hundreds of thousands or millions of communications devices. Management of such networks is new for utilities—indeed, for any company, because the smart grid has a whole new class and scale of devices to manage.
- **Integrating disparate software applications**. For example, data from smart meter network head-end software developed a year ago may need to be sent to a billing system that is three decades old.

The following sections describe these challenges in greater detail and specificity. The concept is that solving the problem begins with understanding the problem.

Smart Grid Data Reliability

The smart grid establishes a new level of required data reliability for utilities. "Reliable" data are data that can be relied upon for their intended uses; the goal is for customers and other data users, such as utility personnel, to have a high level of confidence in the data. Higher reliability is needed for smart grid data because the data will be used for more complex and/or numerous functions. These range from demand response to outage management. One indicator of the significance of this issue is that, as a rule of thumb, a typical, well-run, large-scale smart meter system misses up to 4 percent of the interval usage data it is supposed to record and retrieve each month. For a million-meter system, this totals over 28 million missing data intervals per month. Retrieval of total consumption values from meters—the meter's register read—is more reliable, because only one such read is needed per day rather than 24 (hourly interval data) or 96 (quarter-hourly interval data). Even so, a typical, well-run, large-scale smart meter system can miss more than 1 percent of such data.[8] In fact, the Ontario MDM/R requires that valid data be provided from at least 98% of smart meters.[9] In the USA, each utility has set its own requirement for the amount of missing data that is permissible.

Demand Response

One example of complexity is the use of interval data to create billing determinants for dynamic pricing and other demand response programs, further described in the chapter by Chapter 3. For time-of-use rates, the interval data must be processed into peak and off-peak usage. The 3,000 or so intervals recorded each month per meter must be processed into one total for peak times and a second for off-peak times. In the past, utilities have needed and handled only a single number each month. To deal with this new data volume, utilities implementing smart meters are also implementing MDM systems.

Another complex demand response function is measuring peak demand reduction. This involves calculating a "baseline" load, which is the load the customer would have consumed absent the peak demand event. The reduction is determined by subtracting the actual load from the baseline load. The baseline load is often determined by taking the average load during the same time period (e.g., noon to 6 PM) on the three highest of the ten previous similar days.[10] The baseline must be accurate, because it must comply with regulator-approved tariffs and is used to calculate cash payments. In addition, the methodology must be sound—and auditable—to deal with the issue of customers who may attempt

[8]Pacific Gas and Electric Company, *Advanced Metering Infrastructure, January 2010 Semi-Annual Assessment Report and SmartMeter™ Program Quarterly Report (Updated)*, January 31, 2010 at 35.

[9]IESO *MDM/R Operational Best Practices*, April 2011.

[10]PJM, *PJM Manual 11: Energy & Ancillary Services Market Operations, Revision 45*, June 23, 2010. Available at http://www.pjm.com/~/media/documents/manuals/m11.ashx.

to "game" the system and create artificially high baselines. Finally, because the baseline load changes every day as the calendar advances, this calculation must be done every day—or at least every event day.

Consumer Engagement—Bill-to-Date Calculation

Another instance of complexity is bill-to-date calculation.[11] Customers have said in surveys that they would like to see this information so they can manage their usage and electricity bills better.[12] This function has been piloted and is already in use by some utilities and Texas electricity retailers.[13] Bill-to-date requires daily bill calculations, as opposed to the traditional monthly, with the daily estimated amounts shown on a utility's or retailer's website.

Bill-to-date also requires reconciliation. This process occurs after the actual bill is issued. The purpose is to make sure that the total of the daily estimates calculated prior to the bill issuance equals the actual bill that was sent to the customer. This is done when each bill is issued by taking the actual bill amount and going back and adjusting the daily estimates to match the total bill actually issued. In practice, this means a slight decrease or increase in many of the daily estimates to ensure that the total of the daily estimates matches the actual bill issued. For example, the daily estimates for a given customer may add up to $99.00. The actual bill may then be $100.00. Though the difference is small, the customer's confidence in data provided by the utility depends on the data being accurate. No customer would stand for a utility saying, "Your bill is *approximately* $100.00; please pay this amount." Of course, customers understand estimates during a billing period. But unless told it is an estimate, customers assume the bill is 100% reliable.

The purpose of the daily bill estimate reconciliation process is to avoid customer questions and confusion caused by differences between the total of daily estimates shown on the utility's website, or emailed to customers, and the actual bill amount. The frequency of situations where the final bill does not match the total of the daily estimates depends on the specific utility regulatory situation. Common causes of differences between estimated and actual bills are rate changes and seasonal changes. Often, regulators make rate changes retroactive; since the estimates are done before the retroactive rate change, the estimates will not match the actual bill.

In the case of seasonal changes, winter rates are applied to part of the bill and summer rates to the other part. This occurs for utilities with differences in seasonal rates, which make up perhaps half of U.S. utilities. There are different ways of applying the seasonal changes. Some utilities prorate the usage

---

[11]Chapter 15 further describes the attractive features of "bill-to-date" functionality.

[12]eMeter Strategic Consulting, *PowerCentsDC Final Report*, September 2010 (available at http://www.powercentsdc.org).

[13]*Ibid.; Jeff Read, Reliant's e-Sense Looks Smart*, Smart Meter Marketing, January 7, 2011, available at http://www.smartmetermarketing.com/reliants-e-sense-looks-smart/.

when a bill is issued after a seasonal change, allocating usage to summer and winter based on the number of days in the billing period in each of the seasons. For example, there may be 10 summer days and 20 winter days, in which case the utility would assume 10/30 of the consumption occurred during the summer and 20/30 during the winter period. This proration cannot be done until the final meter read is taken at the end of the billing period. Because this calculation is done based on the proportion of days in each season, the total of the daily estimates will not match the actual bill unless the end-user's consumption is perfectly balanced between the seasons. In the case of proration, a perfect balance occurs only when the average daily consumption in the summer days equals the average daily consumption in the winter days. Accordingly, the vast majority of bills issued after a seasonal change will have a difference between the total of the daily estimates and the actual bill amount. By reconciling the daily estimates—adjusting them to match the actual bill total—utilities can maintain confidence of customers in the bill calculations.

Outage Alerts

Other important issues relating to data reliability involve outage alerts from meters and restoration verification messages. Most smart meters emit a "last gasp" when the power goes out. The last gasp is a radio transmission made possible by a capacitor in the meter that stores energy and can use it after the power goes out to send the outage alert message. Utility outage management systems can use these last gasp messages to identify the location of outages and, with some additional intelligence, sometimes identify the substation or feeder fault location that caused the outage. The additional intelligence requires mapping the individual meters against distribution system elements, such as feeders and substations. For example, if all the alerts are from a single feeder, the outage management system of the utility can conclude that the specific feeder is the source of the problem.

There are a few problems with last gasp messages that must be managed by utility IT systems in order to provide confidence to data users that the outage alerts can be relied upon. The first is that all last gasp messages may not be received by radio receivers in the smart meter communications network. The business rule that solves this problem is straightforward: if the message is not received, the utility cannot assume power is on. The second problem is that meter-generated outage alerts may overwhelm the outage management system. This occurs especially in the case of a large-scale outage usually caused by a major storm; tens or hundreds of thousands of meters can lose power. The ideal solution in these cases—as implemented by utilities such as JEA (formerly Jacksonville Electric Authority)[14]—is to "filter" the outage messages and

---

[14]JEA, *Meter Data Management System*, Presentation at Autovation Conference, Nashville, Tennessee, October 2006.

forward only high-priority alerts. Most utilities define these as alerts coming from hospitals, fire stations, police departments, and other public health and safety authorities. The meters on such facilities are designated "bellwether" meters to signify their high value and priority.

The third problem is false alarms. This problem is perhaps the most insidious, because it causes distribution system operators to have low confidence in the outage alerts from smart meters generally. False alarms are created for many reasons. Meters may report outages when reclosers have actually already restored power. This situation can be solved by programming the meters to report outages only if the outage exceeds the time threshold programmed into the reclosers, but this requires that the system managing the meters maintain the synchronization of values between the reclosers and meters. In other cases, meters report outages caused by crews conducting scheduled work. The utility system can be set up so that such scheduled outages are excluded from reporting, provided the smart meter system, the work management system—the utility CIS often performs this function—and the outage management system are coordinated.

The other source of false outage alarms is meter detection issues. Some utilities have turned off smart meter outage alert systems, because the systems generate too many false alarms. For various and sometimes unknown reasons, smart meters may send outage alerts that do not reflect actual outages. This situation occurs when the meter's outage detection mechanism is triggered by an event—perhaps a voltage sag—that the meter interprets to be an outage. In fact, the event may not have caused an outage. Overall, the IT system must have the flexibility to implement business rules that, over time, can learn that these outage alerts are false and not to forward them to the outage management system. Interpreting meter alerts improves with experience and requires adjustments over time as the utility distribution operations team becomes more familiar with the sensing capabilities of the meters and the like.

The IT system can help if it has the capability to segregate different types of outage alerts or perform other analysis, such as a correlation analysis. If multiple alerts are received at the same time from different meters attached to the same circuit, the likelihood is high that an actual outage has occurred. Another example is based on history; some meters have a history of reporting spurious outage alert data. The right IT application for handling outage data can track such "suspect" meters and exclude alerts from them as likely false positives.

Data Accuracy

Data accuracy is about whether data delivered by smart grid systems are correct. Accurate data are reported data that correctly represent the situation that actually occurred. Examples of accuracy are that a certain number of kWh was consumed during an interval period or that a certain voltage occurred as reported by the smart grid system. Four common data accuracy issues are as follows, and each is described in greater detail later:

- **Sum check**: do the intervals add up to the correct total?
- **Device mis-programming**: were the parameters set properly?
- **False zero values**: is a zero in a dataset an outage, zero usage, or missing data?
- **Meter reset**: did the meter improperly reset its values to zero?

Accuracy differs from reliability. Accuracy is simpler; it is about whether a value is correct. Reliability encompasses the broader context described above: will the data arrive on time, will it be complete, will it represent what it is supposed to represent? An hourly interval smart meter provides an example: from a reliability standpoint, the goal is to receive all of the interval usage data recorded each month from the meter, which would be approximately 720 intervals (30 days times 24 hours). From an accuracy standpoint, the goal is that the interval data accurately reflect the kWh consumed.

These four specific data accuracy issues are further described below.

Sum Check

One test for data accuracy for a smart meter is a "sum check." If the data are accurate, the sum of the intervals will equal the difference between the starting and ending point. An example illustrates this. Let's assume the starting value—the "register" read of the meter—on June 1 is 10,000 kWh and the ending value on July 1 is 11,000 kWh, a difference of 1,000 kWh. If the sum of the 720 intervals differs from 1,000, one or more of the individual hourly values must be incorrect (or the register function has experienced an error, which is also a possibility). Ensuring data accuracy requires an IT system that can perform such a sum check and then provide information that can determine the source of the problem. The solution is not trivial, because the source could be one or more inaccurate interval recordings, an inaccurate register recording, or one or more missing intervals.

Mis-Programmed Devices

Another common source of accuracy errors is improper set-up of meters or devices. Meters for large commercial customers often have a "multiplier." A multiplier is used when only a portion of the current is passed through the meter. This is needed for high-usage commercial accounts, because passing the entire current through the meter would damage it. As an example, if only one-tenth of the current is passed through the meter—this is accomplished using a current transformer/potential transformer combination (CT/PT)—then the consumption recorded by the meter must be multiplied by 10 to deliver accurate total consumption values. If the system is programmed with some other value—say 20—then the reported value will be inaccurate.

The IT system can identify mis-programmed multipliers in different ways. One method is to compare total consumption for similar customer sites; a major discrepancy can indicate a multiplier error. Another common method is looking at consumption history and any major change occurring coincident with the timing of the smart meter installation. A third method is looking at aggregated loads. The usage sum of all the meters on a feeder circuit for a given time period can be compared to the usage recorded at the substation for the feeder as a whole. One source of discrepancy between the two values can be mis-programmed meter multipliers.

False Zero Values

Another potential source of inaccurate data is false zero values. A zero consumption value for a given time interval, for example an hour, may reflect zero consumption, a power outage, missing data, or some combination of the three elements. All three elements are commonly seen in smart meter systems, even though the frequency is often small. However, if 1 percent of the values are zero—a not uncommon situation—this would create over 7 million questionable values per month for a million-meter utility.

Improper Meter Resets

A fourth source of inaccurate data is meter resets. Electric meters have a continuous recording of total consumption, much like the odometer of an automobile that records total mileage. The continuous recording value is called a meter register. Interval data are recorded by taking a snapshot of the register at pre-programmed intervals—every hour, say—and storing the result. However, occasionally a power surge or other disturbance will cause the meter's register to reset. Such a reset will distort the data, because the snapshot taken after the reset will be a lower value than the snapshot taken at the end of the previous time interval. To ensure overall accuracy of the reported data for use in billing and for other purposes, any resets must be identified by the utility's IT system and accounted for. The specific utility business processes will determine how to account for such a situation, but the most common method is to discard the suspect usage interval and estimate the proper value using industry standards.[15]

Data Systems of Record

The smart grid is creating an IT revolution in utilities. Historically, IT has been implemented in "silos," with a single business unit responsible for a function from field device through data collection to data processing in the software

---

[15]Edison Electric Institute, *Uniform Business Practices for Unbundled Electricity Metering, Volume Two, Appendix 5—Interval Data Validating, Editing, and Estimating Rules*, December 5, 2000.

application. To achieve the benefits of the smart grid, communication networks and data must be shared and involve multiple software applications—from billing to demand response to outage management. A key strategy to maintaining data accuracy and reliability in this new environment is identifying and utilizing a system of record for each data element. By definition, the system of record is the system that "owns" the data element and is looked to as the final source of authority as to the accuracy of the record; users and other software applications cannot change the data without the permission of the system of record.

The sections below provide further detail on why utility IT silos developed, why systems of record are needed when multiple applications share the same data, and how to implement a system of record strategy.

Utility IT Silos

Utilities did not have to focus on systems of record in the past, because they generally implemented data applications in IT silos. For instance, the Customer Service Department has been responsible for meters, meter data collection, and processing of meter data in the billing system. Other uses of meter data have been on an exception basis, such as providing periodic outputs of meter data for use in rate design or distribution system analysis. Another example is management of capacitor banks installed on distribution feeders. The Distribution Operations department has been responsible for the capacitor banks, the electronic sensing and switching devices installed on them ("remote terminal units"), the communications network, and the application software that controls the capacitor banks remotely and records the data from them. Figure 11.2 illustrates such systems.

In contrast, the smart grid requires sharing of data, communications networks, and devices by multiple applications. For example, smart meters provide data for monthly billing, consumer web presentment, and outage management—all using the same smart meters and communications network. Inconsistent data cause customer confusion, user concerns, and operational issues. To avoid these problems, multiple applications need to use the same data. Failure to do so can result in costs exceeding tens of millions of dollars when multiple market participants, such as power generators and retailers, are involved.[16]

The Need for a System of Record

Sharing of communications networks and data requires not only new approaches to application integration—connecting the applications to the communications networks and data—but also systems of record. Application integration is

---

[16]Chris King, *Competition at the Meter: Lessons from the UK*, Public Utilities Fortnightly, November 1, 1996.

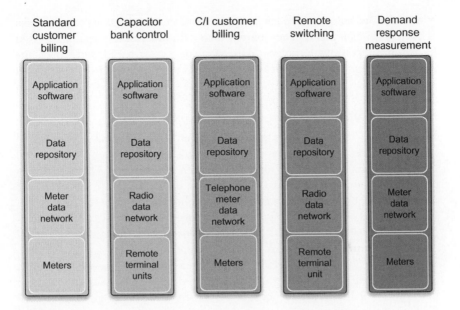

FIGURE 11.2   Utilities historically have operated IT systems in business silos. *(Source: Chris King, MDM Implementation Course, October 2010.)*

addressed below. As a general principle, IT architects prefer to avoid duplicating data, but if every utility application used the same database, all of the applications would be slowed to an intolerable level. The slowing occurs because each time data are processed in the database, it taxes the processing power of the computer server on which the database is loaded. Examples of such processes include sorting, indexing, adding or deleting data records, aggregating data, and creating reports.

The internet provides many examples of how companies use multiple systems to ensure rapid response times and consumer satisfaction. One such example is the Apache Software Foundation distribution network for open source software. When downloading a software package, users are given a choice of multiple locations from which to download the software in order to ensure rapid download.[17] The Apache Software Foundation's strategy ensures that any copy downloaded from any of the duplicate sites is the same as the "master" copy. The system of record approach enables this result.

Both for utilities and in general, a system of record, simply put, is the system that is assigned responsibility for a particular data element and is master of that element. An example would be that the utility customer information system (CIS)/billing system is the master of the monthly electricity bill amount. This amount is utilized in other applications, such as consumer web presentment

---

[17]http://www.apache.org/dyn/closer.cgi.

of the bill and bill comparisons used to inform customers about rate options. If there is ever a discrepancy between the amounts in the databases used by these applications to present data to customers or prepare mailings, the system of record approach dictates that the amount in the CIS is the correct amount.

System of Record Implementation

The system of record concept is straightforward, if not necessarily simple to implement. The first step is designating which system is the system of record for each data element used in various smart grid applications. The next step is ensuring that changes to the data element can be made only in the system of record. In our example, the monthly bill amount can be changed only in the CIS, not in the web presentment software by the customer or in an eBill application where customers can pay their bills online. Third, the system of record approach requires that any changes made to the data element are distributed, or propagated, to the other systems using the data element. In the same example, if the bill is changed in CIS, the new amount needs to be sent out to the customer presentment and eBill applications, so customers using those applications will then see the updated, and correct, amount. Finally, there must be an ongoing synchronization process to check periodically that the different applications are continuing to use the same data. This means that each of the "slave," or subservient, software applications must check back with the system of record to make sure it is continuing to use the proper master data.

Interestingly, with deregulation, defining the system of record also requires specifying the entity in addition to specifying the application. The regulators determine functional and performance requirements at a high level. These requirements then translate into the specific IT system requirements for implementation.

In Texas, the distribution utilities are responsible for meters and metering data, even though retail providers are responsible for bill calculations. Accordingly, if there is a discrepancy between the metered usage in the distribution company's database and the retailer's billing file, the retailer must defer to the distribution company's information.[18]

In Ontario, policymakers felt strongly enough about this issue to mandate, in regulation, the system of record. The province's central MDM/R is the system of record for all meter reading data. The distributors maintain the system of record for all customer data, including any current energy supply contracts with retailers, providing regular updates of any customer changes to the MDM/R.

---

[18]Public Utilities Commission of Texas, *Order Adopting New §25.130 and Amendments to §§25.121, 25.123, 25.311, and 25.346 as Approved at the May 10, 2007 Open Meeting*, May 10, 2007.

In the United Kingdom, the centralized Data Communications Company (DCC) is to perform this role for all 54 million electric and gas customers.[19] The government has concluded that the scope of DCC should be developed in a phased manner. When DCC starts providing its services, its scope will cover secure communications, access control, translation, scheduled data retrieval and central communications, and data management of initial smart grid functions. The UK government says that by limiting DCC's initial scope to these essential functions, an earlier implementation date will be achieved for DCC's services than if registration or other services are included. The government's strategy is to support the rollout plans of suppliers and the early delivery of UK Smart Metering Implementation Program benefits.

The UK government's strategy is a good example of how to implement a complex IT system in the smart grid world. The strategy includes two key elements: a long-term vision and incremental implementation. The vision contemplates the addition of new functionality over time, including support of new technologies. This minimizes the risk of technology obsolescence. The stepwise implementation maintains focus and reliability. Each set of functions is implemented in a phase. During the phase, the functionality is built, installed, and thoroughly tested. Subsequent phases can then focus on the additional functions that can be added later. Examples of utility companies that have used this approach very successfully are Toronto Hydro and CenterPoint Energy, as well as the case studies described later in this chapter.

Communications Management

The challenges associated with communications management for the smart grid are created by two basic aspects. The first is the scope, which includes multiple device types and potentially tens of millions of endpoints. The second is the associated maintenance; communication links must be maintained over wireless and other networks susceptible to occasionally spotty performance. The challenge has some parallels to maintaining large cellular phone networks, except for the variance in devices and lack of current interoperability standards for smart grid data—a key reason the industry is putting so much effort into creating those standards.

Smart Grid Device Tracking

Smart grids include smart meters, in-home displays, smart thermostats, smart appliances, and other equipment. Having successful communications with those devices requires knowing about them. The system needs to know everything about them. This requires an IT system that can track many details, and

---

[19]UK Department of Energy and Climate Change, *Smart Metering Implementation Programme, Response to Prospectus Consultation, Supporting Document 4 of 5, Central Communications and Data Management*, March 2011.

those details change periodically; for examples, when installing new devices, changing the programming (configuration) of devices so they operate or communicate differently, and changing the software (firmware) installed in the device electronics.

The following are several of the items that need to be tracked to enable successful communications:

- Where devices are installed,
- What data they are generating or receiving,
- When such data are generated or received,
- What system, or systems, is ultimately exchanging what data with the device (e.g., the outage management system would receive outage alerts—"last gasp" messages—while usage data would be provided to the billing system),
- What third parties, if any, are authorized to exchange data with the device (e.g., in Texas both the contracted retailer and the distributor), and
- What firmware version is installed in the device.

These data elements are tracked in a database with a data model that can accommodate each element, such as that shown in Figure 11.3.

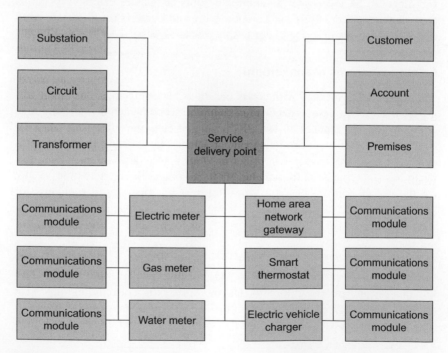

FIGURE 11.3   The data model tracks relationships between a service delivery point and various devices, grid connections, and premises data. *(Source: Chris King, MDM Implementation Course, October 2010)*

To assist in utilizing the data, the data model may also support tracking of the electrical connectivity of devices. For example, the database could track to which distribution substation, feeder, and transformer a smart meter is connected. This would assist in diagnosing the cause of an outage reported by an alert from that meter, because the utility would know which substation, feeder, and transformer was involved.

Device Provisioning

Provisioning in this context refers to establishing the secure logical connection between a home and an account or another device. Linking a new Bluetooth headset with your cell phone is one example of provisioning. It is obviously important that it links with your cell phone, and not someone else's.

In the smart grid, provisioning is important to link a newly installed smart meter with the correct customer account and to link an increasing number of in-home devices to the correct smart meter. This way, for instance, the consumption from the correct meter is displayed on the correct device, so neighbors are not seeing each others' consumption. The value of this and similar applications is described in Chapter 15.

The IT environment challenges associated with device provisioning can be significant. The utilities in Texas struck an innovative approach, building on their existing Smart Meter Texas platform. Energy service providers offer customers programs that often include in-home devices. Rather than duplicating the function, the central Smart Meter Texas service manages all device provisioning between these customer devices and the utility's smart meter.

Communications Maintenance

Successful smart grid communications require that the devices, communications network, and software applications are all working properly. When something fails, communication stops or data become corrupted. Utilities need IT systems that monitor the communications processes. Some system needs to check whether communication is operating successfully with each device on the network and whether the expected data exchanges are happening. For example, most smart meter systems deliver data to the utility's meter data management system at least once a day, during some expected time window. A software application needs to check for this delivery, meter by meter, and report which meters have failed so send data. This is typically done in the meter data management system but could also be done by the head-end communications management software that runs the smart meter network, the so-called advanced metering infrastructure, or AMI, head-end.

The other challenge with smart grid communications is what to do when communications fail. It is one thing to identify, as described above, that a meter has not reported data as expected. It is another thing to determine why the data were not reported and how it can be fixed. The reason could be a meter failure, a failure of the radio in the meter, a failure of the network concentrator unit that normally receives the data from the meter, a failure of the wide area network (WAN) connection to the concentrator,[20] or other things such as a power outage or temporary radio interference.

Exception Management

Some IT system needs to identify the failure—also known as an exception—diagnose possible solutions, and initiate a resolution—all automatically—because manual processes are cost prohibitive in a network of millions of devices. Exception handling is the bane of operations departments, and automated processing is seen by utilities as a major smart grid benefit.[21]

The resolution is typically one of these three approaches:

- **A retry of the communications**, an automated repeat request for the information. Such retries are part of the basic operations of the AMI head-end software. However, retries can be initiated by upstream systems as well, such as the meter data management system or billing system.
- **Simply waiting another day**. Missing meter reads in a daily file of smart meter reads are often resolved the next day in the regular daily meter reading process. This happens automatically in the operation of most AMI technologies.
- **Manual intervention**. A work order is created by the meter data management system, CIS, or work management system—or manually—and a human being takes various steps to diagnose and solve the problem.

Integration of Software Applications

Making the smart grid work requires different IT systems to talk to one another reliably and efficiently. The AMI head-end software needs to talk to the meter data management system, which needs to talk to the billing system. Linking all of these systems can be quite complicated. When multiple systems are involved, and each is individually connected to others, the result is sometimes called "spaghetti integration" as shown in Figure 11.4.

The challenge of application integration is increased by the diversity of the systems. Some are new, others old. For example, a brand-new AMI head-end may be connected to a 20-year-old legacy billing system. Moreover, technical interfaces differ, ranging from modern, XML-based, interfaces to traditional

---

[20]To clarify, the WAN is the communications link between the utility office and communication nodes typically mounted on utility poles in neighborhoods. From there, pole-top nodes communicate via radio frequency or power line carrier to the meters in the areas, this being the LAN. The meters then communicate via a separate radio or power line carrier modem into the home or building to connect to smart devices; this is the HAN.

[21]Southern California Gas Company, *Advanced Metering Infrastructure Chapter II – Summary of AMI Business Case*, Testimony Filed with the California Public Utilities Commission, September 29, 2008.

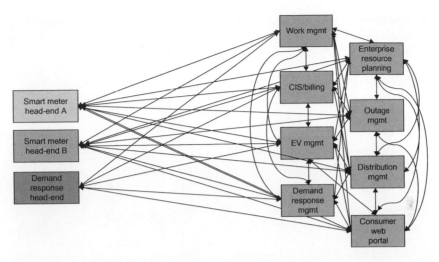

FIGURE 11.4 With the smart grid, multiple software applications must be connected with one another to deliver the desired functionality; a common outcome is "spaghetti" integration. *(Source: Chris King, MDM Implementation Course, October 2010.)*

batch-based file exchanges. In addition, timing requirements may vary substantially, with outage management systems typically using real-time data exchanges, while billing systems operate with nightly file transfers.

ESSENTIAL SMART GRID SOFTWARE PLATFORM

During the implementation of a large-scale advanced or smart metering project, the biggest risk is not related to the meters, the communications, nor even the installation. The greatest potential for getting less functionality than was bargained for or spending more money is a poorly designed information technology (IT) architecture. Getting it right is one of the key steps in ensuring project success. To reiterate, while there certainly are risks associated with investing in smart metering equipment, there are even larger risks of not investing in an integration platform to support the smart metering investment and enable the desired functionality.

Utilities have two overarching goals in providing electricity: maximizing reliability and minimizing costs. These should be the same goals of the IT design for an advanced metering infrastructure. Reliability means timely, complete data, as well as seamless integration with billing, outage, field maintenance, and other essential business systems. Reducing costs means—well—spending less money, of course.

Smart Application Integration Platform

Smart grid functions can be implemented in one of three ways: 1) in a purpose-built smart grid integration platform, 2) in smart grid data communications

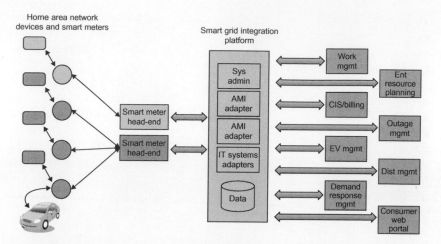

FIGURE 11.5   An integration platform allows sharing of communications networks and data by multiple utility application systems; it also supports multiple meter communications technologies. *(Chris King, MDM Implementation Course, October 2010.)*

software, or 3) in existing utility core business systems. The most reliable, cost-effective architecture is usually the first, as shown in Figure 11.5.

Utilities have found the integration platform architecture to be the most reliable and cost effective for several reasons:

- It centralizes all the smart meter data management activities, linking, for example, data collection with data management and the data warehouse. This results in fewer personnel and less friction between business units than might otherwise be experienced if these functions were separated into different departments,
- It places all of the smart meter data activities in a platform focused on meter or other endpoint data. It has reliable and efficient operation of that platform as its sole mission.
- It effectively and safely manages the typical situation of utilities using multiple networks and technologies, on the one hand, and multiple utility core business systems on the other hand.
- It reduces maintenance costs for both the networks and the utility business systems, because those systems can be operated, maintained, and even replaced on an individual basis without requiring major modifications to the other interconnected systems.

Issues Associated with Other IT Architectures

One alternative is to implement the smart grid platform functions in the head-end of the smart meter communications system. The focus of that system is the operation of the communications network, not managing meters or data.

These systems are optimized for network operation, a very different function from data management. Such an approach makes it much more difficult to integrate multiple communications technologies, a required feature of every utility as technology evolves more and more rapidly.

Another alternative is to implement the smart grid platform functions by modifying core utility business systems. This is expensive because these systems are complex and were built to provide different functions, such as billing. This also adds risk, because the modifications to the legacy systems must happen in parallel with smart grid implementation. In some cases this approach would drive modifications to systems that are scheduled for replacement even before the smart grid or smart meters are fully deployed. These parallel activities mean the system can focus on neither its primary function—say billing—nor the smart metering implementation.

Multiple Operating Companies/Multiple Utilities

The smart grid platform approach supports utilities with multiple operating companies or jurisdictions where multiple utilities and multiple communications networks are involved. This would allow, for example, consistent business rules for validating data to be applied across all operating companies. If there are valid reasons for differences in the business rules, it provides a central location to manage business rules. An added benefit of the platform is that it also allows utilities to work through merger-related consolidation of smart grid communication networks into a single, common system for use by the sister utility companies.

System Interfaces and Interoperability

Every time data are exchanged between one device and another, or between one software system and another, the data must pass through the interface between the devices or the systems. This interface includes numerous layers. The underlying layer is the physical communication: a wire, a radio signal, or something else? Another key layer is the communications protocol, the language the devices or systems speak. A third important layer is the data structure. Standards are used in each layer to enable numerous and diverse devices and systems to work together.

The internet provides an illustration. The physical layer is often a radio connection within the home to a router, then over a cable wire to the cable company's data center, and from there perhaps by microwave radio to the website host's data center in another city. The communications protocol is IP, which stands for Internet Protocol. Web pages use HTML (hypertext mark-up language) as the data structure for the information presented in a browser.

For the smart grid, such standards are also needed for ultimate interoperability of devices and systems. The scope of these standards goes beyond that of this

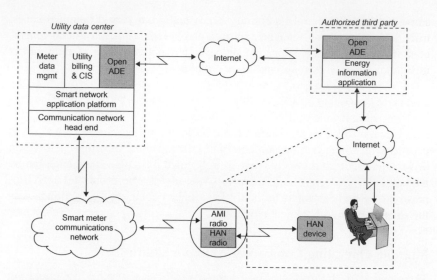

FIGURE 11.6 Two key smart grid standards are the interfaces between utilities and authorized third-party data providers (OpenADE) and between smart meters and HAN devices (OpenHAN). *(Demand Response and Smart Grid Coalition, DRSG Comments to DOE Smart Grid RFI: Addressing Policy and Logistical Challenge, November 1, 2010.)*

chapter, other than to note that the Federal Energy Independence and Security Act of 2007 mandates the development of such standards and adoption by the Federal Energy Regulatory Commission.[22] The National Institute of Standards and Technology (NIST) has responsibility for coordinating the adoption of these standards. Utilities desiring to maximize the value of the smart grid IT infrastructure are embracing these standards where applicable and feasible.

The most important standards are those that apply to interfaces where data are exchanged between one entity and another. One example is sharing data between utilities and third parties authorized by customers via the OpenADE standard ("Automated Data Exchange").[23] Another example is sending data from the utility-owned meter to a customer-owned HAN device via the Open-HAN standard. These interfaces are shown in Figure 11.6.

SMART GRID APPLICATIONS

Utilities implementing the smart grid are faced with also adding the IT application software that ultimately enables delivery of the desired functionality. The range of such software is essentially unlimited. In order to obtain

---

[22]H.R. 6, Energy Independence and Security Act of 2007, Section 1305, "Smart grid interoperability framework."

[23]OpenADE and OpenHAN are under development under the auspices of the National Institute of Standards and Technology. Details are available at NIST's website for smart grid: http://www.nist.gov/smartgrid/.

some of the key, basic functionality described in smart grid business cases, utilities are implementing software to support meter data management, dynamic pricing programs, and, for distributed renewable resources, net metering.

Meter Data Management

Meter data has traditionally been handled by the billing system, which was designed to handle a single meter register read per month from each meter. With quarter-hourly interval data, that data volume is now being increased by a factor of nearly 3,000—not to mention outage and voltage data, tamper alerts, and other data reported by meters. In this new environment, meter data management requires significantly more functionality. To achieve the needed results, utilities implementing smart grids are installing purpose-built meter data management systems (MDMs).

The MDM portion of a smart meter system has three basic functions:

1. Collect meter data,
2. Manage that data,
3. Provide a meter data warehouse.

The MDM essentially acts as a buffer between users of the data and the collection of the data. This allows utilities to use multiple systems without burdening applications with the specifics of how data are collected or how commands are dispatched and, similarly, allows different applications to view the same data in different ways. The "meter data warehouse" is simply a database to store meter usage and other data and is, in fact, the simplest element. A comprehensive meter data management system will provide essential services far beyond a meter data warehouse.

MDM functions reveal an additional layer of complexity when looked at more closely. Collecting meter data involves not only retrieving meter reads, but also sending configuration data to meters and receiving other data such as outage alerts and meter diagnostic flags. Managing smart meter data begins with validation, editing, and estimation (VEE) of meter data. Validation includes various checks such as ensuring the data are coming from a valid meter address. Editing is manual intervention to correct data that have been determined to be incorrect or to insert replacement values for missing data. Estimation is the automatic correction or replacement of data based on various calculation algorithms. VEE is the solution to some of the data reliability and accuracy issues identified earlier in this chapter. Utilities have always performed VEE on interval data; the difference with smart grid is that utilities are now handling interval data for millions of meters rather than hundreds or perhaps a few thousand. Such meters have been used for many years but only for small numbers of large commercial and industrial customers.

Dynamic Pricing Rates

A key source of benefits in the smart grid is dynamic pricing rates. Such programs can account for $2 billion in savings for a utility with less than 1.5 million meters.[24] However, as with the other benefits of smart grid, such savings are possible only if the software application functionality properly supports implementation of the pricing plans. In another example, dynamic pricing benefits in the European Union have been estimated at EUR 53 billion.[25] See the Chapter 3 for additional elucidation of dynamic pricing benefits.

Features of Dynamic Pricing Rates

Dynamic pricing rates include several options. The details of these plans illustrate the numerous capabilities that a smart grid IT application must have in order to bill such rates accurately. Time-of-use (TOU) rate schedules assign different pre-determined rates to pre-defined time periods, and customers pay those pre-determined rates during each time period. For example, during the summer, the rate charged during the afternoon is generally higher than the rate charged at night. The different rates reflect the fact that it is generally more expensive to serve customers during some time periods. TOU rates do not change based on current market conditions. Different TOU rates are set for the summer and the winter seasons.

Critical-peak pricing (CPP) generally describes rates where a very high rate will apply to a customer's usage during CPP events, typically 60 hours per year. In return, the customer gets a small discount during the remaining hours of the year. The CPP event is triggered based on system conditions, such as high temperature. CPP events are for specific hours and are called on a day-ahead basis, and must be tracked in the IT system to enable billing. There are a limited number of CPP events a year. Another name for CPP is peak-day pricing (PDP)

Real-time pricing (RTP) rates are based on prices in the wholesale energy market, for example the PJM day-ahead market. RTP rates apply to every hour of every day, and are subject to change from hour to hour. RTP rates are usually communicated to customers a day ahead. Such rates are also called "hourly pricing."

Peak-time rebate (PTR) is a program that provides the customer a rebate on a per kWh basis for reductions in the customer's usage below a threshold level on days when a PTR event is called. The baseline is specific to each customer and is based on the customer's prior usage for specific days prior to the day of the PTR event. The PTR event is like a CPP event, is called based on system

---

[24]Baltimore Gas & Electric Company, *Application of Baltimore Gas and Electric Company for Authorization to Deploy a Smart Grid Initiative and to Establish a Tracker Mechanism for the Recovery of Costs*, July 13, 2009.

[25]Ahmad Faruqui, Dan Harris, and Ryan Hledik, *Unlocking the EUR 53 billion savings from smart meters in the EU: How increasing the adoption of dynamic tariffs could make or break the EU's smart grid investment*, Energy Policy, October 2010.

conditions, and is called on a day-ahead basis. The number of events per year is either specified or can be within a range, usually up to 15 days, four hours per event. Chapter 12 includes a description of the specific experience in California with CPP pricing.

Software to Support Dynamic Pricing Rates

The IT systems needed to support dynamic pricing vary by specific application. All of the options require bill calculation, of course. All but TOU require some means of notifying customers of prices or events. Such notification systems typically must support automated phone calls, text messaging, and email. PTR requires continuous calculation and tracking of customer baselines for use in bill calculation. RTP requires a link to the wholesale market system to obtain market prices, plus a means of sending prices to customers. This usually involves the internet to display the hourly prices on the utility's website, plus sometimes text messaging and email.

The billing involves very different calculations from those in historical utility billing systems. In some cases, utilities implement software applications outside of their billing systems to perform part of these new calculations. The key such function is framing of interval data into billing determinants. This involves adding up individual interval data values into pre-determined billing periods. For example, in a time-of-use rate, the interval data may be framed into peak, mid-peak, and off-peak values. For a CPP program, the framing may be for critical peak usage quantities. Once the interval data are simplified into a small number of these values—called "billing determinants"—the billing system can then calculate the bill.

Net Metering

Net metering is a method of simplifying the measurement of energy produced by a renewable or distributed energy generator when it is connected to an electric utility distribution system. Net metering is permitted by law for solar, wind, and biomass generators that are generally intended to supply no more than the customer's annual energy usage. The regulation of rates for buying and selling varies significantly from jurisdiction to jurisdiction, so practitioners should take care to review the specific tariffs in the relevant locale. The term "net metering" refers to the measurement of the difference between the electricity supplied by an electric company and the electricity that is generated by an eligible customer-generator and fed back to the electric company during a customer-generator's billing period. In other words, during a single reading period, such as one month, it is the net of the energy supplied by the electric company and used by the customer and the energy produced by a customer and sent back to the electric company through the company's distribution system.

As with dynamic pricing, net metering introduces new functionality requirements for utility IT systems to calculate bills and credits. The terms of utility tariffs typically require a customer to pay the monthly customer charge, regardless of the net energy used. However, for energy billed, the customer only pays for energy that is used, netted against any generation produced by the customer. Any excess generation balance is carried forward to the next month and used in the bill calculation. If any excess generation balance exists at the end of 12 months, the customer is paid a per-kWh amount usually based on the wholesale generation cost, and the cycle restarts. The per-kWh rate paid for the annual true-up is typically about half the total retail price of power, because it excludes transmission, distribution, customer, and other costs outside of generation. The practical effect of this policy is to allow customers to use the utility grid as storage, so that excess energy produced at any given instant can be captured for later use.

Customer generators benefit from net metering by less expensive interconnection with the utility (e.g., only a single standard meter without additional switches). In this way, electricity needs in excess of the renewable output can be obtained from the grid without having to disconnect or shut down a renewable generator. The ease of interconnection allows the customer to use the renewable generator in a grid-connected manner without significant installation or operating expense, thus improving the benefit of having the renewable generator on the grid.

As additional customer generator equipment is installed, power quality problems can emerge. For example, excess solar generation on local distribution grid node can cause voltage problems.[26] Utilities and others are looking to IT applications to solve or prevent such issues. Hawaii is conducting a pilot program. The goal of this demonstration is to solve various issues, including voltage issues related to reverse power flow that stem from photovoltaic solar power generation linked with distribution system circuits.

CASE STUDIES

This section provides case studies of the province of Ontario and the state of Texas. The authors are personally familiar with the implementation of major smart grid IT projects there: the MDM/R in Ontario and the Smart Meter Texas centralized data portal in Texas. These two jurisdictions probably lead the globe in functionality provided via smart meters at commercial scale. Functions available to millions of customers with smart meters in Ontario and/or Texas include next-day online and real-time in-premise energy usage information, dynamic pricing, remote service connect and disconnect, handling of outage alarms from smart meters, and prepayment service.

[26]*Collaborating on Hawaii Smart Grid*, Today's Energy Solutions, May 19, 2011. Available at http://www.onlinetes.com/renewable-energy-manufacturing-Japan-United-States-NEDO-tes-051911.aspx.

Ontario MDM/R

Facing significant shortfalls in generation capacity in coming years, in 2004 the Province of Ontario launched an ambitious program to reduce peak electricity demand through a variety of programs. Central to that objective was the introduction of smart meters and associated time-of-use rates across the entire province by the end of 2010. The deployment of smart meters in Ontario is complicated by extremes in customer density, with very congested urban areas contrasting with rural and extremely remote communities throughout large expanses in the north of the province. Ontario also has over 80 different electricity distributors, ranging in size from 1.2 million customers to several dozen with fewer than 10,000 customers.

After an extensive consultation period, the provincial government announced a two-tiered IT model with some functions remaining the responsibility of the local distribution companies and others being centralized into a new province-wide service. This model achieves three key policy objectives:

- **Scale economies.** The province was faced with having up to 80 separate MDM implementations, one for every local distribution company. As noted, some of these have fewer than 10,000 customers. By having a centralized repository, the province is able to capture the scale economies associated with having a single system and single implementation.
- **Highest service levels.** By concentrating the capability in a single operation, Ontario is able to provide higher levels of service and functionality than individual utilities could afford—especially the smaller utilities. For example, for disaster recovery, Ontario has a full back-up system up and running—at a separate geographic location—to take over operations in the event of an event damaging or destroying the primary facility. Another example is expertise; the centralized operator can afford to hire the best experts, because any additional cost is *de minimus* when allocated across the many users of the system.
- **Consistent and reliable data.** Finally, by having all the utilities use a single system, the province ensures that all customers, regardless of the utility serving them, receive the same, consistent accuracy and reliability of smart meter data.

These benefits of centralization occur as a matter of course in competitive industries. For cellular telephone billing in the United States, for instance, two companies handle almost all of the billing. In contrast, there are hundreds, perhaps over a thousand, billing systems for the 3,000 electric utilities in the United States.[27]

---

[27]Jon Brock, *Business Process Outsourcing: Is It for the Utility Market,* CISWorld.com, August 2003. Available at http://www.cisworld.com/articles/0308_brock3.htm.

Distribution companies are responsible for installing and maintaining smart meters, ensuring hourly consumption data are collected daily, handling customer billing and collecting, answering customer queries, and generally being the agent for conservation programs with the customer. The Independent Electricity System Operator (IESO), the province's electricity grid and market operator, was designated by the government to be the smart metering entity responsible providing the provincial meter data repository (MDM/R) service.

The MDM/R collects consumption data from each distribution company, validates it, and then converts the hourly data into total consumption for each TOU billing period. These TOU billing determinants are then returned to the distribution company for customer billing. Once introduction of TOU rates is complete across the province, the MDM/R will collect and process hourly interval data from approximately 4.5 million smart meters.

Ontario recently introduced a robust feed-in-tariff that is driving significant investments in renewable generation. One of the province's targets is that by 2030 nearly 13% of electricity generated is from wind, solar, and bioenergy.[28] Much of this new renewable generation is expected to come from several hundreds or thousands of small generation facilities at homes and farms, each requiring bidirectional meters that measure and report both energy consumed and generated.

Smart Meter Texas

In 2007, the state of Texas adopted rules to introduce smart meters in support of a number of policy objectives, including improved system reliability (particularly after storms), energy conservation, and enhanced retail services. The smart meter technology standards in Texas include 15-minute interval data collection and ensuring the meter has the communication module necessary to connect to a home area network (also used in businesses).

Facing IT challenges in managing and distributing smart meter data to competitive energy retailers and customers, the four investor-owned utilities in the state (CenterPoint, Oncor, American Electric Power-Texas, and Texas-New Mexico Power) formed Smart Meter Texas.

Smart Meter Texas provides a centralized repository of smart meter data. Distribution companies collect and send smart meter data to the repository, while Electric Reliability Council of Texas (ERCOT) and power retailers receive data from the repository.

Smart Meter Texas (Figure 11.7) also provides centralized services for provision of home area network devices and a default website for customers to access their meter data.

The Smart Meter Texas portal has several consumer benefits. It is compliant with the Americans with Disabilities Act (ADA). As with the Ontario MDM/R,

---

[28]Government of Ontario, *Ontario's Long-Term Energy Plan*, Queen's Printer for Ontario, 2010.

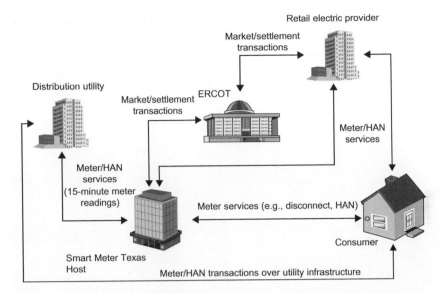

FIGURE 11.7   In Texas, the distribution companies maintain electricity usage data, though the Smart Meter Texas portal also stores the data for retrieval by other authorized users ("HAN" signifies "home area network"). *(Source: Texas PUC Advanced Metering Implementation Team, November 2009.)*

it standardizes functionalities and data access across different distribution utilities within the state. This means consistent content, availability and data format. Moreover, the web portal allows retail electric providers to take full advantage of the smart meters deployed. This includes the ability to develop retail electric services that incorporate dynamic pricing and demand response for residential and small commercial consumers. This standardization ensures that the retailers, customers, and their agents have access to the same basic information in every territory. It also ensures that retailers do not have to develop different systems and processes for providing customer products in each utility territory. In this way, Texas, like Ontario, captures the scale economics inherent in centralized IT approaches.

CONCLUSIONS

Overcoming the IT challenges of the smart grid turns out to be even more difficult than deploying communication networks and millions of meters and other devices. Success requires a transformation of traditional approaches, with the emphasis on sharing data and communications networks and avoiding point solutions that support only a single department or business process. Where utilities have surmounted the obstacles, such as in Ontario and Texas, consumers are receiving the benefits of online and real-time data, pricing choices, and device automation through utility-level and HAN interfaces to service providers or devices. In other cases, consumers are not yet seeing the benefits of smart meters and smart grids, with occasional negative consequences.

The utility industry has been one of the slowest to adopt new IT technology, with many good reasons. Yet the arrival of the smart grid, distributed renewable generation, and electric vehicles is strongly and rapidly increasing the pressure on utilities to modernize and transform their IT systems and strategies.

Chapter 12

How Large Commercial and Industrial Customers Respond to Dynamic Pricing—The California Experience

Steven D. Braithwait and Daniel G. Hansen

INTRODUCTION

Economists have long favored time-varying retail electricity prices, such as time-of-use (TOU) rates, because they reflect correspondingly time-varying marginal costs of generating and delivering electric power, as reflected in wholesale market costs, and thus improve economic efficiency. Electricity costs vary for two primary reasons: 1) consumers' demands for power vary substantially by season, day, and hour due to a combination of normal cycles of personal behavior and business operations, as well as weather differences that affect weather-sensitive uses such as air conditioning; and 2) electricity cannot be easily stored, but must instead be generated at the exact time that it is demanded. These two factors combine to cause electricity suppliers to rely on a mix of generation technologies, including extra reserve capacity, each of which has different capacity and operating costs, to meet these varying demands at reasonable levels of reliability.

Questions are often raised about the extent to which consumers will actually change their electricity usage behavior if they were to face time-varying prices. After all, if usage patterns do not change, then TOU rates will simply change the timing of revenue recovery but leave system costs unchanged. However, numerous studies, beginning in the 1970s, have demonstrated that consumers in general will respond to time-varying electricity prices.[1] Early economic assessments often concluded that even if consumers did respond, the economic benefits to be derived from time-varying pricing would likely fall short of the cost of the more complex metering that is required to bill consumers for consumption at different prices and different times of day. This pessimistic benefit-cost calculus has been altered recently by a combination of falling costs of advanced metering devices, and perceived operational cost savings that may be achieved from widespread installation of advanced or smart metering and smart grid systems, regardless of whether smart pricing is adopted.

If smart meters are in fact installed in the larger context of smart grid development, then the cost barrier to dynamic, time-varying pricing is reduced substantially, opening the way to achieving benefits from consumers' price response, as well as from grid operation savings. Dynamic pricing gives customers access to low-cost power at correspondingly low prices, when available, and occasionally sends high-price signals in hours when generation and transmission costs are highest. Demand response (DR) in the form of load reductions that represent consumers' response to dynamic pricing can provide a reliable and valuable resource at times of generation, transmission, or distribution system capacity constraints, and can thus improve reliability and reduce costs. However, considerable uncertainty still exists about how consumers of various types will accept and respond to smart pricing.

As a result of this uncertainty, a number of electric utilities around the country have initiated pilot programs to test a variety of pricing designs. Numerous papers and reports have summarized the performance of several such pilots targeted at *residential* customers, involving dynamic pricing such as critical-peak

---

[1]See Faruqui and Malko [1], Braithwait [2], and Braithwait and O'Sheasy [3].

pricing (CPP), real-time pricing (RTP), and peak-time rebates (PTR).[2] However, information on the performance of such rates and programs for larger *commercial* and *industrial* customers is much less widely available. Fortunately, several years of load impact evaluations of non-residential DR programs for the three large investor-owned utilities in California—Pacific Gas and Electric (PG&E), Southern California Edison (SCE), and San Diego Gas and Electric (SDG&E)—have produced a wealth of information on how those customers respond to dynamic pricing, such as CPP, as well as to DR program incentives. This chapter summarizes key findings and insights from the most recent statewide evaluations that we have conducted for the California utilities.[3]

The chapter begins with a brief history of the rationale for smart pricing in California, an overview of the dynamic pricing and DR programs offered to large commercial and industrial customers at the California utilities, along with details on the participating customers. It then describes the analysis approach that was used to estimate hourly load impacts, summarizes the empirical findings on customer price response, and offers some perspective on the meaning of the results and future trends for CPP in California. The chapter concludes by commenting more broadly on the potential for DR outside of California, discussing some elements of the tradeoffs between dynamic pricing and DR programs offered by regional ISOs or RTOs.

DYNAMIC PRICING AND DR PROGRAMS IN CALIFORNIA

California's push toward smart metering and dynamic pricing originated in the state's electricity crisis of 2000/2001, which was characterized by periodic supply disruptions, skyrocketing wholesale prices, and frequent load curtailments and rotating outages. At the time, it was generally recognized that if some form of dynamic, time-varying pricing had been in place and customers had reduced load during periods of unusually high prices, then the resulting demand reductions would have gone a long way toward limiting the wild gyrations in wholesale prices and reducing the high electricity costs that ultimately occurred. However, the advanced metering and ratemaking infrastructure required to implement such pricing were not in place at the time.

In view of these conditions, the California legislature in March 2001 agreed to fund the installation of interval meters for customers with maximum demand greater than 200 kW (approximately 25,000 were installed throughout the state, mostly at the three major investor-owned utilities, or IOUs). In parallel,

---

[2]Dynamic pricing is characterized by electricity prices that can vary at short notice, such as with notification of a day ahead or a few hours ahead of the occurrence of high prices. This is in contrast, for example, to basic TOU rates, whose static prices vary by time period, but remain the same for every weekday during a season. In Chapter 3, Faruqui describes several examples of dynamic pricing. Faruqui and Sergici [4] survey a number of residential pilots.
[3]See Braithwait et al. [5–7] for complete descriptions of the CPP, demand bidding, and aggregator DR program evaluations, respectively.

the California Public Utility Commission (CPUC) initiated proceedings that resulted in the three IOUs filing voluntary dynamic price structures, in the form of CPP, along with several DR programs, in 2004 for customers with interval meters.

The CPP tariffs were originally voluntary rates targeted at customers with maximum demand greater than 200kW, who were the first to receive hourly interval meters. The rates were designed in conjunction with existing default tariffs consisting of time-of-use (TOU) demand and energy charges. Customers receive discounts on non-event peak-period prices in return for facing considerably higher peak prices on a limited number of "critical" days. However, following the guidance of the CPUC, the utilities are transitioning toward *default* CPP rates for all customers of size greater than 20 kW of maximum demand, beginning with the largest customers (e.g., those with maximum demand greater than 200 kW) with voluntary opt-out to new TOU rates.[4]

In addition to CPP rates, the utilities also offer a range of DR programs. These generally include three types: 1) traditional emergency, or *reliability-based* programs in which customers receive capacity-based payments in return for agreeing to curtail usage at short notice during infrequent emergencies; 2) demand response programs offered by third-party DR *aggregators* (including the statewide capacity bidding program, or CBP, and individual aggregator contracts); and 3) *demand bidding* programs offered to individual customers (DBPs).[5] CPP events are generally announced on a day-ahead basis, while the DR programs (other than the reliability-based programs) generally offer both *day-ahead* (DA) and *day-of* (DO) notice options.

As of the summer of 2009, enrollments in each of the various rates and programs generally ranged from a few hundred to more than a thousand customer accounts (enrollment in SDG&E's new default CPP rate was nearly 1,600), where the average customer size typically ranged from about 200 to 500 kW. Total enrollment in CPP in 2009 exceeded 2,700 customer accounts, while enrollment in the aggregator programs exceeded 4,000 customer accounts. Enrollment in the DO option of the aggregator programs was generally much larger than for the DA option, with approximately 70% choosing day-of products, presumably due to larger capacity credit payments.

To provide some perspective on the California electricity market, the state has a population of some 36 million residents, who are served by three major investor-owned utilities (IOUs), several large municipal utilities (e.g., Los Angeles Department of Water and Power and the Sacramento Municipal Utilities District), and a number of smaller utilities. PG&E, the largest IOU by area, serves approximately 5 million customers in regions ranging from

---

[4]Designs for real-time pricing are being explored and will likely be implemented in future years.
[5]As PG&E transitions to default CPP, it is renaming the rate "peak-day pricing" (PDP). It is also transitioning DBP to PeakChoice, a flexible bidding program with a variety of notice and payment options.

north of the relatively mild San Francisco Bay area to the hot and dry central valley. Its system peak demand in 2009 was approximately 19 GW. SCE serves a similar number of customers, but with generally warmer weather, reached a peak of about 22.4 GW in 2009. SDG&E serves 1.4 million customers in the southern-most part of the state, reaching 4.5 GW in 2009.

Average bundled electricity prices for the large commercial and industrial customers analyzed in this analysis ranged from $0.11 to $0.14 per kWh across the three major utilities in 2009. These prices, which continue in part to reflect the financial fallout from the 2000/2001 crisis, are substantially higher than U.S. average prices.[6]

CPP CUSTOMERS, TARIFFS, AND EVENT CHARACTERISTICS

This section describes the customers who participated in the California CPP programs in 2009, the nature of the tariffs, and the number and timing of the events called.

Customer Enrollment by Industry Type

In order to assess differences in load impacts across various customer types, program participants were categorized according to seven broad industry types, plus a category for other/unknown. The industry groups are defined as follows according to the applicable two-digit NAICS codes:

1. Agriculture, Mining & Construction: 11, 21, 23
2. Manufacturing: 31–33
3. Wholesale, Transport, Other Utilities: 22, 42, 48–49
4. Retail stores: 44–45
5. Offices, Hotels, Health, Services: 51–56, 62, 72
6. Schools: 61
7. Entertainment, Other Services, Government: 71, 81, 92
8. Other or unknown

Industry types 4, 5, and 7 include customers that are typically thought of as commercial-types, while industry type 2 contains all manufacturing customers. Industry type 3 typically includes a number of water and/or sewage utility customers that often turn out to be quite price responsive.

CPP enrollment by industry type, in terms of numbers of customer service accounts (SAID) and percent of load (using the metric of customers' maximum demand), for each of the utilities is summarized in Table 12.1. Enrollment in CPP at PG&E fell somewhat from 760 accounts in 2008 to 650 customer service accounts in 2009, after having expanded from 337 accounts in 2006 and

---

[6]Data from the Energy Information Administration suggest that average retail prices for commercial and industrial customers in California were 24 to 40% higher than for the United States as a whole.

TABLE 12.1 CPP Enrollment by Industry Type—Customer Accounts and Share of Load, by Utility

Industry Type	Number of SAIDs			% of Max kW		
	PG&E	SCE	SDG&E	PG&E	SCE	SDG&E
1. Agriculture, Mining & Construction Manufacturing	39	24	19	6%	4%	2%
2. Manufacturing	167	221	222	34%	49%	15%
3. Wholesale, Transport, Other Utilities	67	54	266	8%	18%	20%
4. Retail stores	42	35	128	3%	7%	7%
5. Offices, Hotels, Health, Services	127	44	481	25%	7%	36%
6. Schools	159	99	267	14%	13%	9%
7. Gov't, Entertainment, Other Services	49	8	190	9%	1%	11%
8. Other/Unknown			7			0%
TOTAL	650	485	1,580	100%	100%	100%

Source: Braithwait et al. [5].

656 accounts in 2007.[7] The total load of the customer accounts enrolled in CPP, measured as the sum of individual customers' maximum demands, amounted to nearly 400 MW. The Manufacturing; Offices, Hotels, Health, and Services; and Schools industry groups made up the bulk of PG&E's CPP enrollment, measured by the share of maximum demand.

SCE's enrollment in CPP has continued to expand, from just 15 customer accounts in 2006 to 44 accounts in 2007, 201 accounts in 2008, and 485 in 2009. Total maximum demand of customers enrolled in 2009 nearly doubled to approximately 283 MW. Manufacturing and Wholesale, Transportation and Other Utilities industry groups made up the bulk of CPP-participating load at SCE.[8]

---

[7]Enrolled "customer accounts" are defined as the number of "service agreement identification numbers," or SA_IDs. In some cases, a single "customer" will have more than one SA_ID, such as a supermarket or retailer that has multiple facilities at different locations throughout a utility service area.

[8]The relatively smaller number of CPP participants at SCE through 2009, as well as the different mix of industry types (e.g., more manufacturing customers), has largely been due to the presence of a rate, described below, that consists of a notably higher CPP price, but a highly discounted summer peak demand charge, which has attracted a number of very large and flexible customers.

Nearly 1,600 customer accounts out of the original 1,800 customers who were defaulted onto the new CPP rate in 2008 continued to participate in default CPP at SDG&E in 2009, declining to opt out to the new otherwise applicable time-of-use rate.[9] Customers were offered bill protection for their first year, such that their bill under CPP would be no greater than under their otherwise applicable tariff. Since no CPP events were called in 2008, bill protection was extended to 2009, which may have encouraged customers to remain on the rate. Offices, Hotels, Finance, and Services; and Wholesale, Transportation and Other Utilities industry groups accounted for more than half of SDG&E's 611 MW of CPP load.

CPP Tariffs

The CPP tariffs at the three utilities differ somewhat in design and application, both across utilities and across rate classes at the individual utilities. For example, SDG&E's default CPP rate is a commodity tariff that applies only to the commodity portion of customers' bills (e.g., they continue to pay for distribution services under a standard tariff). It specified the commodity prices for on-peak, semi-peak, and off-peak periods on CPP event days and non-event days directly. In contrast, PG&E implements CPP through a surcharge that is applied to energy consumed during the on-peak period on CPP event days, and credits are applied to energy consumed in on-peak and part-peak periods on non-CPP days. Similarly to SDG&E, SCE specifies a CPP price that applies to the generation component of customers' charge for on-peak energy consumption on CPP event days, but offsets that charge through a discount on the summer on-peak demand charge.

In general, the voluntary CPP rates consisted of peak-period prices of approximately $0.80 per kWh on up to 15 event days, with corresponding reductions in non-event energy prices and/or demand charges to maintain expected revenue neutrality. SCE offered a second alternative rate targeted to large flexible customers that set a CPP price of more than $2.00 per kWh in return for a large discount on the summer peak demand charge.

The new default CPP rates (including the one in effect at SDG&E in 2009) typically include event-day on-peak energy prices that exceed $1.00 per kWh, while discounts on non-event energy prices and/or demand charges can be as large as 20 to 30%. Figure 12.1 illustrates the daily pattern of commodity energy prices under the SDG&E default tariff on CPP event days and non-event days, along with a comparable profile for the alternative TOU rate. The CPP on-peak price on event days is approximately 10 times the level on non-event days,

---

[9]Customers of size greater than 20 kW were eligible for the new CPP rate if they met the interval data recorder metering requirement and had been on a demand response program previously. Otherwise, only customers of size greater than 200 kW were assigned to the default CPP rate in the first year.

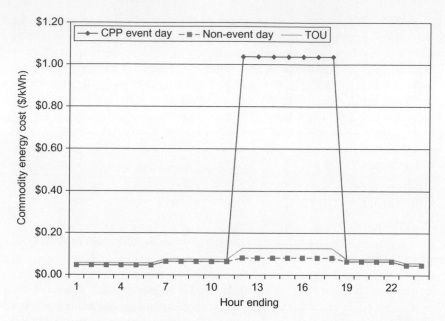

FIGURE 12.1   SDG&E TOU & default CPP price profiles—*Event Day and Non-Event Day.*

while the default CPP prices in all other time periods are less than those for the alternative TOU rate. The latter feature illustrates the potential for customer bill savings under the CPP tariff; customers pay somewhat lower prices most of the time in return for facing very high on-peak prices on a limited number (e.g., up to 15) of summer weekdays.

CPP Events

Table 12.2 lists the CPP event days for each of the utilities in 2009. PG&E and SCE each called 12 CPP events (PG&E's first event was a test event), while SDG&E called 8 events. The utilities often called events on different days, though there was some overlap, particularly in the last week of August. PG&E's events started earliest in the summer and ended earliest in the season, while SDG&E's events did not begin until late August and extended into late September. One of the events occurred on a Saturday.

ANALYSIS METHODS

Estimation of customer load response to event-based dynamic pricing is relatively straightforward compared to, for example, static TOU rates or the effect of modestly changing flat prices over a number of years. In the case of dynamic pricing, hourly load data are typically available for a relatively compact period of time, such as four summer months, and for both days on which consumers

TABLE 12.2 CPP Events, 2009

Date	PG&E	SCE	SDG&E
6/18		1(Test)	
6/29	1		
6/30	2		
7/13	3		
7/14	4		
7/15		2	
7/16	5		
7/17		3	
7/20		4	
7/21	6		
7/22		5	
7/27	7	6	
7/28		7	
8/10	8		
8/11	9		
8/18	10		
8/20		8	
8/27	11	9	1
8/28	12	10	2
8/29			3*
8/31			4
9/1		11	
9/2		12	
9/3			5
9/4			6
9/24			7
9/25			8

*Saturday.

faced relatively low non-event prices, and on days on which they faced high CPP prices. As a result, customers' loads on non-event days may be used as comparison, or control, loads for measuring their load reductions on event days, to the extent that they can be adjusted for weather and day-type effects.

The hourly load data for enrolled customers may be used at a variety of potential levels of aggregation. The most straightforward approach is to aggregate all of the customer loads up to the program level and use that data to estimate overall load impacts. However, that approach hides the potentially rich variation in load response across specific customers or customer types (e.g., across industry types or by location). Largely for that reason, our approach involved using load data at the individual customer level and separately estimating customer-specific regression models for each participating customer. The resulting customer-level estimated load impacts can then be combined or summarized in various ways, including adding up to obtain total utility-level load impacts.

The regression models assume that customers' hourly loads may be explained as functions of weather data; time-based variables such as hour, day of week, and month; and program event information (e.g., the days and hours in which events were called).[10] We also interact event indicator variables with hourly indicator variables to allow estimation of hourly load impacts for each program event in 2009. The resulting equations contain as many as several hundred variables. Automated software procedures allow recovery of key coefficients and their use in post-processing of results. Implicit "reference loads," which represent the load that would have occurred had an event not been called, may be estimated as the sum of the observed load on an event day and the estimated hourly load impacts from the regressions. The estimated load impacts may then be converted to percentage terms by dividing the load impact by the reference load.

ESTIMATED CPP LOAD IMPACTS

Load impacts were estimated for each hour of each CPP event at PG&E, SCE, and SDG&E. The following tables and figures summarize the estimated load impacts at each utility at various levels of detail. We first report overall average event-hour CPP load impacts and percentage load impacts for each of the utilities. Then, for each utility individually, we provide estimates of average hourly load impacts by industry type for the average event, a figure showing the degree of consistency of total load impacts across events, and a figure showing hourly loads and load impacts for the average event. We next present two sets of results specific to the SDG&E default tariff. Finally, we investigate the extent to which CPP load response is concentrated in a relatively few customers.

---

[10]A detailed description of the typical regression equation is provided in an appendix to this chapter.

TABLE 12.3 Average Event-Hour CPP Loads and Load Impacts, by Utility
Average Event

Utility	Customer Accounts	Estimated Reference Load (MW)	Observed Load (MW)	Estimated Load Impact (MW)	% Load Impact	Estimated Load Impact per Customer (kW)
PG&E	642	256	247	8.4	3.3%	13
SCE	476	130	106	24.6	18.9%	52
SDG&E	1,576	419	396	23.3	5.6%	15

Source: Braithwait et al. [5].

Overall Program Load Impacts

Table 12.3 summarizes the number of participating customer accounts, the average event-hour estimated reference and observed loads, and estimated load impacts for the average CPP event at each of the three utilities.[11] Also shown are load impacts as a percent of the estimated reference loads, which ranged from 3 to 19%,[12] and average event-hour load impacts per customer, which were 13, 52, and 15 kW for PG&E, SCE, and SDG&E, respectively. Overall program-level estimated load impacts for 2009 averaged 8.4 MW (3.3% of the reference load) across PG&E's 12 CPP events, 24.6 MW (18.9%) for SCE's 12 events, and 23.3 MW (5.6%) for SDG&E's 8 events.

Estimated Load Impacts by Utility

PG&E

Table 12.4 shows the distribution of estimated reference loads, observed loads, and load impacts (averaged across all event days), in levels and percentages, by industry group, for PG&E. The Manufacturing; Retail stores; and Offices, Hotels, Health, and Services industry types provided the largest load impacts, while Retail stores provided the largest percentage load impacts.

---

[11]Note that the numbers of enrolled customer accounts in Table 12.3 do not match the enrollments in Table 12.1 exactly. Table 12.1 summarizes the characteristics of customers enrolled at the time of *any* event day in 2009, while Table 12.3 shows the average across event days of the number of customers enrolled at the time of each event.

[12]The relatively larger percent load response by SCE customers is due to the price responsiveness of those customers enrolled in the alternative CPP rate option described above, which had a much higher CPP price than the prices offered by the other utilities, in return for a more highly discounted summer demand charge. The rate attracted a number of large manufacturing customers.

TABLE 12.4 Average Event-Hour CPP Load Impacts (kW)—*by Industry Type (PG&E)*

Industry Group	Count	Estimated Reference Load (kW)	Observed Load (kW)	Estimated Load Impact (kW)	% LI
1. Agriculture, Mining & Construction	39	4,021	3,760	261	6.5%
2. Manufacturing	164	87,055	83,351	3,704	4.3%
3. Wholesale, Transport, Other Utilities	67	15,696	15,074	621	4.0%
4. Retail stores	42	11,253	9,802	1,451	12.9%
5. Offices, Hotels, Health, Services	124	85,522	84,105	1,416	1.7%
6. Schools	158	26,765	26,765	0	0.0%
7. Gov't, Entertainment, Other Services	48	25,601	24,642	959	3.7%
Total	642	255,913	247,499	8,414	3.3%

Source: Braithwait et al. [5].

Figure 12.2 reports average event-hour load impacts across the six-hour event period *for each* of PG&E's 12 CPP event days, as well as the average load impact across events. The figure shows considerable variability of load impacts across events. The mean value across events of the average hourly load impacts was 8.4 MW, but load impacts ranged from 4.0 to 12.6 MW, with a standard deviation of 2.4 MW, or 29% of the average value. These values represent percentage load impacts that range from about 1.7% to 4.5% of the reference load, which averaged 256 MW across the event period.[13] The Manufacturing; Retail; and Offices, Hotels, Health, and Services industry types provided the largest load impacts, while Retail stores provided the largest *percentage* load impacts.

Figure 12.3 illustrates the hourly pattern of loads and load impacts, showing estimated reference load, observed load and estimated load impact (right axis) for the average PG&E CPP event. The large portion of CPP load accounted for by office buildings produces a reference load for PG&E that has a typical commercial customer load profile, peaking near mid-day and falling off in the late afternoon. The CPP load response takes a slice off of that load on event days, producing load impacts that are reasonably constant across the event period.

---

[13]The reference load is our estimate of what the CPP customers' load would have been if the event had not been called, and is based on observed event-period loads and the estimated load impacts.

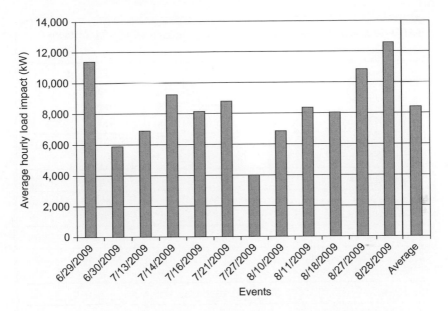

FIGURE 12.2   Average event-hour CPP load impacts by event—*PG&E. (Source: Braithwait et al. [5])*

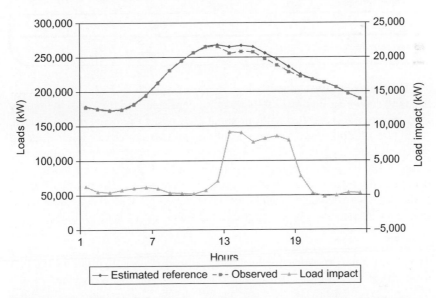

FIGURE 12.3   Hourly load impacts for average CPP event day in 2009—*PG&E. (Source: Braithwait et al. [5])*

SCE

Table 12.5 shows the distribution of average event-hour load impacts across industry types for SCE's average event. Manufacturing customers made up more than half of the total reference load and accounted for the bulk of the load impacts.

TABLE 12.5 Average Event-Hour CPP Load Impacts (kW)—*by Industry Type* (SCE)

Industry Group	Count	Estimated Reference Load (kW)	Observed Load (kW)	Estimated Load Impact (kW)	% LI
1. Agriculture, Mining & Construction	24	3,068	2,677	392	12.8%
2. Manufacturing	217	65,767	48,020	17,747	27.0%
3. Wholesale, Transport, Other Utilities	53	16,791	12,490	4,302	25.6%
4. Retail stores	34	13,602	12,653	949	7.0%
5. Offices, Hotels, Health, Services	44	9,564	8,957	607	6.4%
6. Schools	97	19,961	19,961	0	0.0%
7. Gov't, Entertainment, Other Services	8	1,614	965	649	40.2%
Total	476	130,367	105,722	24,645	18.9%

Source: Braithwait et al. [5].

The average estimated hourly load impacts across SCE's 12 CPP event days in 2009, shown in Figure 12.4, were quite consistent, with an average hourly load reduction of nearly 25 MW, or about 19% of the estimated reference load.

Figure 12.5 illustrates the patterns of the estimated reference load, observed load, and load impacts for SCE's average CPP event day. Note the substantial difference in the reference load compared to that in Figure 12.3 for PG&E. The decline in the reference load and load impacts during the afternoon hours is due to the large share of load of manufacturing customers on SCE's CPP rate, whereas PG&E's CPP rate has a larger share of retail stores and office buildings.

SDG&E

Table 12.6 summarizes event-hour loads and load impacts by industry type for the average SDG&E event. The largest load impacts were provided by the Offices, Hotels, Health and Services; and Wholesale, Transportation and Utilities (largely water utilities) industry groups.[14]

---

[14]Note that the small negative estimated load impact for the "Other/Unknown" industry group indicates that the regression models estimated a higher than expected load on the average CPP event day. This is likely the result of omitted variable bias in the regression model. That is, customer loads are high on event days due to factors not included in the regression model.

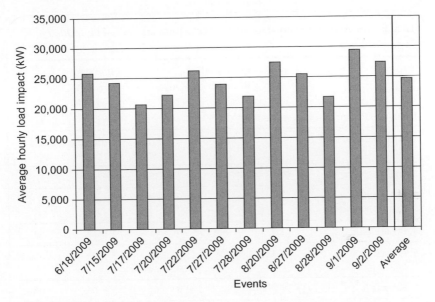

FIGURE 12.4 Average event-hour CPP load impacts by event—SCE. *(Source: Braithwait et al. [5])*

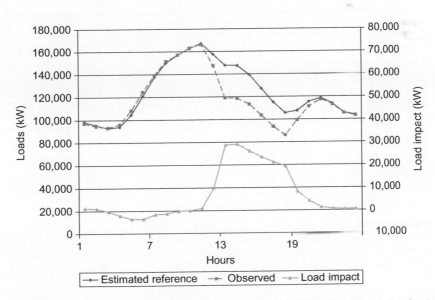

FIGURE 12.5 Hourly load impacts for average CPP event day in 2009—SCE. *(Source: Braithwait et al. [5])*

TABLE 12.6 Average Event-Hour CPP Load Impacts (kW)—by Industry Type (SDG&E)

Industry Group	Count	Estimated Reference Load (kW)	Observed Load (kW)	Estimated Load Impact (kW)	% LI
1. Agriculture, Mining & Construction	19	4,835	3,760	1,075	22.2%
2. Manufacturing	220	59,506	56,533	2,973	5.0%
3. Wholesale, Transport, Other Utilities	265	49,186	42,773	6,412	13.0%
4. Retail stores	128	38,959	36,814	2,145	5.5%
5. Offices, Hotels, Health, Services	480	180,185	172,249	7,937	4.4%
6. Schools	267	41,546	41,546	0	0.0%
7. Gov't, Entertainment, Other Services	190	44,020	41,268	2,752	6.3%
8. Other or Unknown	7	617	653	−35	−5.7%
Total	1,576	418,854	395,595	23,259	5.6%

Source: Braithwait et al. [5].

Like SCE, the average event-hour CPP load impacts at SDG&E were reasonably consistent across the eight events called in 2009, as shown in Figure 12.6[15] Load impacts ranged from 19.8 MW to 29.3 MW across weekday events, with a Saturday event on August 29 producing 19 MW. Load impacts averaged 23.3 MW, or about 5.6% of the CPP reference load. The load impacts were somewhat smaller than average for the Saturday event and the two late-September events. Load impacts were greatest (29.3 MW) on September 3, which was the SDG&E system peak day, as well as the peak day for the state.

---

[15]It should be noted that SDG&E allows joint participation in CPP and the capacity bidding program (CBP) day-of (DO) program type. If CPP and CBP-DO events are called on the same day, customer accounts that are enrolled in both programs continue to face CPP prices on that day, and do not receive energy credits for CBP load reductions. However, the CPUC has ruled that for resource adequacy purposes, capacity-based program load impacts receive a higher priority than those of energy-based programs. Contemporaneous CPP and CBP-DO events were called three times in 2009, on August 27, August 28, and September 3. We estimate that those customer accounts that were enrolled in both programs provided approximately 4 MW of average hourly load impacts. Thus, from a resource adequacy perspective, the estimated CPP load impacts on those three days should be reduced by approximately 4 MW.

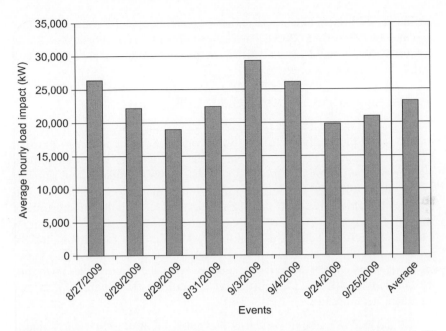

FIGURE 12.6 Average event-hour CPP load impacts by event—*SDG&E. (Source: Braithwait et al. [5])*

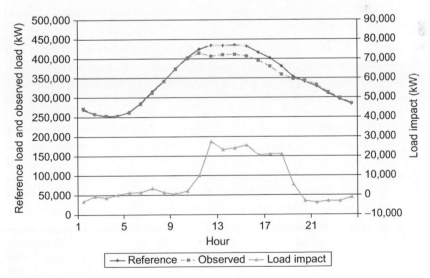

FIGURE 12.7 Hourly load impacts for average CPP event day in 2009—*SDG&E. (Source: Braithwait et al. [5])*

Figure 12.7 illustrates the patterns of the estimated reference load, observed load, and estimated load impacts (right axis) for the average event day. Note that unlike the other utilities, SDG&E's CPP prices apply to seven hours, beginning at 12 noon. Load impacts thus begin an hour earlier than at the other two utilities. Similarly to PG&E, the reference load takes on a typical commercial profile due to the large amount of commercial customer (industry types 5 and 7) load, as shown in Table 12.6.

Special Analyses of SDG&E Default CPP

While the voluntary CPP rates had previously required customers to *opt into,* or choose to participate in the optional rate (i.e., take action to enroll), SDG&E's default CPP tariff provided the first case in which customers were enrolled via an *opt-out* process. That is, eligible customers were automatically enrolled in CPP and were required to take action to leave the rate and return to the alternative TOU rate. In addition, under default CPP, customers were given a *capacity reservation* option, which allows customers to select an amount of electric load (in kW) that they wish to protect from the high CPP price during CPP events, and that they pay for through a fixed monthly capacity reservation charge, similar to a traditional demand charge. These two factors provided us with an opportunity to examine two interesting issues regarding default CPP:

1. Did the load response of customers previously enrolled in SDG&E's voluntary CPP rate differ from that of newly defaulted customers?
2. Does consumers' degree of load response appear to be related to their level of capacity reservation?

We begin by characterizing the differences between the customer accounts that previously volunteered for CPP and those that were transitioned to default CPP beginning in 2008. Tables 12.7 and 12.8 illustrate the differences in the average event-hour load impact, industry group makeup, and price responsiveness of the two groups of customer accounts. As shown in the last column, the overall percentage price responsiveness of the previous CPP volunteers was twice that of the newly defaulted customers (i.e., 10% compared to 5%). However, the difference appears to be due largely to a change in industry makeup, particularly a substantially lower share of load (as shown in the column labeled "% of Max kW") of highly responsive customers in the Agriculture, Mining, and Construction; and Wholesale, Transport, and Other Utilities industry groups among the newly defaulted customers compared to the previous volunteers.

The two groups of CPP customers also differed in their decisions regarding capacity reservation level, which may also have been related to their price responsiveness:

- The previous CPP volunteers accounted for 18.8% of the default CPP customer accounts in 2009.
- Regarding the *capacity reservation level* (CRL), 41.5% of all of the default CPP customer accounts kept the default level of 50% (meaning half of their expected load was not exposed to critical prices during event days).

TABLE 12.7 Characteristics of Customers Previously Enrolled in Voluntary CPP

Industry Type	Num. of SAIDs	Sum of Max kW	Sum of Avg. kWh	% of Max kW	Avg. Size (kW)	Avg. Event LI	% LI
1. Agriculture, Mining & Construction	4	6,740	2,122	6%	1,685	980	43%
2. Manufacturing	28	8,495	4,723	7%	303	349	6%
3. Wholesale, Transport, Other Utilities	107	33,914	13,406	30%	317	3,196	24%
4. Retail stores	25	10,269	7,277	9%	411	163	2%
5. Offices, Hotels, Health, Services	54	35,822	21,986	32%	663	1,591	6%
6. Schools	56	8,811	3,410	8%	157	0	0%
7. Government, Entertainment, Other Services	23	9,274	5,792	8%	403	596	9%
8. Other/ Unclassified	0	0	0	n/a	n/a	0	
TOTAL	297	113,324	58,715	100%	382	6,875	10%

Source: Braithwait et al. [5].

- Of those service accounts that opted to *change* the capacity reservation level, 81.7% selected a capacity reservation level of zero.
- Customers' decision to change their CRL appears to be related to prior participation in the voluntary CPP rate.
- 80.5% of prior voluntary CPP participants changed their capacity reservation level (of which 83% selected zero).
- Only 53.3% of the newly defaulted CPP service accounts changed their capacity reservation level.

Table 12.9 shows observed differences in percentage load impacts by decisions regarding CRL. The 40% of customers who kept the default CRL of 50% (first row) produced average percent load impacts of 3%. In contrast, the nearly 60% of customers (second row) who changed CRL from the default level of 50% (often to zero, as shown in the third row) were three times as price responsive as those who kept the default level (i.e., 9% load impacts compared to 3%).

TABLE 12.8 Characteristics of Customers Newly Defaulted to CPP

Industry Group	Num. of SAIDs	Sum of Max kW	Sum of Avg. kWh	% of Max kW	Avg. Size (kW)	Avg. Event LI	% LI
1. Agriculture, Mining & Construction	15	4,947	2,409	1%	330	95	4%
2. Manufacturing	194	84,202	43,487	17%	434	2,624	5%
3. Wholesale, Transport, Other Utilities	159	86,294	31,756	17%	543	3,216	10%
4. Retail stores	103	32,375	19,383	7%	314	1,982	7%
5. Offices, Hotels, Health, Services	427	184,880	112,958	37%	433	6,346	5%
6. Schools	211	45,605	16,826	9%	216	0	0%
7. Government, Entertainment, Other Services	167	58,301	28,400	12%	349	2,156	6%
8. Other/Unclassified	7	857	570	0%	122	−35	−6%
TOTAL	1,283	497,460	255,789	100%	388	16,384	5%

Source: Braithwait et al. [5].

TABLE 12.9 Differences in Percentage Load Impacts by Subgroups

Customer Type	Percent of SAIDs	Percent Load Impact
Kept default CRL (50%)	42%	3%
Changed from default CRL	58%	9%
Changed CRL to zero	48%	9%

Source: Braithwait et al. [5].

Concentrations of CPP Load Impacts

The methodology of estimating customer-specific regression equations and load impacts also provides the capability to examine the *distributions* of CPP load impacts across individual customer accounts, and to determine the concentration of load impacts among subsets of customers. Table 12.10 summarizes some of the key indicators of the concentration of CPP load impacts across the three utilities.

TABLE 12.10 Concentration of CPP Customer Price Responsiveness

Utility	Share of Customers with LI > 5 kW	Share of %Total LI from Top 5%
PG&E	40%	64%
SCE	59%	55%
SDG&E	35%	74%

Source: Braithwait et al. [5].

The first column in the table reports the percentage of customers who were estimated to provide load impacts of at least 5 kW. The 59% value for SCE (compared to 35 and 40% for SDG&E and PG&E) is consistent with the findings of greater price responsiveness among SCE's CPP customers.[16] The second column shows the share of load impacts provided by the top 5% of CPP customers at each utility, where the customers are ranked according to the size of their estimated load impact. In general, the load impacts are distributed similarly. Relatively large shares of load impacts are provided by a relatively few customers. That is, the top 5% of the customers provide 55–74% of the total program load impacts across the three utilities. Concentration is greatest at SDG&E, likely among the former volunteers, while load impacts are least concentrated at SCE, again reflecting generally broad price responsiveness.

Perspectives on CPP Participation and Load Impacts

In considering how applicable these CPP load impacts at the major California utilities are to other utilities and regions, certain key factors should be kept in mind. First, the results are based on the actions of those customers who volunteered for the program or, in the case of SDG&E's default tariff, elected to stay on the tariff rather than opting out to a TOU rate. The relatively small overall percentage load reductions (e.g., 3–5% in 2010) and the concentration of load impacts among a relatively small fraction of customers are reasonably consistent with previous price-response findings regarding RTP for large customers and CPP for residential customers.[17] The next few years will indicate how willing the defaulted C&I customers at all of the utilities will be to remain on the CPP tariff in the long term.

---

[16]Note that most of SCE's voluntary CPP customers selected the rate option that has the highest CPP price (in return for a discounted summer peak demand charge), and have historically included large and flexible manufacturing and water utility customers who have the ability and financial incentive to reduce load during CPP event hours.

[17]See Goldman et al. [8].

A second factor to keep in mind is that the same pool of customers from which CPP participants are drawn have the alternative option of participating in other demand response programs and receiving financial payments for load reductions on event days.[18] Thus, if a utility outside of California were to offer CPP in the absence of alternative DR programs, it would likely see a different rate of participation and degree of price responsiveness. However, a number of utilities operate in the footprint of organized wholesale markets such as PJM, New York ISO, and ISO New England, which operate a range of DR programs.

To illustrate how the CPP findings in California depend upon the alternative DR programs, consider the following enrollment and load impact information. In 2009, approximately 2,600 customer accounts participated in CPP at the three utilities. In 2010, enrollment grew to about 7,100 accounts with the transition to default CPP at PG&E and SCE.[19] At the same time, aggregator-managed DR programs in 2010 enrolled more than 5,000 customer accounts, and SCE and PG&E enrolled 2,500 customer accounts in their demand-bidding programs (DBP), all from the same pool of large commercial and industrial customers from which CPP draws. That is, approximately equal numbers of customers participated in CPP and in DR programs in 2010.

The estimated load impacts for the aggregator-managed and demand-bidding programs in 2010 were 308 MW and 129 MW, respectively. In contrast, total CPP load impacts in 2009 and 2010 were 57 and 73 MW, respectively.[20] An emergency program, the Base Interruptible Program (BIP), can provide around 800 MW of load reduction during system emergencies. With a total system maximum demand for the three utilities of about 46 GW in 2009, the emergency load relief from BIP amounts to nearly 2% of the system maximum demand. The combined price-responsive load impacts from CPP, DBP, and the aggregator-managed programs amount to a little more than 1% of the system maximum demand; CPP load impacts comprise a relatively small portion of the total.

As a final point of comparison, consider the following differences in the price responsiveness of the CPP and DR program customers, as measured by *percentage* load impacts. As shown in Table 12.3, the CPP customers at PG&E, SCE, and SDG&E in 2009 reduced load during CPP events by 3.3%, 18.9%, and 5.6%, respectively. In 2010, after PG&E and SCE added several thousand newly defaulted customers, the estimated percentage load impacts at PG&E and SDG&E remained about the same, while those for SCE dropped to less than 3%, as the strong price responsiveness of the core volunteers in 2009 was diluted by less responsive newly defaulted customers.

---

[18]In most cases, consumers are not allowed to participate in both CPP and a DR program, where events are likely to be called on the same days. Exceptions are some emergency or reliability-based DR programs that are expected to be called infrequently.

[19]See George et al. [9].

[20]The 2010 load impacts are from George et al. [9].

In contrast to those relatively small percentage load impacts for CPP customers, the comparable values for the aggregator-managed programs were much higher, ranging from 20 to 30%.[21] There are two likely reasons for the larger relative load impacts for the DR programs compared to CPP. First, customers receive a combination of capacity credits for promised load reductions, regardless of whether events are called, and of energy payments for measured load reductions during events. The capacity credits likely attract customers who are willing to commit to reduce load, while penalty provisions for not meeting commitments, as well as energy payments for load reductions, provide strong incentives to perform during events. Second, the third-party aggregators have an incentive to work with enrolled customers to ensure that in aggregate they produce the contractual load reductions agreed upon with the utilities. Our understanding is that in some cases aggregators assist customers with equipment or procedures to automate load reductions when events are called. To the extent that participants in the aggregator programs are inherently more price responsive than non-participants, then that portion of the population is removed from the remaining population from which CPP draws, thus leaving generally less responsive customers to participate in CPP.

LOOKING FORWARD

CPP in California

As noted above, enrollment of large C&I customers in CPP in California increased substantially in 2010, as the utilities transitioned to default CPP rates. Specifically, SCE moved approximately 8,000 customer accounts onto a new CPP rate in the fall of 2009, while PG&E moved around 5,000 accounts to PDP in the spring of 2010. A number of those customers opted out prior to the summer of 2010, leaving 4,100 customers enrolled at SCE (a retention rate of about 50%), and 1,800 customers enrolled at PG&E (a retention rate of about 36%). About 1,500 customers remained enrolled at SDG&E in 2010 from the 2,400 that were originally defaulted (a retention rate of about 60%).[22]

Additional changes are planned for CPP at each of the utilities over the next few years. SDG&E will expand its default CPP rate to medium-sized, customers whose maximum demand lies between 20 and 200 kW as installation of smart meters expands in 2012 and 2013. SCE has proposed to expand default CPP to customers of less than 200 kW and to offer a capacity reservation option to CPP participants. Finally, PG&E plans to expand default PDP to smaller customers in late 2011 as additional smart metering is rolled out. The total pool of medium C&I customer accounts exceeds 200,000. Future load impact evaluations will

---

[21]Percentage load impacts for most DBP customers were in the range of those for CPP (see Braithwait et al. [6]).
[22]Several hundred medium-sized customers were transitioned to default CPP in 2010, adding to the 1,800 large customers in 2008.

track the effect of both expanding enrollment and smaller sized customers on estimated CPP load impacts.

Price-Response Potential Outside of California

In states other than California, increased availability of smart meters through smart grid investments will make dynamic pricing for commercial and industrial (as well as residential) customers much more feasible than it has been historically. Furthermore, interest in demonstrating benefits from smart grid investments is likely to provide incentives for utilities and regulators to explore dynamic pricing.[23] As noted above, the combined load impacts for CPP and DR programs for large C&I customers in California amount to more than 1% of the total system load. Expanding the rates and programs to smaller but much more numerous medium-sized business customers could double that amount. In addition, emergency-based interruptible service programs for large customers and air conditioner load control programs for residential and small business customers add another 2–4% system load-reducing potential.

California differs from other states in a number of ways, including having relatively fewer large energy-intensive industrial customers, in part due to a history of relatively high average electricity prices in recent years; more agricultural customers; milder weather; and a strong historical focus on energy efficiency programs. At the same time, it has many of the same types of office buildings, retail stores, warehouses, and municipal water utilities that are present in all states. Some of the more heavily industrialized states might be expected to generate relatively larger C&I CPP load impacts as a percent of total load. Finally, a study of potential DR load impacts sponsored by FERC concluded that current programs have the potential to reduce total U.S. peak demand by about 4%, while modest expansion of the current "business as usual" case could expand that impact amount to about 9%.[24]

Dynamic Pricing versus DR Programs

Current discussions about the need for and role of price-responsive demand (i.e., customer load reductions in response to dynamic retail prices or DR credit payments) in wholesale markets typically seem to focus unduly on DR bidding programs arranged by DR service providers and run through regional ISO/RTOs. This is in contrast to the seemingly more natural mechanism of dynamic retail pricing offered by utilities and other retail providers, which reflects changes in wholesale market prices. Greater penetration of retail dynamic pricing would obviate much of the need for DR programs. However, there is a long history of barriers to adoption of time-based pricing, which include lack of

---

[23] Why invest in smart grids and smart meters, but continue with not-so-smart pricing?
[24] See FERC [10].

incentives on the part of utilities, concern by regulators about potential bill impacts for some customers, as well as metering costs. These barriers are discussed in more detail in Chapter 3 on the efficiency and ethics of dynamic pricing.

However, DR programs have their own set of issues and problems. Three of the most important are how to *measure* the actual DR load impacts provided, how those load impacts should be *compensated*, and whether *third-party aggregators* should be allowed to recruit utility customers to participate in DR programs in non-restructured states. The measurement issue involves which methodology to use to estimate DR customers' baseline load, or the counterfactual amount that they would have used had they not participated in a DR event. This estimation is necessary because energy not used cannot be measured directly, but instead must be inferred by comparing observed energy consumption to an estimated baseline load to determine the difference.

Various baseline methods have been proposed, tested, or used in different jurisdictions. In a series of baseline analyse conducted using data for several DR programs in California, we have shown that the accuracy of baseline loads, and thus estimated DR load reductions, depends on several factors, including the baseline method used and the type of customer (e.g., industrial or commercial, and degree of inherent load variability), and can often be quite poor.[25] This result raises questions about both fairness of DR payments and credits and the uncertainty of DR resources compared to supply resources.

The DR compensation issue has gained recent prominence due to a March 2011 FERC ruling that requires ISO/RTOs to offer DR programs, and requires a uniform DR compensation structure in which DR customers are to be paid the full wholesale market price, or locational marginal price (LMP), without regard for whether the customers are required to pay for the amount of the load reduction at the generation component (G) of their retail rate (the so-called LMP minus G alternative compensation). Critics contend that full-LMP compensation overpays for DR load reductions, resulting in an effective subsidy that must be recovered through uplift charges to all customers, and that the net benefit test required by FERC's ruling is costly and unnecessary if the subsidy portion of the compensation is removed. A rehearing of the FERC ruling has been requested.[26]

Finally, while requiring ISO/RTOs to offer DR programs, FERC has allowed states to determine whether to allow third-party DR providers to recruit customers of state-regulated utilities. Some states have banned such recruitment, while others are determining whether to do so. Fully regulated utilities often object to their customers being recruited by competitive DR providers,

---

[25]See Braithwait and Armstrong [11].

[26]In view of these issues, ISO New England has encouraged states in its footprint to adopt more dynamic pricing and has recommended DR designs that better mimic dynamic retail pricing. See Chao [12].

on the basis of potential revenue losses and uncertainty about the loads that they must schedule in wholesale markets.

CONCLUSIONS

The new evidence presented in this chapter on the extent to which large commercial and industrial customers will participate in and respond to dynamic price structures such as critical peak pricing (CPP), as well as in price-based DR programs, adds to the results of previous studies of residential customers to confirm that customers on average will respond when peak prices rise substantially above their normal levels, or DR credits are offered, on a limited number of critical days. Two key features of the estimated load impacts—that they range considerably across industry types (e.g., manufacturing, retail stores, and offices) and they tend to be concentrated in a relatively small fraction of participating customers—suggest to utilities and policy makers in other states how load impacts are likely to vary in other settings and what the potential is for increasing customer price responsiveness.

As installation of smart metering and defaulting to CPP continue to reach down to smaller but more numerous business customers in California, load impacts will continue to expand. Outside of California, the expanding installation of smart meters and increasing intelligence of the grid should provide the means and incentives to introduce dynamic pricing to many more customers. Expansion of dynamic pricing will also help to reduce the need for DR programs run by ISO/RTOs, and thus reduce some of the controversial issues that surround DR programs.

REFERENCES

[1] A. Faruqui, J.R. Malko, The residential demand for electricity by time-of-use: a survey of twelve experiments with peak load pricing, *Energy* 8 (1983) 781–795.

[2] S. Braithwait, *Customer Response to Electricity Prices: Information to Support Wholesale Price Forecasting and Market Analysis*, EPRI, Palo Alto, CA, 2001, 1005945.

[3] S. Braithwait, M. O'Sheasy, RTP customer demand response–empirical evidence on how much can you expect, in: A. Faruqui, B.K. Eakin (Eds.), *Electricity Pricing in Transition*, pp. 181–190, Kluwer Academic Publishers, Boston, MA, 2002.

[4] A. Faruqui, S. Sergici, "*Household Response to Dynamic Pricing of Electricity – A Survey of Seventeen Pricing Experiments*," JI of Regulatory Economics (2010) Vol. 38, No. 2, 193–225.

[5] S.D. Braithwait, D.G. Hansen, J.D. Reaser (Christensen Associates Energy Consulting), *2009 Load Impact Evaluation of California Statewide Critical-Peak Pricing Rates for Non-Residential Customers: Ex Post and Ex Ante Report*, Prepared for San Diego Gas & Electric, April 19, 2010.

[6] S.D. Braithwait, D.G. Hansen, J.D. Reaser (Christensen Associates Energy Consulting), 2010 *Load Impact Evaluation of California Statewide Demand Bidding Programs (DBP) for Non-Residential Customers: Ex Post and Ex Ante Report*, Prepared for Southern California Edison, April 1, 2011.

[7] S.D. Braithwait, D.G. Hansen, D.A. Armstrong (Christensen Associates Energy Consulting), *2010 Load Impact Evaluation of California Statewide Aggregator Demand Response Programs: Ex Post and Ex Ante Report*, Prepared for Pacific Gas & Electric, March 29, 2011.

[8] C. Goldman, N. Hopper, R. Bharvirkar, B. Neenan, R. Boisvert, P. Cappers, et al., *Customer Strategies for Responding to Day-ahead Market Hourly Electricity Pricing*, Final report to the California Energy Commission, Lawrence Berkeley National Laboratory (LBNL-57128), August, 2005.

[9] S.S. George, J. Bode, J. Schellenberg, S. Holmberg (Freeman, Sullivan & Co.), *2010 Statewide Non-Residential Critical Peak Pricing Evaluation*, Prepared for San Diego Gas and Electric, April 1, 2011.

[10] FERC Staff Report, *A National Assessment of Demand Response Potential*, Prepared by The Brattle Group, Freeman, Sullivan & Co, and Global Energy Partners, LLC, June 2009.

[11] S.D. Braithwait, D.A. Armstrong (Christensen Associates Energy Consulting), *2008 Load Impact Evaluation of California Statewide Aggregator Demand Response Programs: Volume 2 Baseline Analysis of AMP DR Program*, Prepared for Pacific Gas & Electric, March 30, 2009.

[12] H.-P. Chao, Price-responsive demand management for a smart grid world, *Electricity J.* Vol. 23, Issue 1, January/February 2010, pp. 7–20.

APPENDIX

The basic regression model for measuring CPP load impacts is the following:

$$
Q_t = a + \sum_{Evt=1}^{E} \sum_{i=1}^{24} (b_{i,Evt}^{CPP} \times h_{i,t} \times CPP_t) + b^{MornLoad} \times MornLoad_t
$$

$$
+ \sum_{i=1}^{24} (b_i^{CDH} \times h_{i,t} \times CDH_t) + \sum_{i=2}^{24} (b_i^{MON} \times h_{i,t} \times MON_t)
$$

$$
+ \sum_{i=2}^{24} (b_i^{FRI} \times h_{i,t} \times FRI_t) + \sum_{i=2}^{24} (b_i^{h} \times h_{i,t}) + \sum_{i=2}^{5} (b_i^{DTYPE} \times DTYPE_{i,t})
$$

$$
+ \sum_{i=6}^{10} (b_i^{MONTH} \times MONTH_{i,t}) + b_t^{Summer} \times Summer_t
$$

$$
+ \sum_{i=1}^{24} (b_i^{CDH,S} \times h_{i,t} \times Summer_t \times CDH_t)
$$

$$
+ \sum_{i=2}^{24} (b_i^{MON,S} \times h_{i,t} \times Summer_t \times MON_t) + \sum_{i=2}^{24} (b_i^{FRI,S} \times h_{i,t} \times Summer_t \times FRI_t)
$$

$$
+ \sum_{i=2}^{24} (b_i^{h,S} \times h_{i,t} \times Summer_t) + b^{OTH} \times OTH_t + e_t
$$

In this equation, Q_t represents the amount of usage in hour t for a customer enrolled in CPP prior to the last event date; the b's are estimated parameters; $h_{i,t}$ is a dummy variable for hour i; CPP_t is an indicator variable for program event days; CDH_t is cooling degree hours[27]; E is the number of event days that occurred during the program year; $MornLoad_t$ is a variable equal to the average of the day's load in hours 1 through 10; MON_t is a dummy variable for Monday; FRI_t is a dummy variable for Friday; $DTYPE_{i,t}$ is a series of dummy variables for each day of the week; $MONTH_{i,t}$ is a series of dummy variables for each month; $Summer_t$ is a variable indicating summer months (defined as mid-June through mid-August),[28] which is interacted with the weather and hourly profile variables; OTH_t is a dummy variable indicating an event hour for a non-CPP demand response program in which the customer is also enrolled; and e_t is the error term. The "morning load" variable was used in lieu of a more formal autoregressive structure in order to adjust the model to account for the level of load on a particular day. Because of the autoregressive nature of the morning load variable, no further correction for serial correlation was performed in these models.

---

[27]Cooling degree hours (CDH) was defined as MAX[0, Temperature − 50], where Temperature is the hourly temperature in degrees Fahrenheit. Customer-specific CDH values are calculated using data from the most appropriate weather station.

[28]This variable was initially designed to reflect the load changes that occur when schools are out of session. We have found the variables to a useful part of the base specification, as they do not appear to harm load impact estimates even in cases in which the customer does not change its usage level or profile during the summer months.

Chapter 13

Smart Pricing to Reduce Network Investment in Smart Distribution Grids—Experience in Germany[1]

Christine Brandstätt, Gert Brunekreeft, and Nele Friedrichsen

INTRODUCTION

Around the world a change in electricity generation is desired in order to fight climate change and increase energy security. Consequently renewable energies and distributed generation (DG) are receiving support, and their shares in electricity generation are rising. As described in a number of chapters in this book, both the transmission and distribution networks play a key role when integrating large amounts of distributed and intermittent renewable

---

[1]Acknowledgments: This work has been carried out within the research project IRIN—Innovative Regulation for Intelligent Networks. Financial support by the Federal Government represented by the Federal Ministry of Economics under the 5th Energy Research Programme is gratefully acknowledged. The authors wish to thank the project's research partners and advisory board for helpful comments. Furthermore we are grateful for intensive discussion with representatives from Thüga and useful comments by the editor of this volume. All remaining errors are the responsibility of the authors.

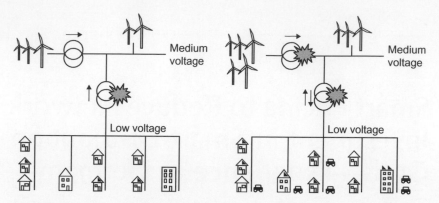

FIGURE 13.1   Distribution network congestion with distributed generation and electric vehicles. *(Source: Authors based on EWE [1,2])*

generation. One of the problems experienced in this context is that the increasing renewable shares may cause congestion in distribution networks.[2] The introduction of large numbers of electric vehicles could create similar problems. An example is illustrated in Figure 13.1. As a result considerable investment is required for expansion and replacement of the existing grid as well as in information, communication, and coordinating technology.

The development of smart energy systems requires significant initial investments. The main cost triggers are the network expansion needs from the integration of distributed and renewable electricity generation and the information and communication technology required for smart grids. Table 13.1 illustrates the investment needs in the United States, United Kingdom, and Germany. With such large costs the need for efficient investment is immediate. Optimizing the use of the existing network improves efficiency in the short run. Long-term efficiency requires coordinated investment decisions. In practical terms investment coordination exploits the trade-offs between the location of generation or demand units and network expansion.[3] DG benefits from investment deferral are in the range of 1% to 15%, depending on location and power factor [3]. In a rural network with long lines, the savings from reduced line losses and operational gains can be over 30% as Sotkiewicz and Vignolo illustrated by a case study in Uruguay [4]. In the context of smart grids, flexible generators or storage can create significant benefits for the network if they are located and operated adequately [5]. The anecdotal evidence suggests that the current framework is suboptimal for efficiently integrating future distributed

---

[2]Other problems may include the intermittency of generation from renewable sources and the lack of dispatchability. These are not the focus of this chapter.

[3]Siting of generators according to free network capacity economizes on line losses and avoids network capacity expansion.

TABLE 13.1 Network Investments and Savings from Smart Grids

Country	Investment Need [Million $/Year]*	Potential Savings [Million $/Year]*	Percentage Savings
US	10,000–15,000[a]	800–2,000[b]	5–20%
UK	4,944[c]	15.5[d]	0.3%
Germany	3,315[e]	n.a.	n.a.

Source: BDEW [8]; EPRI [5]; Ofgem [6]; Li et al. [7].
Investment need refers to estimated additional distribution network investment for DG integration and smart grids. Savings refers to reduction potential from smarter operation and smarter pricing. We used the conversion factors 1 $US = £ 0,64731 and 1 $US = € 0,75464.
[a]*EPRI [5] estimates the transition to smart distribution networks in the US to cost $200–300 billion over the next 20 years. Roughly 40% of this is attributed to accommodating load growth; the other 60% is for technology upgrades.*
[b]*Smart grid benefits of $8–20 billion on deferred transmission and distribution capacity investment over 20 years [5].*
[c]*Network cost for the transition towards a low-carbon energy supply are estimated at £32 billion [6].*
[d]*Li et al. [7] estimated savings in network investment cost in the range of £200 million over 20 years from better locational coordination.*
[e]*Estimated costs of integrating electricity produced from wind turbines and photovoltaics into distribution networks are approximately €25 billion until 2020 [8]. Importantly, this does not include any optimization of network investments or coordination of development plans across network levels or the projected generation developments.*

generation, thereby resulting in higher than necessary cost. Smarter pricing can reduce the investment need. Table 13.1 sums up the figures and estimates for the United States, United Kingdom, and Germany.

Within a liberalized market, decisions are decentralized, thereby requiring a coordination mechanism. Often attention is paid to time-differentiated, dynamic pricing[4] as a means to match demand and generation and to mitigate network usage in peak periods. This chapter focuses on the locational, rather than the time-differentiated dimension of the problem and on the potential this has to increase investment efficiency. The implementation of locational pricing results in socially beneficial investment based on locational coordination of investments as further described in Brandstätt et al. [9].

This chapter examines smart pricing as a means to defer distribution network investment. After a brief overview of the theoretical concepts and international experience with locational pricing in the section "Locational Pricing," an in-depth analysis of the German case is provided in the section "Locational Distribution Pricing in Germany." It is argued that the precise details of smart pricing and smart contracts should be left to the market participants, in particular the network owners, as much as possible. The task for legislators and regulators is to provide market parties with incentives for efficient investment. The section "Conclusion and Policy Recommendations"

---

[4]Chapter 3, among others, discusses dynamic pricing in this book.

describes schemes to implement smart and more differentiated pricing that improve investment coordination. The main message is for flexible application or allowance of already existing rules and regulation. In that case, significant system reform would not be necessary, and it would be left to market parties, especially network owners, to see whether it actually pays off to implement a more differentiated system.

LOCATIONAL PRICING

The cost of electricity supply consists of an energy and a network component. The first remunerates the generation of the electricity used and the latter compensates for the availability of the grid infrastructure. Charges at transmission and distribution, as well as at the wholesale and retail level, are passed on through the different stages and incumbents. The tariff system is what connects different stakeholders across different voltage levels. Locational signals embedded in this tariff structure can thus serve as a means to better coordinate network users and direct them away from congested parts of the network.

Network charges typically include connection and use-of-system (UoS) charges. A one-off connection charge accounts for the establishment of the connection to the network. Ongoing UoS charges recover the running cost of the network such as losses, balancing services, and maintenance. Even though both generators and consumers use the network, in many jurisdictions UoS charges are allocated entirely to consumers [10] (i.e. the so-called generation-load split is 0/100). Generators are charged only for their connection to the network. This can be a market distortion if generators have asymmetrical effects on the network, while the effects are socialized and thus distributed among network users symmetrically.

Final customers usually receive a composite or bundled tariff, including the charges not only for network usage but also for energy consumed. While many countries use uniform pricing and hence do not internalize network conditions, some states, such as the UK, use locational price signals in the network or energy charge. The section "International Experience with Locational Pricing in Distribution Networks" presents country examples where locational differentiation is in use.

The Theory Behind Locational Pricing

Locational signals can be introduced in network and energy charges. They reflect the locational differences of making the infrastructure available for both load and generation. A zonal approach to network or energy price differentiation would, for example, have low prices for generation and demand in remote regions, which would attract further demand. In contrast, prices in densely populated regions with concentrated demand would be high to attract additional generation. More specific signals can be achieved by refining these zones up to branches or even nodes. The challenge is to assess the actual system conditions at a specific location.

Locational Network Pricing

Siting decisions of generators influence the topology of the network. A new network user, be it generation or demand, can either cause or defer considerable investment depending on where in the network it is located.[5] Connection charges typically cover the additional costs of lines, transformers, and other equipment needed to hook up a new user to the grid. Connection charges that reflect the connection conditions potentially influence the siting decision and thus optimize the system.

Basically, connection charges can be either *shallow* or *deep*. With shallow charges, the network user only pays the direct cost of establishing a new connection to the next connection point to the existing grid. Deep charges also include part of the reinforcement that becomes necessary in other, "deeper" parts of the existing network—hence the term. For instance, additional generation at a remote site without corresponding demand may require upgrade of transformers or lines in existing parts of the grid to enable the distribution of the additional electricity. As a rule, deep charges tend to be higher at congested sites, making those locations less attractive.[6]

While elegant and logical, deep charging is not easy to implement. Due to a lack of transparency, an investor may not know at the time of decision the cost variation at different sites. The cost of reinforcement depends largely on the actual condition of the local grid, which is difficult to assess. It is typically not fully disclosed to the network user, and even for the network operator it is not trivial to determine non-discriminating (i.e. fair) deep charges as further described in Brunekreeft et al. [14]. Hence, the benefits of deep charging, namely full cost recovery for the system operator and targeted signals, have to be weighed against substantially higher transaction costs in establishing the charges.

In addition to connection charges, UoS charges can convey locational signals to the investor. However, traditionally this has not been the case. UoS charges were often average, based on each voltage level and further differentiated by the extent of use. This did not capture all of the effects that network use may have on operation and expansion cost. In order to guide investment, network charges have to reflect the actual condition of the network at a specific site and the impact of the network user. This impact is different for feed-in and take-off of electricity and so should be the charges. Often UoS charges are allocated to demand customers only, as traditionally the same incumbents planned generation in big power plants and the respective network. With the introduction of wholesale and retail competition in electricity markets, often accompanied by unbundling, better coordination through market prices is needed. Even with a 0/100 split where generators do not bear network cost,

---

[5]For a detailed analysis of the potential positive and negative effects, see e.g. Piccolo and Siano [11], Ackermann [12].
[6]See for a more detailed discussion, e.g. Woolf [13].

locational signals can be implemented as long as the sum of the generation charges is zero.

Incremental cost pricing with a long-run perspective is a tool to use to include the expansion cost of the network in UoS charges. In particular it deals with the stepwise cost increase that comes with bulky network investment [15]. Changes in the constellation of network users at a specific location directly influence the respective charges [16]. In other words the charges signal the urgency of network investment. If siting at a certain location defers network investment, charges are low. In contrast, charges are high if new connections cause network reinforcements.

Locational Energy Pricing

In most networks, the largest part of the bundled energy charge on the consumer's bill stems from the energy part, and only a small part originates from the network.[7] In Germany for example energy cost make up about 35% of the final price, the second biggest share are taxes and concession fees with over 30%, and network costs account for only roughly 21% of the final electricity price on average (see Figure 13.2).[8] Therefore locational signals would be strongest if implemented in the energy part of the charge, if not evenly in both parts.

Electricity prices that vary in different zones reflect the scarcity of interconnection between regions. This would encourage generators to connect to the network where prices are high, that is, electricity is scarce. Load is incentivized to connect in regions with excess supply where prices are low. This avoids the need for additional interconnection. Moreover, differences between regions will incentivize network owners to expand in areas where prices are high. In practice, however, it can be difficult to demarcate the price zones, as further described by Björndal and Jörnsten [18]. More precise locational signals originate from nodal spot pricing, also known as locational marginal pricing (LMP) (for further explanations see [19,20]).

A nodal pricing scheme assigns the overall cheapest supply option to the demand units at each node. Nodal prices are calculated by determining the marginal cost for the system of supplying one additional MW of load at each node, while taking loop flows into account [21]. A nodal pricing scheme reflects the topology of the system in detail and thereby takes into consideration losses and congestion. It has been shown that nodal prices send efficient signals for short-term optimization, but insufficient long-term signals. In other words, they send good signals for the optimization of operation [19,21], but since they do not reflect fixed network cost, signals are not sufficient to guide efficient investment decisions [9,14].

---

[7]The specifics, of course, vary from one place to another.

[8]The proportions of network and energy cost are different according to customer group. For industry both the final price (12.29 € ct/kWh) and the share of network cost (~13%) are significantly lower than for household customers (23.42 € ct/kWh and ~25% BNetzA [17]).

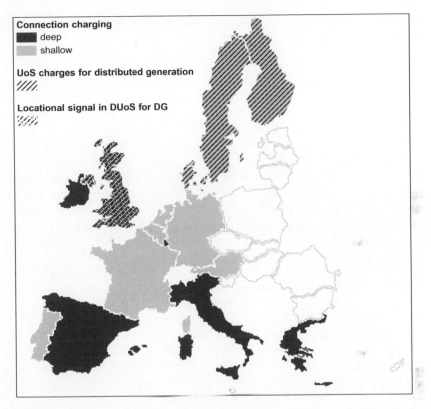

Connection charging
- ■ deep
- ▨ shallow

UoS charges for distributed generation
▨

Locational signal in DUoS for DG
▨

FIGURE 13.2   Composition of a household electricity bill in Germany. *(Source: Authors, based on data from BNetzA [17])*

Nodal spot pricing is often deemed the optimal methodology for network pricing since it gives first best signals for system operation, particularly in terms of congestion management. Indeed, a recent study based on data from U.S. market areas indicates significant benefits for the move towards nodal pricing [22]. The benefits typically outweigh the one-off implementation costs within the first year. However, it seems that this is true for big ISOs but might be problematic for entities within small market areas. For the UK, Green [23] estimates the benefits from moving to locational marginal pricing to be in the range of 1–3% of the generators' revenues. He points out that the effects strongly depend on the specifics of a given market but concludes that the gains might be "worth pursuing". Today however, these price differentiations most often only affect industrial customers since usually retail customers receive flat tariffs from their suppliers. This may change with more advanced technology, more granular information on networks costs, and other functionalities of the smart grid, for example in the introduction and in Chapter 6, the smart grid vision for California, in this book.

International Experience with Locational Pricing in Distribution Networks

In practice, the development towards locational pricing has thus far concentrated mostly on transmission and the wholesale level, while applications in distribution networks are rather rare [14]. However, distribution networks are moving towards smarter systems that efficiently integrate both intermittent generation from renewable and distributed energy sources and a more flexible demand side. These developments increase the necessity for smart pricing.

Currently, most countries apply *shallow* connection charges that convey only few locational signals.[9] However, one can find examples of *deep* connection charges or network charges with locational elements in UoS charges. Yet generation is often exempted from UoS charges and therefore receives too few locational signals. In general, there is a trend towards more flexible, less standardized network charges and negotiated agreements. Figure 13.3 depicts the different approaches for distribution network charging in the EU-15.

An exception is the UK, which uses a more advanced system of cost-reflective locationally differentiated distribution use-of-system charges. The UK abandoned deep charging for connection charges. It was feared that high transaction costs due to negotiations and informational disadvantages might hinder the development of distributed generation [26]. Locationally varying UoS charges were implemented in transmission networks to re-establish the locational signals that disappeared with the elimination of deep charging. They are accompanied by *shallowish* connection charges that cover the connection cost plus a proportion of the reinforcement cost.[10]

The distribution UoS charging methodology in the UK is currently moving towards higher cost reflectivity. To enhance transparency the charging methodologies are published after obtaining regulatory approval[11] [26]. For low, medium, and high voltage levels, a Common Distribution Charging Methodology started in April 2010 [28].

In the distribution reinforcement model, network operators estimate the cost of network development based on the expected growth of DG and load. These are the basis for the calculation of network charges, which are socialized among network users. The model does not feature location-specific components. Yet it differentiates between demand- and load-dominated network areas. In the latter the installation of local generation relieves system stress and avoids network expansion. Accordingly, the charges for distributed generation are negative; in other words DG is rewarded.

---

[9]In some countries connection procedures include a queuing process for connection requests, which can also be interpreted as locational signal [24]. In an area with many other pending connection decisions a request will typically take longer to be fulfilled if the queue is long. Also the cost allocation may depend on the queue position, as for example in the United States [25].

[10]In the future, the connection charges might be revised further and become shallow.

[11]This is a European requirement to strengthen the customer's position [27].

FIGURE 13.3   Distribution charging approaches in the EU-15. *(Source: compiled by authors from various sources. *Shallow charges cover direct connection cost only; deep charges also cover cost for upgrades in the existing network.)*

In the charging methodology for extra high voltage distribution networks, more explicit locational components are to be introduced by April 2012 [29].[12] Network operators can choose between forward cost pricing and long-run incremental cost pricing. The first method calculates average forward-looking charges for different network parts that are not directly connected. Hence the locational signals are limited to distinct sub-networks. The latter method in contrast calculates the impact of new DG and new load connection on the long-run incremental cost of the network for the differential load at each network node. The resulting node-specific charges can be positive and negative. They are scaled by adding a fixed component to ensure cost recovery while complying with regulatory prescriptions on maximum allowed revenue.

---

[12]The implementation was originally planned for September 2011 [30], but the process has been postponed to give network operators more time to develop satisfactory proposals.

The resulting locational signals are strong, administrative, and ex-ante. Although charges may vary over time, they are known at the time of investment and hence create certainty and transparency. Real-world experience with this approach still has to be seen, but the positive effects from such signals are believed to be substantial. Simulations suggest savings in the order of £200 million over the next 20 years for the UK network. With projected investment in distribution networks of £3.2 billion per year over the next 20 years [6], this amount sounds rather low; however, the implementation costs are estimated to be moderate and the gains might still be worth pursuing [7].

Explicit use of locational energy pricing—say, nodal or zonal spot pricing—does not seem to exist at the distribution level at all. All the known examples are at the transmission level. This might suggest that in the past locational signals at the distribution level have not been worth the effort. However, this is changing as more and more generation enters at the distribution level. This changes the paradigm of top-down energy flow and potentially causes congestion in the distribution network. Therefore benefits from steering network demand and generation customers to better use existing network capacity increase. On a theoretical level Pollitt and Bialek [31] advocate locational energy pricing for UK distribution networks in the context of regulating for a low-carbon future. They argue that for distribution networks the differentiation of several price zones might be a reasonable initial step, which captures much of the benefit of more refined locational differentiation.

There are promising developments in the field of demand-side management. Retailers in New Zealand and the United States, for example, offer special tariffs for customers on load control. In the United States, where more and more system operators are offering demand control programs, the federal regulator has recently strengthened the position of providers of demand response [32]. From 2012 providers of demand reduction will be entitled to receive a remuneration equal to the market price for generation, when that reduction balances supply and demand and is cost effective. As the wholesale prices in organized U.S. markets are locationally differentiated, this incentivizes flexible users where they are needed. The United States has therefore made a move towards locational signals within the general pricing system. In addition, at the state level additional regulations may exist that further allow individual solutions. In California, for example, utilities are allowed to offer contracts that ensure the installation of distributed generation at the right time and in the right location [33].[13] On the contrary, the development in New Zealand is rather decentralized and flexible, leaving the decisions with the retailers.

---

[13]Such contracts can also aim at size or physical assurances needed to enable a utility to defer a distribution capacity addition. Interestingly, targeted contracting was seen to obviate the need for additional locational signals in the general tariffs. The contract system is considered to retain most of the efficiencies of locational charges while avoiding "the complications of reversing the long-standing policy of uniform pricing" [33].

As it is the case in New Zealand, most countries have some implicit way to steer investments into distributed generation to reflect the effects on the distribution network and generally maintain uniform charging. So does Germany, as will be described in the next section. Partly for other purposes and partly for exceptional cases, locational pricing can be applied with strong restrictions by current legislation. The next step would then be obvious: to allow more locational signals, lift the restrictions. If the network owners are adequately incentivized to defer unnecessary network investment, they will seek locational signals if allowed.

LOCATIONAL DISTRIBUTION PRICING IN GERMANY

In Germany the share of distributed generation (DG) and generation from renewable sources in general has increased rapidly in recent years. This was mainly triggered by the highly effective feed-in tariffs for renewable generation (RES-E) and combined heat and power (CHP) [34,35]. Like many other countries, Germany is actively promoting smart grid development and more flexible pricing schemes. It has now also become clear that facilitating generation from photovoltaics and wind at the distribution level requires substantial network expansion. Most studies estimate the investment need at around €25 billion until 2020 [8,36], which may be on the conservative side. For comparison, in 2008 the overall expenditure of distribution network operators amounted to €5.57 billion [37].[14] Yet, neither locational network prices nor locational energy prices are implemented explicitly at this time. Nor is there political consensus to move in this direction.

Politicians argue for equal, non-differentiated tariffs to prevent unfair disadvantages for higher price regions, which become less attractive for industry. The argument is that renewable and distributed generation create social benefit, and hence their integration costs, including network investment costs, should be *socialized*. When related to network charging we observe another problem. Varying network charges, especially related to peak usage, are not intuitively understood since the underlying costs are assumed to be constant. However, peak usage drives the required capacity and thereby network costs. Of course opportunity cost considerations are the rationale for differentiated tariffs but often are not perceived as reasonable by customers. Such positions are a major barrier for locally differentiated prices. Looking into existing legislation, however, possibilities exist for introducing locational signals as further described below.

The Challenge: Increasing Renewable Generation in an Inflexible System

As a result of ambitious feed-in tariffs, Germany has a high share of renewable electricity amounting to 16.4% of total gross electricity consumption in 2009 [38]. This share is projected to increase to 38.6% by 2020 [39] and reach

---

[14]This includes operation & maintenance, replacement, expansion, and investments in new control and information technology [37].

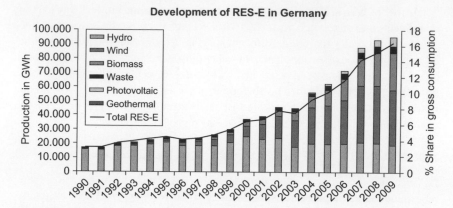

FIGURE 13.4   Development of generation from renewable sources in Germany between 1999 and 2009. *(Source: Brandstätt et al. [53])*

80% in 2050 [40]. Figure 13.4 shows the development of electricity generation from renewable sources in Germany since 1990.

Increasing shares of volatile wind generation are a major driver of network investment in the German transmission networks. The electricity needs to be transported from windy, coastal areas, mostly in the North, to load centers, mostly in the South. Projected costs for network upgrades are in the range of €946 million per year [41]. Main factor is the rapid development of wind both onshore and offshore. Meanwhile in distribution networks too, problems are arising from the growth of distributed generation. The sum of installed electricity generation capacity from renewable sources in distribution networks was around 33.2 GW in 2008 [42] and rose to 40.5 GW in 2009 [43]. Decentralized generation changes the traditional model from top-down electricity flow and requires major changes in the system paradigm and management as described in a number of chapters in this book. The integration costs for generation from renewable energies in the distribution network are projected at €25 billion [8]. A principle cause for the investment need is the boom of photovoltaic installations experienced in the last few years as shown in Figure 13.5.

One problem is the lack of flexibility with generation and demand in the current system. The increasing shares of inflexible and fluctuating renewables can cause regional congestion or voltage problems in distribution networks, especially if reinforcement work is behind schedule. Distributed generation is one of the main drivers of the investment need but hardly receives any market signals nor participates in system management for two main reasons:

- First, distributed generation is often from renewable sources and therefore prioritized under fixed feed-in tariffs and exempted from market prices; and

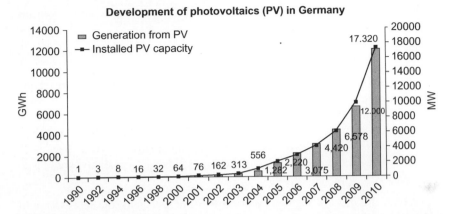

FIGURE 13.5 Development of photovoltaic electricity generation and installed capacity in Germany. *(Source: Authors, based on BMU [38])*

- Second, generators in distribution networks are often too small and not equipped with the technology and characteristics for system management purposes in the balancing markets. Consequently, a large part of distributed generation does not receive the locational signals simply because they are not subject to energy prices.

Furthermore, the price signals for distributed generation that does not receive support from feed-in tariffs are one-sided. Such generators are incentivized with a premium for avoided network charges as a consequence of substituting supply from higher voltage levels. The reasoning is that local generation reduces the reinforcement needed at the transmission level. Consequently the distribution network operator (DNO) pays lower charges to upstream networks and passes the savings through to DG connected to its network. This argument, however, becomes problematic in systems with high shares of DG for at least two reasons:

- First, while DG avoids or reduces network charges in upper levels, it also entails higher network cost in the level it is connected to – namely at the local, low-voltage levels; and
- Second, in systems with high shares of DG, local generation may in times exceed local demand, thus putting an additional burden on the network since it must now balance the load and generation by relying on neighboring distribution and transmission systems.

The premium for avoided network charges accounts only for the avoided network charges at higher voltage levels and is therefore uniform across one distribution network area. This means that the bonus does not relate to site-specific avoided investment needs; the local network situation is ignored. DG may at the same time trigger higher cost in the local network but nonetheless receive a bonus.

Two further issues hinder coordinated investment into generation, demand, and networks:

- First, in Germany generators do not pay for network use, and therefore it is not obvious how to implement effective locational signals; and
- Second, the network operators are obliged to expand the network to accommodate privileged generation from renewable sources, unless this is not economically feasible, which applies only rarely. The issue of the economic feasibility of network investment is clearly critical, but has not been defined so far. As a result network operators are obliged to accommodate as much decentralized generation as is offered, and to invest accordingly, passing on the costs to all network users.

Within the current regulatory framework in Germany the integration of electricity from renewable sources and the respective investment requirements exhibit high regional variance. While distribution network operators in regions with a high share of electricity generation from photovoltaics or biogas face enormous investment, other regions might not be affected in the same way. Currently, network costs are recovered via "postage stamp" tariffs within each network. Tariffs from higher voltage levels are subsequently passed through to lower network levels, but not equalized horizontally with other networks of the same voltage level. Hence, RES-E-driven investment costs cause a regional variation in tariffs.

Importantly, unlike locational charging this does not send locational signals for system optimization. The locational distribution of the burden does not incentivize more beneficial siting decisions for generators that do not contribute to running network cost as feed-in of electricity is presently free of charge (for details see StromNEV [44, §15]). Consequently, consumers alone bear the cost of the network and face higher network cost in regions with high shares of renewable generation. Recently this has led to a proposal to equalize the network tariffs nationwide (see BR 868/10 [45]). The uniform assignment of network cost to demand prevents locational differentiation and is supported by fairness consideration.

Potential for Locational Network Pricing

Currently, the general approach to connection charging in Germany is of the *shallow* variety. However, there are two ways to by pass the shallow approach and have some locational network signals: first, the contributions to connection costs and second, the option in the network ordinance [44, §19.2.3] for individualized network charges and compensation for deferred network investment.

Contributions to Connection and Construction Cost

Network operators may, but do not have to, charge a contribution to the cost of building the network to new connections, which introduces a deep component. The "contributions to connection cost"—in German *Netzanschlussbeitrag*—and the "contribution to construction cost"—in German *Baukostenzuschüsse*—allocate a share of the network connection or expansion cost to the customer. The regulations require that contributions be cost-oriented, non-discriminatory, transparent, and proportionate. Importantly, the contribution to connection cost may only be charged for network investments that are not economically feasible without a contribution from the connecting party. However, the term "economically feasible" has not been defined unambiguously. The contributions to construction costs can be charged generally from every customer to cover the cost of the existing network.

The cost-reflective allocation of the infrastructure cost triggered by a new connection has an important function in steering the capacity demand of customers.[15] It is expected to limit the requested capacity to a realistic demand-oriented value and thereby to contribute to needs-based network expansion while avoiding over-dimensioning. Hence, this element is directly targeted towards efficient network development and is promising development towards locational differentiation.

Despite the requirement for cost-based calculation, certain averaging across network areas and charging based on typical cost of comparable cases are allowed. Theoretically, also the regional differentiation in network sub-areas and the charging of distributed generation are possible,[16] although in practice this does not seem to happen. Network operators rather rely on uniform contributions across their network. It can also be observed that while regulation does not prescribe it, network operators typically only use a standard calculation for contributions to construction cost. This might be motivated, among other reasons, by practical considerations of using a standardized approach. There are two calculation methods that are generally deemed acceptable, one published by the regulator and one by the industry association [47,48]. Both are robust against regulatory scrutiny while other, more flexible individual solutions may be targeted for control. Importantly also, these existing calculation methods rely on the traditional model of unidirectional electricity distribution, from central power plants "down" to end customers. In areas with distributed generation the current calculation even leads to higher contributions for demand, which is inappropriate. Increasing shares of DG and the development of smart grids call for adapted, more sophisticated calculations to take these issues into account.

The assignment of contributions to construction cost is limited to demand customers and non-prioritized distributed generation,[17] although network operators

---

[15]They do not have a financing function. Network operators have to resolve contributions to construction costs they received from demand over 20 years as a cost-reducing factor in the general tariff calculations. Contributions received from generators have to be resolved on a connection-specific basis over 20 years.

[16]For high voltage networks generator connection regulations prohibit the collection of contributions to construction costs [46].

[17]KraftNAV [46] prohibits contribution to building costs for generators bigger than 110 MW and connected to networks of 110 kV and more. KWKG and EEG exempt generators from contribution to building costs.

rarely charge the latter. This de facto exclusion of distributed generation from contributions to construction cost is problematic. As they account for large parts of the investment need, cost-reflective allocation of the resulting investment cost seems justified to internalize those costs in the network customers' decisions. Furthermore, time-differentiated components in contributions to construction costs may be necessary to provide targeted incentives that account for utilization patterns that strongly deviate from the average as, for example, off-peak demand. However, the current connection regulation [49, §11] does not allow for time-differentiated contributions to construction costs in low voltage networks.

In conclusion, the contributions to construction costs give a possibility for more differentiated network charges. Further locational and time differentiation of contributions to construction costs as well as an inclusion of other agents such as smaller customers or distributed generators seems desirable to enhance the effectiveness of the instrument. This requires regulators to give more freedom to network operators and encourage case-specific calculations if they incentivize more efficient network development. These developments towards more individual solutions and negotiated contracts are already known. Experience from New Zealand and the United States indicates that regulated defaults or standard conditions are recommended to maximize benefit from flexible solutions while avoiding negative effects for small generators and customers.

Individual Network Use-of-System Charges

In general UoS-charges are uniform in a network area, but in special cases German network operators are required to offer non-standardized, individually designed network tariffs [44, §19.2]. Standard network charges typically consist of two components, a standing charge based on capacity and a variable charge related to the energy distributed to the respective connection. A coincidence function factors in to which degree the network users contribute to system peak.

This serves cost reflectivity since the capacity demand at system peak is a main driver for infrastructure cost. Individual network tariffs have to be offered if users are expected not to contribute to system peak because their peak demand differs significantly from standard characteristics or for exceptionally large customers.[18] The individual tariff has to reflect the cost savings of deferred network investment, but cannot be less than 20% of the standard tariff [51].[19]

In addition, it may not lead to a substantial increase in the remaining network charges. The contract for an individual network charge is subject to approval by the regulator. If approval fails or preconditions fall away, the standard charges apply. It is important to note, that the individual tariffs can only be lower than the regulated default charges and customers can only improve.

---

[18]Individual network tariffs are generally possible for users with 7,500 utilization hours per year (7,000 h as from 2011) and consuming more than 10 GWh per year [44, §19].
[19]Until 2010 a threshold of 50% applied [50].

If applied accordingly by the network operator, such individual tariffs can be a tool to attract network users with characteristics that are favorable for network development at specific locations. However, the focus of such tariffs on off-peak consumption reduces their effect dramatically as also other characteristics would justify a reduction in network charges. In some cases even consumption during peak times can benefit the network, if for example a lot of photovoltaic feed-in is available during peak periods. Another obstacle for the effectiveness of the tool is its limitation to demand. This results from the fact that feed-in of electricity is presently free of charge [44, §15]; UoS charges are borne by demand only. In principle, this tool could also steer generators if these were subject to use-of-system charges as it is the case for example in the UK.

Recent publications by the German regulator show a tendency to increase the scope for such individual agreements between network operators and customers. The regulator has acknowledged that increasing administrative routine and additional experience justify lowering the preconditions for approval of individual network tariffs at least in some areas [51].

Potential for Locational Energy Pricing

Currently, there are very few locational signals in energy pricing in Germany. There is some discussion at the transmission level to implement two zones, north and south, because wind in the north causes frequent network congestion. Yet, it seems that political consensus towards a zonal approach is still a long way ahead; a discussion on explicit locational energy pricing at the distribution level has not started yet. Nevertheless there are loopholes in the current legislation that would allow locational signals in energy pricing.

Voluntary Curtailment of DG

In 2009, situations with a lot of wind and low demand caused negative wholesale prices as shown in Figure 13.6. This triggered a debate on how to deal with negative prices, and more specifically, whether RES-E, in this case wind, could be curtailed.

FIGURE 13.6   Intra-day prices in Germany on October 3–4, 2009. *(Source: Epexspot.com)*

The RES-E support scheme in Germany is a feed-in system with take-off obligation and RES-E priority in case of network congestion. Moreover, an inflexibility of the system in Germany is that the possibilities for curtailment of RES-E production—are very restrictive by law. Basically, it is not allowed to curtail wind: a conventional plant still produces unless in must-run status or required to operate for system reliability. In other words, all RES-E must be taken into the system with thermal units ramping up and down for system balance and stability.[20]

Clearly then the system constrains the two leverages for the market to operate efficiently: prices and quantities. Prices are fixed through feed-in tariffs—with the result that RES-E suppliers do not respond to market signals—and quantities are fixed by the priority rule and the restrictions on curtailment. The current policy debate is about how flexible the use of RES-E curtailment should be with policy makers typically preferring to maximize renewable generation to meet environmental protection targets.

After heated debate, the authorities loosened the rules for voluntary RES-E curtailment somewhat, as set out in the guidelines by the regulator [52]. Nevertheless, it is still restrictive. Two major obstacles stand in the way of flexible voluntary curtailment of RES-E:

- First, RES-E curtailment is effectively only allowed as a last resort and not for efficiency reasons; and
- Second, for the DNO, if in case of network congestion parties agree on compensated voluntary RES-E curtailment, an obligation to expand the network automatically follows. This practice distorts efficient investment, because it does not allow trading-off between generation and network.

This trade-off is important. A DNO can defer network investment to some extent if allowed to enter into flexible voluntary curtailment agreements. Basically, the network owner would somehow offer compensation for avoided feed-in and foregone revenue and in return save on network investment cost. If parties are properly incentivized this increases overall economic efficiency.

Brandstätt et al. [53] argue that the overall performance of the system would improve significantly by lifting the restrictions on the use of voluntary curtailment agreements, while retaining the priority rule as such. Since generators of RES-E can only improve under this system reform, investment conditions improve, leading to higher installed RES-E capacity. This in turn implies that reduced wind output due to curtailment can actually be offset by higher wind output in all periods in which there is no problem.

This nice property for environmental goals actually creates a problem for locational signals. There might be a perverse incentive to invest more at this particular location, which is what the signals aimed to avoid in the first place. It can be

---

[20]Similar principles apply in many other jurisdictions, where renewable generation is treated as "must take" by the grid operator when available.

expected that if the network owner would expect perverse incentives to occur, he would not compensate the feed-in for curtailment but would support storage options at these locations. The storage would get paid in times of network congestion and would sell to the system if the network is unconstrained. In these cases, the network owner would want to support storage instead of compensating curtailment. Chapter 5 looks into the details of integrating intermittent renewable generation through storage, and Chapter 18 describes the combination of electric vehicles and wind generation. It can be concluded that both the legal possibilities and the awareness to apply curtailment exist. The recommendation is to allow more flexibility in the application of the scope to increase flexibility in the system, through curtailment agreements and investment in storage options.

Remuneration for Location-Specific Flexibility: Call and Curtail Agreements

According to current German legislation, network operators are responsible for system stability. In case of emergencies they are expected to take adequate measures to maintain reliability. Apart from network management, this includes market-oriented measures such as balancing and reserve energy, contracted load interruptions such as demand response, and congestion management [54, §13]. Rewards are given for the provision of reserve capacity and/or via energy prices for produced or curtailed energy, respectively. Again prioritized generation from renewable sources and CHP is exempted and can only be curtailed as an emergency measure in cases of congestion [34, §11] and not for system optimization.[21] Participation in balancing markets is theoretically possible but most, especially small plants, do not qualify for technological reasons. Generation management of prioritized generation is strongly linked to the obligation to expand the network. Whenever renewable or CHP generation is curtailed for network stability, this entails the requirement to expand the network in order to avoid this curtailment in the future. In other words, the system operator cannot avoid network expansion by targeted generation management that avoids network congestion. Generators only receive compensation, if the network operator is liable for the congestion in the sense that it did not sufficiently expand the network. As mentioned in the section "The Challenge: Increasing Renewable Generation in an Inflexible System" it is problematic that regulation does not sufficiently stipulate generation and load management to optimize the system or defer network investment but instead relies on network expansion.

The most interesting of the recent developments in the field of flexibility agreements are so-called call and curtail agreements—*Zuruf- und Abschaltregelungen* in German—that can appear in two different forms. They can be part of an agreement for an individual network tariff or can be established for reliability management [54, §13].

---

[21]Curtailed generators of electricity from renewable sources or CHP in most cases still receive the feed-in tariff.

In the first case the agreement assures that a customer is not consuming during the identified peak period. In practical terms this means that the agreement obliges the customer to reduce its demand when called upon or to allow the network operator to reduce consumption via remote control. A call and curtail agreement only qualifies for an individual network tariff if it refers to the peak periods. If this is not the case, the precondition of atypical usage, generally outside peak periods, would not be fulfilled. In that case the system operator can conclude contractual agreements with generators or load on the provision of balancing power or curtailable load. The parties agree on an individual payment for the provision of such flexibility potentials. Importantly, the usual network tariffs still apply, but are offset against the individually agreed payment. These agreements do not have to go through the regulatory approval process [43].

Both generation management and call and curtail agreements provide the network operator with additional flexibility. Crucially, the management of generation from renewable sources is limited to emergencies. It would be beneficial to allow voluntary curtailment agreements between network operators and renewable generators. Brandstätt et al. [53] show that this would improve system efficiency while also benefiting renewable generators.

Furthermore, flexible agreements in their current form are only possible for customers that individually cause lower network cost. Achieving the full benefit from customer flexibility would require including also smaller customers that can reduce network cost as a group. This would be a form of demand-side management for network purposes. In New Zealand we already observe this in the form of special tariffs for controlled customers, thus rewarding their flexibility potential. In this case customers delegate control to the network operator. However, the example of dynamic retail pricing in the United States [55] indicates substantial system benefits from more cost-reflective pricing even with decentralized control.

Also in Germany further developments towards more dynamic pricing for flexibility, including the network dimension, are within the scope of current smart grid research. The German government commissioned a major research program called "E-Energy—Smart Grids made in Germany" to support research and demonstration of smart grid solutions in six model regions. One part of the research is the implementation of smart grid enabling technologies and functionalities. Among the main objectives of the project is to develop markets that enable the realization of smart grid benefits in a liberalized market [56].

CONCLUSIONS

Experience with locational distribution pricing is still scarce, but shows a trend to more flexible pricing structures. In distribution networks, locational differentiation appears in network tariffs if at all. In the future smarter tariffs can be expected to gain further importance. The energy system becomes more flexible with high shares of renewable generation and flexible demand in smart distribution systems. A flexible tariff structure is necessary to exploit the benefits of smart grids.

Locational pricing can be in the form of locational network pricing or locational energy pricing. The former includes deep connection charges and locationally differentiated use-of-system charges. The latter includes nodal and zonal spot pricing. In between is a field that, in this chapter, is referred to as smart contracts. If network owners are incentivized, they will find innovative contractual ways to steer feed-in and load to defer network investment. In these cases, the authorities do not have to design the markets but rather simply allow flexible use of smart contracts.

For Germany this can result in a further flexibilization of the location-specific connection charges and more individualization of network tariffs. Additionally, adaptations in the current regulatory framework are needed to ensure that network operators have the incentives to use flexible arrangements as an alternative to network expansion.

Several of the mechanisms presented so far could potentially help defer network investments, but to date are not used to their full potential. In principle there are two possible reasons for this. Either the current regulation is too restrictive and does not leave enough freedom to develop smart contracts further, or network operators do not receive adequate incentives to develop the smart contracts that are needed to optimize the system. Consequently some further development of the existing framework is desirable. This chapter discussed four possible targets for loosening the current legislation.

Contributions to construction costs are presently used only to prevent economically in feasible capacity requests. Adjustments to the tool should include application to a broader range of customers, determination in a less standardized way, and addressing prioritized generation. Also more detailed differentiation, for example within one network or according to time patterns, is desirable.

For individual network tariffs, presently only very big network users or those with an uncommon peak behavior are eligible. Other characteristics that can serve to defer network investment, as for example local reliability and voltage support, do not qualify. Also generators and smaller users could contribute to system optimization. For overall efficient network operation it is recommended that network operators be allowed to offer individual tariffs to all users with beneficial characteristics.

Following recent episodes of negative wholesale prices, it is now possible to enter into voluntary curtailment agreements with RES-E suppliers. Application of this option is too restrictive to be effective in deferring network investment. Allowing more flexible use of voluntary curtailment agreements for RES-E suppliers would create further potential without requiring system reform.

The same reasoning applies to the call and curtail agreements that are strongly restricted for renewable generation. This limits the effectiveness of this instrument for the management of smart distribution networks as large parts of generation in distribution grids are exempted. Consequently further flexibility of these tools and the inclusion of renewable generation are highly desirable.

There are two issues for further research. The regulator needs to think about network investment incentives. Without appropriate incentives the network operators will not exploit the structural optimization potential. While network operators are obliged to expand the network to the bitter end, there is not much scope for efficient network development. Also the fact that expansion costs are passed through as long as the regulator approves prevents creativity for investment optimization. Therefore an adjustment of the regulatory framework to provide more incentives for efficient system transformation is necessary.

Another crucial point is market integration. Generation from renewable sources and distributed generation are focal elements in future smart grids and yet do not or not sufficiently participate in the markets.[22] Presently there are only few locational signals, and those that do exist do not reach these critical actors. Additionally, the remuneration for avoided network cost in higher voltage levels that exhibits some form of locational differentiation sends the wrong signals. This remuneration is presently are presently mainly a general support scheme for distributed generation. Small changes, however, could make them reflect the local situation better and could thus help regional networks with locational steering.

REFERENCES

[1] EWE, *Presentation Elektromobilität*, J. Hermsmeier – EWE AG, Bullensee-Kreis, 26 November 2010, Berlin, 2010.

[2] EWE, *Presentation Innovative Regulierung für intelligente Netze – praktische Erfahrungen von EWE NETZ*, T. Maus - EWE NETZ GmbH, IRIN-Workshop, 6 October 2010, Berlin, 2010.

[3] D.T-C. Wang, L.F. Ochoa, G.P. Harrison, C.J. Dent, A.R. Wallace, Evaluating investment deferral by incorporating distributed generation in distribution network planning, *Proceedings of the 2008 16th Power Systems Computation Conference (PSCC'08)*, School of Engineering and Electronics, University of Edinburgh, Edinburgh, UK, 2008.

[4] P.M. Sotkiewicz, J.M. Vignolo, Nodal pricing for distribution networks: efficient pricing for efficiency enhancing DG, *IEEE Trans. Power Syst.* 21 (2) (2006) 1013–1014.

[5] EPRI, *Estimating the Costs and Benefits of the Smart Grid, A Preliminary Estimate of the Investment Requirements and the Resultant Benefits of a Fully Functioning Smart Grid*, Technical Report, 2011.

[6] Ofgem, Ofgem reengineers network price controls to meet £ 32 billion low carbon investment challenge, *Press Release* 26 July 2010, 2010.

[7] F. Li, D. Tolley, N.P. Padhy, J. Wang, Framework for assessing the economic efficiencies of long run network pricing models, *IEEE Trans. Power Syst.* 24 (2009) 1641–1648.

---

[22]Market integration of generation from renewable sources and the adequate design of support schemes are currently the subjects of a big debate. For more information see, e.g., Klessmann et al. [57].

[8] BDEW, *Abschätzung des Ausbaubedarfs in deutschen Verteilungsnetzen aufgrund von Photovoltaik- und Windeinspeisungen bis 2020.* http://www.bdew.de/internet.nsf/id/77103B624B580EF0C125786B0054E259/$file/2011-03-30_BDEW-Gutachten%20EEG-bedingter%20Netzausbaubedarf%20VN.pdf, 2011 (accessed 08.04.11).

[9] C. Brandstätt, G. Brunekreeft, N. Friedrichsen, Locational signals to reduce network investments in smart distribution grids: what works and what not? Utilities Policy (2011), doi: 10.1016/j.jup.2011.07.001. Article in Press.

[10] CEPA, *Cambridge Economic Policy Associates Ltd Review of International Models of Transmission Charging Arrangements, A Report for Ofgem,* 2011.

[11] A. Piccolo, P. Siano, Evaluating the impact of network investment deferral on distributed generation expansion, *IEEE Trans. Power Syst.* 24 (3) (2009) 1559–1567.

[12] T. Ackermann, *Distributed Resources in a Re-Regulated Market Environment.* Ph.D. dissertation, Dept. Electrical Eng., KTH, Stockholm, 2004.

[13] F. Woolf, *Global Transmission Expansion: Recipes for Success,* CMS Cameron McKenna, 2003.

[14] G. Brunekreeft, K. Neuhoff, D. Newbery, Electricity transmission: An overview of the current debate, *Utilities Policy* 13 (2005) 73–93.

[15] F. Li, D. Tolley, N.P. Padhy, J. Wang, *Network Benefits from Introducing an Economic Methodology for Distribution Charging,* A study by the department of Electronic and Electrical Engineering, University of Bath, 2005.

[16] F. Li, D. Tolley, Long-run incremental cost—pricing based on unused capacity, *IEEE Trans. Power Syst.* 22 (4) (2007) 1683–1689.

[17] BNetzA, Jahresbericht 2010, *Annual report of the Federal Regulatory Agency of Germany for the year 2010.* http://www.bundesnetzagentur.de/SharedDocs/Downloads/DE/BNetzA/Presse/Berichte/2011/Jahresbericht2010pdf.pdf?__blob=publicationFile, 2011 (accessed 04.04.11).

[18] M. Björndal, K. Jörnsten, Zonal pricing in a deregulated electricity market, *Energy J.* 22 (1) (2001) 51–73.

[19] W.W. Hogan, Contract networks for electric power transmission, *J. Regul. Econom.* 4 (3) (1992) 211–242.

[20] F.C. Schweppe, R.D. Tabors, M.C. Caramanis, R.E. Bohn, *Spot Pricing of Electricity,* Kluwer Academic Publishers, Norwell, MA, 1988.

[21] S. Stoft, *Power System Economics—Designing Markets for Electricity,* Wiley-IEEE Press, 2002.

[22] K. Neuhoff, R. Boyd, *International Experiences of Nodal Pricing Implementation – Frequently Asked Questions,* Climate Policy Initiative/DIW Berlin, Climate Policy Initiative Working document – this Version February 2011. http://www.climatepolicyinitiative.org/files/attachments/99.pdf, 2011. (Accessed 04.08.11).

[23] R. Green, Nodal pricing of electricity: how much does it cost to get it wrong? *J. Regul. Econom.* 31 (2) (2007) 125–149.

[24] A. van der Welle, J. De Joode, F. Van Oostvoorn, *Regulatory Roadmaps for the Optimal Integration of Intermittent RES-E/DG in Electricity Systems,* Final report of the RESPOND Project, 2009

[25] SGIP, *Small Generator Interconnection Procedures, For Generating Facilities No Larger than 20 MW,* The revisions to this document take effect August 28, 2006, per FERC Order No. 2006-B issued July 20, 2006, FERC Stats. & Regs. 31,221, which was published in the Federal Register July 27, 2006 (71 FR 42587), as amended by the errata issued September 5, 2006, which was published in the Federal Register September 13, 2006 (71 FR 53965), 2006.

[26] Ofgem, *Structure of Electricity Distribution Charges Initial Decision Document* 142/03, November 2003.

[27] EC Directive 2003/54/EC concerning common rules for the internal market in electricity.

[28] Ofgem, *Electricity Distribution Structure of Charges: Distribution Charging Methodology at Lower Voltages*, Ofgem decision document, Nov. 2009.

[29] Ofgem, *Decision on Revised Submission and Implementation Dates for the EHV Distribution Charging Methodology (EDCM)*. http://www.ofgem.gov.uk/Networks/ElecDist/Policy/DistChrgs/Documents1/EDCM%20timelines%20decision.pdf, 2010 (accessed 08.04.11).

[30] Ofgem, *Delivering the Electricity Distribution Structure of Charges Project: Decision on Extra High Voltage Charging and Governance Arrangements*, Ofgem decision document, July 2009.

[31] M. Pollit, J.W. Bialek, *Electricity Network Investment and Regulation for a Low-Carbon Future*, Working Paper EPRG 0721, 2007.

[32] FERC, *Order No. 745 Demand Response Compensation in Organized Wholesale Energy Markets*, 2011.

[33] CPUC, *California Public Utility Commission Proposed Decision of Commissioner Lynch*, 10 January 2003, Paragraph 8.3.2 Discussion: Contracting for Distributed Generation Obviates Need for Deaveraged Tariffs or Incentive Programs at This Time, 2003.

[34] EEG, Erneuerbare-Energien-Gesetz [Renewable Energy Act] 25 October 2008, last changed 12 April 2011, 2009.

[35] KWKG, Kraft-Wärme-Kopplungsgesetz [Combined Heat and Power Act] 19 March 2002 last changed 21 August 2009, 2002.

[36] A. Moser, *Versorgungssicherheit Strom: Energiewirtschaftliche und technische Dimensionen.* [Power supply security: energy economical and technical dimensions], Presentation at Göttinger Energietagung 2011: Aspekte der Versorgungssicherheit Strom und Gas, Göttingen, 13 May 2011.

[37] BNetzA, Bundesnetzagentur, Monitoringbericht 2009 [Federal Regulatory Agency of Germany, Monitoring report 2009], 2009.

[38] BMU, *Development of Renewable Energy Sources in Germany 2010 – Graphics and Tables*, Federal Ministry for the Environment, Nature Conservation and Nuclear Safety. http://www.bmu.de/files/pdfs/allgemein/application/pdf/ee_in_deutschland_graf_tab.pdf, 2010 (accessed 12.05.10).

[39] German Federal Government, National Renewable Energy Action Plan in accordance with directive 2008/28/EC on the promotion of the use of energy from renewable sources. http://ec.europa.eu/energy/renewables/transparency_platform/doc/national_renewable_energy_action_plan_germany_en.pdf, 2010 (accessed 11.11.10)

[40] BMWi, *Energiekonzept für eine umweltschonende, zuverlässige und bezahlbare Energieversorgung*, German Federal Ministry of Economics and Technology, Energy strategy for a sustainable, reliable and affordable energy supply. http://www.bmwi.de/BMWi/Redaktion/PDF/Publikationen/energiekonzept-property=pdf,bereich=bmwi,sprache=de,rwb=true.pdf, 2010 (accessed 11.09.10).

[41] DENA, *dena-Netzstudie II – Integration erneuerbarer Energien in die deutsche Stromversorgung im Zeitraum 2015–2020 mit Ausblick 2025.* [Dena grid study II – integration of renewable energies into German electricity supply 2015–2020 with an outlook to 2025], 2011.

[42] BNetzA, EEG-Statistikbericht 2008. *Statistikbericht zur Jahresendabrechnung 2008 nach den Erneuerbare-Energien-Gesetz (EEG).* [Statistical report for the annual settlement2008 as provided in the feed-in tariff]. Editorial deadline: March 2010. http://www.bundesnetzagentur.de/cae/servlet/contentblob/153014/publicationFile/6555/100427StatistikberichtEEG2008pdf.pdf (accessed 01.04.11), 2010.

[43] BNetzA, EEG-Statistikbericht 2009. *Statistikbericht zur Jahresendabrechnung 2009 nach den Erneuerbare-Energien-Gesetz (EEG)*. [Statistical report for the annual settlement2009 as provided in the feed-in tariff]. Editorial deadline: 28 March 2011. http://www.bundesnetzagentur .de/cae/servlet/contentblob/195642/publicationFile/10358/110318StatistikberichtEEG2009 .pdf (accessed 01.04.11), 2011.

[44] StromNEV, §15 Verordnung über die Entgelte für den Zugang zu Elektrizitätsversorgungsnetzen, 25th Juli 2005 (BGBl. I S. 2225), last changed 3rd September 2010 (BGBl. I S. 1261), 2005.

[45] BR 868/10 Antrag des Freistaats Thüringen Entschließung des Bundesrates zur Herstellung gleichwertiger Lebensverhältnisse im Bundesgebiet durch Vereinheitlichung der Netzentgelte auf Übertragungs- und Verteilnetzebene, Bundesrat Drucksache 868/10, 22 December 2010.

[46] KraftNAV, Verordnung zur Regelung des Netzanschlusses von Anlagen zur Erzeugung von elektrischer Energie [Ordinance on network connection of assets for the generation of electricity], 26 Juni 2007.

[47] VDN, *Einheitliche Berechnungsmethoden für Baukostenzuschüsse* [Uniform calculation methodologies for contributions to construction costs], 19 April 2007.

[48] BNetzA, *Bundesnetzagentur Positionspapier zur Erhebung von Baukostenzuschüssen (BKZ) für Netzanschlüsse im Bereich von Netzebenen oberhalb der Niederspannung (BK6p-06-003)* [Federal Regulatory Agency of Germany, position paper on the charging of contributions to construction costs] (05 January 2009), 2009.

[49] NAV, Verordnung über Allgemeine Bedingungen für den Netzanschluss und dessen Nutzung für die Elektrizitätsversorgung in Niederspannung [Ordinance on the general conditions for network connection and utilization for eletricity supply at low voltage], 1st November 2006 (BGBl. I S. 2477), last changed 3rd September 2010 (BGBl. I S. 1261), 2010.

[50] BNetzA, Bundesnetzagentur, *Leitfaden zur Genehmigung individueller Netzentgeltvereinbarungen nach §19 Abs. 2 S. 1 und 2 StromNEV* [Federal Regulatory Agency of Germany, Guiding document on the approval of individual network tariffs as described in §19 Paragraph 2, sentence 1 and 2 of StromNEV], 2009.

[51] BNetzA, *Bundesnetzagentur, Leitfaden zur Genehmigung individueller Netzentgeltvereinbarungen nach §19 Abs. 2 S. 1 und 2 StromNEV ab 2011* [Federal Regulatory Agency of Germany, Guiding document on the approval of individual network tariffs as described in §19 Paragraph 2, sentence 1 and 2 of StromNEV starting 2011] (document from 29 October 2010), 2010.

[52] BNetzA, *Leitfaden zum EEG-Einspeisemanagement Version 1.0 [Federal Regulatory Agency of Germany*, Guiding document for the curtailment of prioritized generators under the feed in tariff] (document from 03 March 2011). http://www.bundesnetzagentur .de/SharedDocs/Downloads/DE/BNetzA/Sachgebiete/Energie/ErneuerbareEnergienGesetz/ LeitfadenEEGEinspeisemanagement/LeitfadenEEG_Version10_pdf.pdf;jsessionid= D3091117D63764F33395C23C94D68354?__blob=publicationFile, (accessed 08.04.11), 2011.

[53] C. Brandstätt, G. Brunekreeft, K. Jahnke, How to deal with negative power price spikes?—Flexible voluntary curtailment agreements for large-scale integration of wind, *Energy Policy* 39 (2011) 3732–3740.

[54] EnWG, Energiewirtschaftsgesetz vom 7. Juli 2005 (BGBl. I S. 1970, 3621), das zuletzt durch Artikel 4 des Gesetzes vom 7. März 2011 (BGBl. I S. 338) geändert worden ist [German Electricity Act from 7 July 2005, last changed on 7 March 2011], 2005.

[55] L.L. Kiesling, *Deregulation, Innovation and Market Liberalization – Electricity Regulation in a Continually Evolving Environment*, Routledge Studies in Business Organizations and Net-works, 2009.

[56] BMWi, *E-Energy – Auf dem Weg zum Internet der Energie*, German Federal Ministry of Economics and Technology, E-Energy – on the way to an internet of energy, http://www.e-energy.de/documents/BMWI_Brosch_E_Energy_d_16_6_web.pdf, 2010 (accessed 01.04.11).

[57] C. Klessmann, Ch. Nabe, K. Burges, Pros and cons of exposing renewables to electricity market risks: A comparison of the market integration approaches in Germany, Spain, and the UK, *Energy Policy* 36 (10) (2008) 3646–3661.

Succeeding in the Smart Grid Space by Listening to Customers and Stakeholders

William Prindle and Michael Koszalka

INTRODUCTION

This chapter reviews the field deployment experience of programs that seek to deploy smart grid (SG) technologies to support the peak reduction goals of demand response (DR) and the energy savings goals of energy efficiency (EE). It seeks to distill lessons learned, both in program design/implementation and in the process of gaining regulatory approval, and in so doing to provide smart grid stakeholders the benefit of this experience. The chapter's objective is to help make smart grid deployments in the future more successful, in terms of both customer acceptance and regulatory approval.

Smart grid is defined by the Electric Power Research Institute as follows[1]:

The term "Smart Grid" refers to a modernization of the electricity delivery system so that it monitors, protects, and automatically optimizes the operation of its interconnected elements—from the central and distributed generator through the high-voltage transmission network and the distribution system, to industrial users and building automation systems, to energy storage installations, and to end-use consumers and their thermostats, electric vehicles, appliances, and other household devices. (EPRI, 2011)

Smart grid technologies are best viewed in a system context. In such a framework, one might view Smart Grid technologies in terms of what they enable. Smart Grid technologies are expected to enable the following kinds of actions, improvements, and related benefits:

- **Increase customer participation in energy usage.** The smart grid can provide consumers information that helps them modify how they use and purchase electricity. It can provide them choices, incentives, and disincentives in their purchasing patterns and behavior, which in turn can help drive new technologies and markets.
- **Accommodate diverse generation and storage technologies.** These power generation options range from centralized power plants to distributed energy resources (DER) such as system aggregators, grid-scale power projects like wind farms, and building-scale DER such as solar PV or combined heat and power (CHP) systems. Storage systems of various kinds would also be integrated into a mature smart grid system.
- **Enable markets for new products and services.** A smart grid can help enable markets that give consumers greater access to competitively provided energy and related services, from unregulated power purchasing to enhanced information, communication, and control features.
- **Improve power quality.** Smart grid technologies, if deployed in an integrated power grid, can improve the reliability and quality of power supply. With digital technologies increasingly ubiquitous, uninterrupted power supply with consistent voltage, frequency, and related characteristics is increasingly important to individual homes and business operations as well as the productivity of the economy as a whole.
- **Improve utility system asset utilization and operating efficiency.** A smart grid helps manage customer loads and system assets in a more coordinated fashion, such that the system can provide more useful energy services from its total asset base. It also reduces system inefficiencies and operating costs.
- **Minimize outages and system disruptions.** A smart grid can be self-healing to a greater extent than current power grid technologies permit.

---

[1] Hauser & Crandall offer a more extensive discussion of smart grid definitions, goals, and desirable attributes.

It identifies and reacts to system disturbances, using largely automated mitigation methods that enable problems to be isolated, analyzed, and restored with little human interaction. It can use predictive analysis to detect existing and future problems and initiate corrective actions.

- **Improve system security and resilience.** Smart grid designs can resist both physical and cyber attacks. Sensing, surveillance, switching, and intelligent detection, analysis and control software can be built into grid operations to detect and respond to threats. This can make grid systems more resilient, with self-healing technologies that can respond faster and with less impact to human-made and natural incidents.

Another way of looking at these SG benefits from the utility perspective is to list the potential value propositions to the utility, including enabling a host of functionalities, features, and services that are not currently feasible:

- Dispatchable load
- Generation investment deferral
- Reduced need to raise capital
- Increased diversity of the supply portfolio
- Reduction in net power costs
- Non-spinning reserves
- Market price mitigation (wholesale)
- Customer satisfaction improvement
- Cold load pickup during power restoration (dispatch event or power outage)
- Voltage response
- Frequency response
- Reduction in the risk to system reliability
- Load shifting
- Transmission investment deferral
- Distribution investment deferral
- Reduction in emissions
- Reduction in line-losses
- Shorter power outages from faults or other causes

Because several other chapters give more detailed treatment of these issues, we only include this list here.

This chapter focuses on how utilities and other stakeholders can learn from early smart grid deployment experience, fill the gaps in knowledge and program features needed to demonstrate greater net benefit to customers, and advance to the next generation of "smart" smart grid customer offerings. The definitions and goals outlined above embrace a wide range of utility system technologies, from digital switching in transmission systems to digital metering in customer premises, with distribution automation and other technologies in between. In the simplest terms, SG technologies are electronic, capable of two-way operation, capable of some automated functions, and may have some IT-based intelligence capabilities for sensing, reporting, and controlling electricity system components. For the most part, SG technologies are viewed as utility-owned assets, and the only SG technologies that directly touch customers are digital metering devices.

However, it cannot be said that digital meters "touch" customers unless they are used to provide additional information, sensing, or control features. The establishment of a two-way communication system via SG technologies provides the opening for the marketplace to provide new technologies and devices that can aid consumers' management of their energy use and cost. Other chapters, including Chapter 16 go beyond the meter, and Chapter 15 discusses the "set and forget" technology options. We therefore do not provide much detail on the technology content of customer offerings, except to the extent it helps demonstrate customer benefits or the ability of a given program strategy to gain market acceptance and regulatory approval.

This chapter contains the following sections:

- What's the Difference Between DR and EE with Respect to Smart Grid Technology?: Differentiating energy efficiency and demand response with respect to smart grid technology
- How are Customer Benefits Typically Identified and Valued from Smart Grid, DR, and EE?: How customer benefits are defined in smart grid customer offerings
- Regulatory Review Experience with Smart Grid Deployment Proposals: Recent regulatory experience with smart grid deployment
- Utility and Customer Implementation Experience with Smart Grid and Related Deployment Programs: Recent implementation experience with smart grid deployment
- Filling the Gaps: What Smart Grid Designers should Focus on to Better Document Customer Benefits: Filling the gaps in smart grid strategy and program design
- Conclusions

WHAT'S THE DIFFERENCE BETWEEN DR AND EE WITH RESPECT TO SMART GRID TECHNOLOGY?

Demand-side management, or DSM, is the collective term used to describe non-generation demand side (or distributed) resources. Energy efficiency and demand response are subsets of these DSM resources and have their own delineations as well. From a utility perspective, DR and EE are very different resources. Defining precisely the difference between demand-side and supply-side resources is challenging as there are many variations in what constitutes demand response.

> "All KWh are equal, but some KWh are more equal than others"

The National Action Plan for Energy Efficiency's report on the coordination of energy efficiency and demand response contrasts EE and DR in this way:

Energy efficiency refers to using less energy to provide the same or improved level of service to the energy consumer in an economically efficient way; it includes using less energy at any time, including during peak periods. In contrast, demand response entails customers changing their normal consumption patterns in response to changes in the price of energy over time or to incentive payments designed to induce lower electricity use when prices are high or system reliability is in jeopardy [1].

Figure 14.1 is an effort to define the various types of demand response resources. It can be found in many similar forms in the literature. DR is divided into two major types: those that are dispatchable and can be implemented in the ancillary services market, and those that are not dispatchable. Dispatchable DR resources have a known or very precisely predicted result when dispatched as would a generation resource. SG technologies can facilitate greater adoption of these various DR "products" as supply-side operators would refer to them.

HOW ARE CUSTOMER BENEFITS TYPICALLY IDENTIFIED AND VALUED FROM SMART GRID, DR, AND EE?

A recent EPRI report (EPRI 2011) provides an electric industry perspective on valuing the benefits and costs of smart grid technology. It covers the full range of smart grid technologies, including transmission, distribution, and customer-level technologies. This analysis estimates a total cost of $338–476 billion, and total benefits of $1.3–2 trillion, yielding a benefit-cost ratio in the range of 2.8–6.0.

Direct benefits to customers, however, are harder to pinpoint, coming from several different streams, which can be difficult to compare in the same time and value framework. For example, benefits related to enhanced energy efficiency are estimated as shown in Figure 14.2 in the range of $.42–1.76 billion. These benefits are projected for a single year—2030. However, the other major stream of benefits that can be linked closely to customers—reduction in customer service costs such as meter reading, service connection and disconnection, and billing and call center operations—are estimated over 20 years, in Figure 14.3 as $97.8 billion. EPRI also estimates avoided generation benefits of $192 billion to $242 billion in the 2010–2030 period.

Customer-related costs are estimated by EPRI to fall in the range of $24 billion to $46 billion. However, the cost recovery period for these costs depends on their regulatory treatment; utility commissions have to decide how quickly these costs can be recovered, and how the asset value is treated. The customer cost impact of treating AMI costs as any other rate-based asset through a general rate case could result in a longer-term cost recovery period. Special cost recovery riders, as proposed in several AMI initiatives, may have

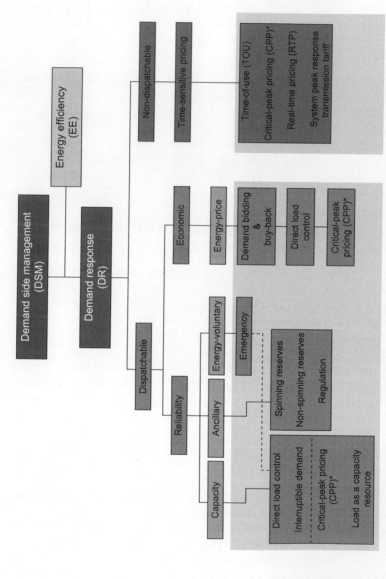

FIGURE 14.1 The relationship between DR and EE in a DSM portfolio. (*Source: NERC* [2])

*NOTE: Dependent on the ISO/RTO Critical Peak Pricing (CPP) may be accepted as Dispatchable Load. It is therefore shown as both dispatchable and non-dispatchable on this graphical representation.

Value of expanded energy efficiency

Type	Billion kWh (2030)		Value $ (@ 7 c/kWh)		Co2 Reduction Mill. Metric Ton		Value @ $50/Ton	
	Low	High	Low	High	Low	High	Low	High
Continuous commissioning	2	9	140M	630M	1	5	50M	250M
Energy efficiency benefits from demand response	0	4	0	280M	0	2	0	100M
Feedback	40	121	280M	847M	22	68	1100M	3400M
Total	42	134	420M	1757M	23	75	1150M	3750M

FIGURE 14.2 Value of smart grid energy efficiency benefits. *(Source: EPRI 2011)*

	$	$/Meter	Potential benefit	Estimated benefit
Meter services	3,909,000,000	801.8642	1.13962E+11	91,169,817,193
Billing operations	187,000,000	38.35984	5,451,763,819	4,361,411,055
Call center	96,000,000	19.69275	2,798,766,453	2,239,013,162

FIGURE 14.3 Advanced metering utility cost reduction benefits. *(Source: EPRI 2011)*

shorter cost recovery periods related to the shorter expected lives of the equipment. These and other regulatory treatment issues, including shareholder returns on rate-based assets, will strongly affect the costs that customers experience in a given year.

It can thus be difficult for regulators to fully capture and align benefits and costs of smart grid investments in a single regulatory proceeding.[2] Costs are known quantities, so customers and their intervenors can see those in filed documents and will experience them soon after the proposal is approved. Benefits, however, typically flow unevenly, both across time and across different customers and customer classes. Deferred generation benefits, as well as some energy efficiency and DR benefits, may appear years later and are not guaranteed to occur, while costs are seen by customers as certainties that are incurred now. So depending on how specific streams of benefits and costs are treated in a given regulatory proceeding,

---

[2]See Chapter 4 for a fuller discussion of equity and other issues revolving around allocation of costs and benefits.

regulators may not find utility smart grid proposals compelling in terms of direct, near-term benefits to customers. These issues are explored further in the section below on regulatory treatment of smart grid deployment proposals.

REGULATORY REVIEW EXPERIENCE WITH SMART GRID DEPLOYMENT PROPOSALS

Smart grid technologies present a set of new challenges to regulators in making decisions regarding smart grid investments. U.S. electricity systems have largely operated at high reliability levels at the transmission level, though distribution system reliability varies. But as long as "the lights stay on" and rates remain relatively low, it can be difficult for policymakers to understand how smart grid technologies can improve the value of electricity systems. It can also be challenging to understand the incremental value of smart grid investments, as they are often implemented in stages, with each stage requiring regulatory approval. But many smart grid benefits come from the combined and cumulative effects of a portfolio of smart grid technologies over time. The Illinois Commerce Commission summarized this conundrum recently (ISSGC, 2010):

The issue of smart grid cost recovery has been a matter of controversy and litigation for several years. Disagreements exist about whether recovery of a utility's smart grid costs should be restricted to the "traditional" rate-base method, or whether a "non-traditional" method (e.g., "rider" recovery) should be used. Some stakeholders are concerned that utility proposals for cost recovery of smart grid investments would lead to significantly higher monthly bills and a shift in the risk of investment from utilities to ratepayers. Others believe that non-traditional cost recovery would be essential to accelerate deployment of smart grid technologies.

Despite these challenges, smart grid technology deployment is proceeding apace. The FERC's 2010 report on AMI and Demand Response (FERC 2011) shows that AMI has been deployed to 8.7% of total U.S. electric meters as of 2010, up from 0.7% in 2006 and 4.7% in 2008. The FERC report, however, also shows substantial unevenness in the penetration of AMI technology:

- Investor-owned utilities (IOUs) lagged the national average, showing a total of 6.6% AMI penetration, compared to penetration rates above 20% for rural electric coops and public power districts.
- States showed a wide disparity of AMI deployment: only five states (AZ, OR, ID, WI, PA) exceeded 20% AMI penetration, while the majority of states (29) showed AMI penetration at less than 5%.

The fact that IOU AMI penetration lags the national average is a concern, because IOUs serve some 75% of total customers. And the unevenness of state action indicates widespread resistance to universal AMI deployment, even as several states move ahead rapidly. These data suggest that unless some of the issues restraining AMI deployment are resolved, the rapid growth of AMI in the last 5 years may slow or even plateau. In fact, the FERC data indicate a slowing in the growth rate in AMI deployment in the 2008–2010 period compared to the 2006–2008 period, from a nearly ninefold increase to just under a doubling. While the more than $4 billion in federal stimulus funding is estimated to support the deployment of some 18 million additional advanced meters, even the availability of federal grant funds has not fully overcome state-level concerns about the net benefits of AMI (see Maryland case below).

Three recent smart grid/AMI deployment case studies serve to highlight the regulatory, perceptual, cost, and process issues that can stymie otherwise promising projects:

Maryland. Baltimore Gas & Electric (BGE's) 2010 proposed AMI deployment plan was initially rejected by the Public Service Commission (PSC); after responding to PSC and stakeholder concerns, a modified proposal was ultimately approved [3]. The original proposal cost $835 million for deployment of over 2 million meters, with an immediate cost recovery surcharge mechanism, and mandatory TOU pricing for all customer classes. Part of the urgency for this proposal was a federal stimulus grant BGE had been awarded, which would have covered some $136 billion of AMI costs; the Department of Energy had asked for an August 2010 approval from the PSC as a condition for grant funding.

The issues raised by the Maryland Office of People's Counsel (OPC) summarize the concerns relating to demonstration of benefits to customers [4]:

- About 20% of the benefits in the original proposal were based on BGE operational savings in meter reading and operations, and in distribution management costs.
- The other 80% of benefits were projected in terms of capacity revenues from wholesale market forward-capacity markets, reduced wholesale prices that would reduce future standard offer rates, reduced energy bills as an effect of dynamic pricing and customer feedback, and reduction in T&D infrastructure costs.
- OPC pointed out that these 80% of projected benefits were subject to considerable uncertainty, and they depended heavily on customer response to price signals and on future wholesale energy and capacity prices. In 2010, wholesale market prices in the PJM power market had shown signs of softening due to the recession; this shift tended to support the OPC argument that such factors are uncertain
- OPC strongly objected to the cost recovery surcharge element of the proposal, asserting that it places all of the risk on ratepayers, with no comparable guarantee of benefits.
- OPC rejected the mandatory TOU pricing element of the proposal, stating that it imposes higher bills on some customers, and that BGE did not include in its proposal any customer-enabling technologies that would help them respond to TOU prices, such as in-home displays or automated appliance control devices.

- OPC also objected to the proposal on the grounds that it would remove consumer protections by enabling remote service disconnects, and also raised customer privacy and security concerns that have not yet been fully addressed by national efforts.

The PSC largely agreed with OPC's objections in its initial order, stating:

The Proposal asks BGE's ratepayers to take significant financial and technological risks and adapt to categorical changes in rate design, all in exchange for savings that are largely indirect, highly contingent and a long way off. We are not persuaded that this bargain is cost-effective or serves the public interest, at least in its current form…
The Proposal is a "no-lose proposition" for the Company and its investors. (MD PSC Order No. 83410. June 21, 2010, pp. 1, 3)

In response to the PSC's initial decision, BGE revised its proposal to remove the automatic cost recovery surcharge, treating the project's costs as a regulatory asset to be recovered through subsequent proceedings, much like any rate case. It also removed the mandatory TOU pricing element, and included an expanded customer education and outreach element. The PSC approved the revised proposal later in 2010. For comparative purposes, the PSC approved Pepco's AMI proposal, which also included a federal grant cost offset, in large part because it did not include a special cost recovery mechanism or mandate TOU pricing. This tends to support the inference that guaranteed cost recovery without risk or cost sharing between customers and shareholders, and the imposition of mandatory time-based pricing, can be significant obstacles to AMI deployment.

OPC also raised the issue that alternative ways to meet utility peak load reduction and system operations goals can be less expensive and less burdensome on customers. While this issue did not become a central point in the BGE case, it has been raised in other states, and can be a critical issue in some situations. For example, in New Jersey the Department of Public Advocate commissioned a study [5] on utility AMI deployment, which focused largely on alternative approaches to meeting the stated goals of the utility AMI deployment proposals. Key points raised in this report include:

- Automated metering reading (AMR) is a less expensive way to obtain a large fraction of the benefits of AMI. Though limited in functionality (may not have hourly reading capability or two-way communication potential), AMR has been deployed by several utilities successfully as based on avoided meter reading and operations costs. If AMR is in place, the benefits of AMI deployment become much harder to justify, as indicated in the Brattle Group's analysis for the Institute for Electric Efficiency (IEE, 2011, p. 13).
- Direct load control (DLC) and related utility load management programs can serve to achieve the peak load reduction benefits of AMI deployment. Utilities have been conducting DLC programs cost effectively for many years.

- The Synapse report further analyzed the dependence of forecast AMI benefits on three key factors: the average load reduction per customer, percentage of customers participating in dynamic pricing, and the long-term persistence of savings. Synapse pointed out that utilities' estimates for customer participant in dynamic pricing were more than double historic participation rates for DLC programs around the country.

Illinois. Commonwealth Edison planned one of the most thoughtfully designed smart grid pilot projects yet conceived, testing a variety of rate designs, information services, and other features in a large sample of some 130,000 customers. Com Ed filed its application with the Illinois Commerce Commission (ICC) for a $360 million pilot program. In late 2008, the ICC approved $274 million in AMI cost recovery charges for the pilot through a special rate rider. Because the ICC had disallowed some of its costs from the 2007 application, the utility initiated an appeal with the Illinois Appellate Court, mainly on matters related to labor costs. This triggered a number of interventions in the appellate court case, including the Illinois attorney general and Citizens Utility Board[3] (CUB), objecting to the ICC's approval of the special cost recovery rider. As was the case in Maryland, consumer representatives objected to the accelerated, automatic cost recovery features of Com Ed's cost recovery mechanism. The appellate court case turned into an expanded version of a full-blown ICC adjudicatory proceeding. In 2010, the appellate court struck down the ICC's 2008 decision, leaving Com Ed's pilot and its entire smart grid initiative in regulatory limbo [6]. In its 2011 session, the Illinois legislature sought to resolve some of these issues through legislation. As of May, however, the governor was threatening to veto the bill, and the situation remained unresolved at this writing.

While this case may reflect the nature of Illinois politics more than the merits of smart grid technology, it does reinforce the core point of the Maryland PSC's treatment of BGE's original AMI application, in which the utility's original special rate rider/surcharge mechanism was rejected as forcing ratepayers to take all of the risk and front all of the costs for smart grid deployment. The lesson that may lie at the core of both of these case studies is that for regulated utilities, especially in states with a history of vigorous consumer representation, smart grid proposals are likely to be more successful if they defer cost recovery to future rate cases or use other methods that give stakeholders more opportunity to advance their views.

Colorado. The Boulder SmartGridCity™ project has been widely heralded as an innovative smart grid experiment. Working with the city of Boulder, one of the "greenest" municipalities in the United States in its commitment to clean energy and environmental protection, Xcel worked with its business partners, including Accenture, Gridpoint, Landis and Gyr, and Current Group, and other stakeholders to design a state-of-the-art program, using fiber-optic technology as part of its AMI deployment. Xcel conducted extensive customer

---

[3]See Chapter 16.

research via surveys and focus groups to help define the kinds of offerings customers would prefer. Fifteen thousand Landis and Gyr advanced meters were ordered for residential installation beginning in 2008; the vision was ultimately to include some 50,000 customers in the program [7].

SmartGridCity™ was designed as an inclusive vision: it was to include new renewable energy sources connected to the grid, such that customers could choose low-emission power pricing plans. It allowed for electric vehicle recharging, and included utility-side upgrades for substations, feeders, and distribution automation hardware and software. In the original plan, customers were to be offered:

- Advanced meters that communicate with home appliances for energy and cost savings
- Energy efficiency incentives integrated with the program
- Online tools for tracking energy use and making changes to fit their lifestyle and efficiency goal
- The ability to automatically manage energy use based on real-time price signals or green-power price signals
- Charging options for plug-in hybrid electric vehicles
- Automatic outage notification
- Automated load management options to reduce peak load and prevent outages
- The ability to choose on-site renewable energy options like wind, solar, and batteries
- Web-based tools to support the above features, including:
  - A control portal: to set use preferences for major appliances and household outlets.
  - An analysis portal: for secure access to usage information and analysis options.
  - An information portal: to educate customers about reducing costs and lowering their carbon footprint.

Customer choice was part of the program design, such that customers would be able to (1) select a plan that works for them, (2) specify their personal plan parameters, and (3) monitor and control their energy use.

Xcel did not initially propose to recover costs from its ratepayers using special rate riders or other mechanisms that resulted in regulatory controversy in Illinois or Maryland. The project was viewed primarily as a research and development initiative, and the company did not feel compelled to require official approval or cost recovery from the Colorado Public Utility Commission (PUC). However, the project ran into cost overrun problems: Xcel originally anticipated that capital costs would be around $15.3 million in early 2008, but in May 2009, it revised the number to $27.9 million, and in 2010 to $42.1 million. PUC staff estimate that total costs will exceed $100 million. As the costs mounted, the PUC ruled that Xcel would be required to apply for a Certificate of Public Convenience and Necessity (CPCN), which will give the PUC authority to regulate the project, including its cost recovery. As that proceeding has unfolded, the City of Boulder has asked to be removed from the proceeding, partly to protect itself from exposure to cost recovery [8].

The PUC in December 2010 approved an Xcel rate increase, primarily to pay for a new unrelated coal-fired facility. But $11 million of this increase is to cover Boulder smart grid costs. How the remainder of costs are to be recovered remains to be soon; possibilities include the City of Boulder and its customers, all Xcel customers, Xcel's business partners, and Xcel shareholders. PUC staff were quoted in 2010 as saying "… it is difficult to ascertain what portion ratepayers should bear, if any."

The program is in operation, with meters installed and many of the planned services available. The company continues to use it as an R&D project, However, the cost issue has placed a shadow over the project. A significant part of the cost overruns appears to stem from unexpected costs for fiber-optic line installation. Xcel's PUC filing in May said that more underground fiber had to be laid than expected, and that installation costs rose due to "having to drill through granite with diamond-tipped drill bits and remove large boulders with cranes and dump trucks…"

SmartGridCity™ turned into a major infrastructure construction effort, using technologies with which there was little local experience. This may in hindsight have been a weakness in the program design. Difficulty in predicting and controlling costs always creates risks in a new project. And yet the project was deliberately intended to test new technologies, so there may have been no other way to find out what the challenges and the costs would be. The difficulty for Xcel was that costs rose high enough to trigger regulatory review. In hindsight, the company might have conducted the project through an unregulated organization structure. Perhaps the principal takeaway is that risks and costs for smart grid technology, as for any new technology, can be uncertain and substantial, and that risk management should be part of the planning process.

Another takeaway from the Boulder project, again benefiting from hindsight, is that many of the customer benefits could have been tested and demonstrated prior to installation of the core infrastructure. Enhanced information and feedback, customer engagement and goal-setting, and even customer device control can be piloted without AMI deployment. Running a series of smaller pilots and field tests for such customer-side elements prior to full construction mode for the core infrastructure might have created an experience base that could have built confidence in the benefits of AMI-based customer offerings. Such a base might help weather later challenges in the technology deployment process.

These three cases, in which AMI deployment has met significant regulatory complications, were not selected to cast a negative light on smart grid technology or its prospects. They were chosen rather to illustrate the challenges that smart grid deployment faces in traditional regulatory structures. Like it or not, in many states these kinds of challenges are likely to emerge, and utilities and other smart grid proponents can learn from these experiences. Key lessons that can be drawn from these cases include:

- **Share risks and costs.** The MD PSC made it clear that the initial BGE submission put too much of the cost and risk burden on customers, and the IL court decision seems to reinforce that point.

- **Demonstrate benefits before seeking cost recovery.** The MD PSC ultimately accepted BGE's revised proposal when it sought to recover costs after deployment. Even though BGE had run an AMI pilot and showed some benefits from that, the PSC did not see a strong enough case for systemwide ratepayer cost recovery prior to deployment.
- **Don't mandate time-based pricing for smaller customer classes.** This may not be an issue in every jurisdiction, but in many states consumer advocates have made strong cases against class-wide pricing plans that involve critical-peak pricing, time-of-use pricing, or other rate structures that tie prices to time of use. This implies that such pricing plans will have to be marketed actively to gain the high market penetration rates needed to fully realize the benefits of smart grid technology, but if this is the path to initial regulatory approval, then that may be the reality smart grid deployment faces in many states. This approach may actually be helpful in the long run, as it could force utilities to learn more about customers, market segments, customer preferences, and the specific bundles of technologies and services that will catch on in various segments.

UTILITY AND CUSTOMER IMPLEMENTATION EXPERIENCE WITH SMART GRID AND RELATED DEPLOYMENT PROGRAMS

Deployment of smart grid technology, especially in the form of advanced metering, is beginning to be seen on a wide scale, as the FERC report cited above indicates. We have reviewed a selection of the ample literature of utility and customer experience with smart grid and related customer offerings from which we seek to distill a few key lessons learned, and to glean insights for the future of smart-grid-based offerings.

Pilot programs and other initial field experience with smart-grid-related customer offerings tend to show that customers reduce energy use as well as peak demand. Most of this experience comes from dynamic-pricing-based offerings, though some also include customer feedback and control options. For example:

- King and Delurey's [9] meta-review found that dynamic pricing produced average savings of 4% of annual energy use. However, reliability-oriented demand response programs reduced energy consumption by only 0.2%, largely because event-driven programs are activated well under 100 hours per year. Information/feedback programs showed stronger savings effects, up to 11%; these typically give customers usage information via the internet or in-home displays but do not involve direct load control. They also found that savings varied greatly, from −5% to 20%, which serves to warn against placing predictive value on these results.

- ACEEE's [10] meta-review of 36 customer feedback programs in North America, Europe, and Asia found that energy savings can range from 3.8% to 12%. Figure 14.4 illustrates the results for the five categories of programs analyzed. The first two categories, enhanced billing and estimated feedback, do not require advanced metering, and so are not dependent on smart grid technology. The three highest categories of savings, however, come from programs that require AMI as a minimum; the highest savings category involves real-time feedback down to the device level. As with King and Delurey, however, the sample size is not large enough to be used for predictive purposes.

- The National Action Plan for Energy Efficiency [1] report on utility rate design included a review of the energy savings impacts of several recent dynamic pricing programs. Those results are summarized in Table 14.1; it shows annual energy savings as high as 7.6%, occurring in the Ontario Hydro One pilot, which combined AMI with in-home displays (IHDs) in half the participating homes. Savings in homes without the IHDs averaged 3.3%, less than half the IHD group. The literature search for this report also reflected a consistent theme in smart-grid-related field programs— relatively few of them measure annual energy savings data. This is understandable to the extent that the purpose of the program was peak demand reduction, not annual energy savings. But given the challenges facing smart grid deployment in terms of documenting customer benefits, it is vital that more data be collected on the annual energy savings impacts of smart-grid-based customer offerings.

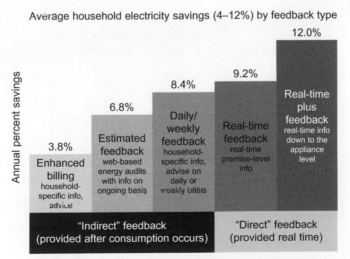

FIGURE 14.4   Energy savings from customer feedback programs. *(Source: ACEEE [10])*

TABLE 14.1 Energy and Demand Savings from Dynamic Pricing Programs

Program	Rate/Price Type	Location	Customer Type/Load Size	Participants	Customer Incentive	Duration	Peak Demand Reductions	Energy Savings
California Statewide Pricing Pilot	CPP	Southern California Edison Service Area	Commercial/ industrial <20 kW	59 in 2004: 57 in 2005; about 33% accepted thermostats	Free installation of smart thermostat that automatically adjusts air conditioning setting in CPP periods	4 months × 2 years: June–October 2004 and 2005	<20 kW: Peak-period energy use fell 4.83%; with thermostats, savings rose to 13%	Savings calculated for peak hours only, not monthly or annual
			Commercial/ industrial 20–200 kW	83 in 2004: 76 in 2005; about 60% accepted thermostats			20–200 kW: Peak-period energy use fell 6.75%; with thermostats, savings rose to 9.57%	
Gulf Power Company— Energy Select	Price-responsive load management with CPP	Gulf Power Company service territory— northwest Florida	Residential	8,500	None—customers pay $4.95/month to participate in the program for the opportunity to save on their electric bill by purchasing electricity at prices lower than the standard rate 87% of the time	March 2000 to present	Summer peak reduction of 1.73 kW/home or 14.7 MW to date

Winter peak reduction of 3 kW/home or 25.5 MW to date | Savings calculated for peak hours only, not monthly or annual |

TABLE 14.1 Energy and Demand Savings from Dynamic Pricing Programs—cont'd

Program	Rate/Price Type	Location	Customer Type/Load Size	Participants	Customer Incentive	Duration	Peak Demand Reductions	Energy Savings
Ontario Energy Board/Hydro One	Regulated price plan TOU rates	Hydro One service area	Residential, farm, small business under 50 kW	500	Real-time in-home display monitors for half the participants	5 months: May– September 2007	Peak load reductions averaged 3.7% With display, impact averaged 5.5%	Annual energy savings averaged 3.3%; with displays, savings averaged 7.6%
Ontario Energy Board—Smart Price Pilot	Regulated price plan TOU; TOU with CPP; TOU critical-peak rebate	Hydro Ottawa's service territory	Residential TOU scheduled to have smart meters installed prior to the start of the pilot	373 participants total: 125 in a critical-peak rebate price group, 124 each in TOU-only and CPP groups	CPP participants: off-peak rate cut to 3.1 cents per kWh to offset critical-peak price TOU with rebate participants: refund of 30 cents per kWh below baseline usage + $75 at end of pilot	7 months: August 2006– February 2007	Peak-load reductions were 5.7% for TOU- only participants, 25.4% for CPP participants	6.0% average annual conservation effect across all customers

(Continued)

TABLE 14.1 Energy and Demand Savings from Dynamic Pricing Programs—cont'd

Program	Rate/Price Type	Location	Customer Type/Load Size	Participants	Customer Incentive	Duration	Peak Demand Reductions	Energy Savings
Community Energy Cooperative—Energy Smart Pricing Plan	Hourly pricing pilot program; air-conditioning cycling added as an option	Chicago	Residential	750 in 2003, rising to 1,100 in 2006	Cooperative provided outreach, education, information materials, high price alerts	2003–2006	Peak reductions up to 25% in first hour; greatest reductions through air-conditioning cycling Peak reductions declined after first hour and over successive high-price days	Summer-month energy usage reduced 3–4%; no annual net usage impact reported

Sources: California Statewide Pilot: George et al. (2006); Gulf Power Company: comments from Ervan Hancock III, Georgia Power Company; Ontario Energy Board: Hydro One (2006); Community Energy Cooperative: Summit Blue Consulting (2004); and NAPEE [11].

These studies make it clear that smart-grid-based programs can save energy as well as reduce peak demand, suggesting that there is no inherent conflict between energy savings and peak reduction goals. What is more relevant to the current debate on smart grid technologies, in our view, is examining the factors that support success in gaining customer acceptance and regulatory approval. Toward that end, we have selected for a more detailed look three smart grid deployment projects, chosen for the key lessons they offer on the positive benefits that smart grid technologies and customer offerings can offer. They are the Grounded Power/Cape Light Compact pilot program in Massachusetts, the DC PowerCents pilot in Washington, DC, and PG&E's Monitoring-Based Commissioning (MBCx) program in California.

Massachusetts. Grounded Power, a startup smart grid company, designed and managed this pilot project for the Cape Light Compact, group of publicly owned distribution utilities in southeastern Massachusetts. The Residential Smart Energy Monitoring Pilot program was launched in 2009 to evaluate potential energy savings from in-home display online energy monitoring systems, and to gain insight on behavioral aspects of energy use [12]. It recruited 100 households in the Cape Cod and Martha's Vineyard areas. Participants had in-home monitoring systems installed for a year, received information and training on them, and had access to an internet-based dashboard, which provided feedback via real-time viewing of current energy use and demand; showed customer savings in kWh, dollars, and CO_2 emissions; and contained learning opportunities for energy-saving behaviors (e.g., unplugging chargers when not in use).

Perhaps most interestingly, the Cape Light pilot did not rely on AMI technology. It installed monitoring devices on the main electrical panel in participants' homes, and then connected the IHD devices to the monitors. The pilot focused primarily on customer feedback and behavior, which are the core of the demonstration of smart grid benefits, without specifically using smart grid technology. This suggests that smart grid pilots should include such tests of customer feedback and behavior early in the smart grid process, to document the benefits and perhaps help build the case for smart grid technology deployment with its costs.

The Cape Light program also engaged customers up front, inviting them to set energy savings goals of their own choosing, and then supporting their chosen goal with specific suggestions for the best mix of hardware measures and behaviors to accomplish the goal. The online features of the program also included the ability to compare participants' experience with others, and even allowed participants to communicate, compare notes, etc. An apparent effect of this approach was to give customers a sense of ownership of the program. Rather than the utility forcing AMI, time-based pricing, and perhaps other services in a one-way fashion, this program was perceived more as customer-driven. This feature was also true of the Boulder Xcel program, but was largely lost in the controversy over the costs of infrastructure development.

By focusing only on customer feedback and behavior, and by giving customers wide control over how they used the program, Cape Light may have paved the way for greater customer acceptance over time, and may ease the way for customers' willingness to absorb future costs of AMI infrastructure. It is also important to point out, however, that Cape Light is a publicly owned utility, and thus has less regulatory oversight than an IOU. Yet the costs of the pilot appear to have been modest, so many IOUs could likely have justified such costs, although no cost data or cost-effectiveness analysis was provided in the evaluation. This lack of cost/cost-effectiveness data may prove to be a shortcoming if this project is used to justify a broader smart grid deployment program.

The Cape Light program was a success by the measures used in the evaluation: participants saved an average of 9.3% of annual electricity costs, corrected for weather and other factors. Recruiting efforts found over 300 willing participants through very limited local media advertising; only 100 were needed. Customers were very satisfied overall, with over 90% interested in continuing participation, and with a reported willingness to pay an average $8/month for future participation. While customer bill savings were not provided, data indicate that monthly savings significantly exceeded $8.

Washington, DC. The PowerCents DC™ program was not overly distinctive in its core content: it tested three dynamic pricing plans in a sample of 900 customers for a year between 2008 and 2009 [13]. These pricing approaches—critical-peak pricing (CPP), peak-time rebate (CPR) and hourly pricing (HP)—have been tested elsewhere. What is most interesting about the DC pilot is its organizational structure: PowerCentsDC™ is structured as a nonprofit organization, with major stakeholders actively engaged, including the incumbent utility Pepco, the Public Service Commission, the Office of Peoples Counsel, the Consumer Utility Board, and the International Brotherhood of Electrical Workers. The pilot also focused intensely on limited-income customers, as this population is often described as not likely to gain the benefits of smart grid technologies.

This combination of utility, regulator, consumer advocate, and labor interests is a notable departure from the more typical proceedings involved in smart grid deployments in regulated IOU service areas. As noted in the Maryland case above, the dynamics of smart grid deployment cases can be much more negative. In this situation, the major stakeholders were not only engaged, but in some cases actively championed the program. One PSC commissioner has presented the program to national audiences such as NARUC (National Association of Regulatory Utility Commissioners) and the U.S. Department of Energy.

PowerCentsDC™ produced positive results in terms of peak load reduction—as high as a 33% average reduction. Customers supplied with smart thermostats reduced peak load even more, as much as 49%, largely because they were able to curb air conditioning usage during peak periods. Customers showed a high degree of satisfaction with the program, even though monetary bill impacts were typically small, as the program was designed to be revenue-neutral. Limited-income customers signed up at higher rates of participation than customers overall, and showed comparable levels of load reduction.

This pilot suggests two key lessons. First, engage key stakeholders and make them part of the deployment planning and implementation process. While PowerCents was a pilot, and no final PSC decision has been reached, it is likely that key parties will be less negative in upcoming proceedings. Second, provide customer-enabling technologies: the smart thermostats not only increased average peak savings but also served to address a common consumer advocate complaint, which is that AMI deployments often come without giving consumers a way to respond. What is not stated in the pilot report, but which may be another key lesson, is that no customer cost recovery was proposed prior to or during the pilot. This supports the notion expressed elsewhere in this chapter that it is preferable to demonstrate customer benefits before asking for cost recovery.

California. This case study focuses not on the residential customer class, but on large commercial customers, as targeted by PG&E's cutting-edge Monitoring-Based Commissioning (MBCx) program. MBCx takes advantage of advanced metering, combined with modern commercial building automation systems, to conduct advanced diagnosis and performance improvement of building energy system operations. PG&E has contracted with third-party energy-efficiency firms to implement the MBCx.

Monitoring-based commissioning refers to the combination of electronic data collection, sophisticated data filtering, and analysis with a comprehensive facility audit. The MBCx Program helps a facility quantify the energy savings from such measures as:

- Detecting faults
- Optimizing schedules
- Adjusting temperature set-points
- Reducing overcooling
- Participating in demand response programs

The MBCx program was designed for large commercial customers with greater than 100,000 square feet of conditioned space or an electric demand of greater than 500 kW. To participate effectively customers must have a BacNET/IP-enabled building management system (BMS), BMS (building management system) metering for utility service point of entry, BMS control and monitoring points for environmental comfort and operating systems, and sufficient IT capacity to support export of energy usage data to third-party firms. Typical program services include:

- Integration of BMS to third-party analysis application
- Comprehensive facility audit and report

- 12 monthly reports that track performance and contain EE
- EE measurement implementation assistance
- Online access to detailed, real-time facility energy usage.

Customers can receive incentive payments for up to 50% of project costs at $0.09/kWh, $1.00/therm, and $100/peak kW.

What is interesting for the purposes of this chapter is that this program reads like a smart grid designer's dream: AMI-based, sophisticated interaction with customers' building automation systems, which allows diagnosis, efficiency improvement, and demand response down to the device level. And yet there was virtually no controversy at the regulatory level on the approval or rollout of this program. It is also interesting that one of the implementing firms, EnerNOC, is better known as a curtailment service provider (CSP), whose business model typically focuses more on wholesale-market demand response programs. But because of the sophistication of its data analytics and other capabilities, EnerNOC is able to seamlessly provide a range of energy efficiency and demand response assistance within a single program design. It is also important to note that PG&E laid a critical foundation for MBCx and other smart grid applications through its early work on automated electronic data exchange. As a leader in the ENERGY STAR Buildings program, PG&E was one of the first utilities to implement the Automated Benchmarking System (ABS) software solution, which is connected to the EPA Portfolio Manager software platform. By providing energy use data electronically, rapidly, and for free, ABS can enable a whole host of customer information-based program solutions.

This case points to the enormous potential for smart grid deployment, and the combined realization of energy efficiency and demand response goals in a single integrated program framework. While most of the literature focuses on the challenges and opportunities in deploying smart grid technology to a fundamentally low-tech, mass-market audience, programs like MBCx (and there are still very few like it) are quietly showing the way forward, working with larger customers who are more tech-savvy and have the digital metering and building automation systems needed to implement sophisticated approaches like this. This suggests that smart grid deployment should start with larger customers first, with mass-market deployment to residential and small commercial customer classes phased in later.

These three case studies imply a number of potential lessons for smart grid deployment:

- **Customer benefits can be demonstrated without AMI.** While this may sound threatening to utilities seeking to deploy AMI as a core part of their smart grid strategy, it also suggests that demonstrating customer benefits in feedback and behavior change, as the Cape Light pilot does, can be achieved relatively quickly and at lower cost than full AMI deployment. Done strategically, as part of a concerted set of pilots, studies, and other activities, such customer-side initiatives can help make the case for wide AMI and other smart grid deployments.

- **Customers prefer to be in charge**. The Cape Light project demonstrates the advantages of giving customers more control of the process—letting them choose their energy savings goals, the energy efficiency measures they want to take on, etc.
- **Use pilots to build the cost-effectiveness case**. A limitation of the Cape Light evaluation was that it was not designed to assess overall benefits and costs. That was not its core design, but other utilities pursuing such pilots would be well advised to try to capture and quantify energy, demand, and other benefits, and to document costs. While some smart grid deployments will only be fully cost effective at scale when all customers are involved, collecting and documenting such data in pilots are necessary parts of the process of building the case.
- **Engage stakeholders proactively**. The PowerCentsDC™ program showed how to gain support from potential opponents by engaging consumer advocates, regulators, and labor interests in the program's design and deployment. While this has not yet translated into final approval for systemwide AMI and dynamic pricing deployment, it can only help with customer acceptance as well as regulatory approval.
- **Give customers enabling technologies**. PowerCentsDC™ included smart thermostats as a customer option for supporting their response to price signals. Not only does this dramatically increase peak demand reductions, and likely energy savings (though energy savings were not measured), it also gives customers concrete tools to deal with dynamic pricing, which was one of the objections Maryland OPC raised in the BGE case. Such technologies thus provide a two-pronged benefit—increased impacts and greater regulatory approval chances.
- **Segment customers strategically**. If one were to read some of the negative press about residential AMI deployment in places like Illinois or Colorado, one could get the impression that smart grid deployment faces big challenges. But the PG&E MBCx program tells a different story, demonstrating most if not all of what smart grid designers dream of, by focusing on the customer segment that is most amenable to the technology and information solutions that smart grid enables. Large commercial and institutional customers are ripe for smart grid solutions, and in many areas comprise a large fraction of electricity load. Even within residential and small business customer classes, there are numerous ways to segment customers from program design and marketing standpoints. For example:
  - The "set and forget" strategies described in other chapters will be most appealing to a certain segment of customers, who want to save money but have little interest or skill for ongoing engagement. They might agree to a simple thermostat/CPP program that they sign up for once and hopefully never notice.

- Another segment, perhaps that targeted by the Cape Light project, would be receptive to an "continuous improvement" approach, involving more choices and more ongoing, active involvement in learning how to manage home energy use.
- It's also important to recognize that some customers may never want to engage in smart grid matters. They may be suspicious of utility motives, resistant to technology, or otherwise disinterested. It may be strategic to allow for some fraction of customers to be "left alone," other than perhaps AMI installation as part of a rollout that the majority of customers come to support.

FILLING THE GAPS: WHAT SMART GRID DESIGNERS SHOULD FOCUS ON TO BETTER DOCUMENT CUSTOMER BENEFITS

Drawing on the field experience with customer acceptance and regulatory approval with smart grid customer offerings across the United States, this section describes the data collection and the program features that should be part of smart grid customer offerings in order to better document and deliver the full range of customer benefits. Filling these gaps will be important in gaining the customer acceptance and regulatory approval that have been lagging in many service areas. These gaps are summarized as follows:

- **Measure annual energy savings.** Far too few smart grid demonstrations and pilots have been able to generate robust data on customer energy savings over annual cycles and longer. Most of these efforts have focused on demand response, dynamic pricing, and related topics. Appropriately, the focus of such programs is on short-term demand impacts, measured as kilowatts of power capacity coincident with the utility system peak. Energy savings measured over a year's period, while not the primary concern of DR and dynamic pricing, are paramount as metrics for the success of energy efficiency programs. There is thus a need for more robust data on the energy AND demand impacts of smart-grid-based customer offerings. This would help utilities, regulators, and other parties develop fuller estimates of the benefits that smart grid technologies can offer.
- **Give customers feedback they can use.** Customer feedback has been shown in many of the programs described above to be effective in producing both peak reductions and longer-term energy savings. But the information has to be designed and delivered to fit the specific needs of customer segments. For a large commercial customer participating in the PG&E MBCx program, data need to be very granular and specific to isolate efficiency and peak demand options. For a typical residential customer, an IHD combined with a user-friendly web portal and bill disaggregation software can point to energy and peak reduction options in a much simpler fashion.

- **Give customers control options they will use.** Residential programs can suffer from overkill, promising too many whiz-bang control features. Many customers will only use a one-time "set and forget" option, while others may be interested in managing home electronics and major appliances as well as heating, cooling, and hot water. Customer segmentation and program design should accommodate these differences. It may require only two to three differentiated offerings to capture these differences and gain wider customer acceptance and program impacts.

However, even if the next generation of customer offerings is developed and perfected to the point of being ready for mass market deployment, utilities and other parties will need to develop program logic models that are persuasive to regulators. This is a critical step: to date, program logic models have had to be relatively simple, with hard-and-fast logic that is easily understood. Such program models typically follow the physical-technological-economic-model (PTEM) structure, in which the program is designed to install or replace a physical device containing a specific technology, offers an economic incentive to the customer for purchasing/installing the device, on the basis that so doing is cost effective under defined economic tests, and that in so doing customers are reliably following a rational pattern of economic behavior.

However, smart grid customer offerings typically depend on customer behavioral responses to price signals and/or usage feedback. Program designers and regulators alike need to develop program models that will allow for robust measurement and evaluation that will pass regulatory review. Substantial work is underway in this regard, and smart grid advocates need to become versed in this emerging area of inquiry [14,15].

CONCLUSIONS

This chapter has sought to show that smart grid technologies can become part of customer offerings that save energy, reduce peak demand, and gain customer acceptance in the market as well as regulatory approval. It has also sought to identify the key challenges that these programs face, in terms of both program design and regulatory review. Based on the field experience we have reviewed, we suggest that a successful smart grid deployment strategy must be comprehensive and long term, allowing for stakeholder engagement, careful documentation of benefits, and a staged deployment approach. We suggest the following principles as guidance for smart grid developers in designing future programs for the best chance of success:

- **Demonstrate benefits before seeking to recover costs.** Use pilots to demonstrate AMI and other smart grid benefits, including a variety of customer offerings, such as dynamic pricing, advanced usage information and feedback, and device control options. Document results, and compile results from other smart grid programs, as part of any proposal for regulatory approval or cost recovery.

- **Segment customer markets.** Start with the large C/I customer classes, where resistance to AMI, dynamic pricing, and active customer-side load control is much lower than for residential customers, and where consumer advocate intervenors are less likely to create regulatory opposition. In some service areas, large C/I customers can provide a large share if not a majority of the potential benefits of smart grid deployment.
- **Share risks and costs.** The BGE/Maryland PSC example illustrates that in many states, proposals that ask for immediate cost recovery from ratepayers are likely to meet stiff opposition. Deferring cost recovery to after installation, and including some shareholder risk and cost sharing, can help improve the chances of regulatory approval.
- **Put the customer in charge.** Give customers choices, including setting their own savings goals, choosing their own preferred usage management strategies, and setting their own preferences for receiving information and feedback. Provide realistic enabling technologies appropriate to each customer segment to allow them to respond to price signals and usage feedback.
- **Be careful about mandatory dynamic pricing.** Many state regulators have resisted imposing mandatory critical-peak pricing or other forms of dynamic pricing, especially on residential and small business customers.
- **Consider the alternatives.** Consumer advocates have pointed out, correctly, that some of the benefits of smart grid technologies do not require AMI or other core technologies. Deployment proposals must carefully address how the proposed plan provides sufficient value to overcome such objections and, depending on the jurisdiction, may require accommodation to include such lower-tech, lower-cost options.

Applying these principles thoughtfully in the context of the specific customer markets and regulatory environment each smart grid developer faces can help America realize the potential for energy efficiency, demand response, and other benefits that smart grid technology promises.

REFERENCES

[1] *National Action Plan for Energy Efficiency, Coordination of Energy Efficiency and Demand Response*, Prepared by Charles Goldman (Lawrence Berkeley National Laboratory), Michael Reid (E Source), Roger Levy, and Alison Silverstein. www.epa.gov/eeactionplan, 2010.

[2] North American Electric Reliability Council, *Demand Response Availability Data System (DADS): Phase I and II.* North American Electric Reliability Corporation, Washington, DC, 2009.

[3] Maryland Public Service Commission, *Case No. 9208: In the Matter of the Application of Baltimore Gas and Electric Company for Authorization to Deploy a Smart Grid Initiative and to Establish a Surcharge for the Recovery of Cost*, Order No. 83410 and Order No. 83531, 2010 (accessed 19.08.11).

[4] P. Carmody, *Dumb Policies and Smart Grids: A Consumer Perspective*, Presentation to University of Florida – Public Utility Research Center, Office of Maryland People's Counsel, Washington, DC, February 3, 2011.

[5] R. Hornby, et al., *Advanced Metering Infrastructure—Implications for Residential Customers in New Jersey*, Synapse Energy Economics, Inc.: Prepared for the New Jersey Department of Ratepayer Advocate, Division of Rate Counsel, Trenton, NJ. 2008.

[6] O. Ghoshal, Smart bet? Illinois court invalides approval of smart grid program. *J. Environ. Energy Law*, 2010.

[7] Xcel Energy, *SmartGridCity*[TM]: *Design Plan for Boulder, Colorado.* http://smartgridcity .xcelenergy.com/media/pdf/SmartGridCityDesignPlan.pdf, 2008 (accessed 19.08.11).

[8] Smart Grid News.com, *Boulder SmartGridCity Cost Overruns: How Bad Is It Really?*. http:// www.smartgridnews.com/artman/publish/Business_Policy_Regulation/Boulder-SmartGridCity-Cost-Overruns-How-Bad-is-it-Really-1868.html, 2010 (accessed 19.08.11).

[9] C. King, D. Delurey, Efficiency and demand response: twins, siblings, or cousins?, *Public Utilities Fortnightly* 143 (3) (2005).

[10] K. Ehrhardt-Martinez et al., *Advanced Metering Initiatives and Residential Feedback Programs: A Meta-Review for Household Electricity-Saving Opportunities*, American Council for an Energy-Efficient Economy: Report no. E105, Washington, DC. 2010.

[11] National Action Plan for Energy Efficiency, *Customer Incentives for Energy Efficiency Through Electric and Natural Gas Rate Design*, Prepared by William Prindle, ICF International, Inc. <www.epa.gov/eeactionplan>, 2009 (accessed 19.08.11).

[12] PA Consulting, *Cape Light Compact Residential Smart Energy Monitoring Pilot Final Report*, Prepared for Cape Light Compact, 2010.

[13] eMeter Strategic Consulting, *PowerCentsDC Final Report.* http://www.powercentsdc.org/ ESC%2010-09-08%20PCDC%20Final%20Report%20-%20FINAL.pdf, 2010 (accessed 19.08.11).

[14] L. Lutzenheiser, et al., *Behavioral Assumptions Underlying California Residential Sector Energy Efficiency Programs*, California Institute for Energy and Environment, Berkley, CA. 2009.

[15] M. Sullivan, *Behavioral Assumptions Underlying Energy Efficiency Programs for Businesses*, California Institute for Energy and Environment, Berkley, CA. 2009.

Chapter 15

Customer View of Smart Grid—Set and Forget?

Patti Harper-Slaboszewicz, Todd McGregor, and Steve Sunderhauf

INTRODUCTION

It may be surprising for consumers to realize that there is a role envisioned for them in the smart grid, but it may also surprise them to realize that their new role does not require changing their lifestyle significantly. The goal is to enable their electrical devices to respond to variable energy prices or other information, following guidelines set up by the consumers themselves. For consumers, the first step is to install a home automation system and gradually transition to a smart home by investing in smart electrical devices, including appliances, or retrofitting existing ones with add-on functionality. Then, using a friendly user interface via their computer, tablet, or smart phone, they can tweak preset rules for how and when their smart appliances and devices will respond in different situations or just stick with the preset responses established by the manufacturers. From that point on, consumers can go about their lives knowing that their smart home is performing their energy management work for them.

The smart home vision is evolving from applying new communication technologies and protocols to energy management. With new wireless communications, which are covered in more detail in Chapter 16 by Hamilton et al., home automation can be offered at a lower price point and can include energy management as well as other features, such as convenience, entertainment, and security.

The chapter is organized as follows: "Emergence of Smart Grid in a World of Rapidly Evolving Technology"; "Smart Grid Vision"; "Consumer Adoption"; "Policy Maker's Burden"; "Home Automation Rides the Smart Grid Wave"; and "Conclusion."

EMERGENCE OF SMART GRID IN A WORLD OF RAPIDLY EVOLVING TECHNOLOGY

The early vision for the smart grid for consumers began with the simple concept that a utility could send price signals to smart thermostats and to consumers. The signals to smart thermostats would be sent using the advanced metering infrastructure (AMI) communication network and a second communication network commonly referred to as the home area network (HAN). Using these two communication networks, a utility would be able to exchange information with smart thermostats, and the goal was to reduce energy use for air conditioning during times of high demand for electricity. The consumer would be notified of pricing either on an exception basis (e.g., when prices were high) or daily using other communication channels such as email, text, and automated phone calls. Many industry dynamic pricing studies have shown that including automation of demand response for large loads, such as central air conditioning, increases the load reduction that residential consumers are able to achieve and makes it more convenient for consumers [1,2]. This type of automation often makes use of sophisticated algorithms to cycle the air conditioning to achieve the load reduction while maintaining comfortable temperatures within the home.

Over time, as utilities, policy makers, and other stakeholders considered the appropriate role for utilities behind the meter, more extensive scenarios began to unfold. It might be possible using the HAN to exchange information with other large loads, such as pool pumps, electric heaters, dryers, and refrigerators. Images of homes with HANs with ten or more devices began to appear on utility websites and in presentations at conferences. New devices were developed, such as in-home display devices, which were portable devices that consumers could use to see energy use in near real time, and at conferences, huge screens showed customer energy use and cost refreshed every few seconds. Regulators and consumer advocates were excited by these displays because direct feedback would be provided to consumers and in the language that consumers understood well: money.

This more extensive vision has not been adopted quickly due to a variety of factors, including high cost, the questionable reliability of HAN in exchanging information with only a few devices joined to the HAN rather than many, security, and the need for standards. In contrast, the use of smart thermostats has been growing along with the use of utility web portals, but penetration of in-home display devices remains low, and it's likely that smart phones and iPads™ will render in-home display devices obsolete before they are offered in significant volumes to customers.

More recently, home automation vendors have begun to recognize that energy management functionality could easily be included in home management systems. The advantage of this approach is that customers could be offered applications that coupled energy efficiency and demand response with other consumer-oriented features, such as convenience, entertainment, and security, thus increasing the appeal of home automation and energy management. Control 4 and Alert Me are examples of two home automation firms now actively pursuing energy management as part of their offerings.[1]

The key to any vision of the consumer smart grid becoming a reality is to sell consumers on the whole concept. A good way to "close the sale" is to interest customers in the benefits of home automation *and* then design systems so that energy efficiency and demand response help justify the cost. Home automation vendors have recognized that the smart grid and new technologies present an opportunity to expand the footprint of their market. They have been active participants in standards organizations to develop new standards to support information exchanges between electrical devices in homes. Demand response vendors, such as Comverge and Tendril, meanwhile, are expanding their product lines to extend beyond energy applications. Essentially then we're seeing a blending of two industries: home automation and demand response. What these companies hope to bring to market are products that consumers value and, at the same time, are complementary to utility demand response and energy efficiency programs.

There is some hard work to be done to move the smart grid into our living rooms by a number of different market participants. Along with supporting standards, so that smart appliances and electrical devices can exchange information with each other and the home management system, it is important to provide opportunities for consumers to save money by reducing peak usage and using energy more efficiently. *At its core, it is about establishing a utility platform that supports greater information to consumers and greater information to the utility and enables the establishment of pricing that more closely tracks the real cost of electricity*, as illustrated in Figure 15.1.

However there is a vital need to educate consumers about the benefits of the new paradigm and the ways to take advantage of it. If the industry gets it "right," consumers will drive the introduction of new products and services that continually improve the ability to benefit and use all of the new capabilities the smart grid will bring.

---

[1]Chapter 16 describes some of the companies and technologies.

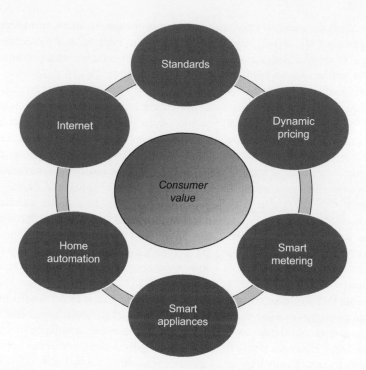

FIGURE 15.1   Smart grid success components for consumer. *(Source: CSC, April 2011)*

Policy makers are facing considerable pressure from those who benefit from the status quo in the energy market and the resulting extraordinary inertia that exists. One only has to watch policy makers struggle with consumer backlash to smart metering rollouts in some areas of the United States to see that initiating change comes with the risk that consumers may not automatically understand what benefits smart metering and home automation are expected to provide.[2] It is easy to maintain the status quo, and it is incredibly difficult to innovate within a tightly regulated industry. Anytime the status quo is changed, some benefit while others have fears that they will lose. Fortunately, this is not a zero-sum game—just because some consumers gain, others don't have to lose. To help avoid negative reactions from some consumers, utilities and

---

[2]See the following ongoing dockets in California and Maine:

- California Proceeding A1103014 at http://docs.cpuc.ca.gov/published/proceedings/A1103014 .htm#top
- Maine: Docket 2010-345 at http://mpuc.informe.org/easyfile/easyweb.php?func=easyweb_ query

At the time of this writing, the California Public Utilities Commission is considering allowing consumers to even opt out of smart metering in a rate setting proceeding. The Maine Public Utilities Commission has already ordered Public Service Company of Maine to allow individual customers to opt out of smart metering but did not allow communities to opt out.

policy makers have to educate consumers on how changes will affect them over time while providing some immediate benefits and reducing the risks of loss. Utilities can improve in this area and need to focus on providing consumers with benefits as soon as practicable.

One golden application that should be offered after smart metering is installed is "bill-to-date," which gives consumers a preliminary estimate of energy spending as they move through the billing month. This incredibly obvious and consumer-empowering application eliminates the surprise of a bad report card—large bill—after it is too late to make any behavioral changes. Consumers understand time and money across all consumer segments. Utilities don't have to wait until dynamic pricing is approved by regulators or consumers invest in home automation—it can be done after installing a smart meter and activating the associated information technology systems. The bill-to-date has proven to be extremely popular[3] with consumers. Providing something tangible sooner rather than later in the smart meter rollout timeline is very important in convincing customers that the smart grid has benefits for consumers *and* utilities. Other early, consumer-facing benefits include the daily presentment of hourly electricity usage data in graphical formats on the utility's website and similar information on the consumer's monthly bill, whether in paper or electronic form. These benefits give consumer's insights into how they are using energy today and, when coupled with energy saving tips on the utility website, can lead to significant energy efficiency gains.

There is some backlash against the rollout of smart metering by a small but vocal percentage of consumers. At the time of this writing, the percentage of consumers asking to opt out of smart metering remains small, but it does complicate the smart meter rollout process for utilities, regulators, and consumer advocates.[4] The backlash is to some extent the result of several factors that the industry needs to understand and manage. Those coming forward with concerns should not be ignored or ridiculed for worrying about such things as smart metering radio frequency transmissions. Rather, it shows that utilities need to introduce smart metering as if it were a new product offering and sell it and continue to sell it. Generally this is accomplished by demonstrating to consumers how this will improve their lives in meaningful ways—the bill-to-date is clearly one example.

At the same time, the industry needs to defend its turf from those that are providing misinformation to consumers. This calls for providing consumers

---

[3]Email exchange between Scott Burns, Smart Energy Product Manager, Reliant Energy and author in February 2011 concerning uptake of bill-to-date by 175,000 customers, which is branded as e-Sense.

[4]The Maine Public Utilities Commission (PUC) ordered Public Service Company of Maine to provide an opt-out plan while the California Public Utilities Commission is still considering a plan filed by PG&E for customers to opt out at a rate setting proceeding. It is noteworthy that the Maine PUC rejected an entire community opt-out plan. The plan approved only allows individual customers to opt out.

with clear information through multiple channels explaining why utilities are deploying smart metering, what the benefits are for consumers, what utilities would like consumers to do, the tools that utilities will be providing to assist consumers, and the overall timeline. If consumers need to wait until certain benefits are available, the industry should explain why but remind consumers their concerns are being heard. Consumers have waited for other products—movies to be released, iPads to go on sale—and consumers were excited even as they waited. The electric utility industry should be doing the same thing for new products and services that are supported by smart metering and the smart grid—building excitement and anticipation for new products, but without over-promising performance. If the industry can market "bill-to-date" in a compelling and memorable campaign that captures consumer attention and generates enthusiasm for this new product, consumer angst about smart metering will lessen and utilities can focus on the task of becoming a trusted partner with their consumers on energy.

While smart metering is being rolled out and policy makers are endlessly debating how best to implement dynamic pricing, appliance and entertainment vendors are moving rapidly ahead with designing smart appliances and smart electronics to enable customers to set up a smart home [3]. The smart components of a smart home all have to "cooperate" with each other as easily (or even easier) than it is to install and start up a new personal computer. And it's up to the vendors to convince customers that a smart home is within their financial means and technical competency and will be a "wow" factor for the neighbors.

Policy makers have another role to play as well: pushing for regulations and standards (or as close as we can get) that require all new appliances, lighting, security, and entertainment components to include smart energy use features. If all appliances and electrical devices had smart energy use features, such as communications and applications for demand response and energy efficiency, then the average cost of including these features will be reduced [4].[5] Imagine being able to avoid rolling blackouts or mitigating high grid prices by simply turning off non-essential electrical devices for all customers in the affected region who had previously volunteered. While some may argue they don't want anyone turning off any of their electrical devices, if some customers volunteer and choose for themselves what's non-essential, avoiding just one set of rolling blackouts or mitigating high grid prices would go a long way toward covering the cost of requiring smart features in all new appliances and electronic devices.

And, of course, life goes on while the energy industry does the hard work of planning and laying the groundwork for the smart grid. Technology is rapidly changing how consumers access the Internet and entertainment, resulting in challenges to entire industries that either must rapidly innovate

---

[5]See http://www.sites.energetics.com/madri/pdfs/lbnl-3044e.pdf.

or fall into the ash heap with the other industries that failed to grasp the need for change. Traditional book stores are facing robust competition from sales of electronic books, for example. But one thing is clear—almost every new innovation introduces new devices that run on electricity, making the smart grid more important than ever for economic growth, quality of life, and innovation.

SMART GRID VISION

Early visions of how consumers would benefit from the smart grid focused on energy applications—helping customers manage their energy spending, participating in programs that rewarded customers from reducing peak energy use, and using energy efficiently and wisely. The emergence of new communication technologies will open up a whole new market for home automation vendors, allowing automation to be offered at lower prices while riding the wave of the smart grid investment and interest. Demand response vendors, seeing this threat from home automation vendors, are changing from energy only to a more expansive view of applications they will offer in their systems.

In fact, home automation systems being deployed today have already begun to include energy applications, piggybacking onto security.

The following is a list of scenarios that consumers might find appealing that combine energy applications with convenience, entertainment, and security, and importantly, are centered on consumer activities. Some are based on offerings available today from home automation vendors, incorporating new trends into existing applications, smart phone applications, and ways of using automation to make it easier to enjoy entertainment options available today.

- Going to bed for the night—Consumers can use an application (app) on their smart phone or iPad to verify that all the doors and windows are closed and locked, lights and entertainment components are off in all unoccupied public rooms, the dishwasher is set to run when energy prices are low, and there are no wet clothes still sitting in the washing machine.
- Leaving for vacation—Another app sets their house to vacation mode, with many of the same functions as an alarm but may add turning lights and TVs on and off to simulate someone being home and adjusting thermostats to reduce energy use for heating and cooling and to readjust the temperature setting for the return to home.
- Leaving home for some or all of a day—An app may check to make sure doors and windows are closed and locked, heating/cooling systems can allow temperature to drift within preset ranges and it's OK to participate in any demand response events; it also checks that lights are out except ones identified to be on for security, the home phone is forwarded to the cell phone, and the security alarm set. It also has an alert if the stove or oven is on, if the washing machine is running, or if wet clothes are in the washing machine.

- Providing periodic updates of the electricity cost in near real time, and identifying which end uses contributed the most to the cost.
- Comparing electricity use to similar households, appliance by appliance, to identify savings opportunities and to spur customers to action.
- Checking which smart appliances or devices are on right now and if the device supports it, whether has been actively used. For example, if the gaming system is on, is someone playing a game right now?
- Generating a reminder that it's time to change the filter on an air conditioner or if one of their appliances needs maintenance or repair.
- Automating lights so that rooms are illuminated when someone is in a room and are dark when rooms are unoccupied.
- Setting up for entertainment: a person selects a movie/show to watch, and the application finds the movie/show and, adjusts the settings on various devices to enable the best entertainment experience, including sound, lighting, and window shades. It may offer family members options on where to screen the show or movie.
- Heating and cooling management—Systems could learn, over time, when rooms need to be cooled and heated rather than operating on times set by consumers.
- Opening and closing window blinds or, in a more advanced scenario, automatically adjust the tinting in windows depending on the time of day or amount of sunshine.
- On common surfaces, such as mirrors, countertops, hallway walls, and refrigerator or microwave doors, useful information may be displayed rapidly to consumers, such as news, weather forecasts, calendars, traffic reports, recipes, and grocery lists, depending on consumer preferences and the location of the surface. It would be similar to setting up a home page on the Internet except the home pages would vary by location and perhaps by which household members are nearby. Messages from the home automation controller could be displayed on these surfaces as well, providing another convenient user interface to the home automation system.

Some consumers are already watching TV shows on their computers, tablets, and smart phones and are using these some of the same devices to stream movies and TV episodes to their TVs. Home management systems could make it easier for consumers to do this by providing virtual controllers for all of the systems as apps on smart phones or tablets, making it more convenient and easier for consumers to manage their entertainment options. At present, not all consumers know how to watch TV on their computer or to send a show available on their computer to a TV so more than one person can watch at the same time. Home automation vendors may offer a dedicated controller to use for this purpose, but as smart phone and tablet penetration increases, it is likely dedicated controllers will give way to devices consumers already own.

With a well-designed system, consumers are expected to be able to save money on their energy bills and enjoy automation conveniences. The key is to design the automation around consumer activities, such as laundry, cooking, keeping their home comfortable, watching TV or movies, working at home, listening to music, reading, playing games, keeping their phones and tablets charged up and ready to use, going to bed, getting up, coming home, leaving home, and washing dishes. Consumer lives are not organized by demand response events, peak hours, or whether one appliance uses more energy than another. Therefore, if we want the smart grid to be adopted by consumers, we need to make it relevant for consumers and easy for all members of the family to use.

Is there any interest in home automation? A report issued by the Smart Grid Consumer Collaborative provides some interesting insights that are relevant here.[6] Based on a study by Best Buy, consumers gain status, at least in the short term, by having the latest gadgets or automation.[7] However, paradoxically, "those who are most motivated by cost savings on their bill are not necessarily the ones willing to pay more for a home energy management system that will allow them to achieve their goals."[8] So, while it may be useful to let customers know what the utility industry is helping consumers to do—reduce peak energy use and use energy wisely—*selling* consumers on the idea of home automation will need to be designed around how consumers live their lives, with energy applications in the background. The energy objectives will still be included as part of consumer applications but they won't be the focus of most of them.

For example, Google is developing an LED light bulb that is shown in a demo blinking on and off in synch with what's going on in a video game [6]. How this will be received by gamers remains to be seen—it may be distracting or it may enhance the game—*but what is of interest is that the focus on Google's use of LED lights communicating with other devices within the home is entertainment, not energy efficiency.*

CONSUMER ADOPTION

Consumers have other priorities besides their electric bill and environmental concerns. While some consumers will make purchases with energy efficiency in mind, to appeal to a broader segment of society the industry needs to think in terms of what consumers think about. For the most part, when consumers spend their discretionary dollars, they are looking for entertainment, convenience, and new features.

---

[6]Smart Grid Consumer Collaborative [5].
[7]Ibid, p. 25.
[8]Ibid, p. 26.

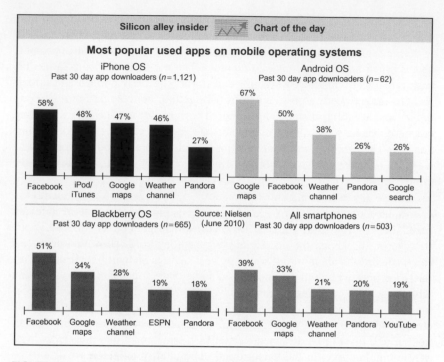

FIGURE 15.2   Most popular smart phone apps downloaded. *(Source: BusinessInsider [7])*

For example, there are thousands of apps available for download to smart phones. As seen in Figure 15.2 the apps most often downloaded across the various smart phone platforms are for entertainment and convenience [7].

Pandora, for example, allows users to create a virtual radio station that plays music from a favorite composer and others like him or her. According to Apple [8], the most popular 20 apps for the iPhone across all categories include:

- Seven games,
- Five entertainment apps,
- Two self-improvement apps,
- Three practical apps,
- Two for style, and
- One energy app (carbon footprint calculator).

Consumers are willing to spend on products that enhance their entertainment and convenience. That provides a good roadmap for home automation positioning: dynamic pricing, demand response, and energy efficiency folded into home automation scenarios will allow consumers to enjoy relatively sophisticated automation for less cost.

The energy industry has not traditionally marketed products to consumers because it didn't have to—almost every household is a utility customer.

In contrast, appliance vendors have been marketing to consumers since the invention of the electric coffee percolator in 1865 by James Nason.[9]

Consider a purchase of a new washing machine and dryer today. One way to assess what features of new washers and dryers are important to consumers is to look at how vendors are marketing them. In almost every ad, styling is either mentioned directly or highlighted by showing matched machines in bright colors against a neutral background. Another highly touted feature is steam, which consumers are told will remove stains and odors, and reduce wrinkles. Some ads, but not all, mention energy efficiency and reduced water use, but more often, ads dwell on larger load capacity.

Early adopters of front-loader washers and dryers may have been strongly influenced by energy efficiency, along with how well the machines clean clothes. But for consumers coming after early adopters, energy efficiency likely plays a smaller role. Consumers are willing to accept energy efficiency as a benefit, but this is partly because the machines also clean clothes better than the older, less efficient designs. Consumers are not forced to choose between energy efficiency and the coolest washing machines on the market—they can have both.

What does this tell us about consumer adoption of home automation and smart appliances? We need to make smart grid technology appealing to early adopters, and the appeal needs to extend beyond practical considerations. While the front loaders offer convenience (more laundry in each load) and clothes are treated more gently (saving money on clothes), they also offer a new stylish design, clothes are tumbled differently (new feature), and they offer some degree of entertainment. Laundry is more fun when you are using colorful appliances that play soft musical tones when you are selecting your cycles and when the cycle is done.

We are fortunate that home automation companies and appliance vendors have experience marketing to consumers and will find ways to make their product offers exciting to consumers, not just practical. Many consumers purchased the new iPad™ without having a real clear idea on how they were going to use it; but they knew they wanted it.[10]

However, the smart grid and home automation are not seen as a positive development by all consumers. Scenarios that provide convenience and security can simultaneously introduce threats to consumer privacy or safety. If doors can be locked using a smart phone app, can they be unlocked by someone with ill intent? Will others be able to analyze energy usage data to tell if someone is home, or home alone?

Clearly, home automation has to provide security along with convenience. Consumers may want to lock doors using automation but they won't want

---

[9]Consumer Appliance Timeline [9].

[10]In less than three months, Apple sold three million iPads (Apple press release on June 22, 2010.) See http://www.apple.com/pr/library/2010/06/22ipad.html.

strangers unlocking their doors by hacking into their system. This will be the challenge for home automation vendors—sell the benefits and manage the risks. Utilities and authorized third-party providers, including energy suppliers, home automation vendors, and so on, also have to plan for and then protect consumer privacy.

From a marketing perspective, there will always be some consumers who prefer the way things are and won't adopt new products in the short or medium run even if they would benefit from adoption. However, consumers who object to the smart grid now may find that some of the benefits are appealing after all, such as tracking their energy spending on a daily basis and being alerted when their energy use suddenly increases.

Consumer activists now opposing smart metering implementations may find that the vast majority of customers appreciate having assistance in managing their energy usage and that the industry has managed security issues to the satisfaction of consumers. After all, security breaches are frequently reported about the Internet, but how many have stopped using it? As a society, online purchases are increasing rather than decreasing, and this is occurring despite the persistent threat of fraud.[11]

The Internet offers benefits to consumers, and these benefits are highly valued. If our industry can position home automation with embedded smart grid features in a similar manner—providing high value—then consumer adoption should follow the pattern of other new technologies. The key is providing high value to consumers and to utilities.

POLICY MAKERS' BURDEN

Policy makers—including utility executives, regulators, consumer advocates, industry pundits, and lawmakers—set policies that try to blend the needs of today with the promises and potential threats of tomorrow. For consumers, the needs of today often loom larger than what may come later. However, some future needs are so great that they must be planned for now. This is the case with planning for future energy needs and capacity. The burden for policy makers is to make the case of why investment needs to be made today and to respond appropriately to those who may prefer the status quo.

For example, many policy makers are overseeing the rollout of smart meters that enable the offering of dynamic pricing. These rollouts are seen by policy makers as essential elements of modernizing the electric grid going forward, such that demand for electricity is more closely linked to the market price of

---

[11]"According to a recent global survey conducted by the Nielsen Company, over 85% of the world's online population has used the Internet to make a purchase, up 40% from two years ago, and more than half of Internet users are regular online shoppers, making online purchases at least once a month [10].

electricity and consumers can respond accordingly. In addition, the smart meter rollouts can help the grid more easily adapt to growing renewable energy supplies, additional penetration of demand response capability, and electric car charging to reduce reliance on the use of gasoline-powered engines, and ultimately, reduce carbon footprints.[12]

A small percentage of consumers are resisting the rollout of smart meters because they perceive no benefit from smart metering, only higher costs, loss of privacy, health effects, outside entities controlling their appliances, and a change in how they are charged for electricity.[13] These concerns need to be addressed directly by utilities and policy makers because while some of the concerns are real (privacy and opening the door for dynamic pricing, for example), others appear to be based on misinformation. The news industry has been dramatically changed by new technology in how, what, and when news is reported, and by whom. Instead of a few deciding what news stories will be published for many to read, the model has changed to "many providing for many." Major news organizations are often in the position of catching up with news that has been widely disseminated by bloggers, Twitter™ and texting, and as a result, stories are often published and checked for accuracy later. For utilities, this allows consumers who are concerned about smart meter rollouts for a variety of reasons to more easily organize and make their concerns heard to lawmakers, consumer advocates, and regulators. Utilities need to consider the perception and perspective of consumers early in their planning and implementation process or be faced with regulatory or legislative roadblocks to their plans.

Another complaint often raised about the smart grid is that some consumers are adamant that they don't want electric utilities controlling when and how they use their electrical devices in their home. The key response to these concerns is that not everyone needs to participate to achieve industry goals, only some. For those who are adamantly opposed to changes in the utility industry, they can decline some or all aspects of home automation, and it won't interfere with the vast majority of consumers who will benefit from going forward.[14] In fact, all customers whether they participate or not will benefit greatly through greater demand response, which will mitigate high energy prices, avoid or defer the need for new infrastructure, and maintain or improve the reliability

---

[12]See Chapter 4 for more details on issues facing regulators in smart metering rollouts.

[13]See the following ongoing dockets in California and Maine:

- California Proceeding A1103014 at http://docs.cpuc.ca.gov/published/proceedings/A1103014.htm#top
- Maine: Docket 2010-345 at http://mpuc.informe.org/easyfile/easyweb.php?func=easyweb_query

[14]At the time of this writing, the California Public Utilities Commission is considering allowing consumers to even opt out of smart metering in a rate setting proceeding. The Maine Public Utilities Commission has already ordered Public Service Company of Maine to allow individual customers to opt out of smart metering, but did not allow communities to opt out.

of supply. Policy makers have made it a best practice that participating in dynamic pricing is either an opt-in or opt-out process for residential consumers, not mandatory. Anyone who doesn't want to be on a pricing plan with incentives to reduce peak energy can find a pricing plan, perhaps with a different supplier, where pricing is not based on when the energy is used. While some fear utilities want to control when consumers use energy, the plans are for voluntary participation in demand response programs, with consumers compensated for their participation.

Policy makers are considering the best method to pay for smart grid investments. In Texas, customers of distribution utilities that are rolling out smart metering are paying a monthly surcharge to cover the costs of smart metering. In Maryland, for example, the investment will be treated as a normal, long-term utility investment, meaning that utilities will not collect any monies from customers until the asset costs are included in the rate base, which will not happen immediately. Many policy makers are supportive of making smart grid investments, but setting policy on how to pay for the investment is challenging. Adding to the cost of utility bills is about as popular as raising taxes.

Dynamic pricing is a price structure change rather than a pricing level change. It introduces a number of new concepts and price signals in the selling of electricity. With these plans, *when* customers use energy, not just how much, influences the amount of the energy portion of the bill, even if the dynamic pricing tariff under consideration is a rebate-only program.[15] After years of debate in the electric industry, it is now clear that for residential customers, dynamic pricing in some jurisdictions will be optional, but in others will be applied on a default basis. This means that residential customers will have a choice in whether to take service under a dynamic pricing rate.

Policy makers can rely on numerous studies that have demonstrated that virtually all residential customers who agree to participate in dynamic pricing programs prefer dynamic pricing to traditional pricing plans. For PowerCentsDC, a dynamic pricing pilot program in the District of Columbia, Figure 15.3 shows that 93% of participants in the program preferred dynamic pricing rates compared to the regular tariff.

Customers with higher and lower incomes, with and without pools, with and without central air conditioning—no matter which way the data have been sliced and diced—reduced peak energy use, as shown in Figure 15.4.[16] Readers should note that this study occurred seven years ago and has been replicated many times since then—simply reconfirming what all of us know: prices affect consumer behavior.

---

[15]Chapter 3 addresses these issues in more detail.
[16]Levy Roger, "California Statewide Pricing Pilot: Lessons Learned," 2006, Slide 10 (Source: Statewide Pricing Pilot, Summer 2003 Impact Analysis, CRA, August 9, 2004, Table 5–9, p. 90).

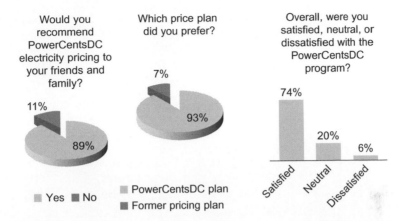

FIGURE 15.3   PowerCentsDC customers preferred dynamic pricing. *(Source: Smart Meter Pilot Program, Inc. [2])*

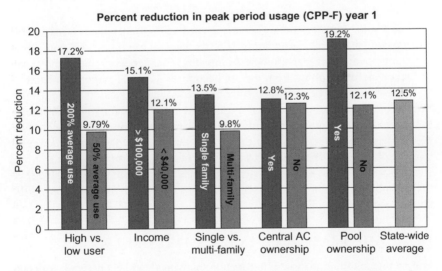

FIGURE 15.4   California pricing pilot results consistent across customer segments. *(Source: Roger Levy, August 9, 2004)*

Providing even more support for the smart home, automation of response—mostly accomplished by smart thermostats or smart switches to date—has increased peak load reduction, and load reduction was greater on the hottest days compared to days with more temperate weather. In Figure 15.5, Power-CentsDC participants on the critical-peak pricing rate using smart thermostats reduced their critical peak energy kWh use by 49% compared to 29% for participants not provided with smart thermostats.

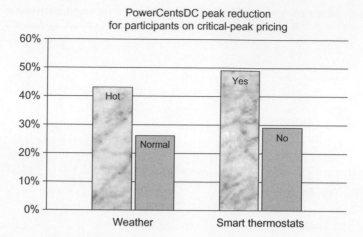

FIGURE 15.5 PowerCentsDC customer load reductions by weather and smart thermostats. *(Source: CSC, April 2011)*

The results show that policy makers are basing their decisions on reasonable expectations that dynamic pricing will bring benefits to virtually all consumers who participate directly and to others indirectly. Chapter 3 covers the direct and indirect benefits resulting from the application of dynamic pricing in detail.

What remains to be seen is whether home automation vendors, which at this point would also include demand response vendors, can convince consumers that investing in home automation will bring monetary rewards through their electric and gas bills and along the way, enhance their lives in meaningful and tangible ways. These services may be coupled with other services in the future, such as home security systems, Internet connections, and/or cable television to become even more attractive for consumers and vendors offering the suite of services.

HOME AUTOMATION RIDES THE SMART GRID WAVE

For a home automation system to function there needs to be a network for communication among devices and a centralized management application, or controller. Network controllers have certain requirements:

- Always be on,
- Stay at the home,
- Be connected to the Internet or an alternative communications pathway.

The controller needs to be in a device that is always on so the controller is always available to perform its home management tasks. There are various devices in the home that could be left on 24/7, such as a desktop computer and routers. However, desktop computers use too much energy to be left on

24/7 compared to a router, which is very rarely turned off intentionally. Another option might be a laptop computer but laptops on the move are not always connected to the Internet, as for example, during air travel. And, for those left at home, having the controller somewhere else may prove to be problematic.

Looking forward, home computers as we know them today may be abandoned altogether in the not too distant future (see Box 1), except perhaps for computers used by gaming enthusiasts—these would not work well as a controller due to their high energy use.[17] Some vendors have produced dedicated devices, such as in-home display devices, but with the penetration of smart phones and tablets occurring at such a rapid pace, dedicated devices are already looking anachronistic [11].

Box 1: Tablets and Smartphones Slow Laptop Sales Growth [12]

"Tablet computers and smart phones are slowing the growth of laptop sales, according to Gartner (IT) released Thursday [3/3/2011]. The research firm projects that laptop sales will grow less than previously expected this year and next year because of the surging popularity of alternative mobile-computing devices....

"We now believe that consumers are not only likely to forgo additional mobile PC buys, but are also likely to extend the lifetimes of the mobiles PCs they retain as they adopt media tablets and other mobile PC alternatives as their primary mobile device," Gartner research director George Shiffler said in a statement.

As smart phones and tablets, such as Apple's (AAPL) iPad, boost their computing power, more and more of these devices are becoming capable of performing the same tasks as laptops. Additionally, laptops are losing their fashion-forward cachet compared to iPads and smart phones, Gartner says."

Almost every home with an Internet connection has a router, and routers are not expensive to replace—this would be within the budget of most households that already own a router. If the controller application were added to the router, it is not expected to significantly raise the cost of the router, especially if is planned that consumers will interface to the controller through other devices. And, of course, the router is connected to the Internet.

The controller needs to be connected to the Internet to access information used by applications that are part of the home automation system. The types of information that may be provided over the Internet include energy prices, comparisons to other "similar" households, weather forecasts, updates to the home automation system, updates to security for home automation systems, updates to applications on smart devices included in the home automation network, and new applications available to download.

---

[17]Gaming computers often include two hard drives, very fast CPU, and a large power supply. This author measured the demand of a gaming computer at 1 kW.

In addition, the home automation network may provide information to utilities or other parties (with the permission of the consumer). For example, a home automation system may order new filters to be delivered to the home on a regular basis during the cooling season, or send information to the manufacturer of a smart appliance when the smart appliance detects a malfunction. In either of these cases, the consumer would need to approve sharing the information with an outside party.

Consumers already take advantage of similar services with regular deliveries of medications, dairy products, books, movies, or other items that are needed or enjoyed on a regular basis. People like the convenience of having items delivered, and in the case of air conditioning filters, the delivery may serve as a reminder that it's time to change the filter.

One manufacturer of smart appliances has already set up a means to extract information from its washing machines if the appliance detects something wrong. After calling a tech center, the consumer pushes a button, and the washer sends data from the appliance directly to the tech team over the phone. A home automation application might send this same information over the Internet to the tech center (again with the consumer's permission); using the information provided, the tech center may be able to diagnose the problem remotely and, through email or text messages exchanges with the consumer, come up with a plan. In some cases, the consumer may simply be provided with instructions for cleaning hoses or filters, or in others, a part may be mailed to the consumer with instructions for installation. Or a service call could be scheduled. In many cases, the applications that might be included in a home automation system have seeds in services already provided or in routines that consumers typically go through on a daily basis.

A weather forecast might be used for managing energy use for cooling or heating to pre-cool or pre-heat when energy prices are lower than during the peak times for heating or cooling. Or, it may allow for cooling to be avoided if a thunderstorm is predicted to move through the area and the high temperatures initially forecast may not occur. A consumer should have the option to specify different criteria to use for planning when heating and cooling will occur based on their preferences. It could be as simple as moving a slide on a smart phone as shown in Figure 15.6, with the focus on comfort or spending. Or more complex options might be provided for the consumer to select what areas of the home always need to be really comfortable and other areas

FIGURE 15.6   Comfort versus spending slide bar. *(Source: CSC, April 2011, with inspiration from PNNL Labs and others)*

where it would be OK for the temperature to drift a bit higher or lower at certain times of the day.

While keeping homes comfortable is likely to continue as a focus of home automation systems, entertainment applications will need to adapt to the increased options for watching TV or movies, listening to music or books, and reading. It used to be that having music in any room, home theater, and watching a movie on any TV in the house, as opposed to the nearest TV to the DVD player, were large parts of home automation. As consumers have shifted to listening to music on MP3 players, iPods™, and smart phones, providing music to every room via a stereo will be a lower priority for consumers. In addition, now that consumers can stream movies or TV episodes to personal devices or large-screen TVs, home automation will likely include applications to make it more convenient and easier for consumers to avail themselves of new options for watching TV and movies.

As consumers bring smart appliances and devices into their homes, each device should be preset to include applications that reduce peak energy use and support the energy-efficient use of the product. Each vendor can differentiate itself by providing innovative ways for its particular product to reduce energy use when prices are high and support energy efficiency, as well as provide other features to take advantage of home automation. Standards come into play here because the home automation controller, housed in the router, must be able to communicate with smart appliances and other devices manufactured by any vendor. In order for the smart grid vision, and indeed the vision of home automation, to catch the fancy of consumers and become the next iPad-like item, consumers can't be restricted to buying only products from GE or Samsung.

When consumers bring a new smart device into their home, the home automation system should "find" the new smart device and ask the consumers if they would like to add it to their home automation network. If they say "yes," then the controller and the new smart device would exchange information so that the controller now knows what type of smart device it is—a dishwasher as opposed to a TV, for example—what applications it supports, and what information the smart device requires to support its applications. For example, a smart refrigerator may be able to reduce energy use during a demand response event by stopping defrosting activity, and allowing the temperatures inside the refrigerator and freezer to drift upwards, but still remain within safe operating temperatures. Perhaps ice making could also be suspended. Thus the smart refrigerator could support demand response events as long as it was notified with the triggering energy price or demand response event. Manufacturers of smart refrigerators would want to select applications carefully that would support reduced energy use but, at the same time, not interfere with the main function of the appliance. In this case, this would most likely mean allowing the light to come on when the door is opened. To support consumer control over their smart devices, if the consumer noticed that ice production had ceased and objected to it, the consumer could use his or her smart phone to change this setting for the smart refrigerator during demand response events.

The example described for a smart refrigerator incorporates two important features of home automation: first, the demand response setting is the default setting, and second, consumers can change the default setting if they wish. If the home automation controller can also use information on the location of individual household members, it is also possible to customize the smart refrigerator's response to a demand response event based on which household member is home at the time. Some home automation systems already feature using the location of individual household members for energy and security applications. Some household members may frequently use ice and others may not, for example.

A smart washing machine and dryer may take a different approach to responding to demand response events than for managing the dishwasher. Consumers may accept the dishwasher running anytime between 11 PM and 6 AM, as long as the dishes are clean when they get up. Laundry is different because it's usually a two-step process, washing and then drying, and more often than not, more than one load needs to be accomplished in one day. The default response for a smart washer and dryer may be simply to alert customers that if they start the load now, finishing the load may extend into the demand response period. Thus, the smart washing machine and dryer may need to know of a demand response event before it happens, if this information is available.

However, there are electric grid emergencies that only require a short period of energy reduction, perhaps seconds or minutes, before the emergency is resolved. The Department of Energy's Pacific Northwest National Laboratory, managed by Batelle, developed a technology to detect a disturbance on the electric grid by monitoring the frequency and voltage of the system.[18] If this technology were included in smart appliances, such as a washer and dryer, then for a few seconds or minutes, the washer and dryer could pause their operations, providing a real benefit to grid operators while not significantly affecting when washing and drying cycles would be completed. During the short response period to the disturbance, a dryer could shut off its electric heating element and tumble the clothes once a minute to prevent wrinkles developing and the washer could just rest.

A smart appliance with technology to detect grid disturbances could alert the controller for the home automation system of the grid disturbance, and perhaps other smart devices could also respond to the grid disturbance, if they didn't have the technology embedded. For example, perhaps lights could be dimmed, a defrost cycle stopped temporarily, or battery charging paused for a few minutes.

---

[18]Grid Friendly™ appliance controller licensed for market development [13].

The grid disturbance and demand response are scenarios of interest to grid operators, generators, energy suppliers, utilities, and other market participants. Both offer potential financial revenue streams. In contrast, consumers are going to be interested in scenarios that address their needs. These scenarios tend to involve more than one smart device and may require more input from consumers to implement, depending on how "smart" the smart devices are. For example, is a sensor, on a door going to be automatically identified as a "door" sensor and will a smart light be automatically identified as being located in the living room as opposed to a bedroom? The goal would be for consumers to provide information about their home in an intuitive manner, which does not require programming skills, which many consumers do not have. Once the controller knows what each device is and where it is located, then the home automation system can offer default versions of applications for consumers to review and possibly tweak to fit their needs.

Consider the "Going to Bed" app discussed earlier in the section "Consumer Adoption." Imagine a household with two children in elementary school, Ted and Lucy, each with their own bedroom, and the parents are using the "Going to Bed" app immediately prior to retiring for the night. Using an iPad, one of the parents activates the app similar to using any other app on their iPad. The app immediately sends out a series of queries to the smart systems/appliances/devices joined to their home area network, as shown in Figure 15.7.

The app would send a signal to the home management system housed in the household router via Wi-Fi to run the "Going to Bed" app. The router would then send signals to each system or appliance through the appropriate protocol such as ZigBee 2.0 or Wi-Fi or another protocol, asking for status information.

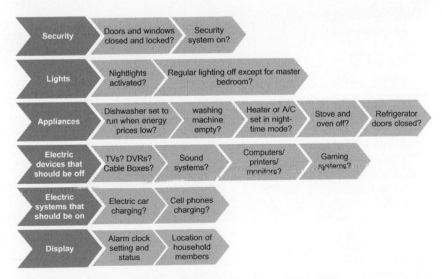

FIGURE 15.7   Queries for "Going to Bed" app. *(Source: CSC, May 2011)*

For example, it would send a request appropriate for the washing machine to try to ascertain if there are any wet clothes left in the washing machine. It will vary from machine to machine how sophisticated and accurate this information might be. One machine may only be able to report whether a cycle is running or not, while another may be able to report on the weight of the clothes in the drum and, if a cycle is running, the time remaining in the cycle. The home management system will use whatever information is returned to provide useful information to the person using the app.

When the home management system has received information back from the queries, it will send the results to display back to the iPad using the Wi-Fi within the home. As example of running the app is shown in Figure 15.8.

From these results, the parents might decide to "Turn off" the light and gaming system in Ted's room, and stay up a few minutes longer to move the clothes from the washing machine to the dryer before going to bed. And, they'll probably decide to close the refrigerator door and let the dog in for the night. The "Going to Bed" app might also feature an alert if anything changes. When Lucy slips back into her room, the parents could be alerted, and if Ted turns the gaming system back on after 30 minutes, they will be alerted that something is amiss.

As the kids grow up, the parents should be able to add, remove, or change features of the "Going to Bed" app depending on their needs. For example, by the time the kids are in high school, the parents may go to bed hours before their kids and may only want to be alerted to a gaming system being on if it's 3 AM. Making these changes should be an intuitive user experience. It shouldn't

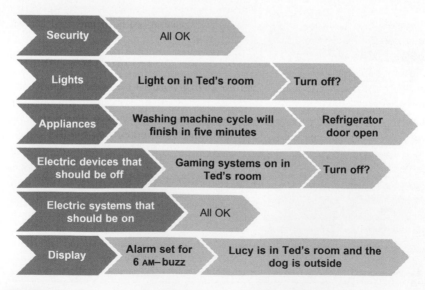

FIGURE 15.8   Results of "Going to Bed" app. *(Source: CSC, May 2011)*

require programming skills on the part of consumers to make adjustments to popular apps or to create a new one. In fact, it should be possible for consumers to go online to download new apps available from other consumers or vendors, similar to downloading apps for smart phones today.

CONCLUSIONS

How the smart grid is manifested in people's living rooms has evolved since the concept was first imagined by the industry. It began with utilities directly controlling large loads like air conditioners and electric heaters and moved forward to price-responsive demand response. Through pilot tests, the industry learned that automating the response of large loads, generally the same ones involved with direct load control programs, increased peak load reduction and made it more convenient for consumers to participate. As shown in test results from PowerCentsDC and verified by many other pilot tests, consumers liked dynamic pricing, preferring it to standard pricing by a high margin. With technology advances, it became feasible to offer all customers the opportunity to participate in demand response programs.

As the smart grid moves into consumer living rooms, the focus will significantly change to marketing to consumers, a very different world than providing electricity over a regulated distribution network. Color and style may trump practical considerations. If society at large wants consumers to invest in home automation that incorporates energy efficiency and demand response, it must be sold to them on the value it will bring. It may be easier to forecast weather than to forecast what products will catch consumer's fancy and become a must-have in their eyes.

Looking at recent spending habits of consumers, it's clear that they spend much of their discretionary spending on entertainment and convenience. In the best of all worlds, new smart grid products should not make consumers choose between energy and entertainment or between energy and convenience. Instead, the new products should roll it all up together and offer customers home automation where they can enjoy unprecedented levels of convenience, and oh by the way, you can save money on your energy bill at the same time and increase the security of your home.

It turns out that the home automation and appliance industries will likely drive consumer interest in and adoption of smart appliances, with utilities and regulators playing important but not leading roles. With energy considerations bundled together with applications for convenience, entertainment, and security, consumers won't have to turn their lives upside down to lower their carbon footprint, any more than people have to understand what a printer driver actually does to use a printer.

In order for home automation and appliance vendors to have success in selling their new products such as smart home, smart washing machines, smart TVs, and so on, many stakeholders need to work together to build the underlying framework: standards, smart metering, dynamic pricing—all are necessary to set the stage for smart homes. Policy makers are in a tough spot, pushing investments in smart metering and changes in the structure of pricing plans in an industry built around exhaustive regulatory hearings, in the midst of a slow recovery from a global recession. There is ample research backing their approach showing that consumers, when given the opportunity, overwhelmingly prefer dynamic pricing compared to standard pricing. The industry should endeavor to reduce barriers to access to dynamic pricing, which is one of the reasons to hope for the ubiquitous deployment of smart metering.

The home automation service offerings are already emerging. What's needed now is to pull together the various strands and position them together on a platform that is inclusive of all smart appliance brands and doesn't require consumers to be able to write code to customize apps for their household. Great strides have been made to reduce the complexity of user interfaces to smart appliances by taking advantage of the talents of people who have experience in designing consumer products. This same set of talents needs to used in the design of home automation scenarios that consider energy, but focus on making life easier, secure, and more fun.

The goal isn't to move utilities into our living room—rather it's to allow consumers to take advantage of some of the same technologies utilities are finding useful in smart metering and monitoring/managing the distribution grid. Home automation systems need to be designed to provide features that inspire consumers to invest in home automation. Consumers should only need to spend a little time to tweak their smart home to meet the needs of their household and then move onto to other things.

Maybe we'll never get to "Set and Forget," but we'll move toward "Set Up and Tweak Occasionally."

REFERENCES

[1] A. Faruqui, L. Wood, *Quantifying the Benefits of Dynamic Pricing in the Mass Market*, http://www.edisonfoundation.net/iee/reports/QuantifyingBenefitsDynamicPricing.pdf, January 2008 (accessed 08.09.11).

[2] Smart Meter Pilot Program, Inc., *PowerCentsDC™ Program Final Report*, September 2010.

[3] GE, *Smart Appliances Empower Users to Save Money, Reduce Need for Additional Energy Generation*, http://www.genewscenter.com/content/Detail.asp?ReleaseID=6845&NewsAreaID=2 (accessed 21.05.09).

[4] C. Goldman, M. Reid, R. Levy, A. Silverstein, *Coordination of Energy Efficiency and Demand Response*, Environmental Energy Technologies Division, Ernest Orlando Lawrence Berkeley National Laboratory, 2010.

[5] Smart Grid Consumer Collaborative, *2011 State of the Consumer Report*, http://smartgridcc.org/wp-content/uploads/2011/02/SGCC+State+of+Consumer+Report.pdf, 2011, http://smartgridcc.org/wp-content/uploads/2011/02/SGCC+State+of+Consumer+Report.pdf (accessed 31.01.11).

[6] C. Lombardi, These smart light bulbs heed iOS, Android devices, *CNET News*, http://news .cnet.com/8301-11128_3-20063928-54.html#ixzz1N73L3Zuu (accessed 08.05.11).

[7] J. Yarrow, K. Angelova, *Chart of the Day: Here's The Most Popular Apps On Android, iPhone, and BlackBerry*, Business Insider, SAI, http://www.businessinsider.com/chart-of-the-day-apps-on-mobile-os-2010-6?utm_source (accessed 10.06.10).

[8] Apple, *All Categories: Most Popular*, http://www.apple.com/webapps/index_top.html (accessed 24.03.11).

[9] *Consumer Appliance Timeline, Trail End State Historic Site*, http://www.trailend.org/ind-timeline.htm (accessed 23.03.11).

[10] *Trends in Online Shopping*, a global Nielsen Consumer Report, http://id.nielsen.com/news/ documents/GlobalOnlineShoppingReportFeb08.pdf, February 2008 (accessed 08.09.11).

[11] K. Fehrenbacher, *Fail: The High-End Home Energy Device Is Toast*, http://gigaom.com/ cleantech/fail-the-high-end-home-energy-device-is-toast/ (accessed 01.02.11).

[12] D. King, *Tablets and Smart Phones Slow Laptops Sales Growth*, DailyFinance, http://www .dailyfinance.com/story/media/tablets-and-smartphones-slow-laptop-sales-growth/19867508/ (accessed 03.03.11).

[13] *Grid Friendly™ Appliance Controller Licensed for Market Development*, Metering.com, http://www.metering.com/Grid/Friendly/appliance/controller/licensed/market/development (accessed 23.03.11).

Chapter 16

The Customer Side of the Meter

Bruce Hamilton, Chris Thomas, Se Jin Park, and Jeong-Gon Choi

INTRODUCTION

Among the key questions in the context of smart grid is: what are the potential achievable benefits of advanced technology to the utility, to customers, and to society as a whole? And more importantly, how can we achieve them?

Smart meter deployments and ongoing pilot projects are revealing just how much transformation is needed in the industry, on both the utility and customer side. The early evidence is that the traditional utility business model alone isn't going to be sufficient to maximize the full potential of smart meter investments. Operationally, achieving the benefits requires dramatic changes in how utilities collect, store, and process information. Functionally this means that utility business rules and practices have to evolve as they become more integrated. While this is certainly a significant challenge to utility management, it's a relatively small problem when compared with the challenges on the customer side of the meter. Customer engagement has traditionally not been the domain of utilities, and it entails a series of challenges that are completely new to the industry.

As explained in Chapter 2, since the early days of electrification, customers have been passive ratepayers because meters only collect usage information for the utility, and do not communicate to the customer the real-time cost of their consumption.[1] Traditional metering only allowed the utility to bill for monthly consumption, using rates that reflect the average cost of electricity. Furthermore, utilities were given regulatory incentives to grow their networks, which meant they had incentives to encourage their customers to use more electricity. As noted in the introduction by Sioshansi, the culture of electricity consumption that has developed encourages customers to plug in without much thought about the cost, until they get a bill at the end of the month. Even then, for many people, the cost of electricity consumption is relatively small compared to their income; so many bills just get paid without a second thought.[2]

Today, there is increasing pressure to find a new model, as the culture of electricity consumption is slowly changing in response to the rising price of natural resources, concerns about energy security, and the environmental cost of electricity generation. Utilities around the world are investing heavily in smart meters with the expectation that consumers will respond to price signals and modify their usage (e.g., by shifting discretionary load to off-peak hours). But not all consumers will be interested in or able to respond to variable prices, as explored in Chapter 15. To truly bring about the full benefits of these new meters, technology beyond the electricity meter and in customers' homes will be necessary. Sometimes called "prices-to-devices," a new class of emerging technologies promises to help customers respond to prices and reduce their consumption easily and painlessly.

The critical question for policy makers, entrepreneurs, manufacturers, and others is how to transmit the price signal beyond the smart meter into the customer's home in a way that will get the response that economists dream of. In other words, what technologies and model of customer engagement will be the most effective?

There are several basic customer engagement models that can begin to bring new technology into everyday usage. In the short to medium term, large commercial and residential buildings will likely see the biggest advances in technology as the business models to deliver new technology emerge first for these customers. However over the long term, many of these same models are expected to be extended all the way down to small residential customers as technology costs decline and energy costs increase.

---

[1] This problem applies in varying degree to virtually all utility customers, but it applies more acutely to residential customers.

[2] This problem isn't limited to just the smallest residential customers; commercial customers have similar limitations in the data they receive about their consumption, even though their electricity bills are a large part of their operating expenses.

This chapter will examine the opportunities for change on the customer side of the meter and the new markets that will be created. We examine the current landscape for demand response, completed and ongoing pilot projects and deployments on the customer side of the meter, technological advances, and customer engagement models that have the potential to realize the promise of smart grid upgrades.

THE CURRENT LANDSCAPE FOR DEMAND RESPONSE

Today most of the focus on the customer side of the meter is on demand response that is operated as capacity resources. System planners see this demand response as primarily an asset to help manage stress on the grid during peak times. However, the emerging smart grid is enabling an entirely new class of demand response, including customer response to prices, more granular balancing opportunities, and overall reductions in energy usage.

The staff of the U.S. Federal Energy Regulatory Commission (FERC) has completed the most comprehensive survey of demand response potential in the United States.[3] FERC staff identified possible sources, and found that the largest driving factor for the future of demand response in the United States is the development of dynamic pricing and the coupling of dynamic pricing with technology, because technology greatly improves customer responsiveness.[4] This is sometimes referred to as sending "prices to devices."

Moving from a centrally controlled power system with passive consumers to a decentralized system with engaged consumers who respond to price signals involves a dramatic change in the way the system is planned and operated. This shift also raises a myriad of questions about how the industry must adapt to enable these developments: Is such a shift even technically possible? How will we get from here to there? How are customers going to find this new technology? And perhaps most importantly, what business models are going to encourage its growth?

Before answering these questions one must examine the current state of customer engagement.

Including the Customer

Policy makers and utilities have been aware of the opportunities that the smart grid can create for customers, and have been discussing these ideas for some time. However, until recently, customers were not a part of this conversation. As a result,

---

[3] Available at http://www.google.com/url?sa=t&source=web&cd=1&ved=0CBkQFjAA&url=http%3A%2F%2Fwww.ferc.gov%2Flegal%2Fstaff-reports%2Fdemand-response.pdf&ei=F2bc-Ta6ULJTRiAKX-qnzDw&usg=AFQjCNENkQhPAN_IlnwIknUkyk4_r51urQ.
[4] Faruqui, Ahmad and Sanem Sergici, "Household response to dynamic pricing of electricity—a survey of the experimental evidence," January 10, 2009.

there has been considerable backlash over some early meter installations.[5] While many in the industry are ready to move past these early problems, we believe they offer a critical lesson. Customers have to be included in the modernization of the system, and the change in technology creates an opportunity to begin engaging customers. However, as discussed in Chapter 15 the idea of getting consumers to become active participants in the market is still novel to many, and is a completely new idea to the average consumer who has been successfully trained to be a passive consumer.

A recent example helps illustrate the point. Commonwealth Edison Company has recently completed a pilot project in Northern Illinois testing in-home devices that allow customers to monitor their energy usage. In the early days of the pilot, only a handful of customers responded to prices, and only a small handful of customers even bothered to enable their new devices.[6] This shouldn't be surprising to many. Consumer Focus, the statutory consumer advocate in the UK, found out in research released in March 2010 that only 26% of people had even read their energy terms and conditions for electric service.[7]

This experience clearly demonstrates that we can't just expect customers to pick up technology and find immediate value. The importance of offering consumers effective communication, education, and choices about prices cannot be overstated. Customers have little knowledge of the cost of electricity and even when they do, it's hard to relate cost to usage, since customers are really only paying for heating, cooling, refrigeration and hair-drying, not for kilowatt hours. Customers are paying for electricity that does useful work in their homes. Successful new technologies will be technologies that help customer maintain the level of cooling, heating, hair-drying, cooking, etc. that they desire, while simultaneously helping them use less energy in the process.

NEW TECHNOLOGIES AND ONGOING PILOTS

Although the FERC Assessment of Demand Response focuses heavily on the role that pricing can play in a more responsive electrical system, it clearly notes that technology can play an even bigger role. Utilities and other actors are currently piloting many different pricing and technology programs for

---

[5]Pacific Gas and Electric (PG&E), a utility in northern California, and Oncor, a Texas utility, both ran into opposition when customers began reporting bills that seemed considerably higher than usual. While both cases were later found to be caused by a confluence of events, the industry and its regulators have begun to learn a valuable lesson: customer engagement is an important and necessary part of making the grid smarter. See http://www.bakersfield.com/news/business/economy/x1305354119/Bakersfields-SmartMeter-trouble-worries-industry.

[6]See http://www.intelligentutility.com/magazine/article/communication-101?quicktabs_11=1.

[7]http://www.consumerfocus.org.uk/files/2009/06/Consumer-Focus-response-to-Smart-Metering-Implementation-Programme-Consumer-Protections.pdf.

customers. This chapter highlights some notable pilots and the resulting business models that can help to get technology into customers' hands.

The ongoing and existing technologies and pilots can be broadly classified into two categories:

- The first is technology that seeks to engage the customer with direct feedback; and,
- The second group of technology allows the customer to "set it and forget it," as further discussed in Chapter 15 by Harper-Slaboszewicz et al.

Direct Feedback Technologies

It turns out that the effect of direct feedback on energy consumption is not a new idea. This issue has been researched by utilities and scientists for more than 30 years. These studies have consistently shown that feedback stimulates changes in behavior, which produce significant energy savings, as illustrated by researchers such as [1]. Table 16.1 summarizes the results of 38 feedback studies that took place over a period of 25 years, and found 21 studies that consider energy savings associated with direct feedback. Among the direct feedback studies included in the report, over half resulted in energy savings of between 10 and 20%. Other papers have found similar conclusions [2].

These findings are intuitive to anyone who's a student of human behavior. Direct feedback is a motivating factor at the gas pump, the grocery store, and elsewhere in the economy, but electricity customers don't receive any significant feedback today, other than an aggregate bill at the end of the month. Thus, electricity consumers don't have the opportunity to respond to changes

TABLE 16.1 Feedback Study Results

Savings	Direct Feedback Studies ($n=21$)	Indirect Feedback Studies ($n=13$)	Studies 1987–2000 ($n=21$)	Studies 1975–2000 ($n=38$)
20%	3		3	3
20% of peak			1	1
15–19%	1	1	1	3
10–14%	7	6	5	13
5–9%	8		6	9
0–4%	2	3	4	6
unknown		3	1	3

Source: *Making it Obvious: Designing Feedback into Energy Consumption*, Sarah Darby [1].

in electricity prices the way they might to gasoline prices or the price of bread and milk. This lack of feedback is so dramatic and persistent that many in the industry are skeptical that residential customers will take direct actions even in the face of significant feedback.

The pilot projects that exist today are investigating two main types of direct feedback mechanisms, briefly highlighted below:

- In-home energy display devices; and,
- Web-based energy usage monitoring.

In-Home Energy Display Devices

The underlying premise identified in the research in this area is that most people have no idea how and when their homes use energy. Their only information about energy use is a monthly bill, which provides only one lump sum of kilowatt hour usage and comes after the fact. This makes it almost completely impossible to correlate quantity and usage patterns. Addressing this fundamental problem has led to the development of a variety of in-home energy display devices that can help customers understand how and when they use energy, so that they have the opportunity to make intelligent choices about their usage.

The first-generation devices to monitor energy use in the home were fairly primitive and oriented more towards the technologically savvy consumers. The Energy Detective (TED), for example, is a device that can be plugged into the wall with appliances plugged into it. It then provides some real-time monitoring of those specific devices' energy use. Other devices such as the Blue Line Power Cost Monitor connect to an existing meter and monitor the whole house's energy use, rather than just that of a specific plug load. As shown in Figure 16.1, pilot projects studying these types of devices have shown fairly encouraging results.

Today literally dozens of in-home devices are available, but their capabilities vary quite substantially. Current in-home energy monitoring devices can be as sophisticated as a full home automation system with sophisticated energy use analysis, or as simple as an energy meter compiling a home's electricity usage over a given day. Two examples of devices currently used are those offered by Control4 and Black & Decker.

Control4 has introduced the EMS100 "home area network" automation system (www.control4.com/energy/products/) see Figure 16.2. The EMS100 is a sophisticated wireless, in-home display (IHD) connected to a wireless thermostat and other appliances in the home. The EMS100 also has control functionality that allows the customer to monitor energy usage and control appliances and other devices. Effectively, the EMS100 combines the functionality of many different devices into one central system. This combined functionality creates comprehensive load management capability that can be used in many different ways.

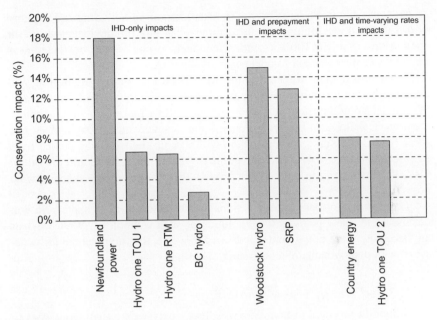

FIGURE 16.1   Conservation impacts by pilot. *(Source: [2])*

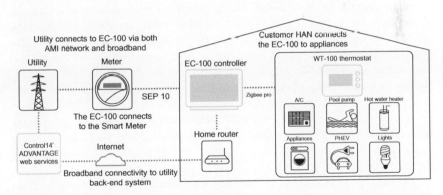

FIGURE 16.2   Control 4 EMS100 Schematic. *(Source: http://www.control4.com/files/energy/ Control4-EMS-Brochure.pdf. Accessed June 6, 2011.)*

At the other end of the spectrum is Black & Decker's EM100B Power Monitor. The EM100B is a relatively simple device, consisting of an in-home display device and an optical sensor that reads the meter. Accordingly, the EM100B is quite a bit cheaper than the Control4 device, and has far less functionality. However, the device is still capable of displaying electricity use, minute by minute, in dollars or kW. It appears that the EM100B's biggest advantage is that it is relatively easy to install, requiring no wiring and working with almost all electricity meters. Customers can buy the device and install it themselves. The EM100B provides feedback to customers, but customers have to take direct action to better manage their usage.

These two devices offer the same core functional energy monitoring and display capabilities; however, they perform these functions in different ways. Which device will customers desire? Which will be more engaging, stimulating bigger changes in behavior? What's the business model for devices like these? Can we assume that customers will pick up their favorite device from the local hardware store, install it, and make use of the information provided? Or, will a third party have to offer a product that limits the customers' upfront costs?

The answers to these questions remain to be seen. In this chapter, we will discuss some likely paths for the evolution of such technology. While devices are becoming more user-friendly and less expensive, the full potential for wide-scale mass deployments remains rather unclear.

Web-Based Energy Use Monitoring

In addition to physical in-home devices, less costly web-based approaches to engaging customers about their energy use have emerged and are evolving rapidly. These tools began as online energy audits sold to utilities as part of online portals. In these audits, the user was asked a number of questions about their home and provided with recommendations about actions that could be taken to reduce energy usage. To get accurate results, these audits often had to ask many questions, some of which required knowledge of the home that's beyond the average homeowner. At their best (such as the Nexus Software [now Aclara] audit) the questions were designed to be intuitive and structured so that the level of detail was flexible. Using the thesis that people don't understand traditional energy bills, Nexus redesigned bills and displayed them online with links to factors such as weather, usage, and the online audit results to provide customers with some context for their energy use.

One challenge is that electric utilities appear to be far less successful than other businesses in migrating customers to online bill payments.[8] There are a variety of reasons for this, but at its core, this trend is reflective of the limited relationships that utilities have traditionally maintained with their customers. As a result, the web portal of utilities has continued to not be a natural starting point for customers interested in their energy use. The possibility created by online energy audits and similar tools to engage customers is hampered by this

---

[8]See for example, *Has the Growth of Electronic Bill Presentment & Payment Reached a Plateau in the Utility Industry?* Available at: http://www.utilipoint.com/issuealert/article.asp?id=3041.

challenge. Utilities are beginning to realize that the smart grid can help them transform the relationship they have with their customers.

As utilities are struggling to redefine their relationship with their customers, other models, including third-party software platforms being promoted by consumer and environmental organizations, are beginning to emerge. For example, Google has developed the web-based PowerMeter, and other companies are working on extensions of the basic audit concept to develop more intensive behavioral change programs.

One example of the latter is the company Efficiency 2.0. This company has developed a consumer-focused Personal Energy Efficiency Rewards (PEER) program see Figure 16.3. PEER "is a multi-channel energy efficiency program

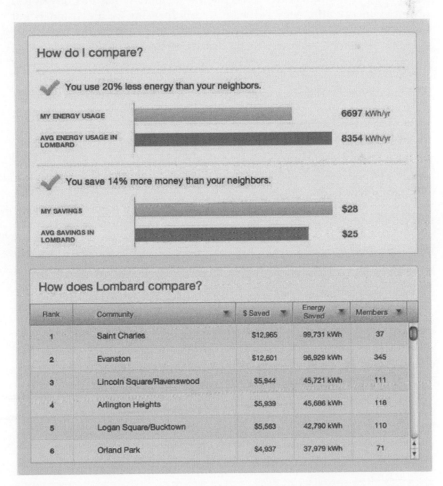

FIGURE 16.3 Efficiency 2.0 PEER comparisons. *(Source: http://efficiency20.com/community. Accessed June 6, 2011.)*

that generates verified energy savings for utilities through best-practice consumer marketing strategies."[9] Customers receive personalized energy savings reports through either print or email, gain access to a program website where they can set goals and compare themselves to others with similar characteristics, and even earn rewards points for saving energy.

Efficiency 2.0's PEER program is a consumer-first approach that uses a comprehensive strategy to generate energy savings. In contrast to a technology-first approach, which markets a specific product installation as a distinct program, a consumer-first approach generates savings by "empowering consumers to adopt a technology and conservation portfolio optimized to meet their specific needs and incentive requirements."[10]

OPower (www.opower.com) is another company that has developed advanced energy efficiency software, and is focusing on a consumer-first approach. Through a patent-pending, customer-engagement approach, OPower leverages behavioral science, customer data analytics, and software to engage customers. The company uses targeted messages to get customers' attention coupled with targeted suggestions for discrete actions to stimulate increased customer efficiency and price responsiveness. Again, the focus on the customer is what's driving this technology. The technology and approach are different, but the goals are the same as those of Efficiency 2.0.

These new web-based developments offer the exciting possibility of bringing the ideas of social networking to energy use.

The introduction of third-party models is also a dramatic shift. Consumer and environmental groups offer credibility on efficiency and savings messages that utilities simply do not have with many customers. The use of social media coupled with the third-party business models could fundamentally change the way that energy messages are delivered. Utilities have always maintained tight control over the messages they send to customers, but the involvement of third parties and the usage of social media channels are beginning to create a much more decentralized messaging system.

New online tools are rapidly evolving, and today it is hard to predict what is going to win out or which business model will be most successful. It could be large players like Google or small start-ups that carry the day, but at this point it's impossible to predict a winner. However, it's likely that the technology will continue to evolve rapidly and be considerably more accurate over the next few years, as increased evidence about customer motivations improves the effectiveness of the messages. More experience with decentralized messages delivered through social media by third parties will also bring significant change to the landscape. Overall, the online energy feedback mechanisms appear to be the most interesting and exciting areas of development for small customers in the short term.

---

[9] EE 2.0.
[10] EE 2.0 Website.

The Future Evolution of Direct Feedback Mechanisms

The fundamentally attractive aspect of web-based energy management tools is the lack of capital costs. The in-home display devices mentioned above are also enticing because some people like the physicality of them, but given the size of the average household energy bill, the cost of the devices is a barrier. By shifting information to the web, there are no hardware costs. Instead, the challenge is to harness the evolution of how people interact with the web. As the mechanism for interacting with the web moves from the computer to the smart phone, to perhaps other devices, such as TVs with built-in web browsing, web tools can adapt to the devices that people turn to for other streams of information, such as their smart phones.

A limiting factor in the development of more innovative web-based residential energy management systems has been the lack of quality energy use data, due to the fact that most residential customers still only have watt-hour meters. Many systems are being designed in anticipation of future increased availability of advanced metering infrastructure (AMI) systems that provide access to hourly or even more granular energy use data. But, a clear expectation gap currently exists. While large deployments are occurring, ubiquitous availability of granular energy use data is still years away for many households. Given this reality, the integration of direct feedback mechanisms with appliance and other in-home devices, further discussed below, may be even further away.

How households will use these new tools also remains to be seen, as most households remain on flat rates where changes in energy use during specific timeframes have a diluted value. Many pilot programs have found that customer do change energy usage when exposed to time-variant rates [3]. But the number of residential households actually on such rates continues to number in the thousands; 10,000 on critical-peak pricing in Bakersfield, California, 23,000 on real-time pricing in Illinois, and so on. While growth in these numbers is expected, compared to the over 100 million households in the United States, change will take time. Furthermore, the experience in states with open access is that many customers exposed to mandatory dynamic rates have elected to switch to flat priced contracts, further complicating the debate. That being said, GE, LG, and other manufacturers have begun to announce the launch of smart household appliances, a clear sign that the appliance industry is preparing for this change.

Prices-to-Devices Technologies

The tools described above rely on active customer engagement and a commitment to make changes. At the other end of the spectrum are the customers who don't have time or the inclination to actively monitor and change their energy use. To reach these customers, another class of technologies is developing. Sometime referred to as "prices-to-devices" or the "set it and forget" model, technology is being developed to automate response to prices and to save customers money without requiring daily usage monitoring and significant behavior change.

In 2007, in a project funded by the U.S. Department of Energy, Pacific Northwest National Laboratories teamed up with regional utilities and other industry partners to perform the first significant tests of smart technology, on the customer side of the meter. There were two separate, but related tests. The Grid Friendly Appliance Project found that appliances can automatically reduce their usage when it's necessary to keep the grid balanced, while the Olympic Peninsula Project found that homeowners were willing to allow devices in their home to respond to price signals, based upon a set of defined criteria.[11] Together, these studies determined that there are no technological barriers to the integration of such devices into the electricity grid. Furthermore, the study found that integrating these technologies into the grid could help avoid about $70 billion in new generation, transmission, and distribution systems in the coming decades.[12]

Still, the development of such devices around the country has been slow. The reasons for this are too many to discuss in this chapter; however, the fact remains that the increased granularity of communications and the control enabled by new smart technology have the potential to increase penetration of these new technologies. The same basic questions exist and are similarly unanswerable today: who will provide the technology, which channels will be used to provide it and what is the business model?

Two technologies for households are discussed below:

- Smart thermostats; and,
- Smart appliances.

Smart Thermostats

The smart thermostat[13] market is evolving. Thermostats control the largest peak-contributing load in most households and businesses: air conditioning. The traditional business model had been a thermostat provided by the utility to the customer in exchange for participating in a program; these are often called air-conditioning cycling programs, and there are large programs like these in existence around the United States. In these programs, the cost of

---

[11]http://gridwise.pnnl.gov/docs/pnnl_gridwiseoverview.pdf.

[12]Id.

[13]These thermostats should be distinguished from programmable thermostats that many energy efficiency programs have promoted. A programmable thermostat allows the user to set different temperature points for different time periods. A smart thermostat is a thermostat that in addition to the functionalities of a programmable thermostat can be controlled remotely. This can be either a one-way signal to the thermostat to change temperature for a set period of time, or a more sophisticated two-way feedback loop.

the thermostat, installation, and incentive payments would be put into the utility rate base after approval by the state PUC. Customers would have no choice about their thermostat model, and utilities would sign multiyear contracts with vendors to provide the thermostats. In some cases the vendors also provided the turnkey operations; in others the utilities managed it themselves. Comverge, Cooper Power Systems, Carrier, and Canon Technologies have been among the market leaders in these programs.

The model of a utility-supplied thermostat has come under pressure. California has some of the toughest building codes in the country, and they periodically ratchet up in terms of performance. In 2007 a proposal was developed to mandate programmable communicating thermostats (PCTs) for all new construction, and subsequently to phase them in over time for existing buildings. These thermostats would receive a radio signal from the utility to allow the utility to control load at critical times. In effect this would have mandated participation in demand response for new homes. In addition, rather than the utility supplying the thermostat, they would be sold through traditional retail channels. So the consumer going to the hardware store to buy a new thermostat, or the developer buying them in bulk for a new subdivision, would buy a thermostat that had built-in demand response functionality.

Days before the California Energy Commission was to have voted on making these thermostats part of the California energy code, news of the proposal found its way to the radio talk shows of controversial conservative commentators and was strongly attacked as being a "Big Brother" proposal to ration energy use and take choices away from consumers. The outcry on consumer privacy issues that ensued forced the Commission to shelve the proposal.[14]

The lesson learned from California is one of the challenges of transformation. The proposal would have vastly increased demand response, transforming it from a specialized utility program, operated at often uneconomical costs, into an embedded functionality of how homes and businesses work. However, customers didn't see the need to change. Utilities and policy makers didn't bring customers along.

Smart thermostats represent a significant opportunity to help reduce usage and better manage costs, but the models to deliver them to customers are still in flux, as we will discuss in more detail.

Smart Appliances

The Association of Home Appliance Manufacturers (AHAM) released a Smart Grid White Paper in December of 2009 which defined the term "smart appliance" as referring to a modernization of the electricity usage system of a

---

[14]http://www.turn.org/article.php?id=670.

home appliance so that it monitors, protects, and automatically adjusts its operation to the needs of its owner. AHAM lists the following six key features that are associated with smart appliances:

- Dynamic electricity pricing information is delivered to the user, providing the ability to adjust demand of electrical energy use.
- It can respond to utility signals, contributing to efforts to improve the peak management capability of the smart grid and save energy by:
  - Providing reminders to the consumer to move usage to a time of the day when electricity prices are lower, or
  - Automatically—shedding or reducing usage based on the consumer's previously established guidelines or manual overrides.
- Integrity of its operation is maintained while automatically adjusting its operation to respond to emergency power situations and help prevent brownouts or blackouts.
- The consumer can override all previously programmed selections or instructions from the smart grid, while ensuring that the appliance's safety functions remain active.
- When connected through a home area network and/or controlled via a home energy management system, smart appliances allow for a total home energy usage approach. This enables consumers to develop their own energy usage profile and use the data according to how it best benefits them.
- It can leverage features to use renewable energy by shifting power usage to an optimal time for renewable energy generation (i.e., when the wind is blowing or the sun is shining).[15]

Appliance manufacturers have been working to reduce the power consumption of their products since California's 1978 passage of Title 24. Title 24 was the first set of regulations to mandate energy efficiency standards. Energy efficiency standards are concerned with how much power an appliance uses, not when it is used. So, the emergence of new smart technologies and price signals generated by organized wholesale markets is beginning to change the way that manufacturers think about energy efficiency. From an electricity grid perspective, when an appliance runs is more important than how much total power it uses. Take refrigerators; for example, on average refrigerators actually only consume about 70 watts. However, the 10- to 15-minute defrost cycle can consume 10 times that amount.[16] Making refrigerators smarter by including a network connection can enable simple adjustments like helping to shift the defrost cycle to low-priced off-peak times.

---

[15]http://www.zpryme.com/SmartGridInsights/2010_Smart_Appliance_Report_Zpryme_Smart_Grid_Insights.pdf.

[16]See: http://www.tendrilinc.com/press-releases/tendril-teams-with-whirlpool-corporation-to-usher-in-new-era-of-smart-appliances/.

Although few smart appliances are available today, the global household smart appliance market is projected to grow to $15 billion by 2015.[17] Many different manufacturers are piloting new technologies in an effort to meet this predicted demand. However, none have yet found the right business model, as we will discuss. Below are some notable examples of activities being undertaken by major manufacturers; these manufacturers are focused on different aspects of the same theme: that automated response to prices will help lower energy costs.

Whirlpool

Whirlpool Corporation markets appliance in nearly every country around the world across all major categories, including fabric care, cooking, refrigeration, dishwashers, countertop appliances, garage organization, and water filtration.[18] The company has announced that all of its electronically controlled appliances will be smart energy compliant by 2015. These appliances will be able to respond to signals from the utility company to optimize operation using dynamic electricity prices.[19] In the short term, the company has even committed to make one million smart dryers by the end of 2011.[20]

Whirlpool's long-term deployment strategy includes short- and medium-term deployment and testing activities to ensure the company is headed in the right direction. In the years leading up to 2015, the company will initiate a 40-home pilot in Houston to study consumer behavior—the pilot will be conducted jointly with Direct Energy, Lennox, Best Buy, and OpenPeak.

The Houston pilot will involve washers, dryers, dishwashers, refrigerators, and water heaters at the 40 homes with products that communicate over a network. It also will install energy monitoring devices so that people can keep track of and adjust energy use.[21]

GE

GE is another worldwide appliance manufacturer looking closely at the smart appliance space. Beginning in November 2009, GE began distributing a type of hot water heater that is the first smart appliance available commercially in the United States.[22] Currently, GE has teamed up with Louisville Gas & Electric Company (LG&E) to create a pilot program that uses smart meters, smart

[17]http://www.zpryme.com/SmartGridInsights/2010_Smart_Appliance_Report_Zpryme_Smart_Grid_Insights.pdf,

[18]http://www.whirlpoolcorp.com/about/overview.aspx.

[19]http://grid2home.com/sg-consumer.html.

[20]http://www.greentechmedia.com/articles/read/whirlpool-plans-to-make-1m-smart-dryers-by-2011/\

[21]http://www.zpryme.com/SmartGridInsights/2010_Smart_Appliance_Report_Zpryme_Smart_Grid_Insights.pdf.

[22]http://www.zpryme.com/SmartGridInsights/2010_Smart_Appliance_Report_Zpryme_Smart_Grid_Insights.pdf.

appliances, and a tiered-pricing program. Participants in this pilot are GE employees living in the LG&E Louisville market, and they were provided with a suite of GE "smart" appliances to replace their standard appliances. Many installations included a refrigerator, range, microwave, dishwasher, and clothes washer and dryer. In addition, LG&E installed a programmable thermostat in the participants' homes. Examples of saving opportunities include:

- The refrigerator can delay the defrost cycle from occurring during peak hours. One participant's energy efficiency efforts with refrigerator usage over 10 weekdays showed over a 20% reduction of energy usage during peak hours.[23]
- Microwave ovens power down slightly by reducing wattage used when operated during peak hours.[24]
- The "smart" dishwasher can delay starting the cycle to off-peak times.[25]

LG

LG Electronics, another major electronics and appliance manufacturer, is planning to incorporate demand response technology in its new generation of smart appliances. The company's smart appliances use patented THINQ™ technology, which allows users to set their operation according to personal preferences and to control or monitor them via websites or smart phones and tablet PC applications. The interface is still evolving, but the goal is to make it easy for customers to make educated decisions about the trade-offs between comfort and convenience. LG is also working on smart grid projects around the world, including a residential energy storage system with Southern California Edison (SCE), a real-time energy pricing project in Germany, and a pilot project on Jeju island in South Korea.[26]

The Future of Smart Appliances

The major appliance manufacturers are all seeking the same core functionalities in their devices, although their technological and marketing approaches are different. Of the physical devices on the residential customer side of the meter, appliances may hold the most value for consumers and the electricity grid. Appliances seem to have the most potential to provide tangible benefits to residential consumers, provided that manufacturers can minimize the cost of installing the smart communication and control technologies in order to make the new functionalities more attractive to consumers. One limiting factor will be the lack of available granular usage information and dynamic pricing programs. As discussed above, the lack of such information and programs will reduce the

---

[23]http://pressroom.geconsumerproducts.com/pr/ge/smart_meter_pilot09.aspx.
[24]http://pressroom.geconsumerproducts.com/pr/ge/smart_meter_pilot09.aspx.
[25]http://pressroom.geconsumerproducts.com/pr/ge/smart_meter_pilot09.aspx.
[26]http://thinkd2c.wordpress.com/2011/01/26/lg-thinq-and-the-smart-grid/.

realized benefits of many smart devices, including appliances. In addition, there are no clear standards for the interconnection of smart appliances yet. This issue is still being debated and should be clarified as the market for such devices begins to grow.

EXISTING MODELS FOR CUSTOMER PARTICIPATION

There are four basic models of customer participation that have emerged for existing demand response and energy efficiency programs:

- Traditional utility model
- CSP (curtailment service provider), or aggregator model
- customer-provisioned model
- Energy services company (ESCO) model

The first model of participation is the traditional model of utility account representatives signing up individual firms to participate in utility-run offerings. In many cases the utilities have offered a site assessment to help businesses identify demand response opportunities and develop a demand response plan. This model functions most clearly in states that have not restructured and where the relationship between the utility and their larger customers has not evolved. In restructured states where commodity electricity is bought from a third-party supplier, these relationships do not exist in the same way.

The second model of participation is through an intermediary, and is the model seen in restructured states. These are either demand response firms ranging from aggregators such as EnerNOC to more focused curtailment service providers such as CPower. These firms take advantage of existing ISO/RTO and utility demand response programs and seek out businesses to participate in them. These aggregators replace the functions provided by utility account representatives and typically bring in the resources of their own centralized control facilities. EnerNOC in particular is well known for its ability to aggregate the usage of onsite backup diesel generation for demand response purposes. Some alternative electric suppliers, perhaps most notably Constellation Energy, also bundle these services in with their energy supply offerings. Unlike traditional utility-offered programs these firms operate in a deregulated landscape. This model is dependent upon the programmatic offerings of utilities and RTOs/ISOs, which often puts CSPs at odds with owners of traditional generation resources.

The third model of participation is through the customer purchasing and provisioning demand response technology and processes for themselves. Today, many large national retail chains have their own internal corporate policies regarding demand response as a way to manage operating costs and that develop and manage participation in programs at a national level. Firms like Wal-Mart and Target carefully monitor the operations of their stores in real time. The same aggregation of information can and is used for demand response purposes. The basic choices to change lighting levels or temperature settings provide consistent demand response, but in different geographical areas the value that can be achieved from those consistent actions varies greatly, as the standardized systems have to intersect with a patchwork of different RTO/ISO and utility demand response programs. This same problem also exists for appliance manufacturers and others looking to provide demand response services. As standards become clearer, and telemetry and control costs decline, large customer opportunities will become clearer, and we anticipate that smaller and smaller customers will be able to provision and maintain their own demand response.

The final model that is in use today is the energy service company or (ESCO) model. The ESCO model is based on the idea that there are savings created by the installation of new more efficient equipment that can be shared between installers and customers. This is sometimes referred to as the shared savings model. There has been tremendous excitement about the ESCO model in the past, but today, despite the success of many companies, the model has not taken off to any large degree. As of yet, property owners and managers have very little real-time information about their energy usage, so it is very difficult to validate that savings were produced by a single technological change. Accordingly, it is difficult to establish benchmarks that define the shared savings mechanisms. While the ESCO model has been successfully deployed, the potential is limited by data access and availability of usage information for many customers.

FUTURE DEVELOPMENTS: INTEGRATING TECHNOLOGIES WITH CUSTOMER PARTICIPATION MODELS

The emergence of more advanced communications and control technologies is creating new opportunities. Instead of a piecemeal approach to demand response, voluntary price response, and energy supply purchases, it appears that an integrated model is emerging. In areas with organized wholesale markets and retail competition, companies will begin to monetize energy savings and load shift through a fully integrated supply model.

New business models combining energy supply with demand response and energy savings opportunities are technically possible today. In effect, such a model combines energy supply with the CSP and ESCO models described above. Such a model will allow suppliers to use savings, from load shifting and efficiency gains, to create models that do not necessarily require customers to pay for efficiency improvement up front. This is a critical metric for widespread success. Programs and offerings need to be designed to minimize the up-front capital required of customers, providing a very short payback period for any investments made by customers.

These new business models are possible, in restructured markets, because a "hedge" against rising prices is a big piece of the value stream for customers. Today, many customers pay suppliers or utilities a flat (or tiered) rate to protect themselves from the volatility of real-time electricity prices. To the extent that physical systems can be used to limit exposure at peak times, the "hedge" is a piece of the current electricity bill that can be leveraged to pay for efficiency improvements or other equipment upgrades. The technology exists for this model to emerge; what's required are specific targeted solutions. These solutions will allow suppliers or other actors to monetize actions taken by physical systems to provide physical hedges that effectively replace, or reduce, the need to purchase financial hedges.

The Illinois Smart Buildings Initiative: Creating Dynamic Efficiency

The Illinois Smart Buildings Initiative (ISBI) aims to develop the fundamental architecture underlying the new model described above. The ISBI is a partnership developed by the State of Illinois Department of Commerce and Economic Opportunity and the South Korean Ministry of Knowledge Economy. Under the ISBI, companies from Illinois and South Korea are working together to enable Chicago skyscrapers to respond to price signals from the PJM market.

In order for an integrated approach to be truly successful, we need to think differently about demand response, energy efficiency, and even energy supply. Instead of optimizing building operation only for tenant comfort, as is the standard today, building operation has to be optimized for both tenant comfort and wholesale electricity market price signals, with tenant comfort being of the upmost importance—referred to as "dynamic efficiency." This may seem like an obvious connection to many, but in the building market, it's a radical operational change. Building owners and managers operate buildings very efficiently for comfort, while minimizing energy usage, but as we've discussed, timing of energy usage is more important than total consumption from an economic perspective, and from an environmental perspective in some circumstances.[27] The Illinois Smart Buildings Initiative is working to connect these piece parts in large buildings in Illinois.

Currently, ISBI partners from Illinois and Korea are beginning a feasibility study of opportunities in a set of candidate buildings and believe that the dynamic efficient operation of building systems is technically and economically possible. The biggest questions that the ISBI seeks to answer relate not to the technological solutions, but to the business process and interaction questions surrounding the sustainability of potential business models.

The advanced metering systems being deployed by utilities are increasing both the granularity of telemetry and control. As noted in this chapter, increasing telemetry and control opens a host of new opportunities for demand response. System operators will have more comfort that changes in demand will be available when they need them; in effect they will understand the

---

[27]In the U.S. Midwest, where the wind is strongest at night, it is possible to shift load to nighttime hours, reducing both the total energy bill and the total carbon footprint of a building, while keeping the total consumption the same.

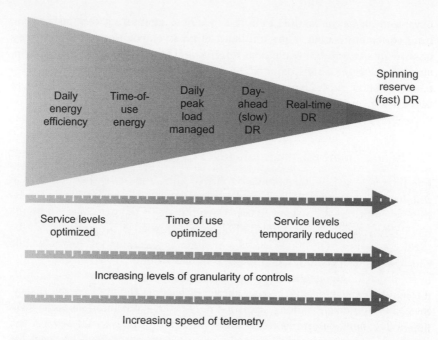

FIGURE 16.4 Linking demand response opportunities to the granularity of telemetry and control. *(Source: Demand Response Research Center at Lawrence Berkeley National Laboratories, M.A. Piette, S. Kiliccote, G. Ghatikar, Linking Continuous Energy Management and Open Automated Demand, November 2008, Presented at the Grid Interop Forum, Atlanta, GA, November 11–13, 2008.)*

reliability implications of demand-side programs, and with this understanding, the system can be planned in a more efficient manner. This is demonstrated in Figure 16.4.

As shown in Figure 16.4 the increasing speed of telemetry and control opens new and diverse markets for demand response participation. The deployment of AMI, and the increasing telemetry and control that it brings, is stimulating a wave of new emerging technologies, as we have discussed in this chapter. The current pace of technological advance has been huge, and the technology that will emerge over the next few years will be radically different from what's available today.

How different? The concept of dynamic efficiency introduced above is an operational solution that optimizes multiple, unrelated inputs in real time. As an example, large commercial and residential buildings today have advanced control and building automation systems.[28] These systems are typically referred to as building energy management systems (BEMS). Today, these systems are

---

[28] Although many have only limited functionality or have been value engineered in one way or another, limiting their functionality.

optimized for tenant comfort, total energy usage, and in some circumstances to limit exposure to high prices and participate in demand response programs offered by utilities and CSPs. But this optimization is constrained. The constraints come primarily from a lack of real-time information about the energy usage of equipment in a building. It's impossible to optimize real-time energy costs without such information.

In many circumstances, building operators have established fixed setpoints for chilled water temperatures, air handling capacity, and other large building system operational parameters. These setpoints are based upon the historic performance of the building and are designed to maintain tenant comfort. This system functions very well and has allowed operators to maintain tenant comfort for many years. However, new predictive methodologies, relying on the thermal mass of a building are emerging.[29] These methodologies create opportunities to maintain comfort using dynamic system setpoints.

Dynamic efficient operation of building systems has benefits for building owners/operators in the form of reduced operating costs, for tenants in the form of reduced energy costs, and for system operators in terms of more effective system planning and real-time operation, not to mention potential ancillary service opportunities that might be available from taking advantage of the momentum inherent in building air handling and chilled water circulation in real time.

More granular communication and control technologies in large buildings offer a host of potential benefits. We expect that these potential benefits offer the clearest path for new business models, and ongoing collaborative efforts like the ISBI offer the best opportunities to realize these models. There are several opportunities, for sustainable models, but in order to realize their true value, suppliers and other actors will have to find solutions that maximize the available telemetry and control.

Can We Relate the Model for Big Customers to Small Customers?

At this point, it's reasonable to ask: how do business models for commercial buildings relate to business models for residential customers? In fact, there are direct correlations; the only difference is the size of the potential load at each site. The models used to operate large loads are equally applicable for small loads, as predictive technologies, similar to those mentioned above, exist for residential air conditioning systems.[30] The only questions are the economics of the systems and the value of corresponding loads involved.

It is important to keep an eye on the continued development, testing, and consumer acceptance of emerging technologies, but it's equally important to keep an eye on the development of new business models. Devices create new

---

[29]Building IQ and Clean Urban Energy are two examples operating as of the publication date of this book.

[30]See Ecofactor, for example.

opportunities for demand response automation, and as the penetration of AMI increases and the available economies of scale for automated response increase, the cost should go down. When combined with appropriate rate designs, such as time-of-use rates, the impact of these distributed energy resources on peak loads could be significant. However, it will take significant evolution of existing business models to create new opportunities to deploy new technology and significantly affect system operation.

CONCLUSIONS

In conclusion, the electric grids of the future will likely look dramatically different from those that exist today. As technology improves and the granularity of communications and control increases, more opportunities for true demand-participation will arise. A truly dynamic demand side is possible with today's technology. However, for this to happen, both customers and system operators will need to perceive and be comfortable with the benefits of these technologies.

Admittedly, this will not occur overnight, and the scale and scope of such a change are not lost on the authors. It will take many years for the industry to be truly transformed by emerging smart technologies on the customer side of the meter. Until that time, traditional demand response and energy efficiency program models will continue to exist and become profitable distribution mechanisms for emerging smart technologies. However, in the long term, customer engagement models for energy efficiency and demand response will look more like smart phone apps and less like the cumbersome regulated models that exist today.

There are a multitude of new ideas emerging that will allow customers to better manage consumption both actively and passively. In the future, a combination of direct feedback and prices-to-devices technologies will likely meet varying customer preferences and needs. Although it's not yet clear how these technologies will find their way to customers, the potential of these technologies is expected to be unlocked in the coming years. As customers become more aware of new electricity usage management capabilities and as technology costs decline, markets will grow and benefits will materialize. Although it's difficult to predict that pace at which these markets will develop, one thing is abundantly clear: to maximize the potential created by utility smart meter investments, we have to go beyond the meter.

REFERENCES

[1] S. Darby, *Making it obvious: designing feedback into energy consumption.* Proceedings of the 2nd International Conference on Energy Efficiency in Household Appliances and Lighting. Italian Association of Energy Economists / EC-SAVE programme, 2000.

[2] A. Faruqui, S. Sergici, A. Sharif, *The Impact of Informational Feedback on Energy Consumption—A Survey of the Experimental Evidence*, 2010. http://www.brattle.com/_documents/uploadlibrary/upload772.pdf (accessed 23.05.11).

[3] A. Faruqui, S. Sergici, *Household Response to Dynamic Pricing: A Survey of Seventeen Pricing Experiments*, Social Science Research Network Working Paper, November 13, 2008.

Part IV

Case Studies & Applications

Demand Response Participation in Organized Electricity Markets: A PJM Case Study

Susan Covino, Peter Langbein, and Paul Sotkiewicz

Chapter Outline

INTRODUCTION

Economics and rapidly changing smart grid technology sit at the heart of the precipitous rise in active participation of demand response (DR) resources in wholesale electricity markets.[1] DR activity in the wholesale electricity market hinges on three things:

- The price of electricity
    - Historically users have paid a flat rate calculated by averaging the cost. Flat rates do not communicate to the user how the marginal cost of producing electricity varies over the course of a day or a year as more expensive generators are called on to meet the growing load. Nor do retail rates communicate to the end-use customer the value of the capacity that is necessary to be procured to ensure resources are available to serve load even on the hottest summer days.
- The value of electricity to the user
    - Since users have not been shown the actual cost of electricity as it changes from hour to hour and from season to season, they have not had the ability to evaluate the value of using electricity relative to the hourly locational cost revealed by the Energy Market. When users are able to see and respond to actual wholesale prices they can then decide to move usage to different times when prices are lower.
- The cost of getting outfitted to meet the requirements for market participation
    - These costs can include the hassle of arranging for electric distribution companies to install interval meters as well as an on-site assessment of load reduction capability and the installation of control devices to deploy that capability.

Transparency of wholesale market prices for capacity, energy, and ancillary services has helped drive DR participation is PJM's markets. As will be discussed in section "Demand Response Participation in the Capacity Market", with the implementation of the RPM Capacity Market, the transparency in the

---

[1] A demand resource has the capability of reducing usage in response to a price signal or direction from PJM Operations during an Emergency Event.

price of capacity to achieve resource adequacy with a loss of load expectation of 1 day in 10 years became immediately apparent, and the value of DR as a capacity resource has driven participation. As one electric distribution company executive stated, "I look at the relative capital costs of DR at $185/kw, combustion turbines fueled by natural gas at $1,850/kw and nuclear generation at $5,500/kw and ask why not install as much DR as possible before building more expensive generation."[2]

Similarly, DR participation in the PJM Energy Market, and its value for energy reductions, tracks the increase in the price of energy in wholesale markets that occurred as natural gas and oil commodity prices and electricity demand rose during the 2006–summer 2008 period as shown in section "DR Participation in the Energy Market (Day-Ahead and Real-Time)". And with the decrease in demand brought on by the recession and falling natural gas commodity prices through 2010, DR participation in PJM's Energy Markets has declined along with the value of energy reductions. DR participation in the Synchronized Reserve Market and in Ancillary Service Markets has also shown steady growth over time, as will be discussed in section "DR Participation in Ancillary Service Markets".

Concurrent with these economic trends and increase in transparency have been the development and evolution of smart grid technologies ranging from interval meters to automated controls such as programmable communicating thermostats. And over time these smart grid technologies have become increasingly affordable. See Chapters 11, 16, and 18 for a more thorough review of smart grid technologies. Not surprisingly as the price of capacity and/or energy has increased and the costs of metering, communications, automated controls and on-site generation have fallen the amount of flexibility in electricity usage increases.

The value of DR participation in PJM's markets should be reflected in the revenues paid to DR participating in each of these markets. As Figure 17.1 clearly shows, DR is receiving the majority of its value as a capacity resource (in gray) to help ensure resource adequacy. DR receives only a small fraction of value for providing energy reductions in the PJM Energy Market, and during 2009 and 2010 revenues from providing Synchronized Reserve have exceeded those revenues from providing reductions in the Energy Market outside of emergency conditions.

The purpose of this chapter is to recount the experience of the PJM Regional Transmission Operator (RTO) in enabling DR participation in its Energy, Capacity, and Ancillary Service Markets and to report the current state of DR participation ranging from self-scheduling an economic reduction in the Energy Market to clearing MW of load reduction capability in the three-year-forward RPM Capacity Market auctions.

Figure 17.2 shows the footprint and vital statistics of the PJM regional market as of June 1, 2011, with the integration of First Energy's American Transmission System, Incorporated (ATSI) and Cleveland Public Power (CPP).

---

[2]BG&E V.P. Wayne Harbaugh, dinner speech at Demand Response Symposium II, May 7, 2007.

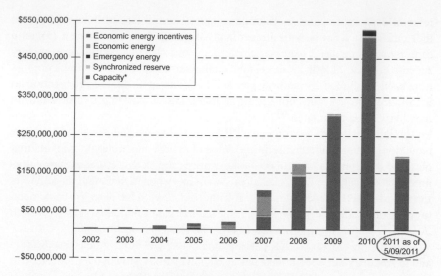

FIGURE 17.1   Estimated DR revenues by PJM market. *(Source: PJM)*

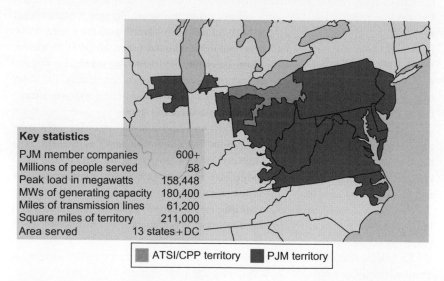

FIGURE 17.2   PJM footprint and key statistics.

Overall, PJM is the largest wholesale power market and centrally dispatched system in the world as of the writing of this chapter, and is vital to the production of almost 20% of U.S. GDP.

The remainder of this chapter is organized around market opportunities for DR and the relative value of DR as reflected by the revenues paid to DR shown in Figure 17.1. Section "Entities that Participate as DR in PJM's Markets" quickly reviews the entities that can participate as DR in PJM's markets. Section "Demand

Response Participation in the Capacity Market" describes DR participation in the RPM Capacity Market with a brief history of DR participation as a capacity resource prior to RPM. Section "DR Participation in the Energy Market (Day-Ahead and Real-Time)" provides a short history of DR participation in PJM's Energy Markets and describes the mechanisms by which DR currently participates in PJM's Day-Ahead and Real-Time Energy Markets. Section "DR Participation in Ancillary Service Markets" summarizes DR participation in Ancillary Service Markets. Section "Administration of DR in PJM's Markets" briefly outlines how PJM administers DR participation in its markets, and section "DR Participation Going Forward: Current Issues related to Demand Response" concludes with an examination of current and emerging issues facing DR participation in PJM's markets.

ENTITIES THAT PARTICIPATE AS DR IN PJM'S MARKETS

When one thinks of demand response (DR), it is natural to envision the end-use customer, whether it is a household a commercial customer, or an industrial customer that is engaged in reducing energy usage. However, these customers (with a few exceptions) are taking electricity service at the retail level and PJM operates at the wholesale level. Consequently, there needs to be a third party that stands between the end-use customer at the retail level, that is an active participant in PJM's markets at the wholesale level. Moreover, this party must be a member of PJM in order to participate in PJM's markets.

There are three types of entities that are PJM members that can participate as DR: Load Serving Entities (LSEs), Electricity Distribution Companies (EDCs), or Curtailment Service Providers (CSPs). LSEs are PJM members that aggregate and/or serve end-use customers at the retail level. LSEs may wish to participate as DR in PJM's markets as a means to hedge exposure to the costs of contracting for capacity to ensure resource adequacy or to engage in load reductions in the energy market to reduce their exposure to periods of high energy prices.

EDCs are PJM members that own or lease distribution networks at the retail level and deliver energy to end-use customers. In many cases EDCs are also LSEs and have the same types of incentives to participate as DR in PJM's markets. EDCs, as the distribution network operator, may also wish to provide DR as a capacity resource to PJM as a by product of maintaining reliability on their distribution networks by ensuring their network assets will not be overloaded on peak days.

Finally, CSPs are PJM members that specialize in aggregating DR at the retail level and providing those DR services to PJM's markets at the wholesale level. A CSP does not have load-serving responsibilities nor does it own distribution assets. A CSP has the incentive to provide DR in PJM's markets based on the economic incentives of payments to the CSP as the PJM member from PJM's markets, and in contracts it can negotiate with the end-use customers at the retail level to split the savings from providing reductions as a capacity or energy resource.

Participation in PJM's markets as DR comes with one caveat that stems of the FERC Order 719 proceeding and compliance filing process at PJM.[3] Order 719 required PJM to also address the need to more clearly delineate federal and state jurisdiction over DR activity in the wholesale market. Given PJM's obligation to administer its tariff in a non-discriminatory manner, the need for a workable way for PJM to recognize and implement decisions by retail regulators about demand response participation in the wholesale market by end-users subject to its jurisdiction became imperative as the level of CSP activity increased. PJM proposed a process for ascertaining and implementing the decisions of Relevant Electric Retail Regulatory Authorities (RERRAs) affecting participation by jurisdictional end-users in PJM's DR market opportunities.[4]

As alluded to in the introduction, it is the PJM member participating in the wholesale market that observes the transparent prices for Capacity, Energy, and Ancillary Service and responds to the incentives just described that has led to increased DR participation in PJM's markets over time, especially in the RPM Capacity Market that is discussed next.

DEMAND RESPONSE PARTICIPATION IN THE CAPACITY MARKET

This section describes how DR participates in the Capacity Market from the registration process to performance obligations and testing requirements. The minimum performance obligation, unchanged since 1991, requires that a participating end-use site be ready, willing, and able to reduce energy usage 10 times per summer season for a maximum of six contiguous hours per event at the direction of PJM Operations. This section also reviews the impact of the new three-year-forward auction for capacity procurement in 2007 known as the Reliability Pricing Model (RPM) on capacity prices and the resulting growth in CSP participation in PJM. The success of the PJM Capacity Market in developing DR has in turn led to the need to develop two new DR products also covered in this section.

Resource Adequacy and the Role of DR

PJM is responsible for resource adequacy planning for the PJM RTO footprint. In carrying out this responsibility, PJM develops load forecasts by EDC zone and RTO-wide, with input from the member EDCs including historical load data. In determining the peak load forecast PJM develops an "unrestricted peak load" forecast that does not account for DR that instead would be included

---

[3]See Wholesale Competition in Regions with Organized Electric Markets, Order No. 719, FERC Stats. & Regs. ¶31,281 (2008) and PJM Interconnection, L.L.C., Initial Filing, Docket No. ER09-701-000 (February 9, 2009).

[4]PJM Interconnection, L.L.C., Initial Filing, Docket No. ER09-701-000 (February 9, 2009).

as capacity (supply) for purposes of meeting resource adequacy requirements that ensure a loss of load probability of one day in 10 years. PJM also is responsible for determining the installed reserve margin (IRM), which along with the forecast peak load determines the resource adequacy requirements for a given Planning Year. PJM's own load forecast can be affected in two ways by an end-user's deployment of load reduction capability: (1) if the end-user simply deploys its load reduction capability during PJM's 5 coincident peak hours, which usually occur late in the afternoon of summer weekdays, then its reduced load will lower PJM's forecasted peak loads; or (2) if the end-use site participates as Load Management in the capacity market, an adjustment will be made to PJM's forecasted peak loads. In either event, the demand response will serve to lower the amount of generation resources needed for resource adequacy.

Brief History Before RPM

Prior to the implementation of the RPM Capacity Market in 2007, PJM operated a Capacity Credit Market that started in 1999. The Capacity Credit Market was a voluntary market for market participants to exchange capacity resources on a daily, monthly, or annual basis. DR participated in the old Capacity Credit Market construct under what was known as Active Load Management (ALM). DR participating as ALM would allow the LSE that had submitted the ALM to reduce its capacity obligation and therefore reduce its share of the cost of maintaining resource adequacy. The assignment of DR as a capacity resource by an LSE in this construct could be done with as a little as a day's notice for the next day, as capacity obligations could possibly change for competitive LSEs as load switched suppliers with in states in PJM that allowed retail competition.

However, prior to the implementation of the RPM Capacity Market, the incentive for LSEs to submit DR for participation in the Capacity Credit Market was relatively small. Table 17.1 shows the annual weighted average prices of capacity from 2002 to 2007 along with the amount of committed DR as ALM as of June 1 of each year from 2002 to 2006.

TABLE 17.1 DR Participation and Capacity Prices Under the Capacity Credit Market and ALM Construct*

	2002	2003	2004	2005	2006
Capacity Price ($/MW-day)	$33.40	$17.51	$17.74	$6.12	$5.73
ALM Quantity (MW)	1,342	1,265	1,412	2,035	1,655

*2007 State of the Market Report, Table 2-97, p. 107 and Table 5-10, p. 246.

As Table 17.1 shows, and will be contrasted below, the price of capacity in the previous construct provided little incentive for DR to participate as a capacity resource as the market was signaling that there was little value as a capacity resource.

RPM Capacity Market

The RPM Capacity Market introduced a paradigm shift in how LSEs could satisfy their capacity obligations and how DR could participate as a capacity resource. RPM created a three-year-forward auction for capacity (known as the Base Residual Auction or BRA) where LSEs had to cover their capacity obligations. This is a paradigm shift from the previous Capacity Credit Market that allowed assignment of resources up to the day before they were needed. The three-year-forward BRA allows both *planned* DR and *planned* generation resources to compete with existing resources. LSEs, EDCs, or CSPs that clear DR in the BRA, have approximately three years to sign contracts with end-use customers and outfit end-use sites before registering those sites in eLRS (eLRS is PJM's DR administrative platform and will be discussed in section "Administration of DR in PJM's Markets").

The participation of DR in the RPM Capacity Market, which utilizes the Reliability Pricing Model (RPM), occurs by means of Load Management, which is the ability to reduce metered load either manually or automatically based on a communication signal. Load Management is broken into two categories: Demand Resource (DR) and Interruptible Load for Reliability (ILR). CSPs, EDCs, and LSEs can aggregate load reduction capability from multiple end-use sites for Load Management. Both DR and ILR must be willing and able to, at a minimum, be interrupted up to 10 times each delivery year for up to 6 hours per interruption as was the case under the ALM and Capacity Credit Market construct.

RPM not only created the opportunity for DR to be offered into the auction on par with generation but also preserved through Interruptible Load for Reliability (ILR) the ability of an end-use site to be certified as Load Management only three months before the beginning of the Planning Year (June 1 to May 31). This alternative path for DR to become a capacity resource in RPM was based on the feedback from large, industrial customers that three-year-forward commitments were unrealistic and risky for them because they couldn't guarantee that they would still be in business when the Planning Year arrived. The difference in value for the earlier commitment was recognized in the related revenue stream from RPM. DR receives the auction clearing price for its three-year-forward commitment, while ILR payments are discounted reflecting the lower risk of committing only three months in advance of delivery.[5]

---

[5]See § 5.8.5 "Zonal ILR Prices," PJM Manual 18: PJM Capacity Market for the rules for calculating Preliminary and Final Zonal ILR Prices. These payments are discounted by the value of the Capacity Transfer Rights, which is the value of transmission import capability into a constrained area.

ILR, however, will expire at the end of the 2011/2012 Planning Year. Anticipation of ILR expiration explains the significant increase in cleared DR beginning with the 2012/2013 Planning Year. The growth in MW of CSPs' zonal portfolios and experience with Load Management have mitigated the concerns that preserved ILR in RPM's original design.

There are three types of Load Management products that can serve as capacity: Direct Load Control (DLC), Firm Service Level (FSL), and Guaranteed Load Drop (GLD). The CSP, EDC, or LSE sends a communication signal to DLC customers, which are typically small customers without hourly interval meters, that controls and cycles usage by end-use devices like central air conditioning units, hot water heaters, and pool pumps. Each FSL participant agrees to reduce usage to a pre-determined level upon notice from the CSP, EDC, or LSE. GLD requires the DR or ILR to make a reduction of a pre-determined amount upon notification from the CSP, EDC, or LSE. Moreover, there are two notification periods that apply to both the FSL and GLD products. Under Short Lead Time notification, DR or ILR must respond within one hour of notification, while under Long Lead Time notification, DR and ILR have up to two hours to respond with reductions.

Participating end-use sites must have metering equipment that provides hourly integrated values. The exception to this requirement is the Direct Load Control product, for which the CSP must submit an engineering study every five years that supports the load reduction value using a PURPA[6] compliant sampling and verifies the number of switches in the field. CSPs may also rely on a matrix of load reduction values developed for Direct Load Control devices on central air conditioners with the assistance of the Lawrence Berkeley National Laboratory in 2006 but the MW of load reduction capability will be significantly discounted in the absence of a timely engineering study.[7]

The applicable EDC or LSE serving the end-use site has the option to review the registrations and report to PJM "other contractual commitments" of the load reduction capability that must be accommodated or may justify denial of the registration. This review enforces a key principle that load reduction capability should be paid only once for performance in any hour. The applicable EDC or LSE serving the end-use site also has the option to review energy settlements submitted by the CSP.

Beginning in 2009 all registered end-use sites that have not responded to an Emergency Event by August 15 must schedule a one-hour test and submit meter data to PJM that demonstrate the capability to reduce load in the amount of MW registered as Load Management. CSPs may choose to conduct the

---

[6]Electric Utility Rate Design Study, REFERENCE MANUAL AND PROCEDURES FOR IMPLEMENTING PURPA, A Report to the National Association of Regulatory Utility Commissioners, March 1979, as amended.
[7]http://www.pjm.com/documents/~/media/documents/reports/20070406-deemed-savings-report-ac-heat.ashx.

mandatory test any time during the summer period but the results will be ignored in the event of a subsequent call for the DR during an Emergency Event.

Market Results

Figure 17.3 shows the growth in DR participation year after year between Planning Year 2005/2006 and Planning Year 2014/2015. It also shows the last two planning years when the ALM and Capacity Credit market construct were in place to provide a contrast to participation under RPM. The Reliability Pricing Model (RPM) was implemented at the beginning of the 2007/2008 Planning Year. Participation has increased almost tenfold since RPM was implemented, from just over 1,600 MW of ALM in 2006/2007 to almost 16,000 MW of DR offering into the 2014/2015 Base Residual Auction that was conducted in May 2011.

The increase in DR participation has been driven by the value of capacity that has become far more transparent under the RPM Capacity Market. Figure 17.4 shows the evolution of RPM Capacity prices since implementation. It is clear,

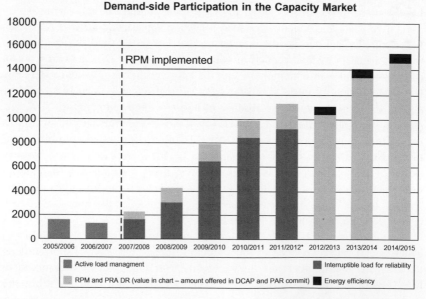

FIGURE 17.3 DR participation in the RPM capacity market.[8] *(Source: PJM)*

---

[8]"2014/2015 Base Residual Auction Results," p. 10, available at http://www.pjm.com/markets-and-operations/rpm/~/media/markets-ops/rpm/rpm-auction-info/20110513-2014-15-base-residual-auction-report.ashx.

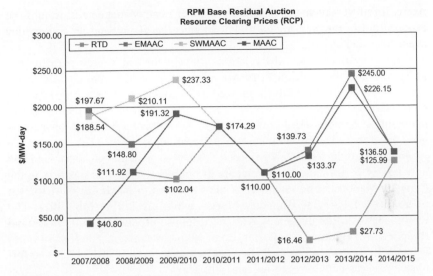

FIGURE 17.4 Evolution of RPM capacity prices.[9] *(Source: PJM)*

contrasting the prices in Figure 17.4 with those in Table 17.1, that price incentives increased DR participation under RPM.

Prices in eastern PJM, which is more constrained than in the west, rose from a few dollars per MW/day before RPM implementation to over $200 per MW/day. The higher capacity prices in constrained areas revealed by the forward, locational RPM auctions have attracted many new CSP market participants. The introduction of locational capacity prices has also increased substantially the efficiency of the Capacity Market. Resource providers, whether of DR or generation, can site resources where they have the most value and by increasing supply in a constrained location will put downward pressure on prices over time by expanding supply. Correspondingly, the end-user customers whose sites are located in constrained areas have a strong incentive to deploy load reduction capability by working with CSPs to reduce their electricity bills.

Expansion of DR Products to Ensure Reliability

The dramatic increase in Load Management as a percentage of the forecasted unrestricted peak load from 1.2% for the 2006/2007 Planning Year to 6.3% for the 2010/2011 Planning Year, as seen in Figure 17.3, prompted PJM to analyze the impact of the limitation on the number of calls and the number of consecutive event hours on reliability. Historical analysis had determined that

[9]"2014/2015 Base Residual Auction Results," p. 10, available at http://www.pjm.com/markets-and-operations/rpm/~/media/markets-ops/rpm/rpm-auction-info/20110513-2014-15-base-residual-auction-report.ashx.

Active Load Management, the predecessor of Load Management, should not exceed 7.5% of the unrestricted peak load given the 10 call and six consecutive event hour limitations.[10]

The analysis explored the relationship between the limited performance obligation and the limit on Load Management as a percentage of the unrestricted peak load. The analysis showed, for example, that the number of consecutive hours of reduction would need to increase from 6 to 10 in order to maintain the number of interruptions at 10 and to permit a Load Management level equal to 8.5% of PJM's unrestricted peak load.[11]

In late 2010 the PJM filed proposed Load Management revisions to add two new sets of performance requirements as options for DR participation in the Capacity Market.[12] The intention of the filing was to maximize participation of qualified DR in the Capacity Market while preserving reliability. The FERC approved the three new demand response products, Limited Demand Response (previously known as Load Management), Extended Summer Demand Response, and Annual Demand Response, in early 2010.[13] The performance requirements for each are set forth below.

Beginning with the Base Residual Auction held in May of 2011 for the 2014/2015 Planning Year CSPs may offer MW of DR that can meet one or more of three different sets of performance requirements.[14] First, PJM's revised market rules have preserved the existing Load Management regime, 10 events for a maximum of six consecutive hours per event between noon to 8:00 PM non-holiday weekdays from June through September, renamed it "Limited Demand Response" and limited the MW quantity that can clear in the BRA. Second, a new product, "Extended Summer Demand Response," must be available on any day from May through October between 10:00 AM and 10:00 PM for a maximum of 10 consecutive hours and will have limitations on the MW quantity permitted to clear in the BRA. Third, a new product, "Annual Demand Response," must be available for a maximum of 10 consecutive hours between 10:00 AM and 10:00 PM from May through October and between 6:00 AM and 9:00 PM from November through April for an unlimited number of events and will have no limitations on the quantity permitted to clear in the BRA. These products and their characteristics are summarized in Table 17.2.

---

[10]PJM Resource Adequacy Planning Department, "Demand Resource Saturation Analysis," May, 2010.
[11]Id. at pg 8.
[12]PJM Interconnection, L.L.C., Initial Filing, Docket No. ER11-2288-000 (December 2, 2010). Stakeholders were unable to achieve required consensus in support of a solution, so the PJM Board of Managers directed staff to file the proposed changes.
[13]PJM Interconnection, L.L.C., 134 FERC ¶61,066 (January, 31, 2011).
[14]Ibid.

TABLE 17.2 DR Products Available Beginning with the 2014/2015 Delivery Year

Requirement	Limited DR	Extended Summer DR	Annual DR
Availability	Any weekday, other than NERC holidays, during June–Sept. period of DY	Any day during June–October period and following May of DY	Any day during DY (unless on an approved maintenance outage during Oct.–April)
Maximum Number of Interruptions	10 interruptions	Unlimited	Unlimited
Hours of Day Required to Respond (*Hours in EPT*)	12:00 PM–8:00 PM	10:00 AM–10:00 PM	Jun–Oct. and following May: 10 AM–10 PM Nov.–April: 6 AM–9 PM
Maximum Duration of Interruption	6 Hours	10 Hours	10 Hours
Notification	Must be able to reduce load when requested by PJM All Call system within 2 hours of notification, without additional approvals required		
Registration in eLRS	Must register sites in Emergency Load Response Program in Load Response System (eLRS)		
Event Compliance	Must provide customer-specific compliance and verification information within 45 days after the end of month in which PJM-initiated LM event occurred.		
Test Compliance	In absence of the PJM-initiated LM event, CSP must test load management resources and provide customer-specific compliance and verification information.		

Source: PJM.

DR PARTICIPATION IN THE ENERGY MARKET (DAY-AHEAD AND REAL-TIME)

This section provides a history of participation of DR in PJM's Energy Markets and a conceptual explanation of the incentives for DR to reduce energy consumption given the differences between the wholesale energy price and retail energy price, and it describes how DR can participate in the Energy Market as Economic Load Response and as Emergency Load Response. When a DR participates as Economic Load Response, then it must change its normal usage pattern in response to hourly Energy Market price signals in order to be compensated. When a DR participates as Emergency Load Response, then it must show how much it reduced its usage in response to the Emergency Event called by

PJM Operations. This section will also report the resulting level of Energy Market participation by DR as well as the benefits to the Energy Market provided by DR participation. The requirements of FERC's recently issued Order 745 are also included in this section.

Energy Market Benefits of Integrating Economic Load Response

The incorporation of DR in the Energy Market allows end-user customers to express their willingness to pay for power, which enhances market efficiency relative to markets where consumer demand is taken as a given regardless of price. As consumers are allowed to respond to prices (and provide additional supply), market competitiveness is enhanced, and potential exercises of supplier-market power are checked as any supplier attempts to raise price will result in a corresponding reduction in demand, making such attempts less profitable. At the level of the end-use customer, the incorporation of DR in the Energy Market provides the opportunity to control electricity expenditures. The empowerment of end-use customers to reduce demand in response to high energy prices provides a mechanism to control the total electric bill while providing reliability benefits in short-term operations.

History

DR participation in PJM's Energy Markets can be traced back to 2002 when following forms of DR participation in the Energy Market were approved by the FERC: (1) DR in response to price, known as Economic Load Response,[15] and (2) at the direction of PJM Operations, known as Emergency Load Response.[16] The FERC Orders authorized these Emergency and Economic Load Response *Programs* to run for three summers, 2002, 2003 and 2004, and to expire on December 1, 2004. The summers of 2003 and 2004 passed with modest growth in the number of sites participating.[17] PJM developed and delivered DR training for market participants, later known as Curtailment Service Providers (CSPs) as described in section "Entities that Participate as DR in PJM's Markets".[18] In October 2004 FERC extended the Emergency

---

[15]PJM Interconnection, L.L.C., 99 FERC ¶61,227 (2002).

[16]PJM Interconnection, L.L.C., 99 FERC ¶61,139 (2002).

[17]167 sites representing 662.7 MW participated in the Emergency Load Response Program and 248 sites representing 468.7 MW participated in the Economic Load Response Program in 2003. 3,873 sites representing 1,557.9 MW participated in the Emergency Load Response Program and 2,466 sites representing 1,644.4 MW participated in the Economic Load Response Program in 2004 according to the 2005 State of the Market Report (pp. 72–75). It is important to note that participating end-use sites could only be registered in one of the programs at a time and that participation as capacity referred to as Active Load Management was a separate program.

[18]The term "Curtailment Service Provider (CSP)" describes the function of aggregating the load reduction capability of end-users for the purpose of participating in PJM's markets. Any member or special member of PJM can chose to fill the role of a CSP.

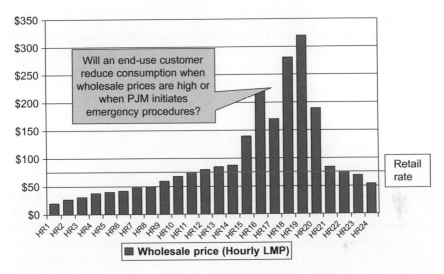

FIGURE 17.5   Incentives for end-use customers to respond to wholesale prices. *(Source: PJM)*

and Economic Load Response Programs until December 31, 2007.[19] PJM later filed market rules for permanently integrating demand response into the Energy Market in late 2005.[20] FERC approval followed in early 2006.[21]

Incentives to Respond to Prices. Figure 17.5 highlights the incentives that end-use customers, LSEs, EDCs, and CSPs, have in responding to prices whether it is for economic reasons, or in an emergency. End-users paying flat, fixed rates at the retail level have little incentive to reduce usage during periods of high prices at the wholesale level or during emergencies as they face the same flat retail rate regardless. However, if the end-use customer works through PJM members that are wholesale market participants and that can directly see and are exposed to wholesale energy prices (Locational Marginal Prices or LMPs), both the end-use customer and the wholesale market participant can work together to effectively create a dynamic rate through the combination of retail and wholesale prices that can benefit both parties.[22]

In Figure 17.5, when wholesale prices are above the retail rate, a reduction in energy consumption by an end-use customer in hour ending 20 will result in a savings of just over $300/MWh in the wholesale market to the LSE that serves the end-use customer. However, the LSE also does not sell that same MW of energy at the flat retail rate of approximately $70/MWh in Figure 17.5. On net the LSE saves the $300/MWh wholesale price (LMP) (power it

[19]PJM Interconnection, L.L.C., Letter Order, Docket No. ER04-1193-000 (October 29, 2004).
[20]PJM Interconnection, L.L.C., Initial Filing, Docket No. ER06-406-000 (December 28, 2005).
[21]PJM Interconnection, L.L.C., 114 FERC ¶61,201 (2006).
[22]See Chapter 3 in this volume for more detailed analysis of dynamic retail pricing.

did not have to buy in the wholesale market), but does not sell the $70/MWh retail energy, thereby realizing a net savings of $230/MWh. This net savings is often referred to as "LMP minus G," where G represents the retail rate. Moreover, since prices during emergency conditions are likely to be very high, the economic incentives to reduce energy consumption coincide with the needs of the system for energy reductions to help maintain reliability. The manner in which the LSE (or CSP) splits the savings between itself and the end-use customer is a contractual matter between the parties and is not overseen by PJM.

Market Mechanics

In PJM's Energy Market, registered DR participates in Economic Load Response by reducing electricity consumption when LMPs are high relative to the retail price paid by the participating end-use site. Economic Load Response enables users for the first time to see and respond to dynamic wholesale electricity prices and to benefit from changing normal usage patterns in response to LMP prices that are significantly higher than the generation and transmission portion of the retail rates they are charged. Participation in Economic Load Response can occur in either the Day-Ahead or Real-Time Energy Markets. All DR—except, DR that purchases electricity pursuant to contracts with Real-Time LMP prices—is eligible to participate in the Day-Ahead Energy Market.

CSPs offering load reductions in the Day-Ahead Energy Market on behalf of registered end-use sites may offer of a quantity of load reduction at a price, as well as shutdown costs and minimum down times for which the reduction must be committed, like the price, quantity, start-up costs, and minimum run times of generators. Load reduction offers that clear in the Day-Ahead Market receive the Day-Ahead LMP. Like generators, registered DR may provide up to 10 price, quantity pairs in eMarket that are used to clear the Day-Ahead Energy Market. Any deviations from Day-Ahead commitments are settled at the Real-Time LMP along with applicable Balancing Operating Reserve charges. Participation by DR in the Day-Ahead Energy Market has been relatively light.

CSPs have two options for participation in the Real-Time Energy Market as part of Economic Load Response: real-time dispatch or as a self-scheduled resource. To be dispatched in the Real-Time Energy Market using the Unit Dispatch System (UDS), a CSP submits an offer for a DR resource that indicates an LMP strike price, a price at or above which the DR commits to make a reduction, the time needed to shut down, and the minimum time the end-use site must reduce load. Likewise, registered DR may provide up to 10 price, quantity pairs in eMarket for use by the Unit Dispatch Software (UDS) as shown in Figure 17.6.

To exercise the self-schedule option, the CSP must provide notification to PJM no less than five minutes prior, and no more than seven days prior, to the demand reduction. The notification must include the start and stop times for the demand reduction as well as the quantity of the demand reduction.

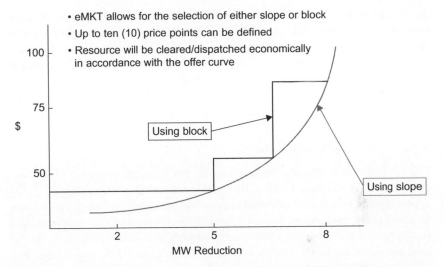

FIGURE 17.6 Example of a DR offer curve in the energy market. *(Source: PJM)*

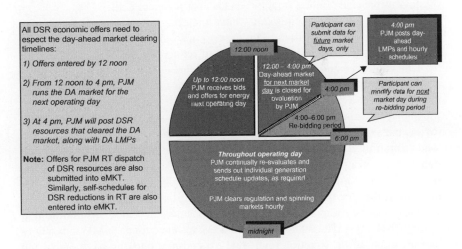

FIGURE 17.7 Options for DR participation in PJM's energy markets. *(Source: PJM)*

Figure 17.7 illustrates how the CSP can utilize the three options for participating as Economic Load Response in the Energy Market: offer into the Day-Ahead Market, offer to be dispatchable by PJM in real time, or self-schedule the load reduction through the notification process in real time.

The CSP may submit a load reduction offer in the Day-Ahead Market by noon of the day before the operating day. PJM evaluates the offer including the shutdown costs and minimum down time. If the offer clears in the Day-Ahead Market, then the CSP will be notified by 4:00 PM of the day before the operating day. If the offer does not clear in the Day-Ahead market, the CSP may submit the same offer or a new offer on behalf of the end-use site for dispatch by PJM in the Real-Time Market. PJM will continually re-evaluate the offer by comparing alternatives throughout the operating day. If PJM dispatches the load reduction in real time, then an email notification will be sent to the CSP. Finally, if PJM does not dispatch the load reduction in real time, the CSP may at the end-use customer's direction self-schedule the load reduction.

Measuring Load Reductions

The load reduction quantity for any hour is the difference between the calculated Customer Baseline Load (CBL) and the metered usage. The CBL is an approximation of what the DR resource would have consumed absent the reduction and is the mechanism that enables DR to effectively be represented as a supply-side resource in the PJM Energy Market software. CSPs must submit load reduction settlements within 60 days of the Load Reduction Event. The 60-day period accommodates end-use sites that rely on EDC meter readings for the settlement of Economic Load Response activity. Many EDCs require 60 days to provide "billing quality" meter data for these settlements.

Participating end-use sites must have metering equipment that provides hourly integrated kWh values. The exception to this rule is the aggregation of registered end-use sites that achieve load reductions through switches that the CSP controls using radio signals or power line communications. The stakeholders developed metering requirements for non-EDC hourly interval meters including pulse recorders in 2008.[23]

The applicable EDC or LSE serving the end-use site has the option to review the settlement data provided by the CSP and to indicate to PJM if the load reduction was the result of the plant's annual shutdown, for example, rather than a change in usage in response to wholesale price signals. Approved settlements are paid by PJM on the basis of the quantity of the load reduction times the difference between the zonal, aggregate, or nodal (as applicable) LMP and the end-use customer's generation plus transmission charge (also referred to as the "retail rate"). The total savings for a DR resource for each MWh of reduction are equal to the avoided retail rate expenditure plus the payment of the LMP minus the retail rate, which adds up to a savings of the applicable LMP.

Settlements in the Energy Market

PJM, in order to pay a settlement or credit the CSP for the end-use customer's Economic Load Reduction activity, must collect the equivalent amount or debit from another wholesale market participant. PJM has put in place a settlement and accounting mechanism that accomplishes both the payment to the CSP

---

[23]PJM Interconnection, L.L.C., Initial Filing, Docket No. ER09-1508-000 (July 28, 2009).

and the collection from the LSE serving the end-use customer. The following hypothetical demonstrates how the settlement and accounting mechanism works.

An end-user site engaging in Economic Load Response would receive the wholesale market price for energy, LMP, less the applicable energy and transmission portion of the retail rate G&T, LMP – G&T, for each MWh of energy consumption avoided below its consumption baseline or CBL. If the G&T rate is $70/MWh and the LMP is $100/MWh, the payment received by the demand responder from the wholesale market for a 1 MWh reduction in consumption is $30. PJM collects the $30 payment from the end-use customer's LSE of record.

The end-use customer engaging in Economic Load Response also avoids paying G&T, which in this hypothetical is $70/MWh for each MWh of avoided energy consumption. Therefore, the demand responding end-use customer obtains a value of $100/MWh for its demand response in a wholesale market context ($30/MWh direct payment plus $70/MWh avoided cost). The LSE has now either saved $100 by avoiding the purchase of electric power at $100/MWh (if satisfying that obligation in the spot market), or has received $100 from the spot market for the excess MW purchased for but not consumed by the end-use customer.

Another way of viewing the settlement for DR resources participating in Economic Load Response is that DR resources are being exposed to the market price of electricity, while still taking service under a regulated retail rate if being served by a traditional EDC/LSE or at a competitive retail rate if being served by a competitive LSE. DR can in this way be exposed to market prices for reductions while remaining hedged for remaining consumption at the retail rate.

There is in fact a subset of end-use customers that pay retail rates that are not fixed, but are so-called "LMP-based" rates. Customers on "LMP-based" rates already buy all or part of their electricity at the Real-Time LMP wholesale price. Since these customers already see wholesale market prices and see directly the bill impacts of LMP-based rates, it was a challenge to develop market rules for them to participate in the Economic Load Response Program. The solution required their load reduction capabilities to be "dispatchable" by PJM in real time. Following dispatch of the DR, PJM assumed the risk of falling LMPs, which ensured that the end-user would realize savings at least equal to the savings calculated on the basis of the "strike price" it provided to PJM. If, for example, PJM dispatched a load reduction with a "strike price" of $100/MWh for four hours and the LMP fell from $150/MWh in hour one to $50/MWh in hours two, three, and four, then the PJM market would make the participant whole for the anticipated savings of $400.

Market Results

The trends in activity in terms of MWh reductions and settlements paid out on a monthly basis are shown in Figure 17.8. As can be seen in Figure 17.8, throughout 2006 and 2007 MWh reductions increased and so did the

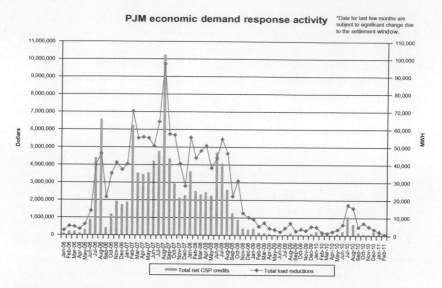

FIGURE 17.8   DR activity in PJM's energy markets 2006–2011. *(Source: PJM)*

dollar value of settlements. Activity and settlements remained fairly steady through the first seven months of 2008 and have fallen off precipitously since then.

The increase in activity coincides with a period of increasing energy prices overall in PJM during the 2006–2008 period. Beginning in September 2008, which also coincides with the recognition that the economy was falling into a deep recession with the financial crisis, PJM energy prices also fell and have remained low relative to levels seen during the 2006–2008 period. Figure 17.9 shows the monthly evolution of PJM load-weighted average LMP from 2006 to 2011.

Moreover, some activity prior to 2008 was likely driven by the incentive payments available to DR in the Energy Market for load reductions occurring at prices above $75/MWh. This incentive was allowed to expire at the end of 2007 per the terms of the PJM Tariff and Operating Agreement.[24]

Economic Load Response participation is shown in Table 17.3. It is important to note when analyzing the payment data that from 2002 through October of 2007 the incentive payment was of the full LMP rather than LMP–G&T for load reductions in hours when the applicable LMP equaled or exceeded $75/MWh. Payment of the incentive required PJM to collect the difference between the LMP and G&T from the end-use customer's LSE, and the G&T portion of the payment from all of the LSEs in the zone on a weighted basis. The allocation of revenue among Economic Load Response activity in

---

[24]See also the denial of a complaint that sought to extend the incentive payments: 121 FERC 61,315.

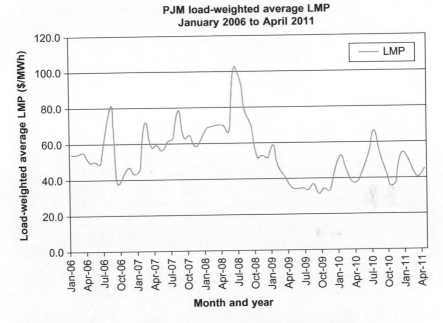

FIGURE 17.9 PJM load-weighted average LMP 2006–2011. *(Source: PJM)*

TABLE 17.3 Annual Revenues Paid to DR in the PJM Energy Market

2002	2003	2004	2005	2006	2007	2008	2009	2010
$801,119	$833,530	$1,917,202	$13,042,865	$18,334,295	$66,987,730	$27,759,624	1,274,675	$3,088,049

the day-ahead, real-time dispatchable, and real-time self scheduled options has ranged on a percentage basis in the low single digits except for the real-time self scheduled option, which consistently exceeds 90%.

Generally, the total revenues paid to CSPs between 2002 and 2009 reflect not only the impact of the incentive payment described above but also the growth in the number of participating CSPs and of registered and active end-use sites shown in Figure 17.10. Even with the decrease in energy prices and therefore more limited opportunities for load reductions, the level of registrants has remained fairly constant.[25]

From the perspective of an individual end-use site, the revenue opportunities for DR participating in Economic Load Response during the high-price years of 2007 and 2008 can be seen in the following example. During the 2007 calendar

---

[25]The dips in registration seen in Figure 17.10 are due to registrations expiring at the end of the delivery year (May 31) and DR participants re-registering just after June 1 of each year.

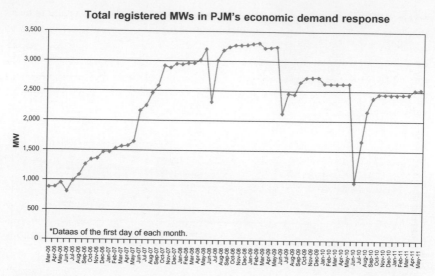

FIGURE 17.10   MWs of DR registered to participate in PJM's energy markets. *(Source: PJM)*

year a 1-MW DR with a retail rate[26] of $75/MWh and a willingness to reduce demand by 1-MW each time LMP was at or above $100/MWh would have collected $44,443 in revenue over and above the savings from avoiding the retail rate during the Load Reduction Events. In 2008 when prices were higher, especially during the first seven months of the year, the same DR acting in the same way would have collected $86,820 in revenue over and above the savings from avoiding paying the retail rate during the Load Reduction Events.

Emergency DR Participation in the Energy Market

DR registered as either Emergency Energy Only or Full Emergency Load Response receives an energy payment for reductions when called by PJM Operations to respond to an emergency event. Response by Emergency Energy Only DR is voluntary, while response by Full Emergency Load Response is mandatory. Full Emergency Load Response resources also receive a capacity payment for their willingness to reduce load at the direction of PJM Operations a maximum of 10 times during the summer period for up to six consecutive hours. The load reduction quantity used to calculate the energy payment associated with these load reductions is measured by comparing the metered load during the full hour before the hour in which PJM Operations calls for Emergency DR with the metered load during each hour of the load reduction event. FERC Order 719 mandated the removal of imbalance penalties for

---

[26]For the purposes of settlement in PJM, the retail rate is the generation and transmission portion of the retail rate also known as G&T.

LSEs whose loads fall below day-ahead schedules when system usage is at its peak and required reserves are short, in order to remove any disincentive to responding to high prices when it helps maintain reliable operations.[27] CSPs cannot easily anticipate precisely when PJM Operations will call for Emergency DR so that the usage level at the site during the full hour before the hour in which the DR is called can be the CBL used for the energy settlement calculation.

Order 745, Demand Response Compensation in Organized Wholesale Energy Markets, Issued 3/15/11

The FERC's recent Order 745 imposes short-term and longer term Economic Load Response compensation requirements on organized wholesale electricity markets. The FERC has determined that DR responding to Energy Market prices should be paid the Locational Marginal Price (LMP) for achieved load reductions but only (1) when the RTO or ISO can use DR to balance supply and demand as an alternative to generation and (2) when dispatch of DR results in benefits, measured by the savings to market buyers from a lower LMP, that exceed the cost of total payments to the CSPs. The Order further requires the RTOs/ISOs to develop a methodology for calculating in advance for each month the LMP threshold at and above which the benefit exceeds the cost. Finally, the RTOs/ISOs must study the feasibility of implementing a tool that determines dynamically the LMP threshold.[28]

If Order 745, as PJM believes, directs organized wholesale markets to pay full LMP when DR responds to PJM dispatch directions in the same way as generation, then CSPs will have fewer options for participating in the Energy Market and will need to alter their current operations significantly. Currently as explained above the CSP can offer the load reduction capability (1) in the Day-Ahead Energy Market or (2) for dispatch by PJM in real time; or it can (3) self-schedule the load reduction through the notification process. Historically, CSPs have used the self-schedule option approximately 98% of the time. The removal of the self-schedule option from the market rules will require CSPs to adjust how they interact with end-users whom they have registered as Economic Load Response.

Resolving Early Issues with CBL Estimation

Estimating what the end-use site would have used but for the demand response activity, known as the CDL, is the single most difficult challenge for DR participation in the wholesale Energy Market. Reviews of some settlements led to questions about whether the Economic Load Response activity was real or

---

[27]See Wholesale Competition in Regions with Organized Electric Markets, Order No. 719, FERC Stats. & Regs. ¶31,281 (2008) and PJM Interconnection, L.L.C., Initial Filing, Docket No. ER09-701-000 (February 9, 2009).

[28]Demand Response Compensation in Organized Markets, 134 FERC ¶61,187 (March 15, 2011).

"happenstance," that is, simply the result of running meter data through the "high 5 of 10" Customer Baseline (CBL) calculation in the market rules. Concerns also arose about stale CBLs resulted from the standard "high 5 of 10" CBL calculation and about end-users whose load varied significantly from day to day. The "high 5 of 10" rule calculated the CBL by averaging the five highest hours corresponding to the load reduction event of the most recent 10 qualifying days.

Concerns were raised in late 2006 about the behavior of some participants in Economic Load Response as expressed in the following question, "Did the submitted settlements (support for an Energy Market payment) reflect a change in the end-user's normal usage pattern in response to hourly Energy Market prices or did gaps in the market rules permit payment of Economic Load Response settlements inappropriately?" Such a gap in the market rules would have threatened the integrity of Economic Load Response by DR in the Energy Market.

PJM stakeholders and staff worked together to develop tariff and manual changes designed to protect the integrity of Economic Load Response in the Energy Market, to permit aggregation of qualified end-use sites in order to meet the required 100-kw threshold, and to enable CSPs to submit up to 10 price, quantity pairs for DR mirroring precisely the offers of generators.[29] This work involved a comprehensive evaluation of a number of CBL methodologies. The criteria for the evaluation included accuracy, bias, and simplicity. The Commission issued an order approving the proposed tariff revisions in mid-2008.[30]

PJM staff subsequently developed a set of objective tools for uniform enforcement of the revised tariff and manuals. PJM flagged end-use sites when CSPs submitted Economic Load Response settlements for less than $5 or more frequently than 70% of the days on a rolling 30-day basis. CSPs with flagged end-use sites must provide PJM with a copy of the contract for curtailment services, a detailed description of the how the end-use site reduces load during an event, and documentation that load reduction activity took place in response to hourly wholesale energy prices. These tools have been effective in aligning CSP activity with PJM expectations and in discouraging the submission of "happenstance" settlements.

DR PARTICIPATION IN ANCILLARY SERVICE MARKETS

Synchronized Reserve Market

Synchronized Reserve (SR) service is necessary to serve load immediately in the event of a system contingency or unexpected need for more power immediately ("Synchronized Reserve Event"). In PJM SR is broken up into Tier 1

---

[29]PJM Interconnection, L.L.C., Initial Filing, Docket No. ER08-824-000 (April 14, 2008).
[30]PJM Interconnection, L.L.C., 123 FERC ¶61,257 (2008).

SR, which is provided by units that are on-line, following economic dispatch, and capable of increasing output within 10 minutes of a call for SR, and Tier 2 SR, which is extra SR capacity committed in excess of Tier 1 capacity to meet SR requirements. The SR Market is the mechanism by which SR is committed.

In the same FERC Order that made the Economic and Emergency Load Response Programs in the Energy Market permanent, the rules were put in place for DR to participate in the SR Market.[31] In order to participate in the SR Market, DR must be capable of dependably providing a response within 10 minutes of a Synchronized Reserve Event call and must have appropriate metering infrastructure that provides data at no less than a one-minute scan rate surrounding a call for SR. DR must also be able to maintain its load reduction in response to a Synchronized Reserve Event for up to 30 minutes. Cleared DR that responds to a synchronized reserve event must provide one-minute meter data from 10 minutes before the event to 10 minutes after the event ends. PJM uses the meter data to assess the performance of the resource by comparing the cleared MW with the difference in metering readings between the time that PJM notifies the CSP of the event (Minute 1) and the beginning of the event (Minute 10) as well as for all of the minutes (up to 30) of the event. PJM does not require telemetry for DR participation in the Synchronized Reserve Market, relying instead on the one-minute meter data discussed above, submitted within 24 hours of the event. This single difference significantly affects the cost of market participation for DR and partially accounts for the continued growth of DR Synchronized Reserve Market participation.

It should be emphasized that DR need not make a demand reduction until a Synchronized Reserve Event is called, but must be ready to do so on short notice. The metering information must be uploaded to PJM's Load Response application (eLRS) within one business day of the event to ensure compliance. The overall participation by DR currently is limited to 25% of the SR requirement in each SR Zone while PJM gains experience with DR participation in this market. To ensure that proper reliability standards are maintained, there are mandatory training requirements for CSPs that desire to bid load reduction capability in the SR Market.

The SR Market also accommodates the participation by industrial sites engaged in "batch load" manufacturing processes. Batch load manufacturing processes include arc furnaces that make steel, which typically draw no power at all for the 10-minute periods between melts. Should PJM call for synchronized reserve during the no-load phase of the process, then it would be impossible for the one-minute metering readings to demonstrate compliance. PJM market rules accommodate "batch loads" that are already down when the event begins by comparing the meter reading at the beginning of the

---

[31]PJM Interconnection, L.L.C., Initial Filing, Docket No. ER06-406-000 (December 28, 2005) and PJM Interconnection, L.L.C., 114 FERC ¶61,201 (2006).

event, which should be zero, with the meter reading 10 minutes after the event ends when the usage should have increased by at least the amount of MW cleared in the Synchronized Reserve Market.

The SR Market is cleared every hour based on the offers that are submitted by various resources. The minimum size of a DR offer is 500 kw or .5 MW. The CSP may aggregate DR to achieve the 500-kw minimum requirement.[32] The market-clearing results establish the SR assignments for each cleared resource in order to meet the SR requirement for each SR zone. Each assigned resource is paid the SR clearing price for providing the service for the hour. Should a DR or generation resource fail to respond to a Synchronized Reserve Event, the penalty imposed is the supply of SR during the same time period in which the resource failed to respond to the call for SR.

Regulation Market

Regulation service provides for the continuous balancing of generation, load, and interchange at a very granular level. Specifically Regulation maintains system frequency at 60 cycles per second (HZ). Regulation is accomplished through the raising or lowering of output by Generation Capacity Resources or the raising or lowering of loads by DR Resources as required. CSPs that offer DR into the Regulation Market must meet all the criteria necessary to provide Regulation in the PJM Manuals including the ability to receive Automatic Generation Control (AGC) signals and must have the appropriate telemetry to the PJM Control Center. As with synchronized reserve, the CSP must be able to aggregate a minimum of 500 kw or .5 MW of qualified load reduction capability to offer a DR resource into the Regulation Market. Current market rules also limit DR to 25% of the Regulation requirement in the RFC region. There are, as with DR participation in the Synchronized Reserve Market, mandatory training requirements for CSPs that desire to offer load reduction capability in the Regulation Market. To date no DR resources have participated in the PJM Regulation Market.

Potential Financial Benefits and Market Results

The ability for a Demand Resource to reduce overall expenditures through participation in the SR Market can be significant. For example, if a 1-MW DR supplied SR in the Mid-Atlantic sub-zone of the Reliability First Corporation (RFC) area in all hours during 2007, that resource would have been paid $63,522 that could have offset its energy and capacity expenditures. In 2008 the same DR would have been paid $45,971.

---

[32]Originally the minimum size was 1 MW and aggregation could not be done, but this was changed by FERC Order 719. See Wholesale Competition in Regions with Organized Electric Markets, Order No. 719, FERC Stats. & Regs. ¶31,281 (2008) and PJM Interconnection, L.L.C., Initial Filing, Docket No. ER09-701-000 (February 9, 2009).

While market rules provide for energy payments related to load reductions achieved in response to a synchronized reserve event, CSPs have not submitted settlements for associated energy payments. Given the infrequency and short duration of synchronized reserve events, the preparation and submittal of an energy settlement for load reduction during a synchronized reserve event are not worthwhile. The incentive for the CSP to participate in the Synchronized Reserve Market comes from the revenue associated with clearing in that market and being ready, willing, and able to reduce load within 10 minutes for a 30-minute period during a synchronized reserve event.

Figure 17.11 shows the additional revenue stream paid to CSPs for the growing number DR resources qualified to provide SR from August 2006 through March 2011. The first qualified DR resource participated in the SR Market on August 17, 2006. The trend in market participation by DR has been increasing and the revenues earned by DR providing Synchronized Reserve have also been steadily increasing.

To date no DR resources have offered into or cleared the Regulation Market. A few DR resources, however, have proved their capability to provide regulation service in the context of Alternative Technology Pilots initiated by PJM in 2010. The financial rewards of DR participation in the Regulation Market have the potential to exceed revenue streams available in the Synchronized Reserve Market. During 2007, for example, 1 MW of DR, provided as regulation service during hours when the clearing price equaled or exceeded \$100/MWh, would have resulted in total compensation of \$50,939 for providing regulation in 337 of the 8,760 total annual operating hours (3.8% of all hours).

Moreover, the participation of DR in the Ancillary Service Market adds competition that drives down the cost of these services for all market participants, increases market efficiency, reserves valuable generation resources for

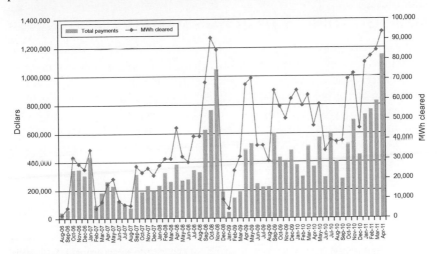

FIGURE 17.11 DR participation in the Synchronized Reserve Market. *(Source: PJM)*

the production of electricity, and reduces the wear and tear of constant ramping, particularly on steam units. For example, the PJM Independent Market Monitor has shown that greater the DR participation in the SR Market, tends to lower the market-clearing price. Figure 17.12 shows clearly the greater the DR participation shown in green, the lower is the market prices shown in blue.

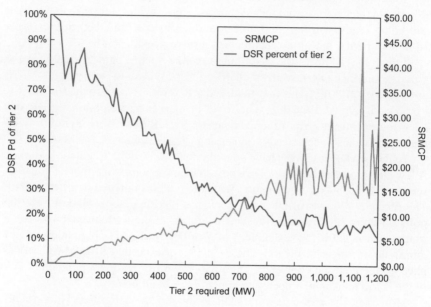

FIGURE 17.12   DR and SR market prices.[33]

ADMINISTRATION OF DR IN PJM'S MARKETS

PJM completed important system enhancements for DR participation in mid-2009. Enhancements to PJM's eMarket system[34] streamlined DR participation in the Energy Market and are known as the eLRS application.[35] eLRS went into production at the beginning of July 2009 following a massive movement of data from the old system. PJM uses the eLRS application to process registrations and settlements for DR, including the calculation of the CBLs used for settlement payments to CSPs. PJM works with CSPs as well as EDCs and

---

[33]2010 State of the Market Report, Figure 6-12, p. 459.
[34]The eMarket system is the software application used by market participants to provide PJM with the offer curve for each resource participating in the Energy Market. PJM uses the pricing information in eMarket to schedule and dispatch offered resources.
[35]The eLRS application is the software application that allows market participants and PJM to adminster DR participation in the markets. The eLRS application administers registrations, notifications of self-scheduled Economic Load Response, and settlements for demand response resources.

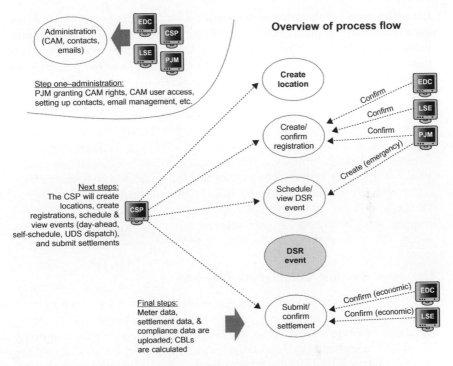

FIGURE 17.13 Flow of information in administering DR in PJM's markets.

LSEs on an ongoing basis to plan, prioritize, and implement eLRS enhancements. eLRS is an xml, IP protocol, web-enabled application that manages DR participation in PJMs markets for approximately 60 CSPs and well over 10,000 end-use sites.

During 2010 eLRS handled over 21,000 site-hours of dispatched Real-Time Energy Market notifications of reductions from over 800 sites and over 71,000 site-hours of self-scheduled notifications of load reductions in the Real-Time Energy Market from over 800 sites. Figure 17.13 shows the data and process workflow handled by eLRS from the creation of registrations, the submission of notifications, and schedules, all the way through to calculation of the CBL and to settlements for load reductions.

DR PARTICIPATION GOING FORWARD: CURRENT ISSUES RELATED TO DEMAND RESPONSE

Price-Responsive Demand

As is being discussed in the PJM stakeholder process, DR is contemplated to evolve into what is being called price-responsive demand (PRD), which accounts for the load reduction activity of DR on the demand side of the market rather than as a supply resource, which is the current paradigm.[36] PRD is not meant to replace DR participation as it currently exists, but is envisioned as providing yet another option for participants on the demand side of PJM's markets that will better integrate the wholesale and retail markets without infringing on the jurisdictional authority of the states and others over the retail side of the market.

Although PRD is the next generation of demand-side participation, DR participation in its current form has become a fixed and necessary piece of PJM's markets. Out of necessity or as a natural evolution of DR participation, there are refinements that are currently under consideration and issues that continually arise that may alter how DR participates in PJM's markets. These range from refinements to calculating customer baselines for settlements in the PJM Energy Market to defining additional DR capacity products based on the ability to respond more frequently in more seasons of the year to the compensation paid to DR in the PJM Energy Market.

Customer Baseline Calculation Revisited

Among the issues addressed by the Load Management Task Force in 2010 was the need for greater clarity in measuring and verifying compliance by DR that commits to reduce load by a prescribed amount during an Emergency Event. Unlike either Firm Service Level or Direct Load Control, measuring the compliance of Guaranteed Load Drop Load Management requires a CBL. While the rules clearly require use of the methodology that will produce the most accurate measurement of the CBL and the resulting achieved load reduction, the CSP must choose from among several methodologies, including Comparable Day, Same Day, Economic CBL, Load Profile, or Regression Analysis. The lack of precise rules for using some of these methodologies led many CSPs to question how to apply some of the methodologies and raised concerns by PJM staff, the Market Monitoring Unit (MMU), and other PJM stakeholders about the integrity of GLD measurement and compliance.

Late in 2010 following a report from the Load Management Task Force, the Market Implementation Committee directed PJM to once again undertake a comprehensive study and evaluation of alternative customer baseline calculations. PJM staff also consulted other organized wholesale electricity markets before settling on 18 different baseline calculations for study and evaluation. KEMA, the consultant retained by PJM to perform the study, used actual, annual meter data from a variety of customers that do not participate as DR to evaluate each of the 18 identified CBL calculations. The results were compared based

---

[36]See PRD under PJM's Issue Tracking feature on its website at www.pjm.com/committees-and-groups/issue-tracking/issue-tracking-details.aspx?Issue=%BCD79A76A-E9E4-4F58-ADF9-88A65F2B61EB%7D.

on accuracy, bias, simplicity, and cost.[37] PRD, being on the demand side of the market and not treated as a supply resource, does not face the same kind of CBL controversy, and PRD's performance is based on how much it consumes alone, rather than how much it consumes relative to a baseline.

Dispatch of DR with a Scalpel Rather than a Sledgehammer

Stakeholder discussion in the Shortage Pricing Working Group, Load Management experience from the summer of 2010, as well as stakeholder discussions about the relationship between reliability and limitations on the performance requirements of Load Management, have resulted in a new stakeholder effort to reconsider the granularity of DR dispatch. The Shortage Pricing proposal filed by PJM included a requirement that CSPs keep PJM informed of MW of Load Management not already reduced in response to market prices and therefore, available to be deployed in response to shortage prices or an Emergency Event. This requirement imposes more sophisticated asset management and reporting obligations on CSPs.[38]

PJM called for Load Management resources six times during the 2010 summer, but in three of these events only the resources in part of one transmission zone were needed. Deploying all of the Load Management in the zone can in some cases exacerbate rather than ameliorate the system condition causing an Emergency Event. The impact of a load reduction downstream of an overloaded transmission line, for example will ameliorate the overload, while the impact of a load reduction upstream may have no salutary effect or worsen the overload.

Load Management experience before the 2010 summer usually called for DR in one or more transmission zones. Compliance has also been measured on a transmission zone basis, meaning that the overall performance of the CSP's Load Management portfolio in each transmission zone determines compliance. Finally, the transmission zone is the most granular information about the location of the Load Management that the registration process currently captures. The stakeholders believe that the newly formed Demand Response Dispatch Senior Task Force should re-examine and rethink Load Management capabilities and develop requirements for sub-zonal dispatch.

CSPs participating in the recent expansion of Load Management performance requirements to include not only Limited Demand Response, but also Extended Summer Demand Response and Annual Demand Response, also

---

[37]Hennessy, Tim and Langbein, Pete, "PJM Empirical Analysis of Demand Response Baseline Methods Results," PJM Markets Implementation Committee, May 10, 2011. http://www.pjm. com/~/media/committees-groups/committees/mic/20110510/20110510-item-09-cbl-analysis.ashx See also "PJM Empirical Analysis of Demand Response Baseline Methods," prepared by KEMA, Inc., April 20, 2011. http://www.pjm.com/~/media/committees-groups/committees/mic/20110510/20110510-item-09a-cbl-analysis-report.ashx.
[38]PJM Interconnection, L.L.C., Initial Filing, Docket No. ER09-1063-006 (June 18, 2010).

considered the impact of automation and portfolios of varied load reduction capability on the ability to deploy more precisely, more granularly, and in a shorter time frame than one or two hours.

The intersection of PJM's operational needs and the capabilities of the DR will continue to evolve. Smart grid investments begun by the infusion of stimulus funds, however, will likely transform the dispatch capability of DR over the next few years.[39] This enhanced capability will aid DR in reaping the available rewards from participation in PJM's markets while also aiding PJM in enhancing the reliability of market and system operations and of the efficiency of PJM's markets.

---

[39] American Recovery and Reinvestment Act of 2009.

Chapter 18

Perfect Partners: Wind Power and Electric Vehicles – A New Zealand Case Study

Magnus Hindsberger, John Boys, and Graeme Ancell

INTRODUCTION

The amount of installed capacity of wind generation around the world is increasing rapidly. The best way of integrating large-scale wind generation into power systems is being considered by regulators, system operators, and transmission owners. Wind generation output is intermittent in nature in contrast to conventional power plants such as thermal or hydro that could be scheduled when the power was needed.[1]

Electric vehicles (EVs) have received less attention to date but may provide a similar challenge in integration. Promotion of EVs includes government subsidies to make them (more) affordable and incentives such as offering free recharge while shopping in the supermarket and rollout of quick charging stations[2] that offer recharging almost as fast as refueling a traditional car. While the latter two initiatives may increase the uptake of EVs, they also promote charging from a convenience point of view, potentially increasing the peak demand for electricity. Increasing peak demand will be costly, as it requires extra capacity in generation, transmission, and distribution networks.

So why are countries worldwide pursuing wind and EVs? This is because wind power and EVs are two key factors in reducing carbon emissions and the dependency on fossil fuels. Both of these are also key objectives of the New Zealand Energy Strategy [1].

New Zealand has great wind resources, and most new generation commissioned is expected to be wind and geothermal generation. The integration of large-scale wind generation poses challenges, even more so where the grid is part of a small island power system such as New Zealand. The integration issues can be managed but the costs of integration can be high.

The uptake of EVs in New Zealand is less certain. There are good benefits from a carbon emission reduction perspective as EVs can be supplied with electricity from largely renewable sources. However, should large numbers of EVs be introduced in New Zealand, high integration costs could be the result as investments in generation, transmission, and distribution need to be brought forward, if EV owners recharge when they want rather than when there is spare capacity in the system.

This chapter will show how getting the right price signals for charging EVs and the use of dynamic demand control (DDC) can considerably reduce the integration costs of both large-scale wind generation and a high uptake of EVs in New Zealand. Their use also provides the opportunity to reduce the magnitude of the perennial problem of large frequency fluctuations on island power systems. While the latter issue may be specific to New Zealand and some other islanded systems, the remainder of the findings should be applicable for most other markets, where wind and EVs are seen as key parts of a future, low-carbon energy system.

This chapter is organized into four sections. The section "Wind Generation and Electric Vehicles in New Zealand" starts by providing a brief overview of the New Zealand power system, its electricity market, and integration issues for

---

[1]The integration issues are discussed in more detail in Chapter 10.

[2]As per May 12, 2011, the CHAdeMO organization reported that 747 of its quick-charge stations had been established worldwide, of which 656 are located in Japan, 87 in Europe, and 4 in the rest of the world (see www.chademo.com).

wind generation and EVs in New Zealand. The section "The Opportunity at Hand" outlines the opportunity to reduce power system operating costs through the use of price and reliability signals. The section "Integration Costs of Wind and Electric Vehicles in New Zealand" describes the results of a study on reducing the integration costs of large numbers of EVs based on price signals. The section "Improving System Reliability by Plugged-In EVs" introduces dynamic demand control and how it can help to increase reliability in systems with a high penetration of wind power, followed by the chapter's main conclusions.

WIND GENERATION AND ELECTRIC VEHICLES IN NEW ZEALAND

New Zealand Power System

The New Zealand power system (Figure 18.1) is made up of two AC island power systems connected by an HVDC bipole link. Given its remote location, New Zealand has no interconnection to other power systems.

The system has a peak demand of around 6,600 MW. The North Island demand ranges between 1,600 and 4,600 MW. The South Island demand ranges between 1,100 and 2,200 MW. The HVDC bipole linking the South Island and the North Island power systems has a current capacity of 700 MW. This capacity will increase to 1,200 MW in 2013.

Generation is largely renewable and generally located far from the load centers, making transmission an important consideration. Table 18.1 shows a breakdown of the installed capacity and how much energy was generated by

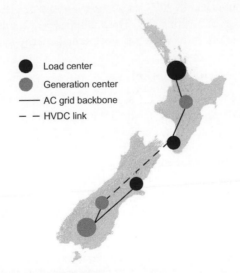

Load center
Generation center
—— AC grid backbone
— — HVDC link

FIGURE 18.1 New Zealand power system. *(Source: Transpower [2])*

TABLE 18.1 Grid-Connected Generation in New Zealand

Fuel Type	Installed Capacity Dec 2009 (MW)	Installed Capacity Dec 2009 (%)	Electricity Generated in 2009 (GWh)	Electricity Generated in 2009 (%)
Hydro	5,378	57%	23,962	57%
Geothermal	635	7%	4,542	11%
Wind	496	5%	1,456	3%
Gas	1,747	18%	8,385	20%
Coal	949	10%	3,079	7%
Oil	156	2%	8	0%
Other	125	1%	578	1%
Total	9,486	100%	42,010	100%

Source: MED [3].

each source in 2009. As seen, renewable generation such as hydro, geothermal, and wind makes up about 70% of total generation. Coal generation has been declining in recent years, being replaced by generation from geothermal, gas, and wind.

New Zealand already generates approximately 70% of its annual electricity from renewable sources, among the highest globally. The New Zealand Energy Strategy to 2050 published in 2007 set a target for 90% of electricity generated from renewable sources by 2025 [1].[3] The additional renewable generation will likely be predominantly from wind and geothermal energy sources.

New Zealand Electricity Market

The New Zealand Electricity Market is a wholesale market that includes offer-based merit order dispatch using locational marginal pricing to determine the lowest cost secure dispatch solution overall (see NZIER [4]).

Nodal pricing takes into account the impacts of both losses and congestion on delivering electricity to a particular point in the power system. Furthermore, the price includes the costs of spinning reserves, as the market model co-optimizes the dispatch of energy and instantaneous reserves.

Generator offers can be submitted up to 36 hours in advance of physical delivery and can be changed at any time until 2 hours before delivery.

---

[3]The 2007 New Zealand Energy Strategy is due to be replaced with a new strategy in 2011. It is expected that the new strategy will maintain the 90% renewable electricity generation target, though with the caveat that only if it does not compromise security of supply.

Generation dispatch occurs every five minutes based on the offers received. Pricing is based on half-hour averages of the prices resulting from dispatch in each five-minute block.

Integration of Large-Scale Wind Generation in New Zealand

New Zealand is regarded as having one of the best wind resources of any country in the world [5]. The country lies across prevailing westerly winds sometimes referred to as the "Roaring Forties." Wind flow in some locations is almost continuous and of relatively high speed, which makes these areas well suited to wind energy development. This is exemplified by West Wind, which is a 143-MW wind generation station near Wellington commissioned in 2009. It is one of the best performing wind generation installations in the world with a capacity factor[4] around 47%. The New Zealand average is approximately 40%, while the international average is ~30%.[5]

Wind generation on the New Zealand power system had an installed capacity of around 530 MW in early 2011.[6] There are further wind generation projects comprising an installed capacity of over 2,400 MW that have already gained regulatory consent or are in the consenting process. During high wind conditions, wind generation could exceed the current demand during some off-peak hours and could equal ~40% of the peak demand. Wind generation, as seen in Figure 18.2, in New Zealand tends not to correlate well with load.

In 2008, the System Operator part of Transpower published a study on the impacts of large-scale wind generation on the power system operation and electricity market operation in New Zealand [6]. The most immediate issue was found to be the effect of unpredictability of wind generation output on pre-dispatch processes. Based on an estimate of future wind generation forecast errors, it was found that these soon would dominate the load forecast error in the scheduling and dispatch processes [7]. At the anticipated level of wind generation, which is around 2,000 MW in 10 years, large forecast errors would be common. Wind generation forecasts six hours ahead of dispatch might have an error equivalent to 10–30% of island demand on a daily basis.[7]

The variability of wind generation output during dispatch was not found to be a major issue. The extent of the predicted variability of anticipated wind generation penetration is similar to the variability of demand [6]. There may be increased requirements for frequency keeping and more occurrences of generation re-dispatch but the effects are manageable.[8]

---

[4]Capacity factor is the amount of energy generated over a period as a percentage of the energy that could have been generated if the power station had been running at its full output for the whole time.
[5]Source: http://www.windenergy.org.nz/nz-wind-farms/operating-wind-farms/project-west-wind.
[6]New Zealand Wind Energy Association; http://www.windenergy.org.nz.
[7]Garrad Hassan, Wind power variability and forecast accuracy in New Zealand, 20 March 2007, Figures 4.8 to 4.10.
[8]Large frequency fluctuations are already experienced in the New Zealand power system, see discussion around Figure 18.9.

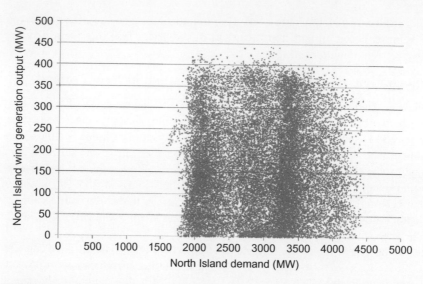

FIGURE 18.2   Correlation between wind generation and demand in the North Island (May 2010 to April 2011).

The probability and size of large sudden changes in aggregate wind generation output are, however, areas of concern. Operational experience with wind generation in New Zealand is very limited. Large sudden changes of Manawatu wind generation output have been observed to occur frequently;[9] for example, changes equal to 66% of installed wind capacity over five minutes [7] as per Figure 18.3.

A doubling of the amount of installed wind generation capacity in the Manawatu region could result in large sudden changes of a size equal to the largest generating unit on the North Island power system. Sudden large changes in regional wind generation output of 33% or greater of installed capacity in the region are likely to occur around 20 times per year. Such large changes over a short period have the potential to cause significant power system frequency excursions.

Integration of Large Numbers of EVs in New Zealand

As in many other places around the world, EVs have been looked upon as an option to meet objectives of reduced carbon emissions and lowering the dependency on imported fossil fuels. The 2007 New Zealand Energy Strategy especially pointed out EVs as one option to meet those objectives [1]. Only a few studies on how large numbers of EVs can be integrated into the New Zealand power system have been published though [8].

---

[9]Most of the existing wind capacity is located in the Manawatu district in the lower North Island region.

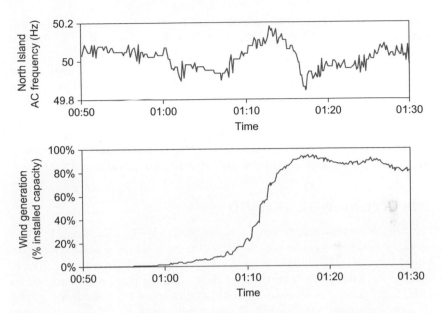

FIGURE 18.3   Observed wind generation output changes and frequency variations on November 15, 2004. *(Source: Adapted from Transpower [7])*

One important variable when assessing the effects of large-scale EV uptake in New Zealand is when the vehicles would appear. There were around 2.6 million registered passenger cars in New Zealand in 2010,[10] a number that is forecasted to grow to 3.2 million by 2040 [8]. Around 200,000 cars are brought into New Zealand each year.

As an extreme case, even if all imported 200,000 cars per year from, say, 2016 were EVs, it would still take more than a decade for EVs to make up the greater part of the passenger vehicle fleet. The actual uptake is expected to be much less initially, so it is more likely to take at least two decades from now before more than 50% of the light vehicle fleet is electric.

Such an uptake will result in about a 12% increase in the electricity demand compared with the total demand in 2008.[11] A 12% increase in electricity generation over a decade or two corresponds roughly to an average annual energy increase of 1% per year. During the 1990s, the annual energy demand growth was around 2%, but it has been growing at a slower rate in the last decade. An additional 1% increase on top of the normal demand increase is not going to pose major challenges to meeting the energy requirements of EVs over the

---

[10]New Zealand Ministry of Transport: http://www.transport.govt.nz/research/newzealandvehiclefleet statistics.
[11]Assuming 1.5 million electric cars, each traveling 40 km each day requiring 1 kWh per 5 km traveled.

coming decades. This is especially the case as the few actual projections that have been made (one example is shown later in Figure 18.4) point towards an even slower uptake.

While the energy balance can be managed, the implications for transmission and distribution planning depend on when the energy is required and hence, what the impact on peak demand will be. It critically depends on when EVs will be charged. For example, if 500,000 EVs (~20% of the current light vehicle fleet) were driven to their respective homes and plugged in at or about 6.00 PM the load on the power system could increase by 1.5 GW in approximately 30 minutes. This represents an increase of more than 20% in peak demand over today.

THE OPPORTUNITY AT HAND

New Zealand's installed wind capacity has around a 40% load factor. In terms of energy only, the entire vehicle fleet of approximately 3 million vehicles could be powered completely by wind power assuming they were EVs if most of the wind projects currently in the consenting process are built. However, infrastructure and economic considerations will likely limit the uptake of wind generation and EVs to well below these figures:

- Large investment in transmission and distribution networks may need to be brought forward to meet the demand from electric vehicles.
- The cost of frequency keeping will increase as the variability of load and generation increases.
- The ability to accurately forecast wind generation and EV charging will be critical to the way in which the technical standards and market arrangements will need to change to accommodate large-scale wind generation in New Zealand. In cases where wind generation forecasting or EV charging is uncertain, the variability will need to be managed during dispatch. Increased variability in dispatch will increase the costs of electricity in New Zealand as additional resources are procured to manage that variability.

There is an opportunity to reduce these costs. Time-of-use pricing signals will provide incentives for owners of EVs to charge their vehicles at times when electricity costs are low and the grid is unconstrained. This topic is further investigated below.

There is another opportunity in New Zealand, and elsewhere, to reduce the effects of the variability of wind generation output and EV charging. This is through dynamic demand control, a technology that has seen significant development in New Zealand since the 1980s. This technology has the potential to significantly reduce the costs of operating the grid as well as neutralizing the variability of wind generation output and EV charging. This technology and how it can be used are described in the section "Integration Costs of Wind and Electric Vehicles in New Zealand."

INTEGRATION COSTS OF WIND AND ELECTRIC VEHICLES IN NEW ZEALAND

This section is based on a detailed computational study undertaken to assess how wind power and EVs may interact in partnership as the title of the chapter suggests. Chapter 19 looks at a similar study in the French grid. This chapter considers how overall integration costs for large-scale wind and EV integration in an island power system can be reduced, and Chapter 19 provides insight into a new way of modeling the effect of EVs on market prices in the French market.

Assumptions

The analysis was based on the Electricity Commission's 2008 Statement of Opportunities [9], which provided an outlook on the New Zealand electricity future out to 2040. It featured five market development scenarios of which two assumed an uptake of EVs. These two, known as the "Sustainable Path" and "Demand Side Participation," scenarios form the basis of the following analysis.

Figure 18.4 illustrates how much the demand forecasts increased above the base forecast for these two scenarios. The forecasted demand for the EVs (i.e., the difference between the full and the dotted line above) is 280 GWh in 2025, growing to 2,660 GWh in 2040 [9]. For 2040, this equals approximately 4% of the total forecasted electricity consumption that year.

The forecast is based on EV uptake rates from the Ministry of Transport's vehicle fleet model and assumptions about the travel distance per car per day.

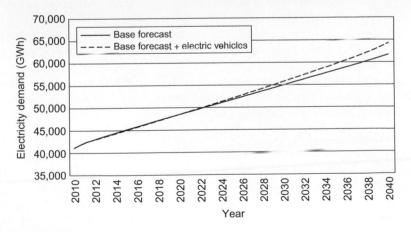

FIGURE 18.4   The demand forecasts from the 2008 Statement of Opportunities. *(Source: Electricity Commission [9])*

It is assumed that EVs are first introduced in 2016 and reach a penetration of 35% by 2040. For the latter year, this corresponds to a situation where 50% of all new car sales are EVs. Overall, the EV forecast used here is consistent with a number of other studies for EV uptake in New Zealand [1,5,8].

The two Electricity Commission scenarios differed considerably with regard to future carbon price assumptions and led to very different wind penetration levels. To supplement these, a third scenario with even higher wind penetration has been analyzed as well in this study. In summary, the three scenarios are:

- High wind scenario: ~25% penetration[12]
- Medium wind scenario: ~15% penetration (Sustainable Path scenario)
- Low wind scenario: ~5% penetration (Demand Side Participation)

Modeling

The analysis was undertaken using the PLEXOS modeling tool.[13] PLEXOS can mimic the Scheduling, Pricing, and Dispatching (SPD) algorithm used as the dispatch engine in the New Zealand Electricity Market, including the co-optimization of energy and instantaneous reserves.

For the analysis, a nodal model of the transmission system was used modeling approximately 170 nodes and the transmission flows in between. The focus has been on the year 2040, which was modeled using hourly time resolution. The fine time resolution was considered essential to capturing the interaction between varying wind power and EV charging. Before that year, the impacts of EVs would be lesser given the relatively slow turnover of the vehicle fleet.

The modeling of both wind power and EV charging used the approach of Rasmussen and Windolf [10]. Historical hourly wind profiles for different locations were used, which ensured that realistic wind variability and geographical correlation were captured. The annual EV demand was converted into a daily energy requirement, which was calculated for each of the 18 regions in the model.

New in this chapter, however, is an analysis of the impacts of that daily energy requirement being recharged in different ways. For this purpose three recharge schemes were formulated to model a range of different behaviors:

- 100% flexible scheme: All EVs are assumed to charge at the optimal time of day based on when the nodal price is lowest. The price is typically lower during night but as more and more wind capacity is added to the system, prices may drop during daytime as well if wind generation is high.
- 50% flexible scheme: Half the EV demand is fully flexible as per above. The remaining half has to be recharged between 8 AM and 4 PM, but will

---

[12]This is measured as wind generation as percent of annual electricity consumption.
[13]Information can be obtained from http://www.energyexemplar.com.

be timed optimally within this period (again based on nodal price). This models people recharging their EVs while at work.

- 24 hours flat scheme: The daily demand from EVs is divided equally among each of the 24 hours, creating a constant demand from recharging.

Overall, this spans from the best case—100% optimized charging—to a case where charging occurs during any hour of the day, including time of peak demand, for convenience reasons. For the two flexible schemes, the recharging is optimized based on the demand and wind output on a day-to-day basis. In some cases, there will be plenty of wind generation and in other cases almost none. The model did not assume demand could be moved from one low-wind day to another windier day.

It was assumed that all the EVs were plug-in hybrid cars, which either could use electricity or gasoline as fuel. The breakeven price for the vehicle owner is $NZ 315/MWh (in wholesale prices) based on a long-term expected gas price of $NZ 2/L.[14] When the nodal price is higher than the breakeven price, gas will be used rather than electricity to power the cars. This happens rarely, typically only during severe constraints during winter peak demand or at times of low hydro storage.

The following key aspects were examined under each scenario:

- Interaction between wind generation and EV recharging;
- System costs impact;
- Cost of recharge;
- Time of recharge; and
- Deferral of generation, transmission, and distribution.

Interaction Between Wind Generation and EV Recharging

The first part of the analysis looks into how the different recharge schemes interact with wind generation and the wholesale electricity price under the medium wind scenario. To illustrate this, one particular day in 2040 was selected, which had significant changes in wind output during the day and a high demand, being in the middle of the New Zealand winter.[15] Wellington was chosen as a location as it has a significant local demand and several wind farms are planned in the area, which has a world-class wind potential.

Figure 18.5 shows the forecasted load profile from one of the Wellington 110-kV substations (Hayward) as well as total modeled wind generation output for the Wellington region that day. Prices are expected to be lower when the Wellington regional net demand (demand less wind generation output) is low. Prices for the day will be lower in the early hours and higher at times of the morning and evening peaks.

---

[14]As per May 1, 2011, 1 NZ$ = 0.8 US$.
[15]New Zealand is a winter-peaking system.

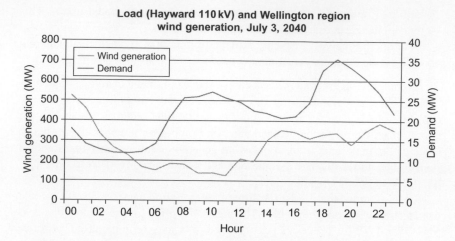

FIGURE 18.5 Example of wind generation and demand data for one day in 2040.

Figure 18.6 contains three charts showing how recharging differed for that particular day for the three recharge schemes as well as the resulting impact on the market price for that location.

In general, it shows that the flexible charging schemes ensure charging occurs in low price hours. Some of the other key observations are:

1. Prices in all three cases generally behave as expected based on the given demand and wind profiles as per the discussion of Figure 18.5.
2. Enough EV demand is moved to the nighttime period under the 100% flexible demand scheme (top chart) to lift the power price slightly above the other cases. As shown later in Table 18.3, it will not make it more expensive for the EV owners compared to the other recharge schemes, but it would lift the revenue for wind generators, making investments in wind generation more attractive or less reliant on subsidies, should that be needed to reach the assumed penetration levels.
3. Prices generally increase at the time of the peak if parts of the EV demand must be met at those hours. This is clearly the case for the 24-hour flat recharge scheme (bottom chart).
4. The cut-off price point for the plug-in cars is reached during hours 18 and 19 for the 24-hour flat recharge scheme (bottom chart) and results in EV demand not being delivered. This simulates that rational plug-in hybrid car owners would refuel with gasoline instead of electricity.

System Costs Impact

Having verified that the recharge schemes worked as intended in the modeling, the total variable system cost for 2040 resulting from the modeling was calculated. This corresponds to the combined cost of fuel, carbon emissions, and

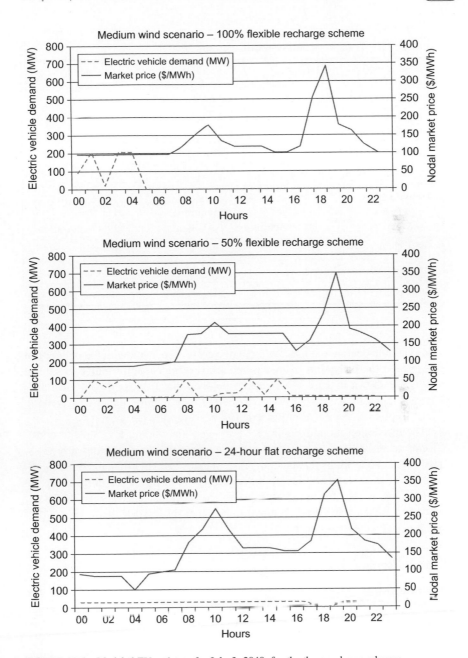

FIGURE 18.6   Modeled EV recharge for July 3, 2040, for the three recharge schemes.

TABLE 18.2 Total Variable System Costs for 2040

System Costs ($million)	Low Wind (5% Penetration)	Medium Wind (15% Penetration)	High Wind (25% Penetration)
No electric vehicles	$1,601	$1,364	$1,198
100% flexible scheme	$1,775	$1,487	$1,255
50% flexible scheme	$1,799	$1,499	$1,283
24-hours flat scheme	$1,827	$1,540	$1,303
System Costs (% of No Electric Vehicles)	Low Wind (5% Penetration)	Medium Wind (15% Penetration)	High Wind (25% Penetration)
100% flexible scheme	110.8%	109.0%	104.7%
50% flexible scheme	112.3%	109.9%	107.1%
24-hours flat scheme	114.1%	112.9%	108.7%

other variable operational and maintenance costs associated with the delivery of energy and the provision of instantaneous reserve capacity that year.[16] The results are shown in Table 18.2. It can be seen that for less flexible recharge schemes the costs are higher, as occasionally they require higher cost peaking plants to provide the electricity needed to recharge the EV fleet.

The discussion around Figure 18.4 showed that adding EV demand corresponds to a 4% increase in demand over the base case without any EVs. Table 18.2 shows the percentage increase in costs is higher than the 4% percentage increase in demand. This is because the majority of the demand in the no-EV case is met by renewables with no fuel costs (as per Table 18.1), while any increase in demand above this is met by fossil technologies with a significant fuel and carbon cost.

Cost of Recharging

The next part of the analysis looked into the costs of recharging from an EV owner's perspective. Table 18.3 shows that more wind lowers the average annual cost of recharging the EVs. Also, the average cost is reduced if recharging is flexible, so it can be optimized based on the wind generation each day.

Comparing the three charts in Figure 18.6 illustrated that flexible charging could lift prices in off-peak hours, but this table shows that the benefits from not

---

[16]Capital costs are not considered here, so one cannot conclude that 25% wind penetration is better overall for society than 5% or 15%.

TABLE 18.3 Average Recharge Costs

Average Recharge Costs ($/MWh)	Low Wind (5% Penetration)	Medium Wind (15% Penetration)	High Wind (25% Penetration)
100% flexible scheme	$65.12	$46.11	$21.25
50% flexible scheme	$74.01	$50.71	$31.75
24-hours flat scheme	$84.52	$65.87	$39.17

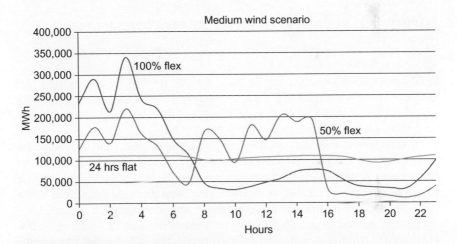

FIGURE 18.7   EV charging times.

charging in more expensive peak hours more than counter the increased costs off-peak. As indicated, investors in wind generation will benefit from this increase in price though, as it typically coincides with times with high wind generation. Overall, there are synergies between wind power and flexible recharged EVs.

Time of Recharge

The time at which EVs are recharged is a critical factor in reducing market costs, as Figure 18.7 shows when the model decided to recharge the EV demand over the full year for the medium wind scenario.

The daily profile demand on the New Zealand power system is generally similar to the demand profile shown in Figure 18.5. Demand tends to peak in the morning (hours 8–10) and in the evening (hours 18–20). The 100% flexible scheme predominantly recharges during nighttime but also between the morning and evening peaks. The 50% flexible scheme must recharge half during work hours, and the rest is mainly nighttime recharge. The 24-hour flat scheme draws power uniformly over the course of the day unless the price of electricity exceeds the breakeven price, which can be seen in the slight reductions in power drawn between hours 8–10 and hours 18–20.

The key observation is that even in the flexible cases the model finds it optimal to recharge sometimes during the evening peak.[17] This will be driven by a combination of localized high wind generation and transmission constraints limiting export capability from those locations. While this is optimal from a transmission point of view, no consideration has been given to distribution network capability, which may require significant upgrades if recharging is happening during peak hours.

IMPROVING SYSTEM RELIABILITY BY PLUGGED-IN EVS

This section will describe how EVs can help improve system reliability or maintain reliability at lower costs, including reliability issues related to increased wind power penetration. The key to this is the use of DDC for controlling when and how quickly plugged-in EVs are recharged.[18] DDC technology has been given much less attention than smart meters and other smart grid technologies so it will be explained in more detail in this section.

In comparison with the typical smart grid technologies, DDC is based on a simple concept, which has been around for a long time. As will be seen, it is potentially a very powerful addition to how to operate the power system; either as a means of improving reliability or providing the same level of reliability at lower costs. In DDC, the information to control the system is contained in the actual frequency that the grid is operating at. This is illustrated in Figure 18.8 assuming a 50-Hz system.[19] It is based on the fact that the "system health" is signaled by the frequency. If the frequency is below 50 Hz, demand is exceeding generation (case 1), and if the frequency is above 50 Hz, generation is exceeding demand (case 3).

Frequency has to be kept close to 50 Hz (case 2) at any time to prevent damage to electrical equipment and potentially blackouts. During normal operation, generation is adjusted continuously to match changes in demand throughout the day. However, during abnormal system conditions caused by outages, large changes can happen very quickly

- Outage of large generator: leads to frequency dropping
- Outage of major industrial load: leads to frequency increasing
- Outage of transmission line: leads to frequency dropping in the receiving end while frequency increases in the sending end

---

[17]Similar behavior is observed for the other wind scenarios, though there is a tendency that higher wind penetration levels cause the model to recharge during the evening peak more often.

[18]Vehicle-to-grid (V2G) operation, where power is sent from a fully or partly charged battery back to the grid, is not considered. Only the speed of recharging is altered.

[19]Other systems uses 60 Hz, but for ease of explaining, 50 Hz has been assumed in the following.

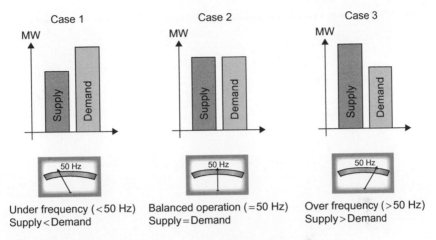

FIGURE 18.8 Power system frequency as an indicator of system health.

Frequency drops are most critical, as it takes time to start up generators to cover the loss of generation or transmission.

Frequency Fluctuations in New Zealand

The New Zealand power system experiences large frequency fluctuations compared to large continental power systems. This is due to the lower inertia of the power system and the size of experienced disturbances. Frequency is managed within a normal target band of 49.8 Hz to 50.2 Hz, which is a large range compared to frequency variations on a continental power grid. Figure 18.9 shows the variation in frequency for the North Island and South Island power systems over a period of 50 minutes in the early hours of December 16, 2010, which is a typical day. Over this time, the frequency in each island moved over +/– 0.1 Hz.

Variations of this magnitude and speed would be rare in a large power system such as Europe but are common in smaller systems. In New Zealand most excursions are within the normal band, which is 49.8 to 50.2 Hz, and frequency typically returns to within the normal band within 2 minutes.

Momentary frequency excursions of up to 51.5 Hz and down to 48 Hz are quite common. These excursions can be caused by the loss of a large load such as a reduction line at the Tiwai Point Aluminium Smelter or the loss of a large generating unit. The loss of a reduction line, around 180 MW, is large compared to the load on the South Island power system, which ranges between 1,100 and 2,200 MW. The largest generating units in the North Island are around 400 MW in size, which compares to North Island load that ranges between 1,600 and 4,600 MW.

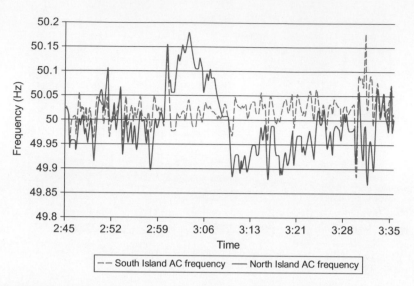

FIGURE 18.9   Variation in power system frequency in the New Zealand power system. *(Source: Internal Transpower information)*

The System Operator in New Zealand manages the frequency fluctuations described above through the procurement of three ancillary services [11][20]:

- Instantaneous reserves: Fast acting reserves from partially loaded generating plant, certain hydro generating units operating in tail water depressed mode and interruptible load are procured to arrest the decline in frequency for the loss of large generating units or an HVDC pole. The cost of instantaneous reserves currently varies between NZ$66 and NZ$72 million per year.
- Frequency keeping: A generating station referred to as the frequency keeping station is dispatched in each island to manage frequency. These stations will adjust output to maintain frequency to within the target normal band, which is 49.8 Hz to 50.2 Hz. The cost of frequency keeping currently ranges between NZ$50 and NZ$100 million per year.
- Over-frequency reserves: Certain generating units are provided with over-frequency protection systems that operate in a coordinated manner when there is a sudden increase in frequency due to the loss of a large load. The cost of over-frequency reserves is currently around NZ$0.7 million per year.

Using DDC for Frequency Control

DDC can help stabilize the network at the target frequency by adding or removing load in response to real-time measurements of the system frequency. The concept is illustrated in Figure 18.10.

---

[20]The specification of ancillary services varies from power system to power system.

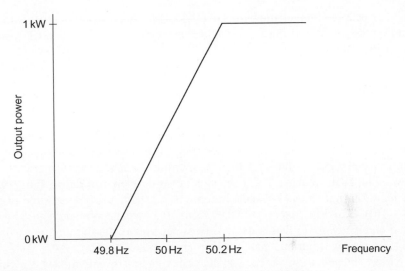

FIGURE 18.10   Response characteristics for a 1-kW DDC-controlled load.

As seen, when the system frequency varies over a control range—49.8 Hz to 50.2 Hz in this example[21]—the consumer load varies from zero to its maximum value of 1 kW. If the system frequency is 49.8 Hz or lower, then there will be no demand from the DDC-controlled load on the power system, whereas if the system frequency is greater than 50.2 Hz then the full 1-kW load will be added to the system.

The consumer may independently turn this load off with a switch, but it may not turn on immediately when wanted—as it depends on the system frequency. As the main use of the response is during system events, such as outages of power plants or major transmission lines, delays experienced when turning on a load are infrequent and only for short durations until generation has been adjusted.

Load types that work well with DDC, apart from recharging of EVs, could be washing machines, water heaters, pool pumps, and refrigerators (i.e., types of demand where the service is not affected by power being switched off for shorter periods of time). In practice, a system as big as the one in New Zealand could have several million DCC controllers, which would contribute to providing reliability.

DDC can either be built into the EV car chargers and household appliances themselves or into a "smart box" at the metering point that controls different appliances through a home area network, office network, and so on. The signal is available to anything connected to the power system. As such, there is no need

---

[21]Both the range of control (0.4 Hz here) and the center frequency (50 Hz) can be varied as required for the particular application.

for two-way communication for DDC to work effectively, but there would still be benefits of two-way communication as discussed in Section "Towards a Smarter DDC."

DDC Development Trends

Early Days

The concept behind DDC is far from new. It was developed by Fred Schweppe, an American engineer, in the late 1970s [12]. However it has evolved little subsequently, partly due to difficulties within the regulatory framework in rewarding DDC systems. It did see significant development during the 1980s in New Zealand, where islanded system made it particularly interesting (e.g., [13–16]). Based on that work, DDC technology has been manufactured and sold commercially to hundreds of small-scale applications around the world in sizes ranging from less than 1 kW to greater than 200 kW.

More recently, there has been an increasing interest, exemplified by the Grid Friendly appliances work by Pacific Northwest National Laboratory in the US [17], demonstration-scale trials by RLtec in the UK [18], and a study about how DDC can be used to stabilize frequency on the Danish island of Bornholm [19].

Towards a Smarter DDC

The simpler system explained so far can be enhanced in a system with two-way communication. With two-way communication available, the System Operator would have the ability to continually alter the response curves of individual loads to end up with a system that can restore system frequency to 50 Hz after a major outage, regardless of the actual uncontrolled load.

Such a system is shown in Figure 18.11 where 10 loads (numbered 1–10) are shown with different response curves. This ensures fast proportional control that stabilizes the system and slow integral control that ensures the frequency is restored to the desired frequency, here 50 Hz, over time.

Response curves can also be altered to take into account distribution network constraints—such as real-time monitoring of transformer temperatures—only allowing responses that do not cause any risk of overloading assets.

Another benefit of having data access through two-way communication comes from enabling the System Operator to monitor the overall supply-demand balance. Even if the system frequency is 50 Hz, the system risk is very different depending on whether all DDC-capable demand is utilized or only a minor part.

Lastly, knowing who has contributed to restoring frequency allows for better rewarding. Without this information, all DDC customers would get the same overall discount. There will be more system benefits arising from those who plug in the EVs at any time they are not using it compared to those who only plug it in when the battery level is below 20% and they want it recharged.

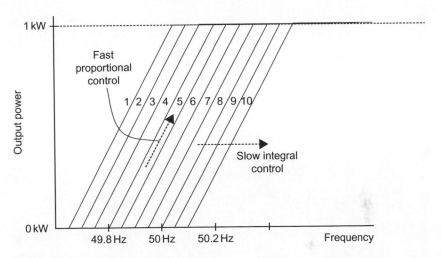

FIGURE 18.11 Example of response curves for a set of DDC controllers (here 10) to restore system frequency at 50 Hz.

Smart—Without the Smart Grid

Price-based optimization of demand for EVs as described in the section "The Opportunity at Hand" will assure optimal economic allocation of resources where the marginal willingness to pay matches the marginal cost of production. It does, however, rely on smart grid infrastructure in the form of smart meters connected with two-way communication that send the price signal to the consumers and subsequently settle accounts according to the response.

Smart grids are expensive to build, and for many developing countries, they will not exist for years. Even among the larger developed countries, there might be rural areas that are considered too expensive for a smart grid rollout. A DDC can achieve some of the outcomes seen in the section "The Opportunity at Hand" but without the smart grid infrastructure.

A laboratory-scale experiment has been undertaken at Auckland University showing that two different "energy flows" can coexist in the same network. Here a wind generator subject to a gusting wind flow was simulated. This is shown in the upper chart of Figure 18.12.

The system had a DDC controller, which acted as if charging an EV battery. The use of this is shown in the lower chart of Figure 18.12, while the frequency either with or without this DDC controller is shown in the middle graph.

It can be seen that the system is capable of keeping frequency to a narrower band while the same amount of energy is recharged over the time period. Thus it is feasible to transmit fluctuating power through a grid and capture it by types of load, which can be shifted, elsewhere in the system. A grid-scale experiment on this technique is needed to show whether it is scalable.

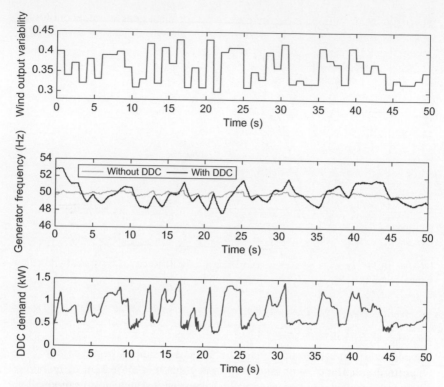

FIGURE 18.12 Experimental results with a DDC system. Upper trace: Generation change, Middle: System frequency, Lower: DDC demand.

While the experiment looked at intermittency in the shorter term (seconds), it will work for longer time horizons as well, say 24 hours. The new paradigm is that connected DDC devices are not guaranteed power at any particular time, but over a 24-hour period they will get some fractional energy guarantee. Some types of intermittent generation will be able to give a rather high energy guarantee (e.g., solar and tidal), while for wind it will be lower.

Economics of DDC

The single most important driver for these developments with DDC, and other grid management systems, is reducing the operating cost of the grid. Some of the possibilities are outlined in this section with a view to seeing how it might be possible to make system improvements using DDC.

At present frequency keeping involves using essentially one power station operating at partial load so that it can pick up load when required or reduce power when needed. For a generator operating at less than full output, this means a considerable opportunity cost. Also for most generation technologies, especially thermal generation, running at reduced output means running at reduced fuel efficiency and higher emissions.

These costs range from NZ\$50 to NZ\$100 million per year in the New Zealand system alone. Using DDC the power stations could run at full power, and excess generation would be used in water heating and other 'energy' loads. Thus this simple DDC controller allows matching the generation to the load and controlling the frequency to be 50 Hz. When the load changes the frequency transiently changes from 50 Hz, but as the frequency is restored by choosing the appropriate response the load is also matched to the generation.

In New Zealand, the System Operator annually spends between NZ\$66 and NZ\$72 million to procure instantaneous reserves to cover the loss of a large generating unit or the loss of an HVDC pole.[22] Some of that is provided by industrial and distribution load reductions, but the majority is from generators. As for frequency keeping, generators providing instantaneous reserve will have an opportunity cost for not generating as well as an extra fuel cost—or higher water use per MWh generated for hydro power plants—for not operating at their most efficient point.

Using DDC would allow most of the NZ\$66 to \$72 million per year to be saved. These are costs that could be paid to equipment providers to build in a chip that makes water heaters, refrigerators, electric car chargers, etc. DDC capable.

A calculation of the amount of MW that could be obtained from domestic refrigeration in New Zealand shows that about 140 MW of controllable capacity can be obtained longer term should all be converted into DDC.[23] This amount, 140 MW, is more than enough to cover all the frequency reserves currently needed and some of the instantaneous reserves. Adding a large fleet of EVs to the system will add to the controllable load, allowing more of the savings in provision of ancillary services to be obtained.

An estimate in the UK indicates that a typical under-counter fridge-freezer could "earn" around £30 during its life from allowing thermal generators to run more and run more efficiently.[24] The DDC chip allowing this is expected

---

[22]The loss of both poles (bipole outage) is not covered by the reserves provision. Covering the loss of a bipole would require up to 1,200 MW of capacity from 2013 while ~400 MW normally is enough for a single pole failure or loss of the largest generating unit, To prevent the whole country from blackout in case of a bipole outage, an AUFLS (automatic under-frequency load shedding) system is in place. In such a case, up to 33% of the demand will be shed on the importing island, not gradually as in DDR, nor by appliance level, but with whole distribution feeders being cut completely to save the remaining system from a total blackout.

[23]The calculation is based on data on refrigeration load and number of appliances per household from BRANZ [20] and household projections from Statistics NZ. In 2021, there will be approximately 1.9 million households, and even assuming a 40% efficiency improvement of new refrigerators and freezers, it will have an average demand of about 140 MW.

[24]See http://www.dynamicdemand.co.uk/pdf_economic_case.pdf.

to be significantly cheaper, providing a good business case for producers of refrigerators.

CONCLUSIONS

New Zealand's island power system currently experiences large frequency fluctuations. These fluctuations are expected to increase with the connection of large-scale wind generation and the charging of large numbers of electric vehicles unless there is some coordination of these resources. There is also a great opportunity to reduce wind and EV integration costs by coordinating the charging of EVs with the variability of wind generation output. This chapter has explored two ways in which this can be achieved.

First, time-of-use pricing signals will provide incentives for owners of EVs to charge their vehicles at times when electricity costs are low and the grid is unconstrained. This reduces market integration costs for both wind generation and EVs and defers the need for transmission and generation investment. Prices can be adapted to reflect constraints on the distribution network, allowing further deferment of investment.

Higher wind penetration levels lower the cost of recharging electric vehicles, especially if recharging is flexible. Flexible recharging increases the value of intermittent generation into the power system. The cheaper costs resulting from flexible charging will allow incentives that can be paid to electric vehicle owners to ensure smart charging.

Second, in New Zealand and elsewhere there is an opportunity to reduce the effects of the variability of wind generation output and electric vehicle charging through DDC. This technology has the potential to significantly reduce the costs of operating the power system by providing a low-cost alternative to generation-based frequency control and instantaneous reserves.

ACKNOWLEDGMENTS

The authors would like to thank Kevin Lao, Transpower, and Sarah Smith, Transpower, for the valuable contribution to the modeling behind the results presented in the section "The Opportunity at Hand," which is a large part of the work reported on in here. Also, the authors are grateful for the comments and guidance from the editor, which helped to improve the chapter significantly.

REFERENCES

[1] MED, *New Zealand Energy Strategy to 2050 – Powering our Future*, Ministry for Economic Development, Wellington, New Zealand, www.med.govt.nz/upload/52164/nzes.pdf, 2007 (accessed 09.09.2011).

[2] Transpower, *Annual Planning Report 2011*, Transpower New Zealand Limited, http://www .transpower.co.nz/n3652,381.html, 2011 (accessed 01.04.11).

[3] MED, *Energy Data File 2010*, Ministry for Economic Development, Wellington, New Zealand, http://www.med.govt.nz/upload/73585/EDF%202010.pdf, 2010 (accessed 09.09.2011).

[4] NZIER, *The Markets for Electricity in New Zealand*, Report to the Electricity Commission by New Zealand Institute of Economic Research (NZIER), February 2007.

[5] MED, *New Zealand's Energy Outlook – 2009/2010 Edition – Changing Gear Scenario, Ministry for Economic Development*, Wellington, New Zealand, http://www.med.govt.nz/upload/70162/April%202010%20Energy%20Outlook%20-%20Changing%20Gear%20Scenario.pdf, 2010 (accessed 09.09.2011).

[6] Transpower, *Wind Generation Investigation Project*, Transpower New Zealand Limited, http://www.systemoperator.co.nz/studies-agreements, 2008 (accessed 23.02.11).

[7] Transpower, *Manawatu Wind Generation: Observed Impacts on the Scheduling and Dispatch Processes*, Second revision, Transpower New Zealand Limited, http://www.systemoperator.co.nz/f1689,28051615/man-wind-gen-impact-rpt-28-sept-05.pdf, 2005 (accessed 23.02.11).

[8] CAENZ, *Electric Vehicles – Impacts on New Zealand's Electricity System*, New Zealand Centre for Advanced Engineering, Christchurch, New Zealand, 2010.

[9] Electricity Commission, *2008 Statement of Opportunities*, Electricity Commission, now New Zealand Electricity Authority, Wellington, New Zealand, http://www.ea.govt.nz/document/12466/download/industry/ec-archive/soo/2008-soo/, 2008 (accessed 09.09.2011).

[10] I.M. Rasmussen, M.H. Windolf, Wind power integration in New Zealand – a scenario analysis of 15–25% wind power in the electricity market in 2025, Master's Thesis, Technical University of Denmark, August 2008.

[11] Transpower, *Procurement Plan – 1 December 2010 to 30 November 2011*, Transpower New Zealand Limited, http://www.systemoperator.co.nz/Procurement-Plan, 2010 (accessed 23.02.11).

[12] F.C. Schweppe, Frequency adaptive, power-energy re-scheduler, US Patent 4,317,049 filed, September 1979.

[13] J.T. Boys, J.M. Elder, M.K. Forster, J.L. Woodward, Low-cost AC generating system for small hydro plant, *Trans. NZIE* 8 (3/EMCh) (1981) 75–86.

[14] J.M. Elder, J.T. Boys, J.L. Woodward, Integral cycle control of stand-alone generators, *IEE Proc. B* 132 (5) (1985) 260–268.

[15] J.M. Elder, J.T. Boys, Self excited induction generators, *IEE Proc. C* 131 (2) (1984) 33–41.

[16] J.L. Woodward, J.T. Boys, Method of governing a generator and/or apparatus for governing a generator, US Patent 4,563,630 filed, January 1987.

[17] PNNL, *Pacific Northwest GridWise Testbed Demonstration Projects – Part II. Grid Friendly Appliance Project*, Pacific Northwest National Laboratory, http://www.pnl.gov/main/publications/external/technical_reports/PNNL-17079.pdf, 2007 (accessed 09.09.2011).

[18] Carbon Trust, *Creating Dynamic Demand in the Energy Marketplace, RLtec Investment Case Study by Carbon Trust*, http://www.rltec.com/sites/default/files/Carbon%20Trust%20J7653_RLtec_03.pdf, 2011 (accessed 09.09.2011).

[19] Z. Xu, J. Ostergaard, M. Togeby, Demand as frequency controlled reserve, *IEEE Trans. Power Systems* (99) (2010).

[20] BRANZ, BRANZ report SR 221, *Energy Use in New Zealand Households – Final Report on the Household Energy End-use Project (HEEP)*, www.branz.co.nz, 2010.

Chapter 19

Impact of Smart EVs on Day-Ahead Prices in the French Market

Margaret Armstrong, Amira Iguer, Valeriia Iezhova, Jérôme Adnot, Philippe Rivière, and Alain Galli

INTRODUCTION

In 2007 the leaders of the European Union (EU) endorsed an ambitious integrated climate and energy policy that aims to combat climate change and increase the EU's energy security while strengthening its competitiveness. To kick-start the process, they set a series of demanding climate and energy targets that are known as the 20-20-20 targets.[1] These call for

- a reduction of at least 20% in EU greenhouse gas emissions below 1990 levels,
- 20% of EU energy consumption to come from renewable resources,
- a 20% reduction in primary energy use compared with projected levels, to be achieved by improving energy efficiency.

---

[1] See http://ec.europa.eu/europe2020/targets/eu-targets/index_en.htm.

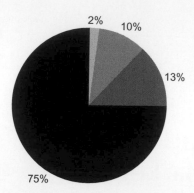

FIGURE 19.1   Energy Mix in France: 75% nuclear, 13% hydro-power, 10% fossil fuel thermal, and 2% renewable. *(Source: RTE [1])*

Compared to many of its neighbors France has been slow in developing renewable sources. At present, electricity in France is 75% nuclear, 13% hydro-power, and 10% fossil-fuel thermal, with only 2% from renewable energy sources other than hydropower (Figure 19.1). To remedy this situation the government plans to introduce large quantities of wind power.

Other changes will also affect the French power generation fleet over the next 15 years: the nuclear plants built in the 1970s and 1980s are reaching the critical 40-year mark[2] and will require more maintenance (some might be decommissioned); two new third-generation nuclear plants have been ordered, and several new gas-powered turbines are being constructed to meet the demand in peak and semi-peak periods. In addition since the recent nuclear problems in Japan, there have been calls for a reduction on the reliance on nuclear power. So there is considerable uncertainty on the supply side.

The introduction of electric vehicles is one of the measures envisaged for reducing greenhouse gas emissions and also air pollution levels in cities. Plans call for the number of EVs[3] to rise from a few thousand at present to 2 million in 2020 and to 4.5 million in 2025 out of a total of about 20 million vehicles [2]. This will increase electricity consumption. At a recent forum on EVs, the French electricity distribution authority [3] stated that charging 1 million EV simultaneously would have a significant impact because they would draw between 3,000 and 6,000 MW, that is, the equivalent of two to four of the new third-generation nuclear reactors. A quick calculation shows how much will be required to charge the 2 million EVs that will be on the roads in 2020, and likewise for the 4.5 million EVs in 2025 (Figure 19.2). As the government also plans

---

[2]The original lifespan of these plants was planned to be 30 years. After extensive testing, this was extended to 40 years, with the possibility of further extensions out to 50 or 60 years.
[3]The French government and the car manufacturer Renault are focusing on battery-operated EVs rather than plug-in hybrid EVs.

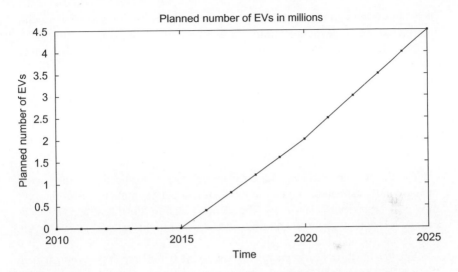

FIGURE 19.2 Evolution of the number of EVs from 2010 until 2025. *(Source: Batut [2])*

to introduce smart meters into homes by 2017 to optimize electricity use at the distributional level, these vehicles will be recharged at night after the evening peak in consumption is over.

The question is whether the introduction of wind power and the new nuclear plants will increase the supply fast enough to cope with the increasing numbers of EVs. At present even though there is surplus capacity in France and in Europe in general [1, p. 12], occasional periods of stress occur, resulting in price spikes. For example, on October 19, 2009, the market price was fixed at the technical ceiling (3000 €) over a period of four hours.[4] On that occasion there was surplus capacity in neighboring countries, and so the high prices attracted more sellers the following day, reducing the pressure on the system. But the French grid operator [1, p. 14] has warned that after 2015 the existing surplus capacity in Europe may well be squeezed between the resumption of demand (which had dropped in Q4 2008 due to the financial crisis) and declining generation capacity.

The objective of this chapter is to evaluate the combined impact that the introduction of EVs and the changes in generation fleet would have on day-ahead electricity prices over the next 10–15 years. We are interested in statistically predicting their impact by simulating the offers to buy and sell electricity on the day-ahead market. By "statistical" we mean that the histograms of the simulated prices should be realistic; we are not attempting to predict prices on particular

---

[4]The results of the inquiry held by the French electricity regulator, the CRE, were made public on November 20, 2009, and are available on its website: www.cre.fr/en/.

dates. It is important to be able to reproduce realistic peaks because they provide the price signal to encourage utilities to construct peaking plants.

This chapter focuses on weekdays because electricity consumption is much lower on weekends, and consequently the effect of the changes will first be felt on weekdays.

Two scenarios for recharging the electric vehicles [4] are considered:

1. recharging the batteries at night from 9 PM onwards,
2. recharging the batteries at night and drawing power from the EVs during the day when the demand is high.

The first scenario called grid-to-vehicle (G2V) would lead to a substantial increase in the demand at night, but would not affect day time prices or the evening peak hour from 7 PM to 8 PM. The second scenario illustrates the vehicle-to-grid (V2G) option where smart grids use the power in the batteries to supply the grid during the daytime peak period.

After the liberalization of energy markets in 2000, the French government gave the grid operator, RTE, the responsibility of evaluating the risk of not having enough electricity over the next 10–15 years. The first step in addressing this risk consists of identifying the trends in the demands and in the evolution of the generation fleet. Based on the latest studies [1,5] and the government's environmental proposals, Iezhova [6] identified three possible evolutions for the nuclear fleet (High, Reference, and Low), two for wind power (High and Low), and two for thermal plants (High and Low), that is, a total of 12 scenarios (see Table 19.1). Figure 19.3 shows the three scenarios for the evolution of the nuclear generation fleet (top), the two for thermal power (middle), and the two for wind power (lower), over the next 15 years. In this chapter we consider the two most extreme cases denoted by HHH and LLL. These correspond to the high option for nuclear plants, wind power, and thermal plants, and to the low option for all three.

The next sections review existing methods for predicting the effect of changes, such as the introduction of EVs and of wind power, on the electricity system as a whole and on day-ahead prices for electricity.

TABLE 19.1 Summary of Twelve Scenarios: Ref = Reference Scenario, H = High, and L = Low

Scenario	1	2	3	4	5	6	7	8	9	10	11	12
Nuclear	Ref	H	L	Ref	H	L	Ref	H	L	Ref	H	L
Wind		H	L	H	L	H	L	H	L	H	L	H	L
Thermal		H	H	H	H	H	H	L	L	L	L	L	L

Source: Iezhova [6].

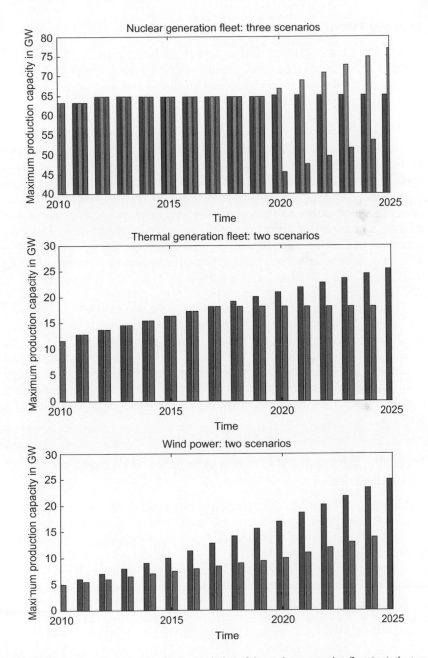

FIGURE 19.3   The three scenarios for the revolution of the nuclear generation fleet (top), the two for thermal power (middle), and the two for wind power (lower). Note the drop in the low nuclear scenario in 2021 is caused by the decommissioning of several nuclear power plants. *(Source: Iezhova [6])*

Effect of EVs on the Power System as a Whole

Kempton and Tomic [7] were among the first to study the effect of V2G power transfers. They concluded that it would not be a viable source of baseload power but that it could be economically profitable in three markets: spinning reserves, regulation of the voltage, and peak production. Denholm and Short [8,9] and Tomic and Kempton [10] further explored the possibilities of the V2G concept. Scott et al. [11] computed the life-cycle cost of a plug-in hybrid electric vehicle and compared it to that of a fuel-efficient conventional vehicle, for two cases: (1) if the host utility and the grid only have to make minor adjustments to accommodate the plug-in hybrid EVs and (2) if additional generation capacity is required. The results of their comparison are favorable in the first case. In addition to reducing greenhouse gas emissions and dependency on foreign oil, it improves the economics of the electricity industry, which has spare generation and transmission capacity at night at present. They note that unless additional generation is built before the surplus generation capacity is fully used, the system will experience problems providing the required power in the future.

Rousselle [12] carried out a detailed study of the impact of EVs on the electric system in France sponsored by the French grid operator, RTE. She developed a model for simulating the recharge times and the state of charge in the battery, in order to estimate the total load curve. After analyzing the impact of different load curves on the electric balancing system, she proposed several solutions to lower these impacts. The total load curves that she found were reported by Batut [2]. Rousselle did not study the impact of EVs on electricity prices.

Predicting the Impact of Changes in Supply and Demand on Electricity Prices

The approaches that are used for predicting the impact of changes in supply and demand for power on day-ahead prices depend on (1) the size and complexity of the system, (2) the way in which prices are determined, and (3) the information that is available to the public.

Chapter 18 by Hindsberger et al. presents a case study on New Zealand, which has a small, self-contained electrical system, whereas in this chapter we consider a much larger system with interconnections to six neighboring countries. In New Zealand most electricity is generated in the South Island whereas the demand is concentrated in the North Island, so transmission constraints arise. Electricity prices are determined by co-optimizing energy and instantaneous reserves every five minutes after taking into account the different transmission costs and other technical constraints. Hindsberger et al. used the Plexos modeling tool that mimics the algorithm used in New Zealand. In their case only 170 nodal points were required to model the electrical system.

TABLE 19.2 Reduction in the Day-Ahead Electricity Price due to the Introduction of Wind Power

Authors	Country	2005	2006	2007	2008
Saenz de Miera et al. [14]	Spain	7 €/MWh	5 €/MWh	12 €/MWh (1st half)	
Weigt [16]	Germany		10 €/MWh	17 €/MWh	19 €/MWh (1st half)
Sensfuss et al. [15]	Germany		7.8 €/MWh		

Source: Saenz de Miera et al. [14], Weigt [16], and Sensfuss et al. [15].

Because of the size of the system in France[5] and its interconnections with other countries, it would not be possible to implement an overall optimization of this type. Other approaches have to be used.

Studies on EVs rarely consider their impact on electricity prices. One exception is Hadley and Tsvetkova [13] who determined the marginal generation type in 12 regions in the United States, and hence the impact on wholesale prices. That is, they were effectively using the full merit order. Similarly, most studies on the impact of wind power on day-ahead prices [14–16] assume that:

- the demand is inelastic
- the supply curve is equal to the full merit order.

Saenz de Miera et al. [14] evaluated the impact on day-ahead prices in Spain, whereas Weigt [16] and Sensfuss et al. [15] both studied the impact in Germany. Table 19.2 summarizes the average drop in the electric price found by these three studies over different periods from 2005 until mid-2008. Despite some differences, there is an overall agreement between the values obtained.

Pfluger et al. [17] studied the impact of solar power[6] from Africa on day-ahead prices in Italy using the same agent-based simulation approach as Sensfuss et al. [15]. One particularly interesting part of their study is a comparison between simulated prices and the observed prices. Their simulation model delivers realistic predictions except during peak hours and in the early morning. As the authors noted, "*the price in peak hours turns out to be problematic as the very high prices in these hours cannot be explained fundamentally by variable costs used in PowerAce.*"[7] In fact, the simulated prices during peak periods on

---

[5]The generation fleet as of May 2011 consisted of 58 nuclear plants, 20 coal-fired plants, 8 gas-fired plants, 18 peaking plants, 39 lake hydropower stations, and 14 run-of-river generators. Source: http://clients.rtefrance.com/lang/an/clients_distributeurs/vie/prod/parc_reference.jsp.

[6]The idea was to set up solar panels in the Sahara and transport the power generated by direct current cable to the southern tip of Italy.

[7]PowerAce is the name of their agent-based simulation model.

weekdays rarely exceeded 80 €/MWh, whereas the actual peak prices ranged between 110 €/MWh and 130 €/MWh.

Two reasons can be advanced to explain the discrepancy: first the authors averaged sets of 50 simulations, which effectively smoothe out any peaks and troughs. More importantly, as Sensfuss et al. [15] noted, most electricity is traded via bilateral contracts (i.e., OTC) rather than through organized markets. For example, in France about 10% of the power consumed is traded via the day-ahead market. In fact, producers and major consumers do not wait until the day before delivery to start the process of buying and selling electricity. They spread their transactions over time by using futures markets and financial derivatives. For example, buyers often purchase part of their requirements months ahead of time through the futures and forward markets, and then as the delivery date approaches, they make adjustments by purchasing more or selling off unwanted power through the day-ahead market. Last-minute adjustments are made through the intra-day market. So the key question is whether or not the aggregate offers to sell power on the day-ahead market are a scaled-down version of the full merit order.

There are two ways of tackling this question: one is by computing the merit order and comparing it to the aggregate offers. This requires a detailed knowledge of the costs of all the generators and being able to evaluate the opportunity costs for hydropower stations and the nuclear plants,[8] which is difficult to do. An alternative way of testing this hypothesis is by studying the evolution of the aggregate curves over short periods (e.g., a few consecutive days) when the same power plants should be operational and hence when the merit order should be constant. If the offers reflect the full merit order, then they would remain stable over these periods [20]. Armstrong et al. (2011b) showed that the offers to sell power at a given time of day change radically from one day to another, particularly when the system is stretched. This means that the curves are not a scaled-down version of the full merit order. This is why a method for simulating the offers to buy and sell power on the electricity exchange, and hence the prices, was developed.

The primary aim of this chapter is to present a new method for simulating the aggregate offer to buy and sell power on the day-ahead market without these restrictive assumptions. The proposed method uses three types of data that are available hourly 365 days per year: (1) the aggregated offers to buy and sell electricity in France, (2) the total consumption, and (3) the actual production for each class of power plant. As these data are available at exactly the same times, whatever factors (weather, etc.) affected one set also influenced the others. While part of the data has been available since the liberalization of the electricity market in 2001, some of it has only become public more recently.

---

[8]As the dates for routine maintenance and for loading fuel rods are fixed well in advanced, nuclear plants have to decide how much to produce at any given time in much the same way as hydro plants.

As the information on the production per class only became available to the public from November 2006 onward and as this study started early in 2010, there were only two full years that could be used as the reference: October 2007 to September 2008, or October 2008 to September 2009. During the fourth quarter of 2008, the electricity consumption in France dropped sharply because of the financial crisis. So that year is atypical. This is why we used the year October 2007 to September 2008 as the reference year. In the future it will be possible to base the study on several reference years, which will be more representative.

This chapter is structured as follows. The next section presents our method for simulating the aggregate offers to buy and sell electricity on the day-ahead market. It takes account of the evolution in the structure of the generation fleet on the supply side, and two factors on the demand side: the expected increase in consumption and the quantities required by the EVs. The results of the simulations are presented in the following section. Our conclusions and perspectives for future work are given in the last section.

APPROACH

Having demonstrated that the aggregate offers to sell electricity are not a scaled-down version of the full merit order, and that the demand on the bourse is not inelastic, we need a method for simulating realistic curves of the aggregate offers to buy and sell electricity in order to simulate day-ahead prices at future dates. In fact we need matched pairs of curves corresponding to the same market and weather conditions. The key point is to link the overall supply and demand to the aggregated offers in the auction market. Figure 19.4 summarizes our approach.

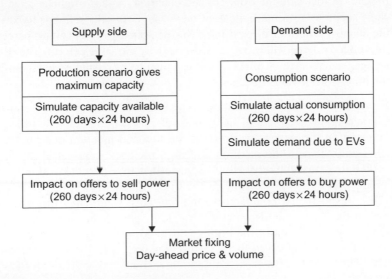

FIGURE 19.4   Diagram showing the approach proposed.

Simulating the Supply Side

For each of the generation scenarios, the capacity that will be available is simulated for each type of power plant (nuclear, fossil-fuel, run-of-river hydro, pumped hydro, peaking plants and wind power) at some date in the future. This is carried out in four steps:

1. Select the reference year (or years) for which the aggregate offers, the total consumption, and the actual production for each class of power plant are available. Here it is the 52-week period from Monday October 1, 2007 until Friday September 26, 2008.

2. Each of the scenarios for the generation fleet specifies the maximum capacity available for each class of generator in the future. Using a procedure similar to that used by the RTE [5], we generate multiple realizations of the production for each class of power plant (nuclear, thermal, etc.) for 24 hours per day for 260 working days per year from 2011 until 2025. Details are given in Armstrong et al. [18].

3. The aggregate offers to sell electricity are split into three tranches according to the marginal cost of production:
   a. low marginal costs, which corresponds to nuclear power, run-of-river hydro, and wind power
   b. mid-range marginal costs, which correspond to conventional thermal plants
   c. high marginal costs (or high opportunity costs), which correspond to peaking plants and pumped hydro.

4. It is assumed that producers will apply the same strategy in the future as at present. In that case the offers to sell electricity in each tranche have the same shape up to a multiplicative factor as during the reference year. If the quantity of power offered increases or decreases by a certain percent, the volume offered will increase or decrease by the same percent. The overall aggregated curve of offers to sell can then be reconstituted by summing the offers for the three tranches.

Figure 19.5 illustrates this concept. The solid black and grey curves were the original aggregate offers to buy and to sell power in the reference year. In this illustration the thresholds between the low and medium marginal cost tranches and between the medium and high tranches were set at 20 euros and 100 euros, respectively. Now suppose that in a particular simulation the production capacity available in the three tranches rises by 10%, 50%, and 20%, respectively, compared to the reference year. Then the volumes offered in the first slice also increase by 10%, giving the first segment of the dotted curve. Similarly the volumes offered in the second tranche rise by 50%. These are added to those in the first slice, and so on for the third tranche. The dotted curve represents the new simulated aggregate offers to sell power.

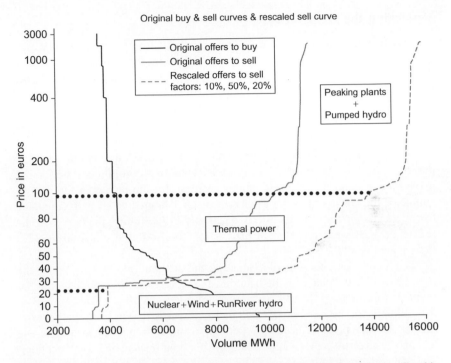

FIGURE 19.5 The simulated aggregate curve of offers to sell electricity. The quantity offered in each tranche is rescaled to take account of the simulated production in that production class.

The choice of the threshold prices separating the three different production tranches is important. In general utilities are reluctant to give out information on the marginal costs of different types of generators. In France these can be deduced from the strike prices of VPP (virtual power plants) auctioned by the EDF every three months. These are derivative contracts that were designed to allow EDF's competitors to access power without divesting the power plants.

The strike prices are set to be equal to the average marginal cost of nuclear power plants for baseload VPP and to thermal power plants for peakload VPP, respectively. During the reference year the strike price of baseload VPP was 9 € throughout the year whereas that of peak load VPP varied from 64 € to 85 C, with an average of 73 €.

As the marginal cost for nuclear plants was 9 € on average, it was below 9 € for some plants and above for others. So the limit between the lower and middle tranches was set at 20 € because we assume that nuclear power plants were offering to sell their production at prices between 0 € and 20 € per MWh. Similarly as the average strike price for thermal plants was 73 € the threshold between the middle and upper tranches was set at 100 €.

Simulating the Demand Side

The procedure for simulating the demand is split into two stages. The first stage simulates the ordinary demand (i.e., excluding the part due to EVs). In the second stage the demand due to the EV was evaluated and added to the ordinary demand. In the first stage, the typical consumption pattern for France was computed for each 1-hour time slice on weekdays by averaging the consumption for three consecutive years starting in October 2005 and ending in September 2008, after rescaling to have the same average value in winter. (Otherwise they would not be comparable.) This gives a smoothed typical pattern. The variability around the mean was also computed. Then using these statistics and the expected annual increase for the next 15 years [1] we simulated the consumption for France for each 1-hour time slice. To compute the aggregate curve of offers to buy electricity in the future, the curves in the reference year were rescaled in proportion to simulated consumption for France as a whole.

The second step consists of evaluating the demand due to the EVs for the two scenarios for recharging the batteries. In the first scenario, the batteries are recharged at a rate of 2.5 kW per hour from midnight till 7 AM and from 9 PM till midnight. In the second, V2G scenario, power is drawn from the batteries at a rate of 1 kV per hour from 9 AM until 4 PM, and consequently to compensate for this the batteries are recharged at 3.2 kW per hour instead of 2.5 kW per hour. The total quantity required is computed by multiplying the rate by the total number of EVs in service and is then added to the offers to buy if the batteries are being charged. Otherwise it is added to the offers to sell. In both cases it is treated as a price inelastic order.

RESULTS FOR THE YEAR 2020

The day-ahead prices in 2020 were simulated for the two recharging scenarios:

1. if EVS are introduced and the batteries are charged late at night
2. if EVS are introduced and the batteries provide power to the grid at peak periods as well as being charged late at night (i.e., the V2G case).

for the most favorable and the least favorable scenarios for the generation fleet (HHH and LLL), as a function of the number of EVs in service in 2020. It is felt that the French government's objective of having 2 million EVs on the roads by 2020 may be too ambitious.

In each case, the offers to buy and sell electricity were simulated, and the simulated day-ahead prices (and volumes) were obtained from the intersection of the two curves. A total of 50 simulations were run for all 24 hours per day for 260 weekdays. The results are presented for the 5th and 12th hours, that is, from 4 AM to 4:59 AM and from 11 AM to 11:59 AM. These are denoted H5 and H12. The first one is off-peak; the second corresponds to the midday peak in France. Figures 19.6 and 19.7 present the simulated prices at H12 and H5,

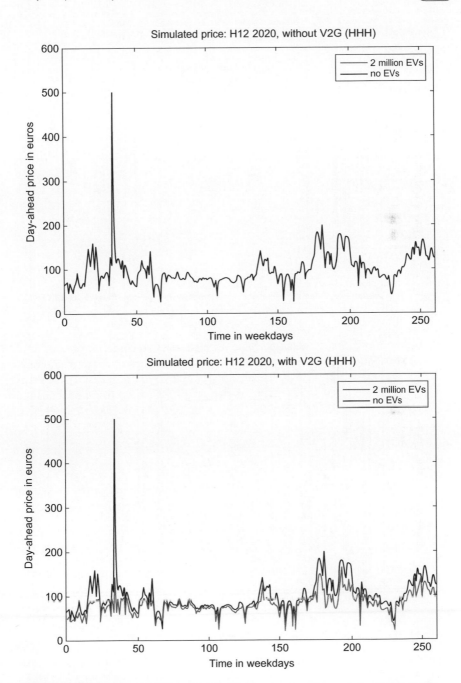

FIGURE 19.6 The simulated prices at H12 for the most favorable scenario (HHH) for one particular simulation out of the 50 that were generated, with the first case above and the second (V2G) below. The solid lines correspond to the case where there are no EVs, while the grey one corresponds to the case where there are 2 million EVs.

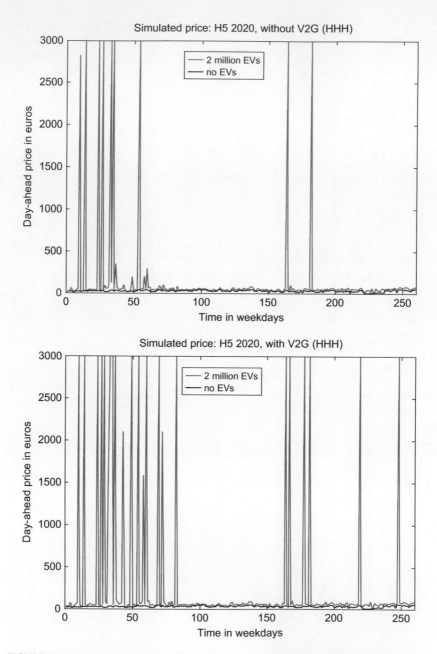

FIGURE 19.7 The simulated prices at H5 for the most favorable scenario (HHH) for one particular simulation out of the 50 that were generated, with the first case above and the second (V2G) below. The solid lines correspond to the case where there are no EVs, while the grey lines correspond to the case where there are 2 million EVs.

respectively, for the 260 weekdays in 2020, for the most favorable scenario (HHH) for one particular simulation out of the 50 that were generated. The upper figure corresponds to the first scenario without the V2G option, while the lower one is for the V2G case. The solid line in each figure corresponds to the case with no EVs, whereas the grey line corresponds to the case where there are 2 million EVs.

Looking at these four figures it can be seen that

- The two sets of simulated prices (Figure 19.6) are identical at H12, as expected, because no power is being drawn from the batteries. So only one curve is apparent.
- The simulated prices for 2 million EVs (dotted line, Figure 19.6) are *lower* than those for no EVs (solid line) at H12, as expected because power is being supplied from the batteries of the EVs to the grid.
- As is usual for electricity, price spikes occur when the system as a whole is stressed. One price peak occurred at H12 in the simulation.
- The simulated prices for 2 million EVs (grey line, Figure 19.7) are *higher* than those for no EVs (solid line) at H5, as expected.
- The striking feature in Figure 19.7 is the number of days when the price hits the technical ceiling (3000 €), which is a sign of market curtailment (see Box 19.1 for details). This corresponds to a situation of stress—usually because the demand is much greater than the supply, but occasionally because the supply is much greater than the demand.

While plotting the simulated prices for a single realization provides interesting insights, the statistics over all 50 simulations give a more complete view.

Box 19.1: Explaining How Market Curtailment Occurs

Figure 19.8 shows the offers to buy and sell electricity in black and red, respectively, on one day in the reference year Oct 2007–Sept 2008, together with three other sets of offers to buy electricity in gray. These were obtained by successively increasing the amount to be bought. This is why the curves move to the right. The day-ahead price in the reference year was 32 euros. As more and more power is required, the price rises, first to about 50 euros, then to about 100 euros. In the third case, the two curves no longer intersect, so no market fixing is possible. This is called curtailment. It would happen if more power is required to charge the batteries of the EVs than is available. Exactly the same problem occurs if too much power is taken from the batteries during peak hours for sale, except that the offers to sell increase, rather than the offers to buy.

It is assumed that buyers and sellers continue to use the same strategies as in the reference year. In practice they would probably change them.

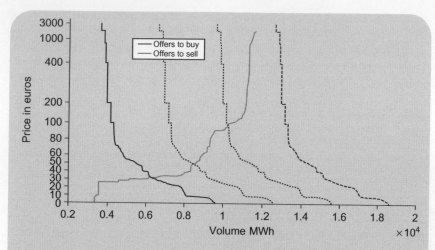

FIGURE 19.8 The original offers to buy and to sell power in the reference year, plus three sets of offers to buy obtained by successively increasing the amounts. Note that market curtailment occurs in the last case.

Figure 19.9 presents the average simulated day-ahead price for off-peak periods (Figure 19.9a) and for peak hours (Figure 19.9b) for summer and winter, as a function of the number of EVs, for the pessimistic generation case LLL. The solid line corresponds to the V2G case. Note that the average drops during peak periods in summer and winter, but that it rises rapidly during peak hours both in winter and in summer. The average price is above 150 euros in off-peak periods in winter and above 300 euros in summer for the V2G case. The results for the optimistic generation scenario HHH are very similar to these. As the striking feature in Figure 19.7 is the number of days when the price hits the technical ceiling (3000 €), we also computed the average number of days when this occurred in the simulations as a function of the number of EVs (Tables 19.3 and 19.4).

Impact on the Average Price

During peak periods the average price drops in the V2G scenario as the number of EVs increases, because more and more power is drawn from the batteries of parked cars. The average price drops from about 85 euros to about 75 euros for 2 million EVs. In off-peak periods additional power has to be put into the batteries to compensate for the drawdown during peak hours, so the average price is higher in the V2G scenario. In winter, the average price rises to above 300 euros both in the optimistic and the pessimistic generation scenarios. This raises three questions. First would users accept paying more in off-peak periods than in peak hours? Second, how much power would smart meter operators be prepared to offer to sell from the batteries, knowing

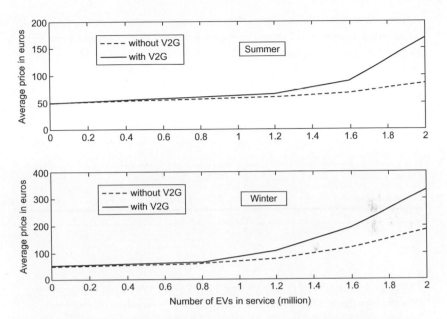

FIGURE 19.9a   The average off-peak price in summer (above) and winter (below) as a function of the proportion of power needed for the EVs that transits via the bourse. The solid line corresponds to recharging scenario N°2 (with V2G), while the dotted one corresponds to recharging scenario n°1.

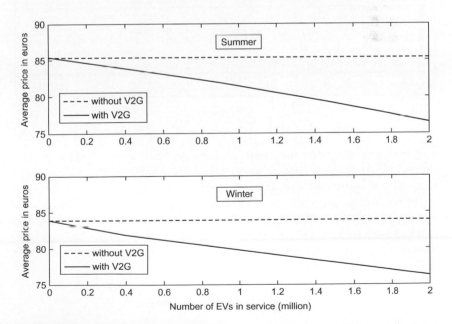

FIGURE 19.9b   The average peak hour price in summer (above) and winter (below) as a function of the proportion of power needed for the EVs that transits via the bourse. The solid line corresponds to recharging scenario N°2 (with V2G), while the dotted one corresponds to recharging scenario n°1.

TABLE 19.3 Number of Hours During Each Period When No Market Fixing Would Have Been Obtained, as a Function of the Number of EVS in Service. Scenario 1 for Recharging Batteries, Scenario LLL for the Generation Fleet

Not V2G, LLL	0	0.4 M	0.8 M	1.2 M	1.6 M	2 M
Peak Winter	0	0	0	0	0	0
Peak Summer	0	0	0	0	0	0
Off-peak Winter	0	0.02	0.78	5.62	20.48	46.5
Off-peak Summer	0	0	0	0.06	0.94	4.40

TABLE 19.4 Number of Hours During Each Period When No Market Fixing Would Have Been Obtained, as a Function of the Number of EVS in Service. Scenario 2 for Recharging Batteries, Scenario LLL for the Generation Fleet

V2G, LLL	0	0.4 M	0.8 M	1.2 M	1.6 M	2 M
Peak Winter	0	0	0	0	0	0
Peak Summer	0	0	0	0	0	0
Off-peak Winter	0	0.12	2.54	16.30	50.62	105.54
Off-peak Summer	0	0	0.04	0.62	5.28	24.44

that the more they offer the lower the price will be? They would probably optimize their offering strategy so as to maximize their profits, to the detriment of EV users and consumers in general. Third, the increase in price will have a flow-on impact on the balancing market (*marché d'ajustement*, in French) run by the grid operator, RTE, because in order to promote stability and security of supply on the grid the prices are pegged relative to the day-ahead market. Higher prices on the day-ahead market will therefore flow through the balancing market to utilities.

Interpreting the Results on "Market Curtailment"

When the French government gave the grid operator, RTE, the responsibility of evaluating the risk of not having enough electricity over the forthcoming

10–15 years, its objective[9] was to ensure that the estimated shortfall would be kept to socially and economically acceptable levels. The criterion chosen is that the annual loss of load expectation (in hours) should remain under three hours per year. Tables 19.3 and 19.4 show that market curtailment occurred increasingly often in the simulations in off-peak periods with and without the V2G option. In many cases the number of hours exceeded the acceptability threshold prescribed by the French government (3 hours per year). These results show that if bidders continued to use the same strategies as in 2007–2008 for offering power on the bourse, market curtailment would become increasingly common.

In a companion paper [19] it was shown that the market reacts quite quickly to high prices. In the case studied, the day after the price spike, larger quantities were offered for sale, albeit at higher prices, from neighboring countries, which become exporters rather than importers. So as long as these countries have excess power or are able to source it cheaper elsewhere, then it is likely they will continue to divert power to France in order to optimize profits. However, as the European Union is pushing to reduce CO_2 emissions from transport and to rid urban areas of gas-powered vehicles, Europe could reach a point where there is no surplus capacity and market curtailment could become a reality. Even if market players adapt their bidding strategies to the changing situation, the difference in consumption between peak and off-peak periods will be attenuated. There will be far less need for additional peak producers. Instead additional baseload capacity will be required.

CONCLUSIONS

Most studies on the impact of wind power and of the introduction of EVs on electricity prices have assumed that the supply curve was equal to the full merit order and that demand is inelastic. Armstrong et al. [19] demonstrated that neither of these assumptions holds for the French day-ahead market. The major contribution of this chapter is to propose a method for modeling the offers to buy and selling power on the wholesale market in the absence of such restrictive hypotheses. Instead a method has been developed based on a dataset of the observed aggregated offers to buy and sell power on the bourse over a 12-month reference period from October 2007 to September 2008.

The underlying hypothesis is that buyers and sellers will continue to use the same strategies when making offers in the future as they do today. One advantage of this approach is that it generates pairs of aggregate offers that correspond to a given set of conditions (availability of power stations, weather, fuel prices, etc). In the long run further work is required to analyze bidding behavior, in particular to determine how it changes during periods of stress.

---

[9]RTE, General Adequacy Report (2009) p. 12.

Three factors that will be important in the future but that were not taken into account in this study because they are too recent are

- the market coupling between the French and German electricity markets that took effect on November 9, 2010, and so less than 12 months of data were available. Later on, it would be interesting to repeat the study using the new aggregate curves.
- the stochastic nature of wind power. When this study started, the average wind availability (22%) was given in the latest RTE reports [1,5], but information on its day-to-day variability only started to be published in January 2011. As with the aggregate curves after market coupling, it will be very interesting to repeat the study when a longer time series is available.
- a new law called the *Loi Nome* which has been passed by the French parliament but has not yet been implemented (as of June 1, 2011). It will allow retail sellers of electricity to buy what they require directly from wholesalers through long-term contracts. This will dry up the liquidity on the market.

The next 10 years will be a dynamic and exciting period for the electric system in France. Many changes will come into effect: smart meters, 2 million new EVs, and large quantities of wind power, in addition to the *Loi Nome*, a capacity market, and the changes in the generation fleet that were listed earlier. The electric system as a whole and the market may have trouble adapting to so many changes in such a short period of time.

ACKNOWLEDGMENTS

We would like to thank EpexSpot (formerly Powernext) for allowing us to use the aggregate offers to buy and sell electricity on the French day-ahead market. Thanks also to Audrey Mahuet, Florence Very, and Aymen Salah Abou El Enien of EpexSpot for answering our questions and to Philippe Vassilopoulos for his comments and suggestions.

The opinions expressed here are those of the authors.

REFERENCES

[1] RTE, *Update on the Generation Adequacy Report on the Electricity Supply-Demand Balance in France.* http://www.rte-france.com/uploads/media/pdf_zip/publications-annuelles/generation_adequacy_report_update_2010.pdf, 2010, pp. 16 (accessed 28.05.11).

[2] J. Batut, *RTE: Point de vue de Jacques Batut, présenté au Forum sur les véhicules électriques, organisé par le CRE,* le 12 octobre 2010, http://www.smartgrids-cre.fr/index.php?rubrique=dossiers&srub=vehicules&page=17, 2010 (accessed 28.05.11).

[3] ERDF, *L'Impact sur la courbe de charge, présenté au Forum sur les véhicules électriques, organisé par le CRE,* le 12 octobre 2010, http://www.smartgrids-cre.fr/index.php?rubrique=dossiers&srub=vehicules&page=5, 2010 (accessed 28.05.11).

[4] A. Iguer, *Impact sur le prix de l'électricité suite à l'introduction des véhicules électriques en France,* Internal Report, CERNA et CENERG, Mines-ParisTech, August 2010, pp. 55.

[5] RTE, *Generation Adequacy Report on the Electricity Supply-Demand Balance in France*, 2009 ed. http://clients.rte-france.com/htm/an/mediatheque/telecharge/generation_adequacy_report_2009.pdf, 2009, pp. 172 (accessed 28.05.11).

[6] V. Iezhova, *Impact sur le prix de l'électricité de l'introduction des centrales au cycle combiné gaz*, Internal Report, CERNA et CENERG, Mines-ParisTech, August 2010, pp. 55.

[7] W. Kempton, J. Tomic, Vehicle to grid power fundamentals: calculating capacity and net revenue, *J. Power Sources* 144 (2005) 268–279.

[8] P. Denholm, W. Short, A Preliminary Assessment of Plug-In Hybrid Electric Vehicles on Wind Energy Markets, Technical report, National Renewables Energy Laboratory, NREL/TP-620-39729 April 2006.

[9] P. Denholm, W. Short, *An Evaluation of Utility System Impacts And Benefits of Optimally Dispatched Plug-In Hybrid Electric Vehicles*, Technical report, National Renewables Energy Laboratory, NREL/TP-620-40293 July 2006.

[10] J. Tomic, W. Kempton, Using fleets of electric-drive vehicles for grid support, *J. Power Sources* 168 (2007) 459–468.

[11] M.J. Scott, M. Kintner-Meyer, D.B. Elliott, W.M. Warwick, *Impacts Assessments of Plug-In Hybrid Vehicles on Electric Utilities And Regional U.s. Power Grids: Part 2: Economic Assessment*, Pacific Northwest National Laboratory, http://energytech.pnl.gov/publications/pdf/PHEV_Economic_Analysis_Part2_Final.pdf, 2007.

[12] M. Rousselle, *Impact of the Electric Vehicle on the Electric System*, Master's Thesis KTH, https://eeweb01.ee.kth.se/upload/publications/.../XR-EE-ES_2009_018.pdf, 2009, pp. 97.

[13] S.W. Hadley, A. Tsvetkova, Potential impacts of Plug-in hybrid electric vehicles on regional power generation, *Electricity J.* 22 (10) (2009) 56–68.

[14] G. Saenz de Miera, P. del Rio Gonzalez, I. Vizcaino, Analysing the impact of renewable electricity support schemes on power prices: the case of wind electricity in Spain, *Energy Policy* 36 (2008) 3345–3359.

[15] F. Sensfuss, M. Ragwitz, M. Genoese, The merit order effect: a detailed analysis of the price effect of renewable electricity generation on spot market prices in Germany, *Energy Policy* 36 (2008) 3086–3094.

[16] H. Weigt, Germany's wind energy: the potential for fossil capacity replacement and cost saving, *Appl. Energy* 86 (2009) 1857–1863.

[17] B. Pfluger, F. Sensfuss, M. Wietschel, Agent-based simulation of the effects of an import of electricity from renewable sources in Northern Africa into the Italian power market, *Internat. Energiewirtschaftagung an der TU Wien*, 2009, pp. 15.

[18] M. Armstrong, V. Iezhova, A. Galli, An Empirical Analysis of the Impact of Wind Power on Day-Ahead Electricity Prices in France, Working report, CERNA, Mines-ParisTech, 2011b, pp. 50.

[19] M. Armstrong, V. Iezhova, A. Galli, *Assessing the Impact of Wind Power on Day-Ahead Electricity Prices in France*, Presented at the 34th International Conference of the IAEE to be held in Stockholm, 19–22 June, 2011.

[20] I.M. Rasmussen, M.H. Windolf, *Wind Power Integration in New Zealand—A Scenario Analysis Of 15–25% Wind Power in the Electricity Market in 2025*, Master's Thesis from the Technical University of Denmark, August 2008.

The topic of the smart grid, examined from a number of perspectives by the contributing authors of this volume, offers exciting opportunities, savings, benefits, and a host of features and functionalities that are simply beyond the capabilities of the existing grid. Moreover, taken together, the smart grid has the potential to usher the electric power sector into the information age, at last, as pointed out by Guido Bartels in the preface of the book.

But as one looks at the road ahead, it is abundantly clear that many remaining obstacles have to be overcome and pitfalls avoided—technical, regulatory, financial, and behavioral, to name a few—if we are to achieve the full potential benefits of the smart grid. In this sense, the smart grid today resembles the proverbial glass, it appears half empty to some, while others will see it as half full. The former view is mainly focused on the enormous costs, the many technical, implementation, and integration hurdles. The latter view is mainly focused on the potential gains that could be had if we persevere and manage to get it almost right.

Fereidoon P. Sioshansi
Menlo Energy Economics

Index

Page numbers in *italics* indicate figures, tables, and footnotes.

Printed and bound by CPI Group (UK) Ltd, Croydon, CR0 4YY

13/10/2024

01773514-0003